W9-COG-022

# THE ATLAS OF THE CIVIL WAR

# THE ATLAS OF THE CIVIL WAR

Edited by Pulitzer Prize-Winning Historian
James M. McPherson

SP

Skyhorse Publishing

First published in the United States in 2005 by Courage Books, an imprint of Running Press Book Publishers
First Skyhorse Publishing edition 2022



Skyhorse Publishing books may be purchased in bulk at special discounts for sales promotion, corporate gifts, fund-raising, or educational purposes. Special editions can also be created to specifications. For details, contact the Special Sales Department, Skyhorse Publishing, 307 West 36th Street, 11th Floor, New York, NY 10018 or info@skyhorsepublishing.com.



Visit our website at www.skyhorsepublishing.com.

10 9 8 7 6 5 4 3 2 1

Library of Congress Cataloging-in-Publication Data is available on file.

Print ISBN: 978-1-5107-5640-3
Ebook ISBN: 978-1-5107-5670-0

Printed in Malaysia

# Contributors

EDITOR:
James M. McPherson
George Henry Davis Professor of American History
Princeton University, New Jersey
Pulitzer Prize-winning author of *Battle Cry of Freedom: The Civil War Era*, Vol. VI in the *Oxford History of the United States*. His other books include *The Struggle for Equality, Marching Toward Freedom, Ordeal by Fire: The Civil War and Reconstruction, Abraham Lincoln and the Second American Revolution* and *What They Fought For, 1861–1865*

Stacy D. Allen
Historian
Shiloh National Military Park,
Shiloh, Tennessee

Edwin Cole Bearss
Special Assistant to Director
National Park Service (Military Sites)
Arlington, Virginia

Albert Castel
American Civil War Historian
Emeritus Professor of History,
Western Michigan University

William M. Fowler, Jr.
Professor of History
Northeastern University,
Boston, Massachusetts
Editor of the *New England Quarterly*

D. Scott Hartwig
Supervisory Park Ranger
Gettysburg National Military Park,
Gettysburg, Virginia

Lawrence L. Hewitt
Professor of History
Department of History and Government
Southeastern Louisiana University,
Hammond, Louisiana

Frank O'Reilly
Historian and author
Fredericksburg, Virginia

William G. Piston
Associate Professor
Department of History,
Southwest Missouri State University,
Springfield, Missouri

Dr. William Glenn Robertson
Professor of History
U.S. Army Command and General Staff College,
Leavenworth, Kansas

# Contents

# Prologue

The Civil War was the most violent and fateful experience in American history. At least 620,000 soldiers were killed in the war, 2% of the American population in 1860. If the same percentage of Americans were to be killed in a war fought in the 1990s, the number of American war dead would exceed five million. An unknown number of civilians, nearly all of them in the South, died from causes such as disease, hunger or exposure inflicted during the conflict. As a consequence, more Americans died in the Civil War than in all of the country's other wars *combined*. The number of casualties incurred in a single day at the battle of Antietam (September 17, 1862) was four times the number of Americans killed or wounded on the Normandy beaches on D-Day, June 6, 1944. More Americans were killed in action that September day near Sharpsburg, Maryland, than were killed in combat in all the other wars fought by the United States in the 19th century.

How did this happen? Why did Americans fight each other with a ferocity unmatched in the Western world during the century between the end of the Napoleonic Wars in 1815 and the beginning of World War I in 1914? The origins of the American Civil War lay in the outcome of another war fought by America fifteen years earlier: the Mexican War. The peace treaty signed with Mexico in 1848 transferred 700,000 square miles of Mexican territory to the United States. However, the dramatic victory of American forces in the Mexican War fulfilled the prediction made by the philosopher Ralph Waldo Emerson in 1846 at the war's outset: "The United States will conquer Mexico, but it will be as the man swallows arsenic, which brings him down in turn. Mexico will poison us."

The poison was slavery, which many Southern politicians wanted to introduce into the new territories; anti-slavery Northerners wanted to keep slavery out of them. In the House of Representatives, they had the votes to pass the Wilmot Proviso (offered by Congressman David Wilmot of Pennsylvania) stating that slavery should be excluded from all territories acquired from Mexico. In the Senate, Southern strength defeated this Proviso. South Carolina Senator, John C. Calhoun, introduced instead a series of resolutions affirming that slaveholders had the constitutional right to take their slave property into any United States territory they so wished.

These opposing views set the scene for a crisis when gold was discovered in California in 1848. Eighty thousand gold seekers poured into the region in 1849. To achieve some degree of law and order, the Forty-niners organized a state government and petitioned Congress for admission to the Union as the thirty-first state. As California's new constitution prohibited slavery, this request met with fierce resistance from Southerners. The crisis escalated when the American President, Zachary Taylor, encouraged the huge territory of New Mexico (embracing the rest of the cession from Mexico) also to apply for statehood without slavery.

Pro-slavery Southerners threatened to secede from the Union if they were denied their "right" to take slaves into these territories. "If, by your legislation, you seek to drive us from the territories of California and Mexico," Congressman Robert Toombs of Georgia informed Northern lawmakers, "*I am for disunion.*" The controversy in Congress became so heated that Senator Henry S. Foote of Mississippi flourished a loaded revolver during a debate, and his colleague Jefferson Davis challenged an Illinois congressman to a duel. In 1850 the American nation seemed held together by a mere thread, with armed conflict between free and slave states an alarming possibility.

But cooler heads prevailed. The Compromise of 1850 averted a showdown. This series of laws admitted California as a free state, divided the remainder of the Mexican cession into the territories of New Mexico and Utah, and left to their residents the question as to whether or not they would have slavery. (In fact, both territories did legalize slavery, but few slaves were taken there.) At the same time, Congress abolished the slave trade in the District of Columbia, ending the shame – in Northern eyes – of the buying and selling of human beings within sight of the White House and the Capital. But the Compromise of 1850 compensated the South with a tough new fugitive slave law that empowered federal marshals, backed by the army, to recover slaves who had escaped into free states. It thus postponed, but did not resolve, the sectional crisis.

During the 1850s, polarization between North and South intensified. The fugitive slave law embittered Northerners compelled to watch black people – some of whom had lived in their communities for years – being forcibly returned in chains to slavery. Southern anxiety grew as settlers poured into those Northern territories that were sure to join the Union as free states, thereby tipping the sectional balance of power against the South in Congress and the electoral college. In an attempt to bring more slave states into the Union, Southerners agitated for the purchase of Cuba from Spain and the acquisition of additional territory in Central America. Private armies of "filibusters," composed mainly of Southerners, even tried to invade Cuba and Nicaragua to overthrow their governments and bring these regions into the United States – with slavery.

Nothing did more to divide North and South than the Kansas-Nebraska Act of 1854 and the subsequent guerrilla war between pro- and anti-slavery partisans in Kansas territory. The region that became the territories of Kansas and

Nebraska was part of the Louisiana Purchase, acquired by the United States from France in 1803. In 1820, the Missouri Compromise had divided this territory at latitude 36° 30', with slavery permitted south of that line and prohibited north of it. Regarded by Northerners as an inviolable compact, the Missouri Compromise lasted for 34 years. But in 1854, Southerners broke it by forcing Stephen A. Douglas of Illinois, Chairman of the Senate Committee on Territories, and leader of the Northern Democrats, to agree to the repeal of the ban on slavery north of 36° 30' as the price of Southern support for the formal organization of Kansas and Nebraska territories.

Douglas capitulated under Southern pressure, even though he expected it to "raise a hell of a storm" in the North. It did. The storm was so powerful that it swept away many Northern Democrats and gave rise to the Republican party, which pledged to keep slavery out of Kansas and all other territories. One of the most eloquent spokesmen for this new party was an Illinois lawyer named Abraham Lincoln, who believed that "there *can* be no *moral right* in the enslaving of one man by another." Lincoln and other Republicans recognized that the United States Constitution protected slavery in the states where it already existed. But they intended to prevent its further expansion as the first step toward bringing it eventually to an end. The United States, said Lincoln at the beginning of his famous campaign against Douglas in 1858 for election to the Senate, was a house divided between slavery and freedom. "'A house divided against itself cannot stand.' I believe this government cannot endure, permanently half *slave* and half *free*." By preventing the further expansion of slavery, Lincoln hoped to "place it where the public mind shall rest in the belief that it is in the course of ultimate extinction."

Douglas won the senatorial election in 1858. But two years later, running against a Democratic party split into Northern and Southern halves, Lincoln won the presidency by carrying every Northern state. This was the first time in more than a generation that the South had lost effective control of the national government. Southerners saw the writing on the wall. A substantial and growing majority of the American population lived in the North. The pro-slavery forces had little prospect of winning any future national elections. Thus, to preserve slavery as the basis of their "way of life," during the winter of 1860–1861 the seven lower-south states seceded one by one. Before Lincoln took office on March 4, 1861, delegates from these seven states had met at Montgomery, Alabama, adopted a Constitution for the Confederate States of America, and formed a provisional government with Jefferson Davis as president. As they seceded, these states seized the national arsenals, forts, and other property within their borders – with the significant exception of Fort Sumter in the harbor of Charleston, South Carolina. When Lincoln took his oath to "preserve, protect, and defend" the United States and its Constitution, the "united" states had already ceased to exist.

*The inauguration of Jefferson Davis as President of the Confederate States, at the State House, Montgomery, Alabama, February 18, 1861.*

Secession transformed the principal issue of the sectional conflict from the future of slavery to the survival of the Union itself. Lincoln and most of the Northern people refused to accept the constitutional legitimacy of secession. "The central idea pervading this struggle," Lincoln declared after war had broken out in 1861, "is the necessity that is upon us, of proving that popular government is not an absurdity. We must settle this question now, whether in a free government the minority have the right to break up the government whenever they choose." Four years later, looking back over the bloody chasm of war, Lincoln said in his second inaugural address that one side in the controversy of 1861 "would *make* war rather than let the nation survive; and the other would *accept* war rather than let it perish. And the war came."

*James M. McPherson*

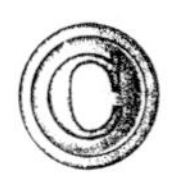

# Key to Maps

## ARMY TERMS

Although the strength of the army units varied widely, the following is a general guide to the terms used in the atlas:

**Army:** Any force operating in a theater. The strength of any army could vary from 10,000 to over 100,000

**Corps:** Composed of two or three divisions, a corps' strength varied from between 15,000 and 20,000

**Division:** Usually composed of two or three brigades amounting to 5,000 men

**Brigade:** Composed of two or more regiments and totaling between 1,200 and 3,000 men

**Regiment:** Composed of 10 companies of 50–100 men each

## ARMY COLORS

Confederate

Union

Confederate HQ

Union HQ

## ARMY HIERARCHIES

LEE MEADE — Army Commander

LONGSTREET — Corps Commander

STONEYMAN — Division Commander

ARMISTEAD BIRNEY — Brigade Commander

elts — Element (or part of unit)

## ARMY SYMBOLS

LONGSTREET
Army Corps

STONEYMAN
Army Division

ARMISTEAD — Army units of varying strength

BIRNEY
Cavalry units of varying strength

## ARMY MOVEMENTS

First position Later position

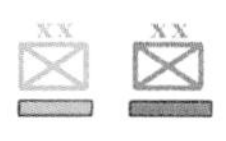

Corps & Divisional arrow

Corps & Divisional Dispersal or End Movement

Corps & Divisional retreat

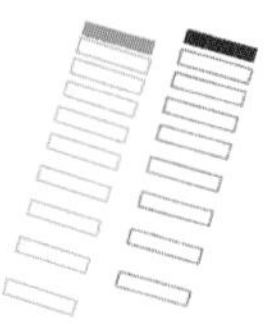
Unit movements

Unit retreat

Skirmish Line

## GENERAL SYMBOLS

Urban Area

Town/Settlement

Farm or Building

Station

Church

Bridge

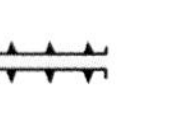
Pontoon Bridge

Ford

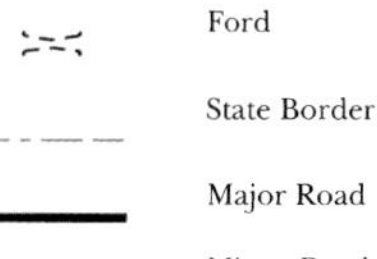
State Border

Major Road

Minor Road

Railroad

## PHYSICAL FEATURES

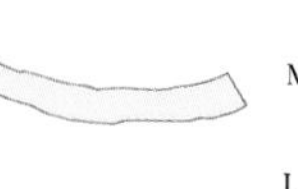
Major Rivers/Estuary

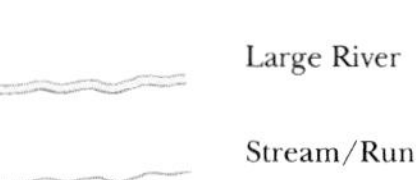
Large River

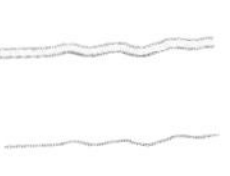
Stream/Run

Forest or Wood

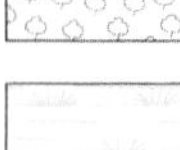
Marsh or Swamp

Agriculture

Seminary Ridge
Hill and Mountain Feature

## GENERAL MILITARY SYMBOLS

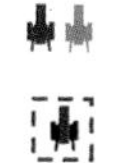
Artillery

Artillery Reserve

Sail or sail-steam ship

Gunboat/Ironclad

Ships (various types)

Fort/Battery

Encampment

Fortification Line

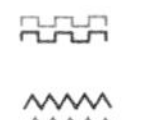
Siege Line

Battle Site

Mine

Siege/Explosion

## TYPE STYLES

V I R G I N I A — State

C L A R K E C O U N T Y — County

GETTYSBURG — City

FALMOUTH — Town

LEWISVILLE — Settlement

*Peach Orchard*
*Cemetery Hill*
*Devil's Den* — Small Physical Feature

## ABBREVIATIONS

P.O. — Post Office

C.H. — Court House

R.R. — Railroads

STN. — Station

MT. — Mountain

# Eastern Theater

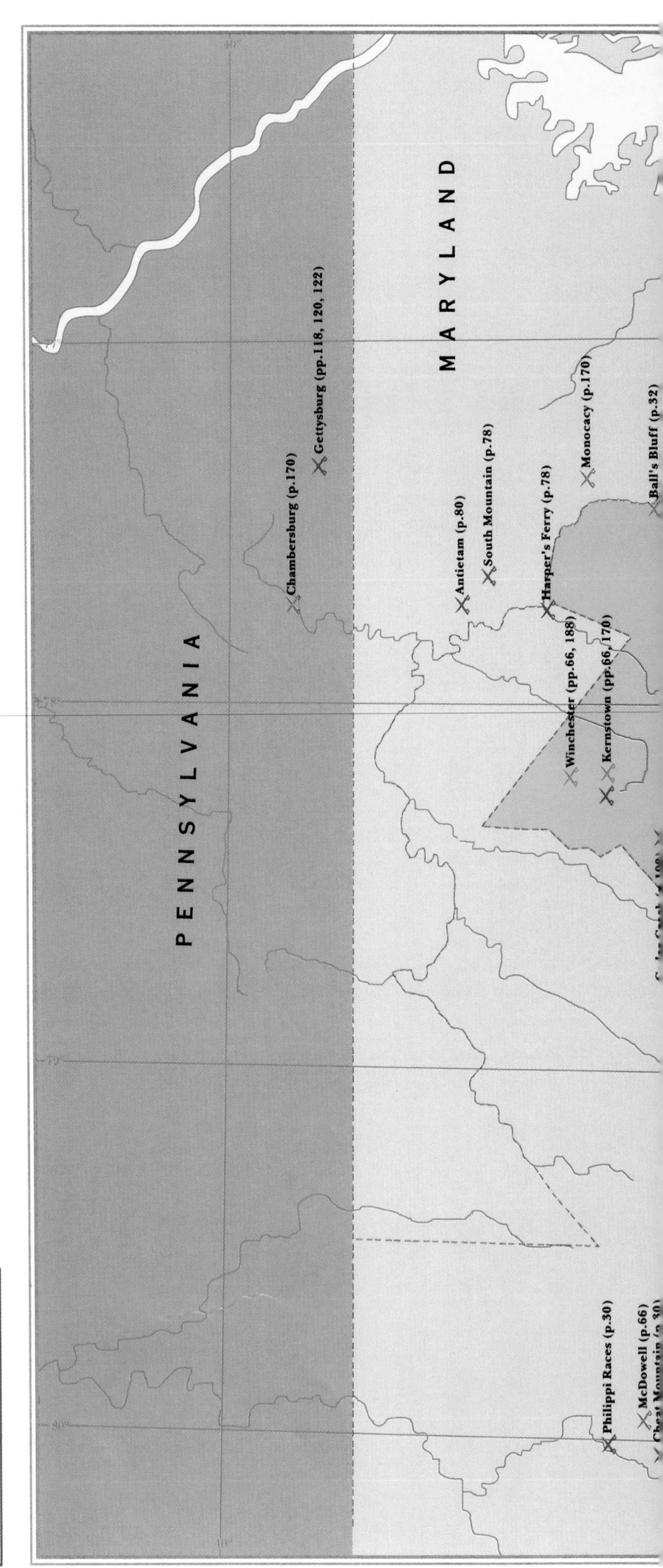
PENNSYLVANIA
MARYLAND
Chambersburg (p.170)
Gettysburg (pp.118, 120, 122)
Antietam (p.80)
South Mountain (p.78)
Harper's Ferry (p.78)
Monocacy (p.170)
Ball's Bluff (p.32)
Winchester (pp.66, 188)
Kernstown (pp.66, 170)
Philippi Races (p.30)
McDowell (p.66)

Union territory
Confederate territory
border slave states
Union victory
Confederate victory
inconclusive

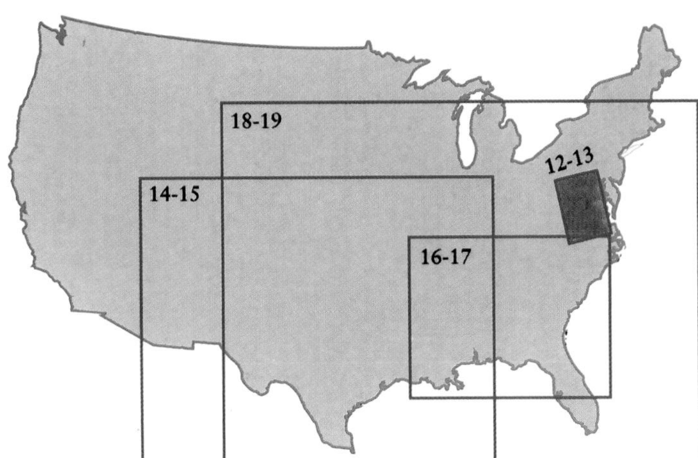
18-19
14-15
12-13
16-17

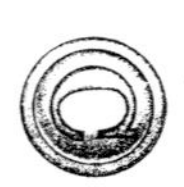

Chesapeake Bay
Potomac River
Rappahannock River
York River
James River
Appomatox
VIRGINIA
NORTH CAROLINA
Jackson's turning movement (p.74)
Chantilly (p.76)
New Market (p.168)
Gorveton (p.76)
First Manassas/Bull Run (pp.32, 76)
Cross Key (p.68)
Bristoe (p.138)
Port Republic (p.68)
Catlett's Station (p.74)
Piedmont (p.168)
Brandy Station (p.116)
Cedar Mountain (p.74)
Fredericksburg (p.82)
Waynesboro (p.190)
Chancellorsville (pp.112, 114)
Mine Run (p.138)
The Wilderness (p.152)
Spotsylvania (pp.154, 156)
Lynchburg (p.168)
Trevilian Station (p.166)
North Anna (p.158)
Yellow Tavern (p.166)
Gaines' Mill (p.70)
Cold Harbor (pp.158, 160)
Mechanicsville (p.70)
Seven Pines/Fair Oaks (p.64)
Oak Grove (p.70)
Savage's Station/White Oak Swamp (p.72)
Glendale (p.72)
Darbytown Road (p.186)
New Market Heights (p.186)
Drewry's Bluff (p.164)
Malvern Hill (p.72)
Deep Bottom (p.186)
Battle of the Crater (p.184)
Petersburg (pp.162, 184, 210)
Weldon Railroad (p.184)
Fort Stedman (p.210)
Hatcher's Run (pp.184, 210)
White Oak Road (p.210)
Poplar Springs Church (p.184)
Five Forks (p.210)
Globe Tavern (p.184)
Reams Station (p.184)
Yorktown (p.64)
Williamsburg (p.64)
Big Bethel (p.32)
N
0 10 20 KM
0 10 20 MILES

# Western Theater

ILLINOIS
MISSOURI
New Madrid
Island No 10
(p.54)
Fort Donelso
Fort Henry (p.4
Franklin (
TENNESSE
Spring Hill (p.19
ARKANSAS
Fort Pillow (pp.54, 180)
Memphis (pp.54, 180)
Shiloh (pp.50,52)
Corinth (pp.54, 88)
Iuka (p.88)
Mississippi River
Brice's Crossroads (p.180)
Tupelo (p.180)
Elkin's Ferry (p.150)
Arkansas Post (p.104)
Forrest's and Van
Dorn'sRaids (p.90)
MISSISSIPPI
Chickasaw Bluffs (p.90)
Vicksburg
(pp.56, 90, 104-9)
Champion's Hill (p.106)
Jackson (p.106)
Big Black River (p.106)
Raymond (p.106)
Meridian (p.146)
Port Gibson (p.104)
Port Hudson (p.110)
LOUISIANA
L. Ponchartrain
Mobile Bay
New Orleans (p.56)
Basataria Bay
Fort St Philip (p.56)
Fort Jackson (p.56)
Union territory
Confederate territory
border slave states
Union victory
Confederate victory
inconclusive
18-19
16-17
12-13
14-15

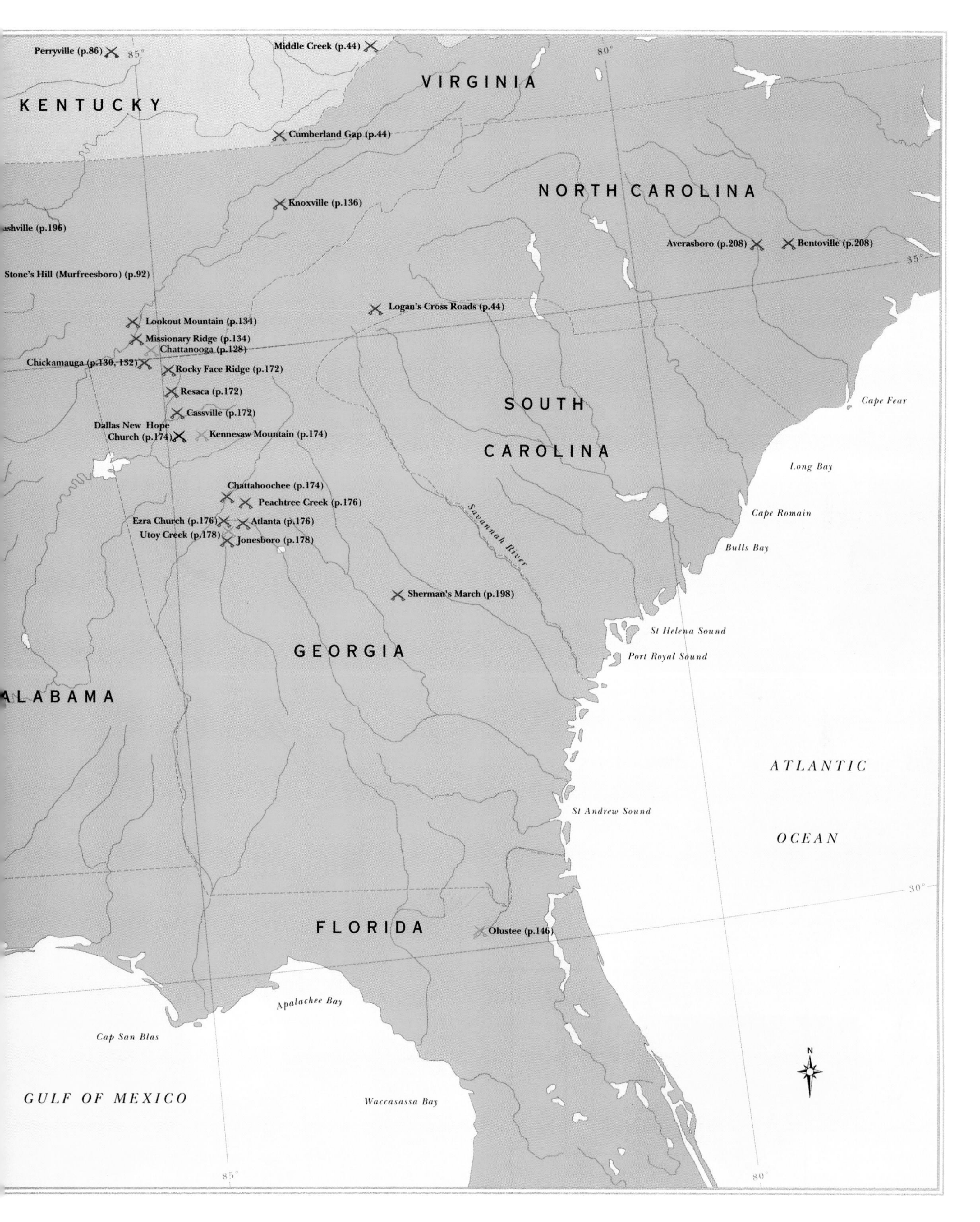

Perryville (p.86)
Middle Creek (p.44)
VIRGINIA
KENTUCKY
Cumberland Gap (p.44)
Knoxville (p.136)
NORTH CAROLINA
ashville (p.196)
Averasboro (p.208)
Bentoville (p.208)
Stone's Hill (Murfreesboro) (p.92)
Logan's Cross Roads (p.44)
Lookout Mountain (p.134)
Missionary Ridge (p.134)
Chattanooga (p.128)
Chickamauga (p.130, 132)
Rocky Face Ridge (p.172)
Resaca (p.172)
Cassville (p.172)
SOUTH CAROLINA
Cape Fear
Dallas New Hope Church (p.174)
Kennesaw Mountain (p.174)
Long Bay
Chattahoochee (p.174)
Peachtree Creek (p.176)
Ezra Church (p.176)
Atlanta (p.176)
Utoy Creek (p.178)
Jonesboro (p.178)
Cape Romain
Savannah River
Bulls Bay
Sherman's March (p.198)
St Helena Sound
Port Royal Sound
GEORGIA
ALABAMA
ATLANTIC
OCEAN
St Andrew Sound
FLORIDA
Olustee (p.146)
Apalachee Bay
Cap San Blas
GULF OF MEXICO
Waccasassa Bay
N

# Trans-Mississippi Theater

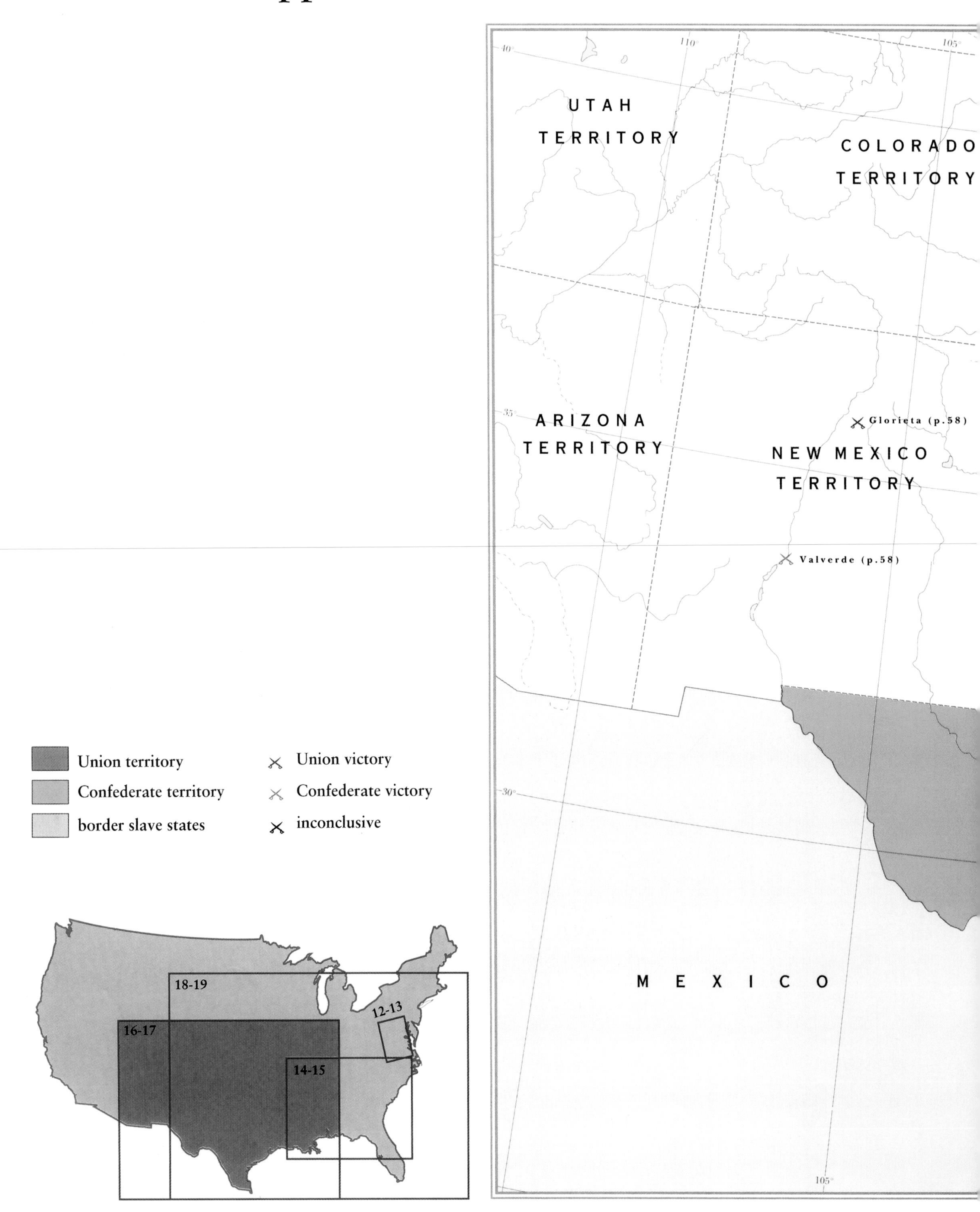
UTAH
TERRITORY
COLORADO
TERRITORY
ARIZONA
TERRITORY
NEW MEXICO
TERRITORY
Glorieta (p.58)
Valverde (p.58)
MEXICO
Union territory
Confederate territory
border slave states
Union victory
Confederate victory
inconclusive
18-19
16-17
12-13
14-15

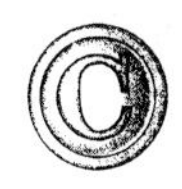

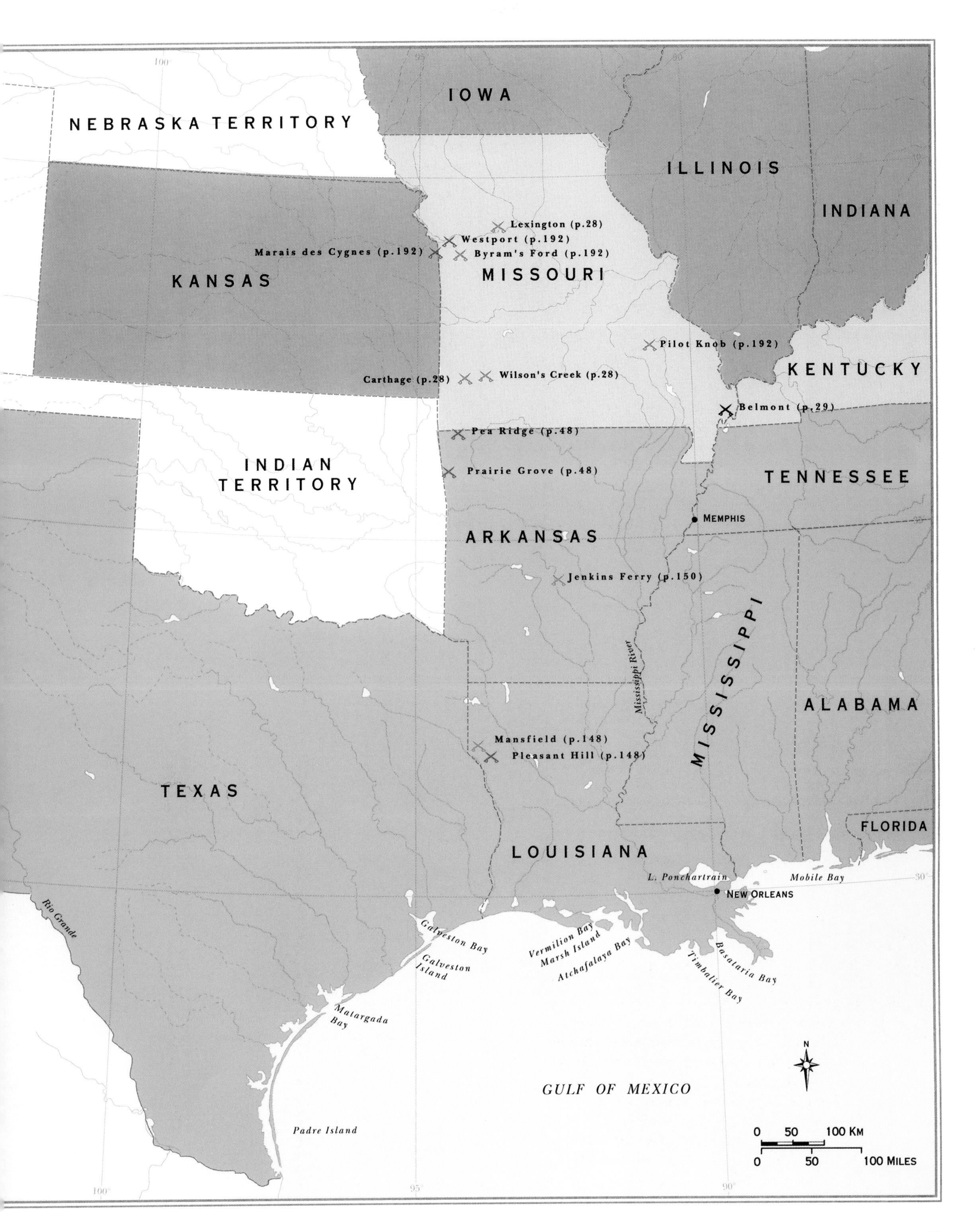

IOWA
NEBRASKA TERRITORY
ILLINOIS
INDIANA
KANSAS
MISSOURI
Lexington (p.28)
Westport (p.192)
Marais des Cygnes (p.192)
Byram's Ford (p.192)
Pilot Knob (p.192)
KENTUCKY
Carthage (p.28)
Wilson's Creek (p.28)
Belmont (p.29)
Pea Ridge (p.48)
Prairie Grove (p.48)
INDIAN TERRITORY
TENNESSEE
Memphis
ARKANSAS
Jenkins Ferry (p.150)
Mississippi River
MISSISSIPPI
ALABAMA
Mansfield (p.148)
Pleasant Hill (p.148)
TEXAS
FLORIDA
LOUISIANA
L. Ponchartrain
Mobile Bay
New Orleans
Rio Grande
Galveston Bay
Galveston Island
Vermilion Bay
Marsh Island
Atchafalaya Bay
Timbalier Bay
Basataria Bay
Matargada Bay
GULF OF MEXICO
Padre Island
N
0 50 100 Km
0 50 100 Miles

# Coastal War

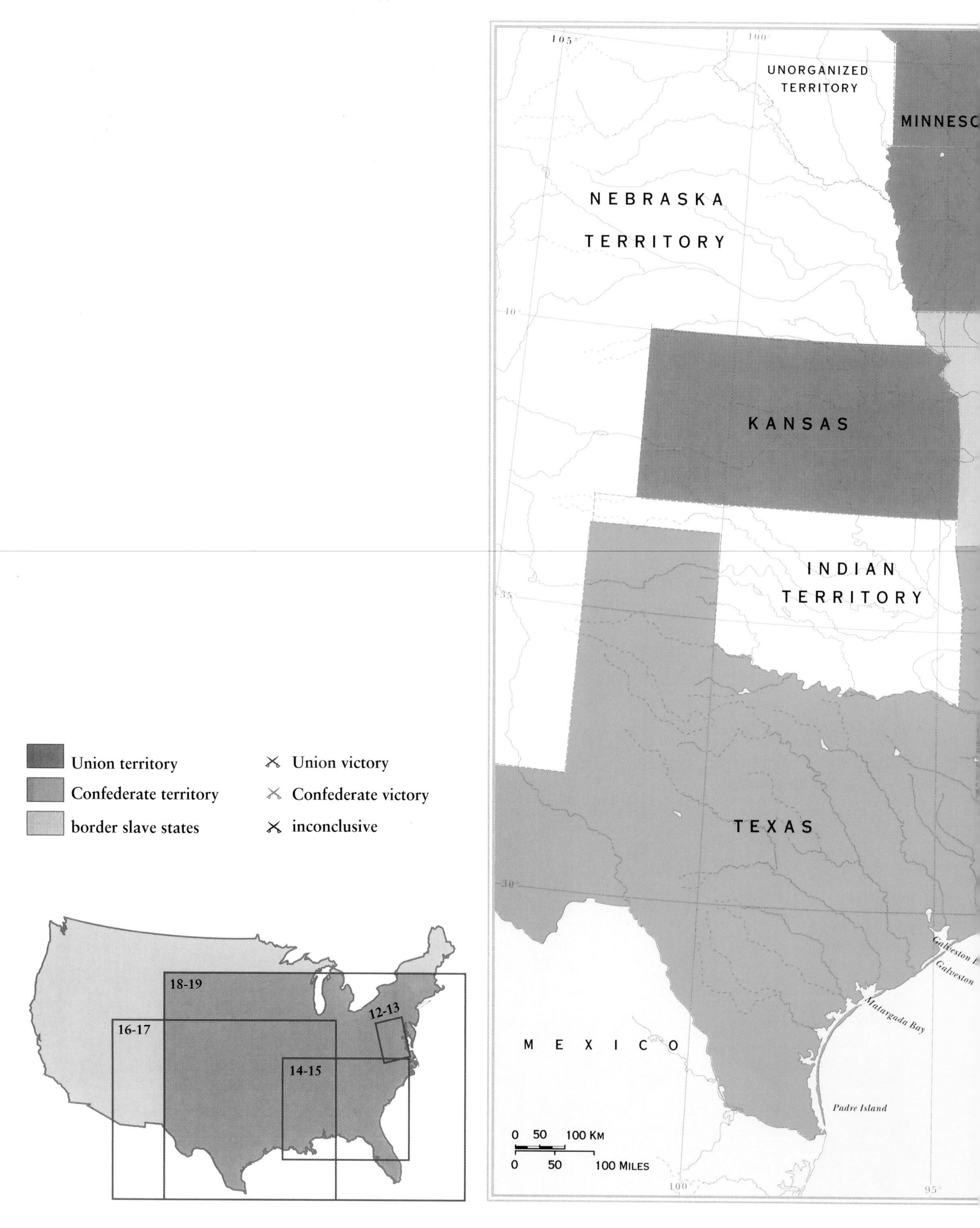

UNORGANIZED TERRITORY
MINNESO
NEBRASKA TERRITORY
KANSAS
INDIAN TERRITORY
TEXAS
MEXICO
Galveston
Matagorda Bay
Padre Island
0 50 100 KM
0 50 100 MILES
Union territory
Confederate territory
border slave states
Union victory
Confederate victory
inconclusive
18-19
16-17
12-13
14-15

WISCONSIN
MICHIGAN
CANADA
Lake Huron
Lake Ontario
Lake Michigan
Lake Erie
NEW YORK
VERMONT
NEW HAMPSHIRE
MASSACHUSETTS
Cape Cod
RHODE IS.
CONNECTICUT
NEW JERSEY
Long Island
OWA
PENNSYLVANIA
OHIO
ILLINOIS
INDIANA
MARYLAND
Delaware Bay
DELAWARE
WASHINGTON, DC
WEST VIRGINIA
Chesapeake Bay
James River
Cape Charles
Hampton Roads (p.62)
VIRGINIA
MISSOURI
KENTUCKY
Albemarle Sd
Roanoke Island (p.60)
New Bern (p.60)
Pimlico Sd
Cape Hatteras
Hatteras Inlet (p.34)
NORTH CAROLINA
Fort Macon (p.60)
Cape Lookout
TENNESSEE
RKANSAS
Onslow Bay
SOUTH CAROLINA
Fort Fisher (p.206)
Cape Fear
Long Bay
Mississippi
Savannah River
Cape Romain
CHARLESTON
Bulls Bay
Fort Sumter (pp.26,126)
Fort Wagner (p.126)
St Helena Sound
ATLANTIC
Port Royal Sound
Port Royal Sound (p.34)
Fort Pulaski
MISSISSIPPI
ALABAMA
GEORGIA
St Andrew Sound
OCEAN
JACKSONVILLE
OUISIANA
Fort Gaines (p.182)
FORT PICKENS
FLORIDA
Fort Morgan (p.182)
L. Ponchartrain
NEW ORLEANS
Mobile Bay (p.182)
Apalachee Bay
Mobile Bay
Cap San Blas
FORT JACKSON
Marsh I.
Atchafalaya Bay
Barataria Bay
Timbalier Bay
Waccasassa Bay
Tampa Bay
GULF OF MEXICO
N
Cape Romano
Biscayne Bay
Cape Sable
Florida Bay
Florida Keys
Strait of Florida
Bahama Islands

*Abraham Lincoln, photographed by Mathew Brady on February 27, 1860, the day before he delivered his Cooper Union speech. Lincoln was later to state that, "Brady and the Cooper Union speech made me President."*

# 1861: The Coming of War

When Abraham Lincoln took the oath of office as the sixteenth – and, some speculated, the last – president of the *United* States on March 4, 1861, he knew that his inaugural address would be the most important such speech in American history. On his words would hang the issues of union or disunion, peace or war. His goal was to prevent the eight slave states that had not yet seceded from doing so, while cooling passions in the seven states that had seceded, hoping that in time their old loyalty to the Union would reassert itself. He pledged in his address not to "interfere with the institution of slavery where it exists." Referring, however, to Fort Sumter and three other minor forts in the seceded states, he pledged to "hold, occupy, and possess the property, and places belonging to the government" – without defining exactly what he meant or how he would do it. In his eloquent peroration, Lincoln appealed to Southerners as Americans who had shared with other Americans four score and five years of national history. "We are not enemies, but friends," he said.

> *Though passion may have strained, it must not break, our bonds of affection. The mystic chords of memory, stretching from every battlefield and patriot grave to every living heart and hearthstone all over this broad land, will yet swell the chorus of the Union when again touched, as surely they will be, by the better angels of our nature.*

Lincoln hoped to buy time with his inaugural address – time to demonstrate his peaceful intentions and to enable Southern Unionists (whose numbers he overestimated) to regain the upper hand. But the day after his inauguration, Lincoln learned that time was running out. A dispatch from Major Robert Anderson, commander of the U.S. army garrison holding Fort Sumter, informed him that his supplies would soon be exhausted: the fort must be resupplied or evacuated. The majority of Lincoln's cabinet advised him to evacuate the garrison to avoid provoking a shooting war. But Lincoln feared that withdrawal would give the Confederacy a moral victory, confer legitimacy on its government and probably lead to diplomatic recognition by foreign powers. Having pledged in his address to "hold, occupy, and possess" national property, could Lincoln afford to abandon that policy during his first month in office? If he did, he would go down in history as the president who consented to the dissolution of the United States.

Lincoln finally arrived at a solution that would place the onus of starting a war – if there was to be a war – on the other side. He decided to send an unarmed ship with supplies to Sumter, and to hold troops and warships outside the harbor with authorization to go into action only if the Confederates used force to stop the supplies. He would also notify Confederate officials in advance of his intention. This shifted the decision for war or peace to Jefferson Davis. In effect, Lincoln flipped a coin and said to Davis: "Heads I win; tails you lose." If Confederate troops fired on the supply ships, the South would stand convicted of starting a war by attacking "a mission of humanity" bringing "food for the hungry men." If Davis allowed the supplies in, the American flag would continue to fly over Fort Sumter. The Confederacy would lose face at home and abroad, and Southern Unionists would take heart.

Davis did not hesitate: he considered Fort Sumter to be Confederate property. By ordering Confederate artillery to open fire against the fort on April 12, before the supply ship arrived, he started the biggest war in American history. The attack triggered an outburst of war fever in the North. "The town is in a wild state of excitement," wrote a Philadelphia diarist. "The American flag is to be seen everywhere. Men are enlisting as fast as possible." Because the tiny United States army – most of whose 16,000 soldiers were stationed at remote frontier posts – was inadequate to quell the "insurrection," Lincoln called on the states to supply 75,000 militia. The free states filled their quotas immediately: more than twice as many men volunteered than Lincoln had called for. Recognising that the 90 days' service – to which the militia were limited by law – would be too short a period, Lincoln on May 3 issued a call for three-year volunteers. Before the war was over, more than two million men would serve in the Union army and navy.

The eight slave states still in the Union rejected Lincoln's call for troops. Four of them – Virginia, Arkansas, Tennessee, and North Carolina – seceded and joined the Confederacy. Forced by the outbreak of war to choose between the two sides, most residents of those four states chose the Confederacy. As a former Unionist in North Carolina remarked: "The division must be made on the line of slavery. The South must go with the South." When news of Sumter's surrender reached Richmond, a huge crowd poured into the state capitol square and ran up the Confederate flag. "I never in all my life witnessed such excitement," wrote a participant. *The Times* of London's correspondent described crowds in North Carolina with "flushed faces, wild eyes, screaming mouths, hurrahing for 'Jeff Davis' and 'the Southern Confederacy.'" No one in those cheering crowds could know that before the war ended at least 260,000 of the 850,000 soldiers who fought for the Confederacy would lose their lives, together with 360,000 Union soldiers, and that the slave South they fought to defend would be utterly destroyed.

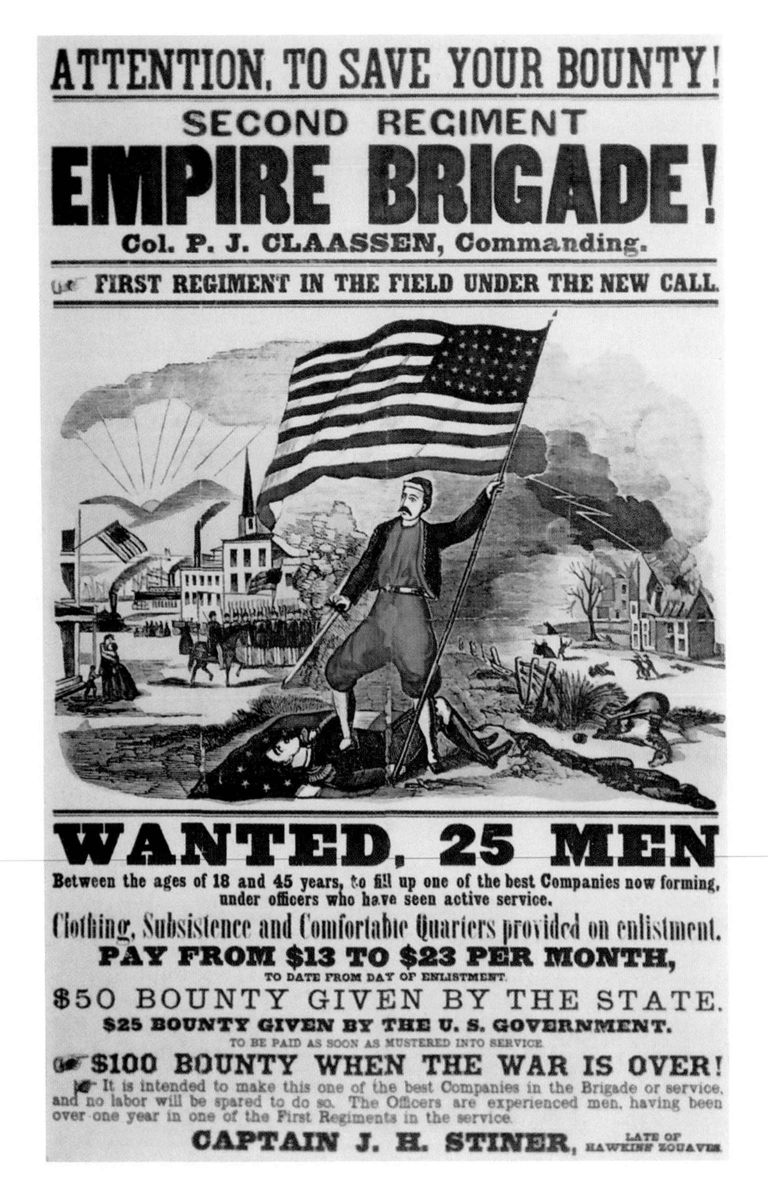

ATTENTION, TO SAVE YOUR BOUNTY!
SECOND REGIMENT
EMPIRE BRIGADE!
Col. P. J. CLAASSEN, Commanding.
FIRST REGIMENT IN THE FIELD UNDER THE NEW CALL.
WANTED, 25 MEN
Between the ages of 18 and 45 years, to fill up one of the best Companies now forming, under officers who have seen active service.
Clothing, Subsistence and Comfortable Quarters provided on enlistment.
PAY FROM $13 TO $23 PER MONTH,
TO DATE FROM DAY OF ENLISTMENT.
$50 BOUNTY GIVEN BY THE STATE.
$25 BOUNTY GIVEN BY THE U. S. GOVERNMENT.
TO BE PAID AS SOON AS MUSTERED INTO SERVICE.
$100 BOUNTY WHEN THE WAR IS OVER!
It is intended to make this one of the best Companies in the Brigade or service, and no labor will be spared to do so. The Officers are experienced men, having been over one year in one of the First Regiments in the service.
CAPTAIN J. H. STINER, LATE OF HAWKINS' ZOUAVES


*One of the many recruitment posters which were very successful in swelling the ranks of the Union army*

The four border slave states – Maryland, Kentucky, Missouri, and Delaware – were badly divided by the war. All four remained officially in the Union, but probably one-third of their white residents sympathized with the Confederacy, and many men from these states joined the Confederate army. Clashes between Union troops and local pro-Confederate crowds in Baltimore and St. Louis took dozens of lives in April and May, 1861. Guerrilla warfare, as well as pitched battles, plagued Kentucky and especially Missouri throughout the war. More than any other state, Missouri suffered from a civil war within the Civil War; its bitter legacy was to persist for generations.

The war itself produced a fifth Union border state: West Virginia. Most of the delegates from the part of Virginia west of the Shenandoah Valley had voted against secession. A region of mountains, small farms, and few slaves, its economy was linked more closely to nearby Ohio and Pennsylvania than to the South. Delegates who had opposed Virginia's secession from the Union returned home determined to secede from Virginia. With the help of Union troops, they accomplished their goal. Through a complicated process of conventions and referendums – carried out amidst continuing raids and skirmishes – they created in 1862 the new state of West Virginia, which entered the Union in 1863.

With a population of nearly 23 million compared with 9 million (3.5 million of whom were slaves) in the Confederacy, the Union states had considerable superiority in military manpower. In economic resources, Northern superiority was even greater. But even when fully mobilized, the North's superiority in resources and population did not guarantee success. The Confederacy had come into being firmly in control of 750,000 square miles – a vast territory, larger than all of western Europe. To win the war, Union forces would have to invade, conquer, and occupy much of that territory, cripple its people's ability to sustain a war of independence, and destroy their armies. Britain had been unable to accomplish a similar task in the American War of Independence, even though it enjoyed a greater superiority of resources over the United States in 1776 than the Union enjoyed over the Confederacy in 1861.

To "win" the war, the Confederacy needed neither to invade nor conquer the Union, nor destroy its armies; the South needed only to stand on the defensive and hold out long enough to convince Northerners that the cost of victory was too high – as Britain had concluded with respect to the American colonies in 1783. As the military analyst for *The Times* of London wrote in 1861: "No war of independence ever terminated unsuccessfully except where the disparity of force was far greater than it is in this case. Just as England during the revolution had to give up conquering the colonies, so the North will have to give up conquering the South."

While the railroad, telegraph, and naval steampower had radically altered wartime logistics, communications, and navies, the organization and tactics of armies had changed little since the Napoleonic Wars half a century earlier. But Civil War experience brought some changes in tactics dictated by the widespread adoption of the rifled musket. This innovation was only a decade old, dating to the perfection in the 1850s of the "minie ball" (named after French army Captain Claude Minie, its principal inventor), a cone-shaped lead bullet whose base expanded upon firing to "take" the rifling of the barrel. This made it possible to load and fire a muzzle-loading rifle as rapidly (two or three times per minute) as the old smooth-bore musket. The rifle had greater accuracy and four times the effective range (400 yards or more) of the smooth-bore. By 1862, most infantrymen in the

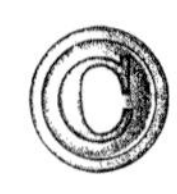

Civil War were equipped with rifles; by 1863 many Union cavalrymen had breech-loading repeating carbines – an innovation accelerated by the Civil War – and by the last year of the war some Union infantry also had repeating rifles.

Tactics adjusted fitfully to the greater lethal range and accuracy of the rifled musket. Generals continued to order old-fashioned close-order assaults, which resulted in enormous casualties from 1862 onward. The defensive power of the rifle became even greater when troops began digging into trenches from 1863 on. Massed frontal assaults became almost suicidal. Soldiers and their officers learned through hard experience to adopt skirmishing tactics, taking advantage of cover and working around the enemy flank rather than attacking frontally.

After the fall of Fort Sumter, one of the first acts of the Lincoln administration was to proclaim on April 19, 1861, a naval blockade of the Confederate states. Initially, this blockade was more of a policy than a reality, for the navy had few ships on hand with which to enforce it. The task was formidable: the Confederate coastline stretched for 3,500 miles, with two dozen ports and another 150 bays and coves where cargo could be landed. The U.S. navy, which had been converting since the 1840s from sail to steam, recalled its ships from distant seas, took old sailing vessels out of mothballs, and bought or chartered merchant ships and armed them in an effort to create a blockade fleet overnight. Eventually the navy placed several hundred warships on blockade duty. But in 1861 the blockade was so porous that nine out of ten vessels slipped through it on their way to or from Confederate ports, bringing in munitions and other matériel vital to this agricultural society with its slender industrial base.

The Confederacy, however, inadvertently contributed to the blockade's success by adopting in 1861 a foreign policy that has been called "King Cotton diplomacy." Cotton was important to the British economy because textiles were at the heart of British industry, and three-fourths of Britain's suply of raw cotton came from the American South. Southerners therefore believed that Britain would recognize the Confederacy's independence and use the British navy to break the blockade. Southerners were so firmly convinced of King Cotton's leverage that they kept the 1861 crop at home rather than try to export it through the blockade, hoping thereby to compel the British to intervene. But the strategy backfired: bumper crops in 1859 and 1860 had piled up a surplus of raw cotton in British warehouses, thus delaying until 1862 the "cotton famine" on which Southerners had counted.

In the end, the South's 1861 voluntary embargo of cotton cost them dearly. By 1862, when the Confederacy needed to export cotton in order to pay for imported war matériel, the Union blockade had tightened to the point that the slow sailing ships with large cargo capacity could not get through. The sleek, fast, steam-powered "blockade runners" – that became increasingly prominent – had a smaller cargo capacity and charged high rates because of the growing risk of capture or sinking by the Union navy. Although most of these runners got through, the blockade by 1862 had reduced the Confederacy's seaborne commerce to the point where both the Southern armies and the homefront began to suffer serious shortages. As a naval power that had relied on blockades in past wars, and expected to do so in the future, Britain refused to challenge the legitimacy of the Union blockade.

Lacking the capacity to build a naval force at home, the Confederacy hoped to use British shipyards for the purpose. Through a loophole in the British neutrality law, two fast commerce raiders built by a private firm in Liverpool made their way into Confederate ownership in 1862. Named the *Florida* and the *Alabama*, they roamed the seas for the next two years, capturing or sinking Union merchant ships and whalers. Altogether, these and other Confederate commerce raiders destroyed or captured 257 Union merchant vessels and drove at least 700 others to foreign registry. The U.S. merchant marine never recovered. But this Confederate achievement, though spectacular, made only a tiny dent in the Union war effort, especially when compared with the 1,500 blockade runners captured or destroyed by the Union navy, or the thousands of ships that were deterred from even attempting to beat the blockade.

In 1861 neither side anticipated how long and destructive the war would become. Southerners expected the Yankees to give up after Confederate armies had whipped them in one or two battles. Northerners were likewise confident of success, expecting that after one or two Union victories the Southern people would come to their senses, throw out their secessionist leaders, and return to their old allegiance.

The General-in-Chief of the United States Army was Winfield Scott, a veteran of the War of 1812, and commander of the army that had captured Mexico City in 1847. A Virginian who had remained loyal to the Union, Scott evolved a military strategy based on his conviction that there were many Southerners ready to be won back to the Union. The main elements of his strategy were the naval blockade and a combined army–navy expedition to take control of the Mississippi River, thus surrounding the Confederacy. In Scott's words, this would "bring them to terms with less bloodshed than by any other plan." The Northern press gently ridiculed Scott's strategy as "the Anaconda Plan," after the South American snake that slowly squeezes its prey to death.

Most Northerners believed that the Confederacy could be overcome only by victory in battle. Virginia emerged as the most likely battleground, especially after the Confederate government moved its capital to Richmond in May, 1861. "Forward to Richmond," clamored Northern newspapers. And forward toward Richmond marched the main Union army, only to be defeated on July 21 on the banks of the Bull Run 25 miles southwest of Washington. Victory in the battle of Manassas, as Southerners called it (the Union name is the battle of Bull Run) exhilarated Confederates and confirmed their feelings of martial superiority over the Yankees. But this overconfidence had its negative side: some Southerners thought the war was won. Northerners, by contrast, were jolted out of their expectations of a short war. A new mood of reality and grim determination gripped the North. Congress authorized the enlistment of up to a million three-year volunteers. Hundreds of thousands of new recruits enlisted during the next few months. Lincoln called General George B. McClellan to Washington to organize the new troops into the Army of the Potomac.

An energetic, talented officer only 34-years old, small of stature but great with an aura of destiny, McClellan soon won the sobriquet in the Northern press of "The Young Napoleon." He organized and trained the Army of the Potomac into a large, well-disciplined, and well-equipped fighting force. He seemed to be just what the North needed after the dispiriting defeat at Bull Run. When the 75-year-old Scott stepped down as General-in-Chief on November 1, McClellan succeeded him.

But as winter approached and McClellan did nothing to advance against the smaller Confederate army whose outposts stood only a few miles from Washington, his failings as a commander began to manifest themselves. He was a perfectionist in a profession where nothing could ever be perfect. His army was perpetually *almost* ready to move. McClellan was afraid to risk failure, so he risked nothing. He consistently overestimated the strength of enemy forces confronting him (sometimes by a multiple of two or three) and used these faulty estimates as a reason for inaction. This caution and defensive-mindedness that McClellan instilled into the Army of the Potomac persisted for almost three years. In the meantime, the Confederate Army of Northern Virginia acquired a new commander in 1862, Robert E. Lee, who seized the initiative in the Eastern theater and thereby reversed a momentum toward Union victory that had built up during the first five months of 1862.

*The Battle of First Manassas (Bull Run), July 21, 1861, from a lithograph by Kurz and Allison*

# Fort Sumter APRIL 12-14, 1861

On December 20, 1860, South Carolina seceded from the Union, an act which led by February 2, 1861, to the secession of six more states; the formation of a Confederate government; and a confrontation in Charleston Harbor that had momentous consequences.

Posted at Fort Moultrie, one of four Federal forts in Charleston Harbor, were two companies of artillery commanded by Major Robert Anderson. Concerned that Moultrie could not be defended, Anderson on December 26 transferred his men to Fort Sumter, a three-tiered masonry work at the harbor's entrance. While the fort was prepared to resist an attack, South Carolina forces took possession of Moultrie and the other forts, and positioned batteries to command the sea approaches to Sumter.

On January 9, a cannon on Cumming's Point opened fire on the merchant steamer *Star of the West*, sent by the Union to supply and reinforce Sumter, and she turned back. In February, when the Confederate Government was established – with Jefferson Davis as President – the Confederacy assumed responsibility for the defense of Charleston Harbor. Brigadier General P.G.T. Beauregard was placed in command.

In his inaugural address as President of the United States, Lincoln vowed to "hold, occupy, and possess" Federal property in the South. Thus, on April 9, a ship carrying supplies for Fort Sumter sailed from New York. On April 11, apprised of Lincoln's plans, Beauregard demanded that Anderson evacuate Fort Sumter. Although aware that lack of supplies would force him to evacuate within a few days, Anderson refused. After a further exchange, Anderson replied that unless he received instructions or supplies from Washington by noon, April 15, he would evacuate.

However, on April 12 at 4.30 am, following Anderson's refusal to leave the fort immediately, the battery at Fort Johnson fired a signal round which exploded high above Fort Sumter.

By daybreak Fort Sumter was under heavy fire from more than 70 Confederate guns, but it was 7am before the first of the fort's cannon responded. The bombardment continued for 34 hours, until 2.30pm on April 13, when Anderson surrendered. Although the fort had suffered considerable damage, no lives were lost.

The surrender of Fort Sumter united a previously divided North behind President Lincoln in his mission to preserve the Union.

*Major Robert Anderson (left), a Kentuckian by birth, commanded a garrison of one hundred officers and other ranks. Lacking powder, guns, and men Anderson managed to hold out against the Confederate bombardment for 34 hours, finally surrendering only when further resistance seemed both fruitless and impracticable.*

1. **Dec 26, 1860:** Anderson evacuates Fort Moultrie, transferring his men to Fort Sumter.
2. **Dec 27, 1860:** Confederate troops take possession of Forts Moultrie, Johnson and Pinkney.
3. **Jan 9, 1861, dawn:** Federal supply ship, *Star of the West*, is fired on by Morris Island battery, and turns back to New York.
4. **April 11:** Gen. Beauregard demands fort's surrender. Anderson refuses. **April 12, 3.20am:** Confederates reject Anderson's response.
5. **April 12, 4.30am:** Fort Johnson fires a signal round which explodes above Fort Sumter. Forts Johnson, Moultrie and others open fire on Sumter.
6. **April 12, 7am:** Sumter responds.
7. **April 13, 2.30pm:** Anderson surrenders.

Magnolia Cemetery
N.E. R.R.
S.C. R.R.
Drum Island
Hobcaw Po
ASHLEY R.
Wappoo Creek

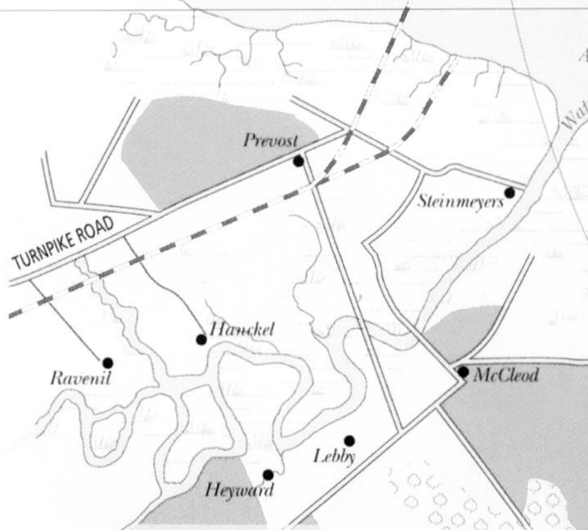
Prevost
Steinmeyers
TURNPIKE ROAD
Hanckel
Ravenil
McCleod
Lebby
Heyward

*Charleston, (below) looking toward the harbor and Fort Sumter prior to the opening of hostilities.*

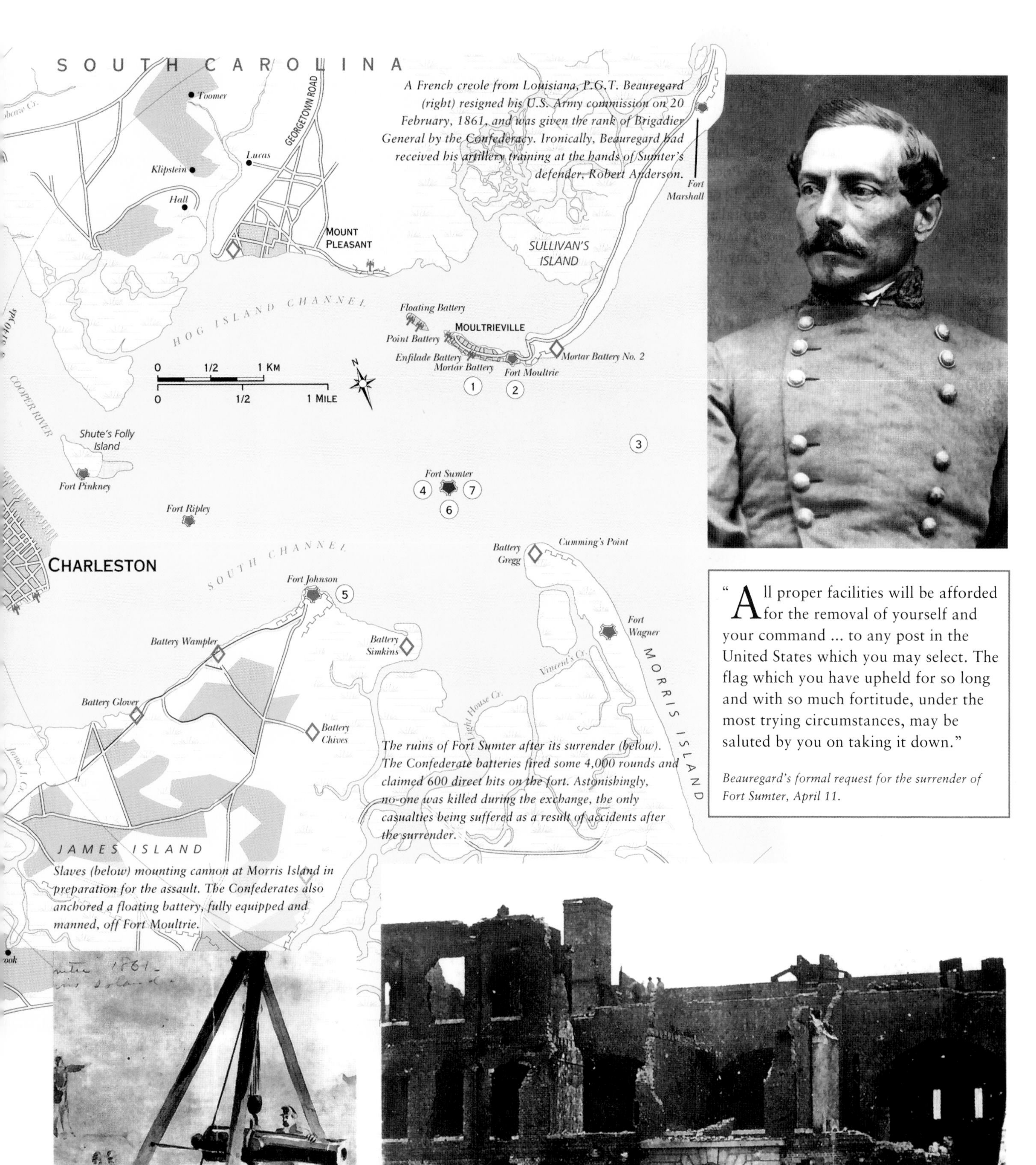
SOUTH CAROLINA
Toomer
GEORGETOWN ROAD
Lucas
Klipstein
Hall
MOUNT PLEASANT
SULLIVAN'S ISLAND
Fort Marshall
HOG ISLAND CHANNEL
Floating Battery
MOULTRIEVILLE
Point Battery
Enfilade Battery
Mortar Battery
Fort Moultrie
Mortar Battery No. 2
0 1/2 1 KM
0 1/2 1 MILE
N
COOPER RIVER
Shute's Folly Island
Fort Pinkney
Fort Ripley
Fort Sumter
1
2
3
4
5
6
7
CHARLESTON
SOUTH CHANNEL
Fort Johnson
Battery Gregg
Cumming's Point
Fort Wagner
MORRIS ISLAND
Battery Simkins
Battery Wampler
Vincent's Cr.
Battery Glover
Light House Cr.
Battery Chives
JAMES ISLAND

*A French creole from Louisiana, P.G.T. Beauregard (right) resigned his U.S. Army commission on 20 February, 1861, and was given the rank of Brigadier General by the Confederacy. Ironically, Beauregard had received his artillery training at the hands of Sumter's defender, Robert Anderson.*

"All proper facilities will be afforded for the removal of yourself and your command ... to any post in the United States which you may select. The flag which you have upheld for so long and with so much fortitude, under the most trying circumstances, may be saluted by you on taking it down."

*Beauregard's formal request for the surrender of Fort Sumter, April 11.*

*The ruins of Fort Sumter after its surrender (below). The Confederate batteries fired some 4,000 rounds and claimed 600 direct hits on the fort. Astonishingly, no-one was killed during the exchange, the only casualties being suffered as a result of accidents after the surrender.*

*Slaves (below) mounting cannon at Morris Island in preparation for the assault. The Confederates also anchored a floating battery, fully equipped and manned, off Fort Moultrie.*

# Clashes in Missouri JULY 5 – NOVEMBER 7 1861

The war in Missouri in 1861 pitted Federal forces, under Brigadier General Nathaniel Lyon, against the pro-secessionist State Guard, led by Governor Claiborne Fox Jackson and Major General Sterling Price. Although Missouri had not seceded, Lyon drove the state legislature from the capital at Jefferson City on June 15. Two days later, Lyon routed the State Guard at Boonville, then sent a force southwest to cut off their retreat.

From Springfield, a Federal force of 1,100 advanced to a point just north of Carthage. There, on July 5, they were defeated by 4,000 Confederates marching south under Governor Jackson. The Federals retreated to Springfield. Jackson continued south to Cowskin Prairie, where Price had gathered an additional 1,200 men from across Missouri. The Missourians withdrew into Arkansas, uniting with Confederate and Arkansas State troops under Brigadier General Ben McCulloch. Lyon started south on July 3 and arrived at Springfield on July 13 with a force of over 5,000 men, but lacked supplies for offensive operations.

On August 1, McCulloch led the combined Confederate forces north toward Springfield, but rain canceled his plans to attack Lyon on August 9. The aggressive Lyon, however, launched a surprise dawn attack on the Confederate camp at Wilson's Creek on August 10. Lyon was killed and the defeated Federals retreated to Rolla.

McCulloch withdrew to Arkansas, but Price advanced with the State Guard toward Lexington. The Confederates besieged from September 12 to 20, finally capturing the 2,800 defenders, with minimal losses. Lack of supplies forced Price to retire back to Springfield, and the year ended with Federal forces controlling most of the state.

In a separate action in eastern Missouri, Brigadier General Ulysses S. Grant sailed down the Mississippi from Cairo, attacking and nearly defeating an equal Confederate force under Brigadier General Gideon H. Pillow at Belmont. When Confederate reinforcements from Columbus drove Grant back to his gunboats, he returned to Cairo.

> "The rebels have been elated and emboldened while our troops have been depressed, if not discouraged... It may be said of these victims, 'They have fallen, and to what end?' "
>
> *The* Chicago Tribune *after Grant's return to Cairo.*

*Union artillery under the command of Capt. James Totten repel a Confederate cavalry charge during the battle of Wilson's Creek. The Confederates quickly rallied and succeeded in breaking the Union line.*

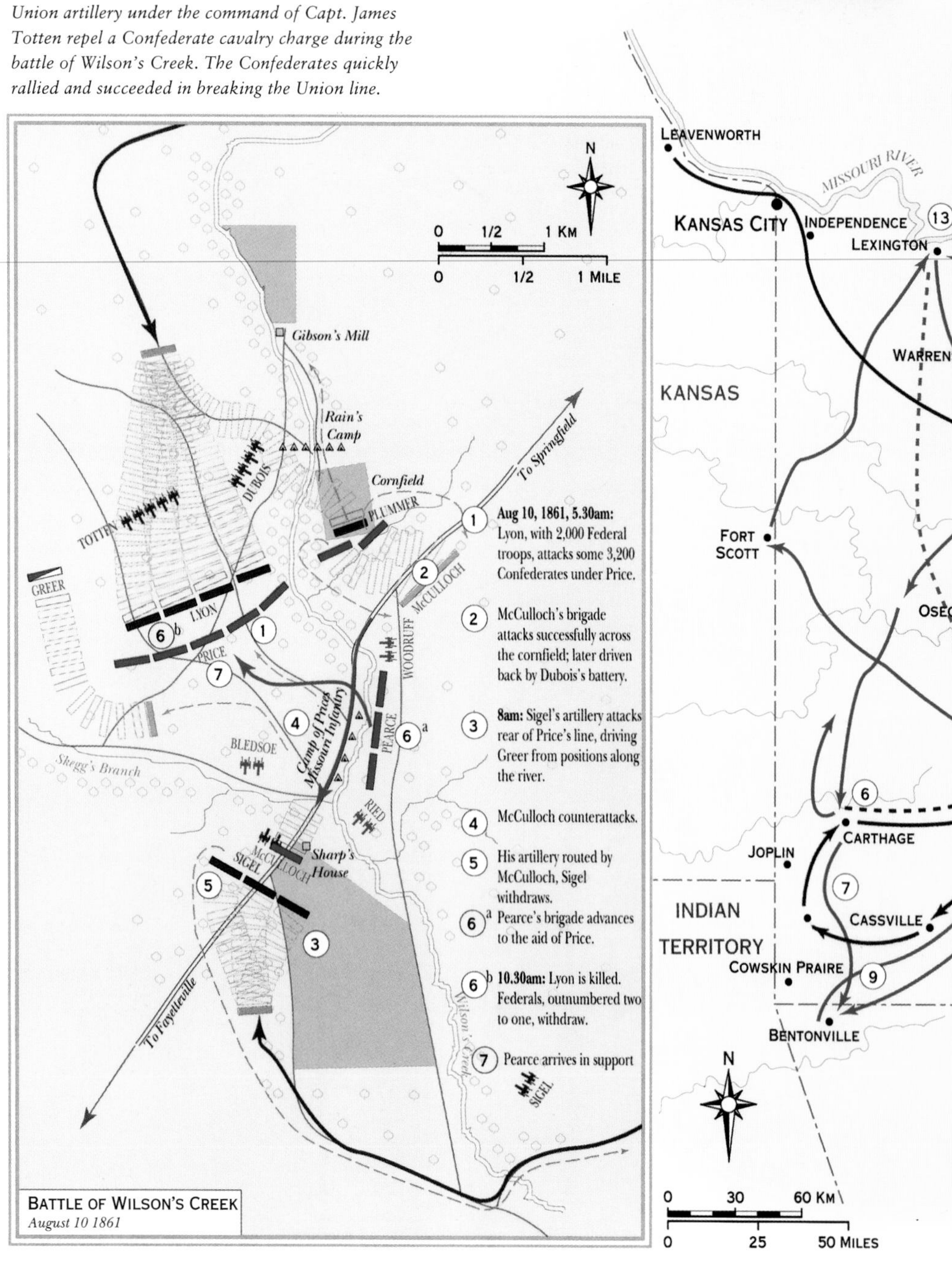

N
0 1/2 1 KM
0 1/2 1 MILE
Gibson's Mill
Rain's Camp
Cornfield
To Springfield
TOTTEN
DUBOIS
PLUMMER
GREER
LYON
PRICE
McCULLOCH
WOODRUFF
Camp of Price's Missouri Infantry
PEARCE
BLEDSOE
Skegg's Branch
RIED
Sharp's House
SIGEL
To Fayetteville
Wilson's Creek
1 Aug 10, 1861, 5.30am: Lyon, with 2,000 Federal troops, attacks some 3,200 Confederates under Price.
2 McCulloch's brigade attacks successfully across the cornfield; later driven back by Dubois's battery.
3 8am: Sigel's artillery attacks rear of Price's line, driving Greer from positions along the river.
4 McCulloch counterattacks.
5 His artillery routed by McCulloch, Sigel withdraws.
6a Pearce's brigade advances to the aid of Price.
6b 10.30am: Lyon is killed. Federals, outnumbered two to one, withdraw.
7 Pearce arrives in support
BATTLE OF WILSON'S CREEK
August 10 1861
LEAVENWORTH
MISSOURI RIVER
KANSAS CITY
INDEPENDENCE
13
LEXINGTON
WARRENSB
KANSAS
FORT SCOTT
OSECOL
6
CARTHAGE
JOPLIN
7
INDIAN TERRITORY
CASSVILLE
COWSKIN PRAIRE
9
BENTONVILLE
N
0 30 60 KM
0 25 50 MILES

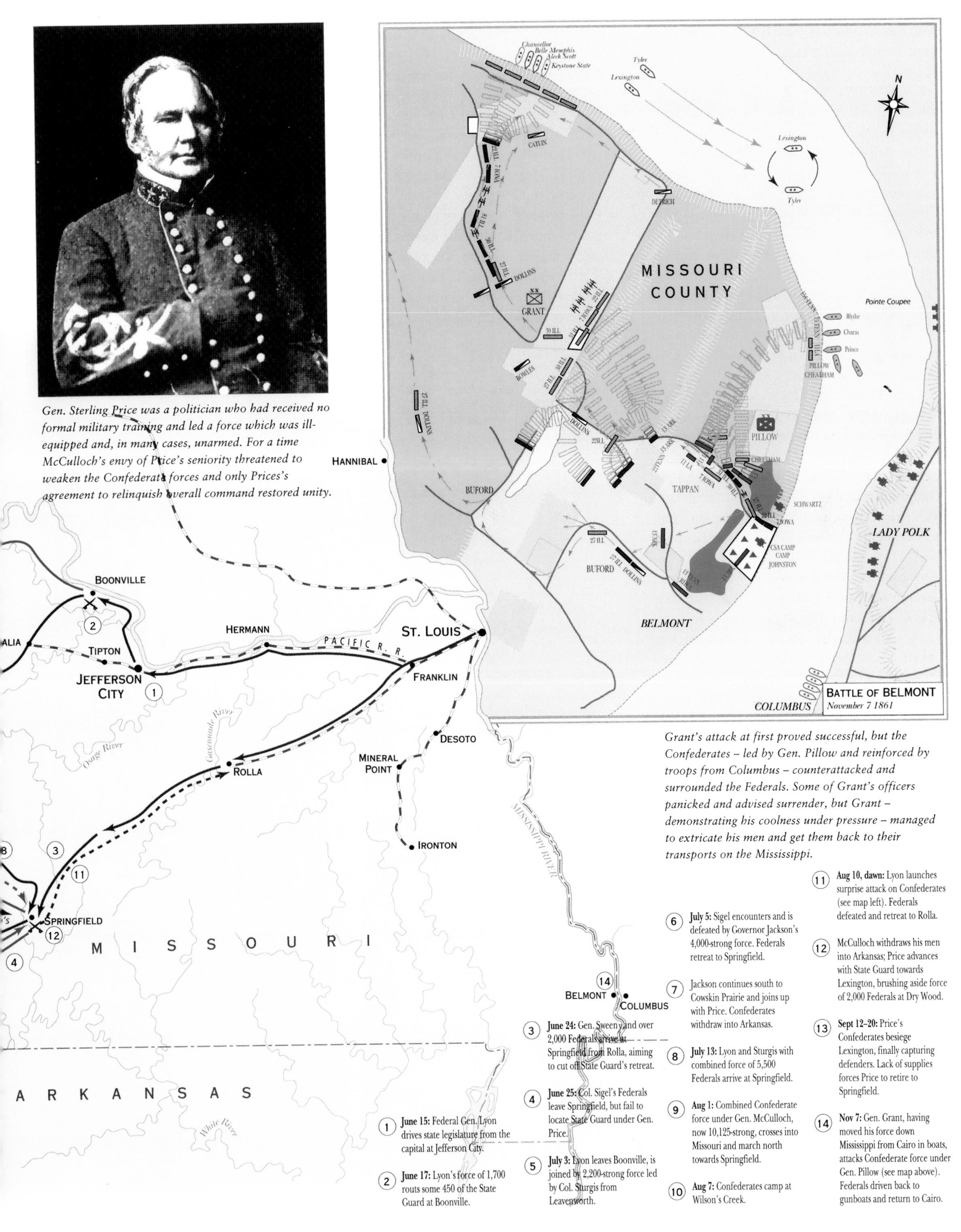

MISSOURI COUNTY
Pointe Coupee
LADY POLK
BELMONT
COLUMBUS
BATTLE OF BELMONT
November 7 1861
GRANT
PILLOW
CHEATHAM
TAPPAN
BUFORD
DOLLINS
SCHWARTZ
CATLIN
DETRICH
BOWLES
CSA CAMP
CAMP JOHNSTON
Lexington
Tyler
Blythe
Charm
Prince
HANNIBAL
BOONVILLE
HERMANN
ST. LOUIS
PACIFIC R. R.
TIPTON
JEFFERSON CITY
FRANKLIN
DESOTO
MINERAL POINT
ROLLA
IRONTON
SPRINGFIELD
Osage River
Gasconade River
MISSISSIPPI RIVER
White River
BELMONT
COLUMBUS
MISSOURI
ARKANSAS


*Gen. Sterling Price was a politician who had received no formal military training and led a force which was ill-equipped and, in many cases, unarmed. For a time McCulloch's envy of Price's seniority threatened to weaken the Confederate forces and only Prices's agreement to relinquish overall command restored unity.*

*Grant's attack at first proved successful, but the Confederates – led by Gen. Pillow and reinforced by troops from Columbus – counterattacked and surrounded the Federals. Some of Grant's officers panicked and advised surrender, but Grant – demonstrating his coolness under pressure – managed to extricate his men and get them back to their transports on the Mississippi.*

(1) **June 15:** Federal Gen. Lyon drives state legislature from the capital at Jefferson City.

(2) **June 17:** Lyon's force of 1,700 routs some 450 of the State Guard at Boonville.

(3) **June 24:** Gen. Sweeny and over 2,000 Federals arrive at Springfield from Rolla, aiming to cut off State Guard's retreat.

(4) **June 25:** Col. Sigel's Federals leave Springfield, but fail to locate State Guard under Gen. Price.

(5) **July 3:** Lyon leaves Boonville, is joined by 2,200-strong force led by Col. Sturgis from Leavenworth.

(6) **July 5:** Sigel encounters and is defeated by Governor Jackson's 4,000-strong force. Federals retreat to Springfield.

(7) Jackson continues south to Cowskin Prairie and joins up with Price. Confederates withdraw into Arkansas.

(8) **July 13:** Lyon and Sturgis with combined force of 5,500 Federals arrive at Springfield.

(9) **Aug 1:** Combined Confederate force under Gen. McCulloch, now 10,125-strong, crosses into Missouri and march north towards Springfield.

(10) **Aug 7:** Confederates camp at Wilson's Creek.

(11) **Aug 10, dawn:** Lyon launches surprise attack on Confederates (see map left). Federals defeated and retreat to Rolla.

(12) McCulloch withdraws his men into Arkansas; Price advances with State Guard towards Lexington, brushing aside force of 2,000 Federals at Dry Wood.

(13) **Sept 12–20:** Price's Confederates besiege Lexington, finally capturing defenders. Lack of supplies forces Price to retire to Springfield.

(14) **Nov 7:** Gen. Grant, having moved his force down Mississippi from Cairo in boats, attacks Confederate force under Gen. Pillow (see map above). Federals driven back to gunboats and return to Cairo.

# Western Virginia JUNE 3– SEPTEMBER 13 1861

Western Virginians had little in common with the rest of the state, and actively opposed secession. Steps were taken to organize a loyalist Virginia government in Wheeling, and to raise Virginia units to fight for the Union.

The Baltimore & Ohio Railroad crossed western Virginia; to control it was a strategic imperative for the Union, and by late May, Federal troops had occupied Grafton, a key junction. The Confederacy responded by sending soldiers into the region to recruit and threaten the railroad. These forces clashed at Philippi on June 3. Federal columns made a night march and converged at Philippi. Suprised and routed in a skirmish known as the "Philippi Races," the Confederates fled southward.

Both North and South responded vigorously, sending in reinforcements. Confederate Brigadier General Robert B. Garnett divided his small army, taking the larger force to Laurel Hill to guard the road to Philippi. Lieutenant Colonel John Pegram, with 1,300, held Rich Mountain on the vital Staunton-Parkersburg Turnpike. In early July, Major General George B. McClellan sent a force from Philippi to threaten Garnett at Laurel Hill. McClellan marched against Pegram at Rich Mountain. On July 11, Brigadier General William S. Rosecrans's column climbed Rich Mountain and overwhelmed the small Confederate force in the pass. With the Federals in his rear, Pegram was forced to abandon his position during the night.

The Federal victory at Rich Mountain rendered Garnett's Laurel Hill position untenable, and he evacuated on July 12. Garnett, believing McClellan was at Beverly, sought to extricate himself from the area by way of the Cheat River Valley. The pursuing Federals overtook Garnett at Carrick's Ford on the 13th; Garnett was killed and his small army escaped to Monterey.

The Confederate government's response was prompt: on July 28, General Robert E. Lee left Richmond for western Virginia. Aiming to defeat Federal forces occupying the Tygart Valley and East Cheat Summit, Lee sent three columns down the valley; one

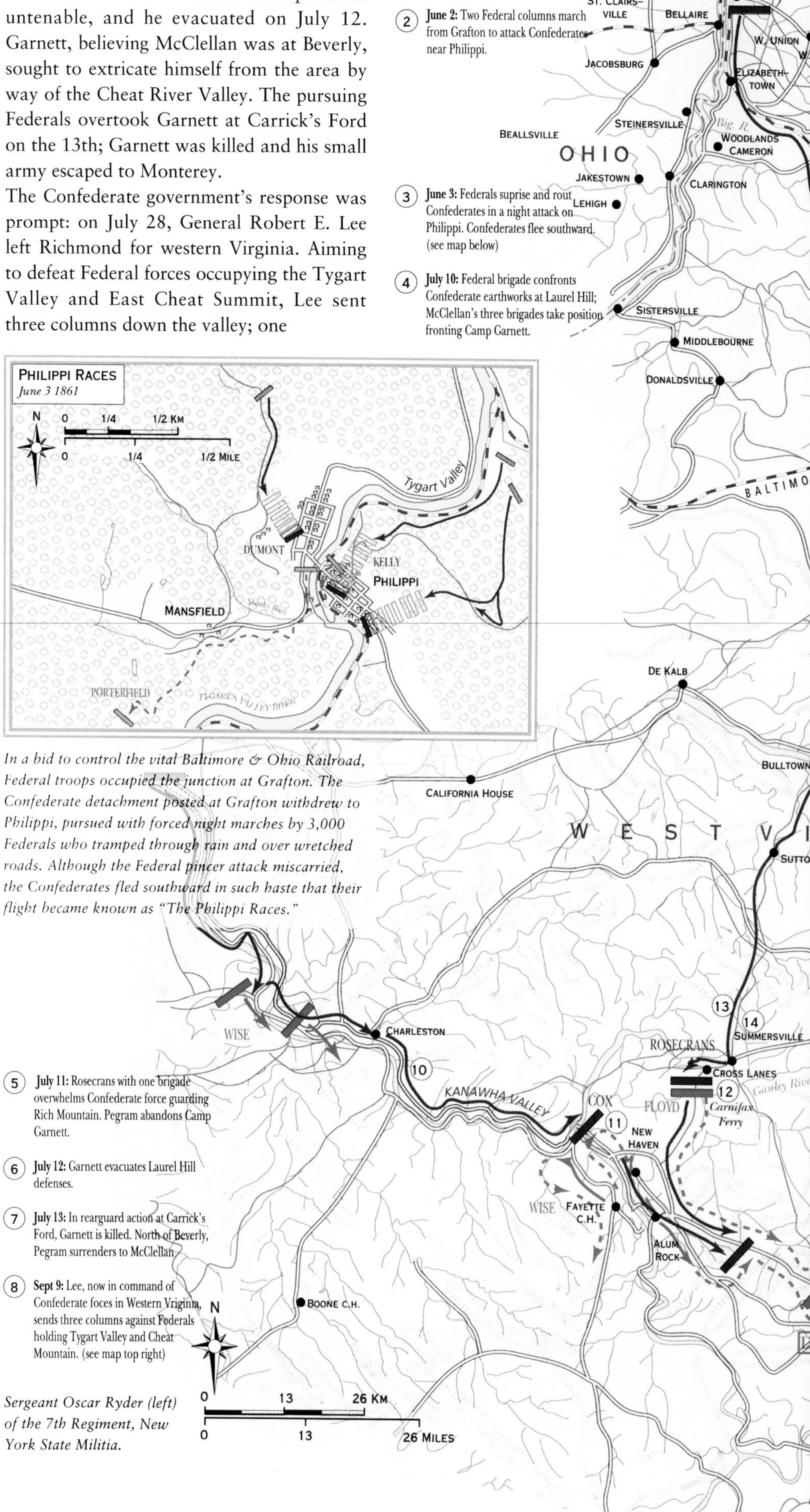

1 May 30, 1861: Union troops arrive by rail from Wheeling and occupy Grafton
2 June 2: Two Federal columns march from Grafton to attack Confederates near Philippi.
3 June 3: Federals suprise and rout Confederates in a night attack on Philippi. Confederates flee southward. (see map below)
4 July 10: Federal brigade confronts Confederate earthworks at Laurel Hill; McClellan's three brigades take position fronting Camp Garnett.
5 July 11: Rosecrans with one brigade overwhelms Confederate force guarding Rich Mountain. Pegram abandons Camp Garnett.
6 July 12: Garnett evacuates Laurel Hill defenses.
7 July 13: In rearguard action at Carrick's Ford, Garnett is killed. North of Beverly, Pegram surrenders to McClellan.
8 Sept 9: Lee, now in command of Confederate foces in Western Vriginia, sends three columns against Federals holding Tygart Valley and Cheat Mountain. (see map top right)
PHILIPPI RACES
June 3 1861
N
0 1/4 1/2 KM
0 1/4 1/2 MILE
Tygart Valley
DUMONT
KELLY
PHILIPPI
MANSFIELD
PORTERFIELD
BURLINGTON
FARMINGTON
W. ALEX
BRIDGEPORT
WHEELING
ST. CLAIRSVILLE
BELLAIRE
W. UNION
JACOBSBURG
ELIZABETHTOWN
STEINERSVILLE
BEALLSVILLE
WOODLANDS
CAMERON
OHIO
JAKESTOWN
CLARINGTON
LEHIGH
SISTERSVILLE
MIDDLEBOURNE
DONALDSVILLE
BALTIMOR
DE KALB
BULLTOWN
CALIFORNIA HOUSE
W E S T V I
SUTTON
WISE
CHARLESTON
ROSECRANS
SUMMERSVILLE
CROSS LANES
KANAWHA VALLEY
COX
FLOYD
Carnifax Ferry
NEW HAVEN
WISE
FAYETTE C.H.
ALUM ROCK
BOONE C.H.
N
0 13 26 KM
0 13 26 MILES


*In a bid to control the vital Baltimore & Ohio Railroad, Federal troops occupied the junction at Grafton. The Confederate detachment posted at Grafton withdrew to Philippi, pursued with forced night marches by 3,000 Federals who tramped through rain and over wretched roads. Although the Federal pincer attack miscarried, the Confederates fled southward in such haste that their flight became known as "The Philippi Races."*

*Sergeant Oscar Ryder (left) of the 7th Regiment, New York State Militia.*

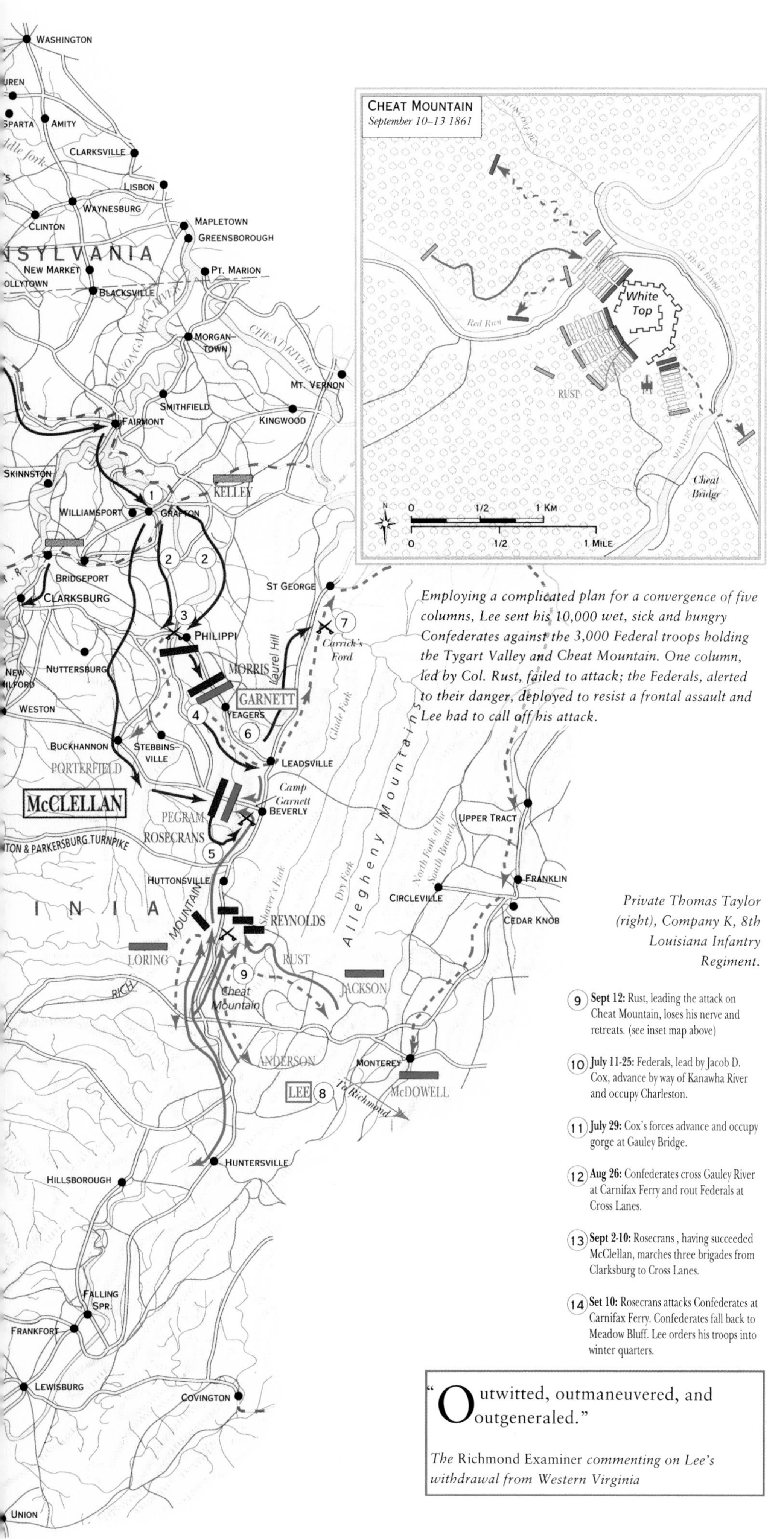

CHEAT MOUNTAIN
September 10–13 1861
White Top
Red Run
RUST
Cheat Bridge
0 1/2 1 KM
0 1/2 1 MILE
WASHINGTON
AMITY
CLARKSVILLE
LISBON
WAYNESBURG
CLINTON
MAPLETOWN
GREENSBOROUGH
NEW MARKET
PT. MARION
BLACKSVILLE
MORGANTOWN
CHEAT RIVER
MT. VERNON
SMITHFIELD
FAIRMONT
KINGWOOD
SKINNSTON
KELLEY
WILLIAMSPORT
GRAFTON
BRIDGEPORT
CLARKSBURG
ST GEORGE
PHILIPPI
Carrick's Ford
NUTTERSBURG
MORRIS
Laurel Hill
GARNETT
WESTON
YEAGERS
BUCKHANNON
STEBBINS-VILLE
LEADSVILLE
PORTERFIELD
McCLELLAN
Camp Garnett
BEVERLY
PEGRAM
ROSECRANS
UPPER TRACT
FRANKLIN
HUTTONSVILLE
CIRCLEVILLE
CEDAR KNOB
REYNOLDS
LORING
RUST
Allegheny Mountains
JACKSON
Cheat Mountain
ANDERSON
MONTEREY
McDOWELL
LEE
To Richmond
HUNTERSVILLE
HILLSBOROUGH
FALLING SPR.
FRANKFORT
LEWISBURG
COVINGTON
UNION
9 Sept 12: Rust, leading the attack on Cheat Mountain, loses his nerve and retreats. (see inset map above)
10 July 11-25: Federals, lead by Jacob D. Cox, advance by way of Kanawha River and occupy Charleston.
11 July 29: Cox's forces advance and occupy gorge at Gauley Bridge.
12 Aug 26: Confederates cross Gauley River at Carnifax Ferry and rout Federals at Cross Lanes.
13 Sept 2-10: Rosecrans, having succeeded McClellan, marches three brigades from Clarksburg to Cross Lanes.
14 Set 10: Rosecrans attacks Confederates at Carnifax Ferry. Confederates fall back to Meadow Bluff. Lee orders his troops into winter quarters.


*Employing a complicated plan for a convergence of five columns, Lee sent his 10,000 wet, sick and hungry Confederates against the 3,000 Federal troops holding the Tygart Valley and Cheat Mountain. One column, led by Col. Rust, failed to attack; the Federals, alerted to their danger, deployed to resist a frontal assault and Lee had to call off his attack.*

brigade established a roadblock on the west summit of Cheat Mountain, while Colonel Albert Rust's force prepared to overwhelm the Federals on the east summit. On September 12, Rust, who was to trigger the attack, lost his nerve. Lee's Cheat Mountain campaign unraveled, and he recalled his troops.

Following a Confederate victory at Cross Lanes in late August, Rosecrans with 6,000 men left Clarksburg to support Federal forces that had thrust up the Kanawha Valley. On September 10, Rosecrans attacked and defeated 2,000 Confederates in earthworks at Carnifax Ferry on the Gauley River. The Confederates retreated south to Meadow Bluff.

With the advent of cold weather, Lee gave up all thoughts of offensive operations and ordered his troops into winter quarters. The Confederacy had lost trans-Allegheny Virginia, which in 1863 would become West Virginia.

> "Soldiers! I have heard there was danger here. I have come to place myself at your head and share it with you. I fear now but one thing – that you will not find foemen worthy of your steel."
>
> *Gen. George McClellan, upon his arrival at Philippi*

*Private Thomas Taylor (right), Company K, 8th Louisiana Infantry Regiment.*

> "Outwitted, outmaneuvered, and outgeneraled."
>
> *The* Richmond Examiner *commenting on Lee's withdrawal from Western Virginia*

# Virginia JUNE 10 - OCTOBER 12 1861

The transfer of the Confederate capital to Richmond guaranteed that Virginia would become a major battleground of the Civil War. Large numbers of Federal troops soon began to concentrate on Virginia's northern border and at Fort Monroe in Hampton Roads. The first clash occurred on June 10, 1861, near Fort Monroe, when Major General Benjamin Butler sent a 4,400-strong Federal force to attack a Confederate outpost at Big Bethel. Sharply repulsed by Colonel D. H. Hill's 1,400 defenders, the Federals withdrew in confusion. Losses were slight on both sides, totaling only 76 Federals and 11 Confederates, but the action received inordinate attention because it was the first significant clash of the war.

Activity now shifted to northern Virginia, where in mid-July Brigadier General Irvin McDowell led 37,000 Federal troops from Washington toward the rail hub at Manassas Junction. Facing him behind Bull Run was General Beauregard with 21,000 men. McDowell expected Brigadier General Robert Patterson to hold General Joseph Johnston's 11,000 Confederates in the Shenandoah Valley. Unknown to McDowell, Patterson failed to restrain Johnston, whose troops began to join Beauregard by rail. On July 21, McDowell feinted toward Beauregard's front while turning his left flank. Initially successful, McDowell's attack pushed Confederate defenders southward until they rallied on Henry House Hill. There, the timely arrival of Confederate reinforcements combined with Federal fatigue to collapse McDowell's line by mid-afternoon. Shielded by a battalion of regulars, the Federal volunteers withdrew in great disorder. Fortunately for McDowell, the Confederates were too disorganized to pursue, permitting the Federals to escape to the safety of the Washington defenses. McDowell lost 2,896 men while Beauregard's and Johnston's commands suffered a total of 1,982 casualties.

The final major clash of the year in Virginia occurred on the upper Potomac River at Ball's Bluff on October 21. Ordered to make a demonstration across the Potomac in support of another Federal probe, Colonel Edward Baker incautiously led his brigade into a trap sprung by Confederate Brigadier General Nathan Evans. Forced to fight with the river at his back, Baker was killed and his brigade routed. Again, Federal casualties totals greatly exceeded Confederate. Clearly, the road to Richmond would be neither short nor cheap, and both sides began to prepare for a long war.

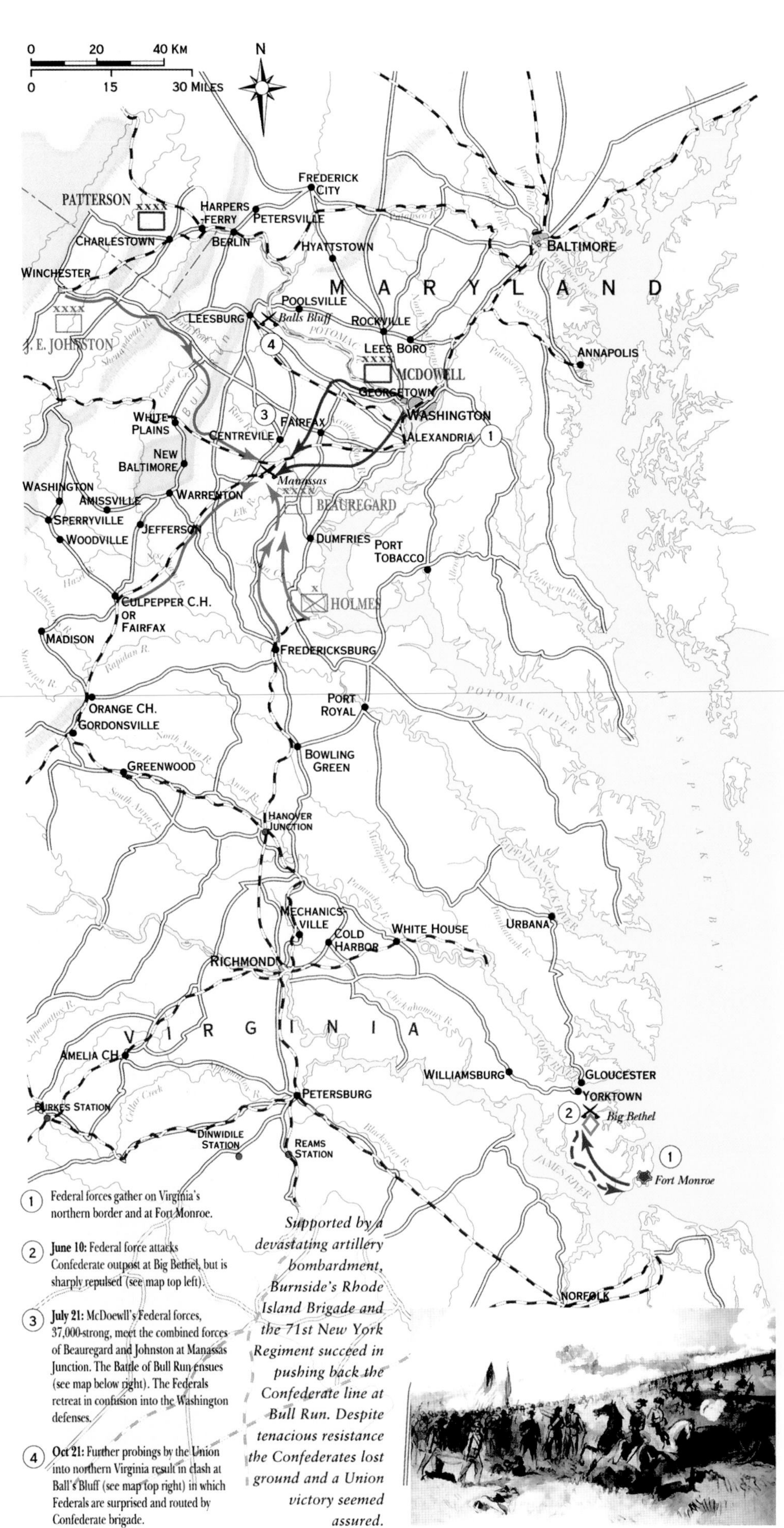
0 20 40 KM
0 15 30 MILES
N
PATTERSON
HARPERS FERRY
FREDERICK CITY
PETERSVILLE
CHARLESTOWN
BERLIN
HYATTSTOWN
BALTIMORE
WINCHESTER
M A R Y L A N D
POOLSVILLE
J. E. JOHNSTON
LEESBURG
Balls Bluff
ROCKVILLE
LEES BORO
ANNAPOLIS
MCDOWELL
GEORGETOWN
WHITE PLAINS
CENTREVILLE
FAIRFAX
WASHINGTON
ALEXANDRIA
NEW BALTIMORE
WASHINGTON
AMISSVILLE
WARRENTON
Manassas
BEAUREGARD
SPERRYVILLE
JEFFERSON
WOODVILLE
DUMFRIES
PORT TOBACCO
CULPEPPER C.H. OR FAIRFAX
HOLMES
MADISON
FREDERICKSBURG
POTOMAC RIVER
PORT ROYAL
ORANGE CH.
GORDONSVILLE
BOWLING GREEN
GREENWOOD
HANOVER JUNCTION
MECHANICSVILLE
COLD HARBOR
WHITE HOUSE
URBANA
RICHMOND
V I R G I N I A
AMELIA CH
WILLIAMSBURG
GLOUCESTER
YORKTOWN
PETERSBURG
BURKES STATION
Big Bethel
DINWIDDIE STATION
REAMS STATION
Fort Monroe
JAMES RIVER
NORFOLK
1 Federal forces gather on Virginia's northern border and at Fort Monroe.
2 June 10: Federal force attacks Confederate outpost at Big Bethel, but is sharply repulsed (see map top left).
3 July 21: McDoewll's Federal forces, 37,000-strong, meet the combined forces of Beauregard and Johnston at Manassas Junction. The Battle of Bull Run ensues (see map below right). The Federals retreat in confusion into the Washington defenses.
4 Oct 21: Further probings by the Union into northern Virginia result in clash at Ball's Bluff (see map top right) in which Federals are surprised and routed by Confederate brigade.

*Supported by a devastating artillery bombardment, Burnside's Rhode Island Brigade and the 71st New York Regiment succeed in pushing back the Confederate line at Bull Run. Despite tenacious resistance the Confederates lost ground and a Union victory seemed assured.*

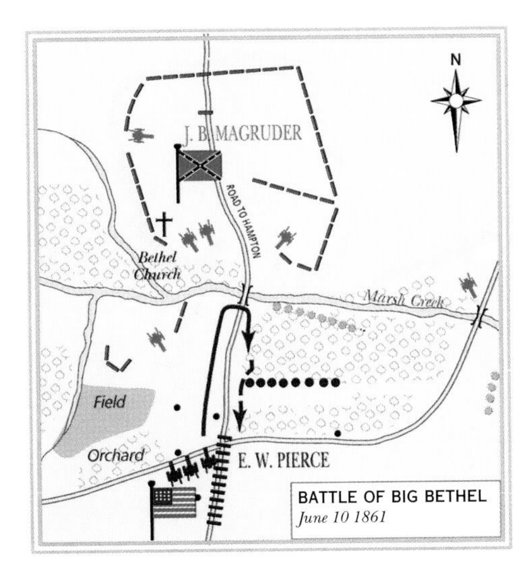

N
J. B. MAGRUDER
ROAD TO HAMPTON
Bethel Church
Marsh Creek
Field
Orchard
E. W. PIERCE
BATTLE OF BIG BETHEL
June 10 1861


*A strong Federal force marched from Fort Monroe through the night to attack Confederate positions at Big Bethel. The Federal attack was hesitant and confused. After about an hour, the Union force retired in confusion, having suffered nearly ten times as many casualties as they had inflicted (see map above).*

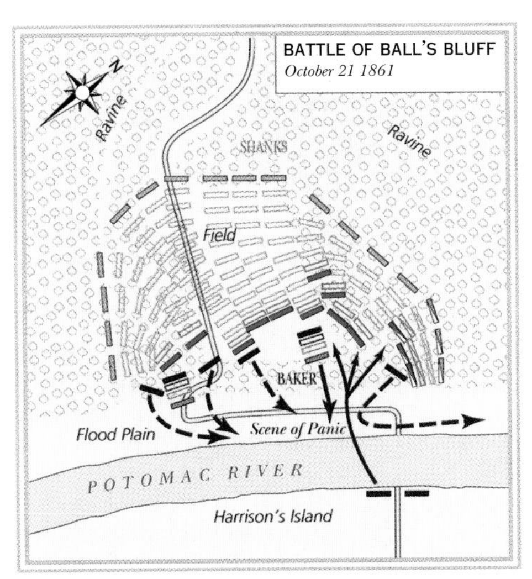

BATTLE OF BALL'S BLUFF
October 21 1861
N
Ravine
SHANKS
Ravine
Field
BAKER
Flood Plain
Scene of Panic
POTOMAC RIVER
Harrison's Island


*Hoping to dislodge the Confederates from Leesburg, McClellan ordered a demonstration across the Potomac River. Baker's brigade crossed the river, but ran into a Confederate brigade at the top of Ball's Bluff which drove them in disorder down the bank and into the river (See map above).*

*Confederate Col. D.H. Hill (above) whose command repulsed Butler's assault at Big Bethel. An almost fanatical supporter of secession, the North Carolinian was an impetuous but courageous commander. His spirited defense at Big Bethel resulted in 76 Union casualties and eight Confederate.*

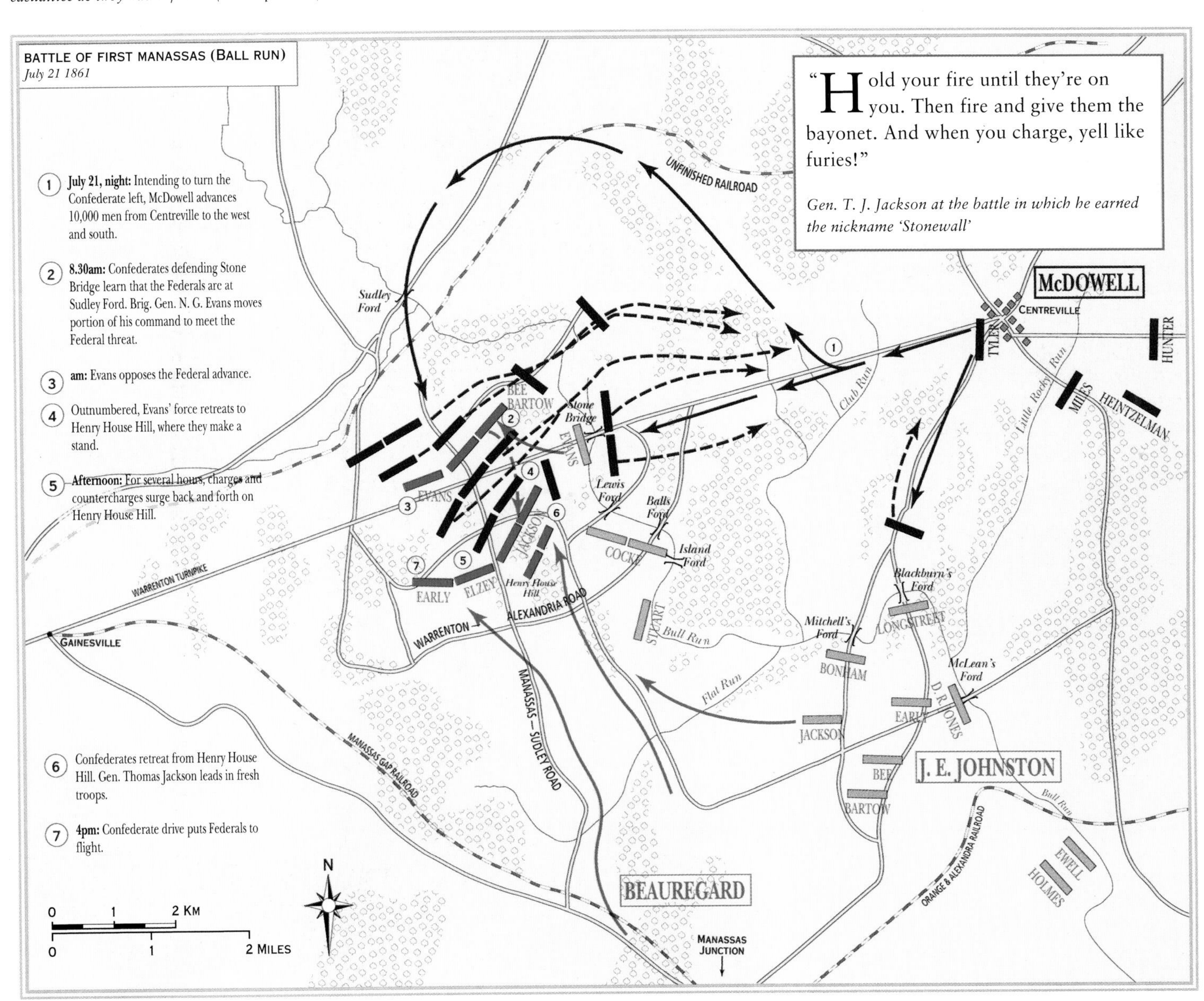

BATTLE OF FIRST MANASSAS (BALL RUN)
July 21 1861
1 July 21, night: Intending to turn the Confederate left, McDowell advances 10,000 men from Centreville to the west and south.
2 8.30am: Confederates defending Stone Bridge learn that the Federals are at Sudley Ford. Brig. Gen. N. G. Evans moves portion of his command to meet the Federal threat.
3 am: Evans opposes the Federal advance.
4 Outnumbered, Evans' force retreats to Henry House Hill, where they make a stand.
5 Afternoon: For several hours, charges and countercharges surge back and forth on Henry House Hill.
6 Confederates retreat from Henry House Hill. Gen. Thomas Jackson leads in fresh troops.
7 4pm: Confederate drive puts Federals to flight.
"Hold your fire until they're on you. Then fire and give them the bayonet. And when you charge, yell like furies!"
Gen. T. J. Jackson at the battle in which he earned the nickname 'Stonewall'
UNFINISHED RAILROAD
Sudley Ford
McDOWELL
CENTREVILLE
TYLER
HUNTER
MILES
HEINTZELMAN
Little Rocky Run
Cub Run
BEE
BARTOW
Stone Bridge
EVANS
Lewis Ford
Balls Ford
Island Ford
COCKE
JACKSON
EARLY
ELZEY
Henry House Hill
WARRENTON TURNPIKE
WARRENTON
ALEXANDRIA ROAD
GAINESVILLE
STUART
Bull Run
Blackburn's Ford
Mitchell's Ford
LONGSTREET
BONHAM
McLean's Ford
D. R. JONES
EARLY
Flat Run
JACKSON
BEE
BARTOW
J. E. JOHNSTON
MANASSAS–SUDLEY ROAD
MANASSAS GAP RAILROAD
ORANGE & ALEXANDRIA RAILROAD
Bull Run
EWELL
HOLMES
BEAUREGARD
MANASSAS JUNCTION
N
0 1 2 KM
0 1 2 MILES

# South Atlantic Coast AUGUST 27 – NOVEMBER 7 1861

While the Federal navy made plans for a systematic blockade of southern coasts, they also reacted to specific threats. In July 1861 Confederate vessels, operating from behind the North Carolina Banks, captured two Federal merchantmen. This action, coupled with a fear of what Confederate raiders might accomplish if allowed to operate in these waters, convinced Lincoln it was necessary to control Albermarle and Pamlico Sounds. This could best be accomplished by seizing Ocracoke Inlet, guarded by Forts Hatteras and Clark.

A naval squadron, commanded by Commodore Silas Stringham, accompanied by some 1,000 soldiers led by General Benjamin Butler, sailed from Hampton Roads on August 26, arriving off the Confederate forts the following day. Butler's troops landed to the north of the forts, while Stringham maneuvered his ships to within range of the forts and opened fire on Fort Clark. Unable to withstand the bombardment, the Confederates abandoned Clark and took refuge in Fort Hatteras. Butler's forces, meanwhile, became marooned by heavy surf on the beach, and were unable to participate. The next day Stringham bombarded Hatteras, forcing its surrender.

While the battles resulted in a Federal victory, the lack of coordination between the naval and military elements was appalling.

To blockade the South Atlantic coast effectively, the Federal navy required a port in the area from which to support its ships; Port Royal was selected. On October 16, 1861, a squadron commanded by Commodore Samuel F. Du Pont left New York for Norfolk where they embarked 1,200 troops, led by General Thomas Sherman. On October 29 the force sailed for Port Royal. On November 7, the squadron steamed to a point midway between the two forts. For over four hours the ships fired at the forts with devastating accuracy. With their cannon destroyed, Fort Walker's defenders had to abandon their posts, and Du Pont took possession. Fort Beauregard also surrendered. Sherman's troops were landed, and Port Royal secured.

"The *Susquehanna* opened the ball, and in a few minutes the entire fleet concentrated its fire on Fort Hatteras... The bombardment was continued without intermission, when, at half-past eleven... a white flag was displayed on the fort... General Butler... demanded an unconditional surrender. These terms after a Council of War, were accepted."

*New York Illustrated News, September 16, 1861.*

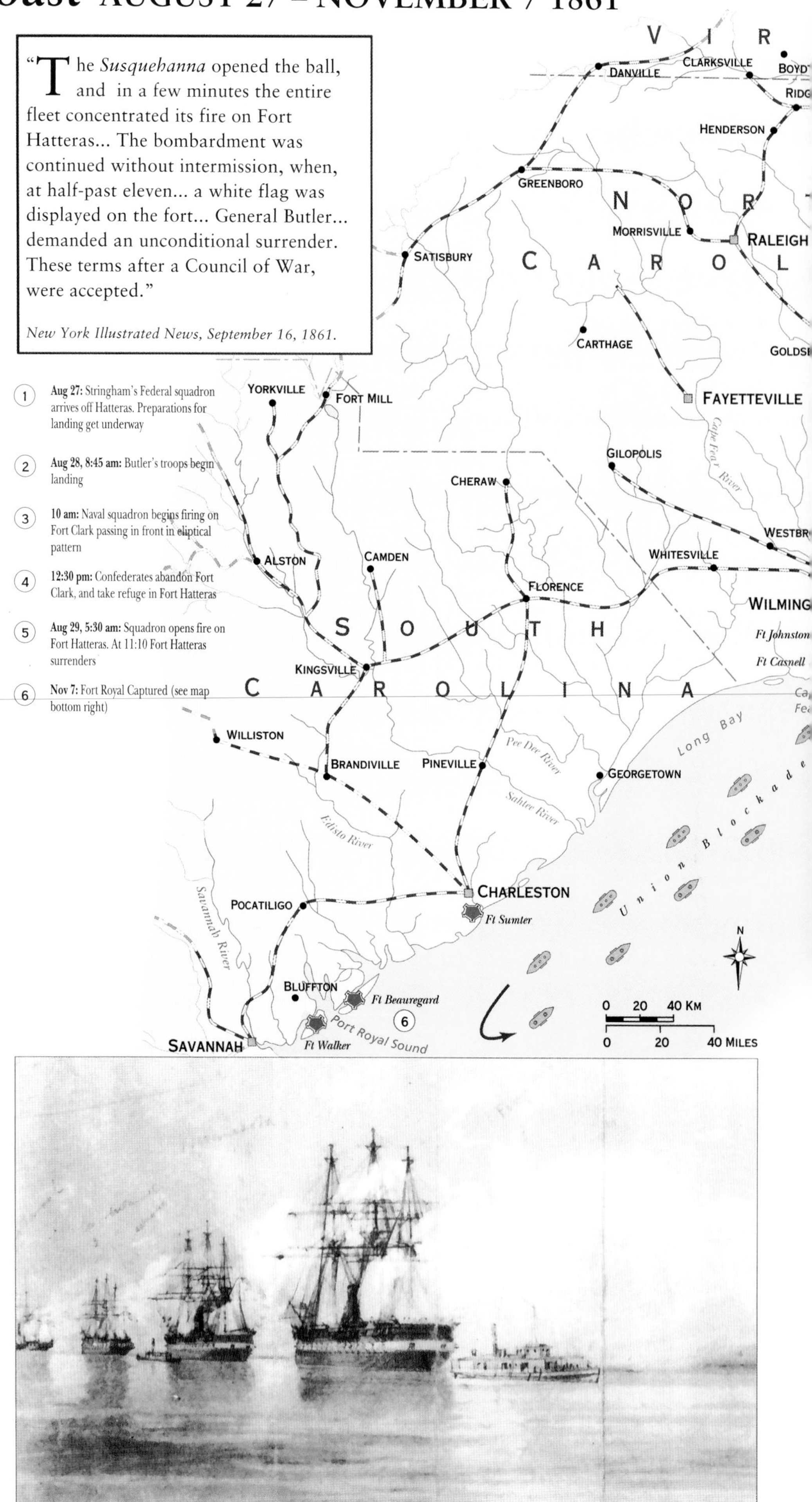
1 Aug 27: Stringham's Federal squadron arrives off Hatteras. Preparations for landing get underway
2 Aug 28, 8:45 am: Butler's troops begin landing
3 10 am: Naval squadron begins firing on Fort Clark passing in front in elliptical pattern
4 12:30 pm: Confederates abandon Fort Clark, and take refuge in Fort Hatteras
5 Aug 29, 5:30 am: Squadron opens fire on Fort Hatteras. At 11:10 Fort Hatteras surrenders
6 Nov 7: Fort Royal Captured (see map bottom right)
V I R
NORTH CAROLINA
SOUTH CAROLINA
DANVILLE
CLARKSVILLE
HENDERSON
GREENBORO
MORRISVILLE
RALEIGH
SATISBURY
CARTHAGE
FAYETTEVILLE
YORKVILLE
FORT MILL
GILOPOLIS
CHERAW
Cape Fear River
WESTBR
WHITESVILLE
ALSTON
CAMDEN
FLORENCE
WILMING
Ft Johnston
Ft Casnell
KINGSVILLE
WILLISTON
BRANDIVILLE
PINEVILLE
Pee Dee River
Santee River
GEORGETOWN
Long Bay
Union Blockade
Edisto River
Savannah River
POCATILIGO
CHARLESTON
Ft Sumter
BLUFFTON
Ft Beauregard
Port Royal Sound
Ft Walker
SAVANNAH
N
0 20 40 KM
0 20 40 MILES

*Stringham's squadron commences its bombardment of Fort Hatteras on August 29 (right). A total of 158 guns pounded the fort, forcing its surrender when their fire was directed at the magazine. In itself, the victory was not a startling event but it became important as a booster of flagging Northern morale.*

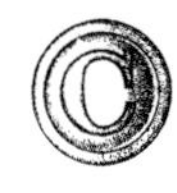

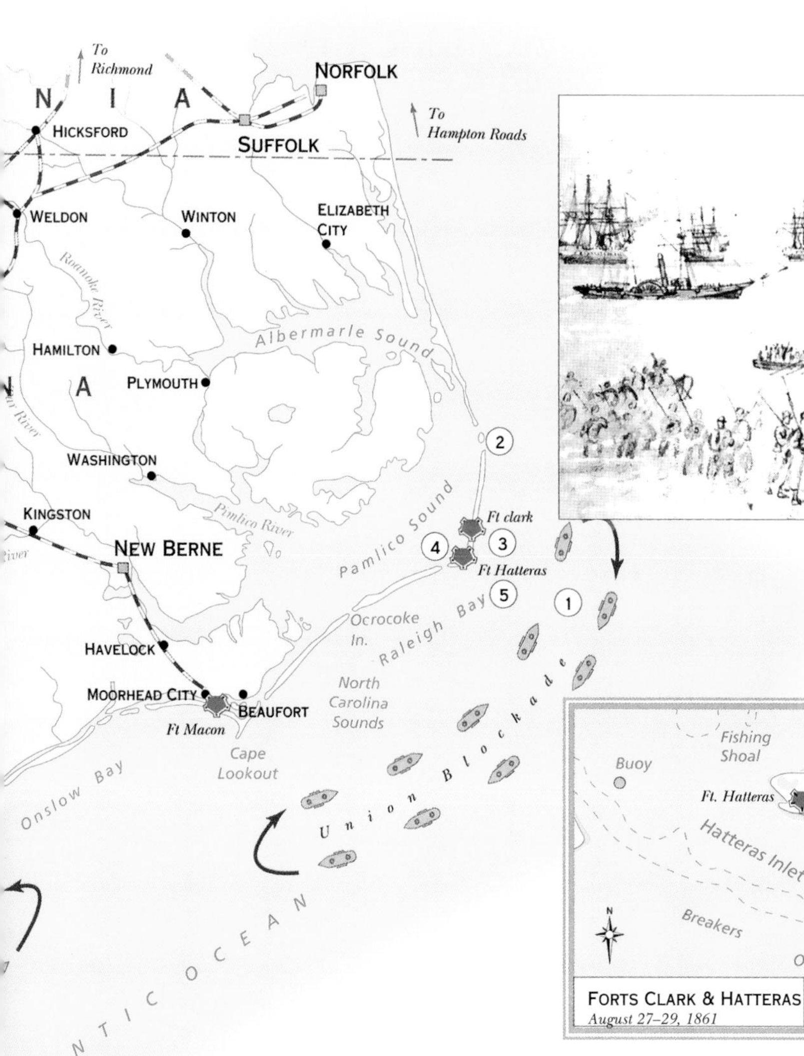

To Richmond
NORFOLK
To Hampton Roads
HICKSFORD
SUFFOLK
WELDON
WINTON
ELIZABETH CITY
Roanoke River
HAMILTON
Albemarle Sound
PLYMOUTH
WASHINGTON
Pamlico River
KINGSTON
NEW BERNE
Pamlico Sound
Ft clark
Ft Hatteras
Ocrocoke In.
Raleigh Bay
HAVELOCK
MOORHEAD CITY
BEAUFORT
Ft Macon
North Carolina Sounds
Cape Lookout
Onslow Bay
Union Blockade
ATLANTIC OCEAN
Fishing Shoal
Buoy
Ft. Clark
Landing beach 3 miles from Hatteras
Ft. Hatteras
Breakers
Hatteras Inlet
1600 yards
1300 yards
U.S. Fleet
Outer buoy
FORTS CLARK & HATTERAS
August 27–29, 1861


*Benjamin Butler's Union assault force is landed at Hatteras Inlet, August 28 (above). Poor conditions resulted in only 300 of the 1,000-strong contingent being landed, rendering useless the army's contribution ration.*

*Using tactics which would have been impossible before the advent of steam power, du Pont ordered his ships to steam back and forth in an oval pattern past the twin forts of Port Royal. As they passed the shore batteries each ship could deliver a full broadside while presenting only a moving target. The plan was executed on November 7 with such deadly precision that both forts surrendered within four hours of the attack's commencement (see map below).*

*Having occupied Fort Clark without opposition, the Federal flotilla concentrated its attention on Fort Hatteras. An intense artillery duel between the Federal ships and the Confederate batteries began on August 29. After suffering severe damage, Fort Hatteras surrendered at 11:10 am (see map above).*

*Joint operations between the army and the navy were notorious for the rivalry which they inspired. Brig. Gen. Thomas West Sherman (above) commanded the land forces at Port Royal Sound, a campaign which was remarkable not only for its success but also for the unprecedented cooperation betwen the two services.*

*Port Royal under fire from Du Pont's fleet (below). Despite the tightening of the Federal blockade, nine out of every ten blockade-runners still reached port safely.*

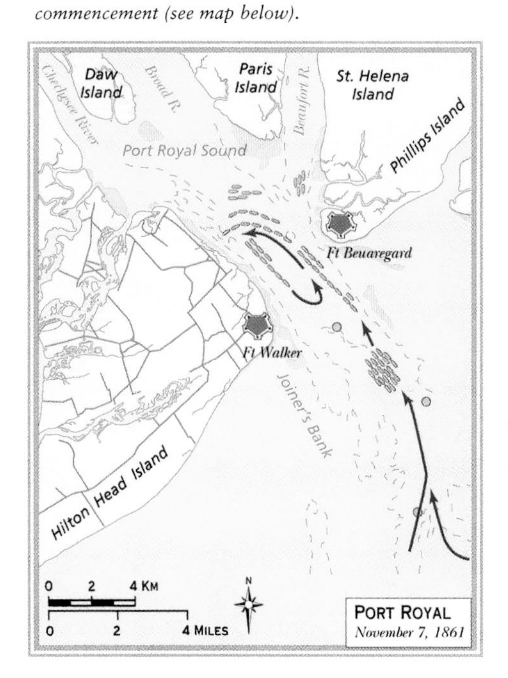

Daw Island
Paris Island
St. Helena Island
Chechesee River
Broad R.
Beaufort R.
Port Royal Sound
Phillips Island
Ft Beuaregard
Ft Walker
Joiner's Bank
Hilton Head Island
0 2 4 KM
0 2 4 MILES
N
PORT ROYAL
November 7, 1861

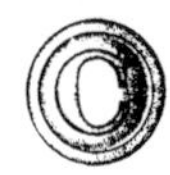

# 1862: A War For Freedom

DURING THE CHRISTMAS season of 1861, the mood in Richmond was buoyant, while gloom prevailed in Washington. Despite the Union's naval triumph at Port Royal, the balance of military victories in 1861 lay with the Confederacy, while a diplomatic showdown between Britain and the United States threatened to add a second war to Abraham Lincoln's problems. In November a Union warship had stopped the British steamer *Trent* on the high seas and taken from it the Confederate diplomats James Mason and John Slidell, who were on their way to Europe to seek British and French diplomatic recognition of the Confederacy. Uproar ensued in London as the British government branded this action an outrageous violation of British neutrality and freedom of the seas. War loomed between Britain and the United States until the Union government backed down the day after Christmas and released Mason and Slidell. "One war at a time," was Lincoln's sage philosophy. But the letdown following this affair added to the despondency that gripped Washington in the early weeks of the new year.

The *Trent* affair had provoked a brief panic in Northern financial circles. Banks suspended specie payments (the backing of their notes in circulation with gold and silver) and Secretary to the Treasury, Salmon P. Chase, found it difficult to sell bonds to finance the war. General McClellan fell ill with typhoid fever. From General Henry W. Halleck, commander of Union forces in Missouri and western Kentucky, came negative reports on the prospects for an advance in that theater. "It is exceedingly discouraging," said Lincoln. "The people are impatient; Chase has no money; the General of the Army has typhoid fever. The bottom is out of the tub."

Beginning in February 1862, news from military fronts transformed the moods in North and South. The first reports came from western Tennessee, where the joint efforts of Union army and naval forces won significant victories on the Tennessee and Cumberland rivers which flow through Tennessee and Kentucky, emptying into the Ohio River just before it joins the Mississippi. The unlikely hero of these victories was General Ulysses S. Grant, who had failed in several civilian occupations since resigning from the peacetime army in 1854. Rejoining the army when war broke out, he had demonstrated a quiet efficiency and determination that won him promotion from Illinois colonel to brigadier general, and command of a small but growing force based at Cairo, Illinois, in the fall of 1861. When Confederate units entered Kentucky in September, Grant moved quickly to occupy the mouths of the Cumberland and Tennessee rivers. Unlike McClellan, who had known nothing but success in his career and was afraid to risk failure, Grant's experience of failure made him willing to take risks, having little to lose. He demonstrated that willingness dramatically in the early months of 1862.

Military strategists on both sides understood the importance of these navigable rivers as highways of invasion into the South's heartland. The Confederacy had built forts at strategic points along the rivers, and begun to convert a few steamboats into gunboats to back up the forts. The Federals also converted steamboats into "timberclad" gunboats – so called because they were armored with oak bulwarks that protected their machinery but did not impair their speed and shallow draft for river operations. Northern shipyards also built a new class of ironclad gunboat designed for river warfare. Carrying 13 guns, these flat-bottomed, wide-beamed vessels drew only six feet of water. Their hulls and paddlewheels were protected by a sloping casemate sheathed in iron armor up to two and half inches thick.

In February, 1862, when the first four of these strange-looking but formidable craft were ready, Grant struck in a lightning campaign that captured Fort Henry on the Tennessee River and Fort Donelson on the Cumberland. The fall of these forts opened the rivers to Union forces. On February 25, Northern troops occupied Nashville, the first Confederate state capital to fall to the enemy. The center of the Southern defensive line stretching from the Mississippi River to the Appalachian foothills was breached, forcing Confederate theater commander Albert Sidney Johnston to retreat, yielding all of Kentucky and much of Tennessee to Union occupiers.

Johnston concentrated what was left of his army at Corinth in northern Mississippi. He brought reinforcements from New Orleans and Pensacola on the Gulf Coast, leaving these areas dangerously exposed to Union naval forces. But the urgent need to defend the upper Mississippi Valley and, if possible, to drive the enemy out of western Tennessee, received top Confederate priority. Grant concentrated his army at Pittsburg Landing, 20 miles north of Corinth, for a drive on that strategic Southern railroad junction. But Johnston struck first, launching a surprise attack on Grant's forces on April 6 that precipitated the bloody two-day battle of Shiloh. Johnston was killed in the battle and the Union army almost driven into the Tennessee River before receiving reinforcements overnight and counterattacking the next day, snatching victory from the jaws of defeat and driving the Confederates back to Corinth.

Though Shiloh was a Union victory, it was a costly one. The heavy Federal casualties, and the suspicion that he had been caught napping the first day, temporarily damaged

*Jefferson Davis, photographed by Mathew Brady in 1861.*

Grant's reputation. Nevertheless, Union victories in the western theater continued throughout the spring. On April 24, a Federal naval force under Captain David G. Farragut (soon to be promoted to become America's first admiral), fought its way past the forts guarding the Mississippi below New Orleans, and compelled the surrender of the Confederacy's largest city and principal port. General Halleck took overall command of Union troops in western Tennessee, and drove the Confederates out of Corinth at the end of May. Meanwhile, the Union gunboat fleet steamed down the Mississippi, virtually wiping out the small Confederate river navy in a spectacular battle at Memphis on June 6. At Vicksburg, 200 miles below Memphis, the Union gunboats from upriver joined part of Farragut's fleet that had come up from New Orleans, taking Baton Rouge and Natchez along the way. However, the heavily fortified Confederate bastion at Vicksburg proved too strong for the firepower of the fleets. But the dramatic succession of Union triumphs in the western theaters – including a decisive victory at the battle of Pea Ridge in northwest Arkansas on March 7 and 8 – convinced many Northerners that the war was nearly won. This conviction was reinforced by news from the North Carolina sounds, where an expeditionary force under General Ambrose Burnside won a string of small victories, and occupied several ports along the North Carolina coast from February to April 1862. "Every blow tells fearfully against the rebellion," boasted the leading Northern newspaper, the New York *Tribune*, on May 23, 1862. "The rebels themselves are panic-stricken, or despondent. It now requires no very far-reaching prophet to predict the end of this struggle."

The *Tribune*'s confidence stemmed from the situation in Virginia as well as from victories in the West. But even as the editorial writer penned these words, affairs in Virginia were about to take a sharp turn in the Confederacy's favor. Within three months the Union, so near an outright victory in the spring, would be back on the defensive.

*The battle of Shiloh, painted by T.C. Lindsey*

In the western theater, broad navigable rivers pointed like arrows into the heartland of the Confederacy, facilitating Union invasions. But in Virginia half a dozen small rivers flowing west to east lay athwart the line of operations between Washington and Richmond, thus providing the Confederates with natural lines of defense. McClellan persuaded a reluctant Lincoln to approve a plan to transport the Army of the Potomac down the Chesapeake Bay to the tip of the Virginia Peninsula. This route would shorten the distance to Richmond, and give the Union army a seaborne supply line secure from harassment by Confederate cavalry and guerrillas.

It was a good plan – in theory. The logistical achievement of transporting 100,000 men and their equipment, animals, and supplies by sea to the jump-off point near Yorktown, was impressive. But then McClellan's failings began to surface. His caution, his exaggeration of enemy forces facing him, and his bickering with the Lincoln administration about reinforcements, bogged down his army. This enabled the Confederates to delay him a month at Yorktown while reinforcements poured into the Peninsula to defend Richmond. Eventually, McClellan's siege tactics forced the Confederates to retreat up the Peninsula to a defensive line only six miles east of their capital.

Despite McClellan's sluggishness, it seemed in May 1862 only a matter of time before his siege tactics would compel the surrender of Richmond. Then the remarkable team of Robert E. Lee and "Stonewall" Jackson turned the war around. Lee had been one of the most promising officers in the pre-war U.S. army. After war broke out at Fort Sumter, Lee's fellow Virginian, Winfield Scott, wanted Lee to become field commander of the principal Union army. Although Lee had opposed secession, sadly he resigned from the U.S. army after the Virginia Convention passed an ordinance of secession on April 17, 1861. "I must side either with or against my section," he told a Northern friend. "I cannot raise my hand against my birthplace, my home, my children." Together with three of his sons and a nephew, Lee joined the Confederate army. After a frustrating stint as field commander of a small army in western Virginia, followed by efforts to shore up Confederate defenses along the South Atlantic coast, Lee returned to Richmond in March 1862 to serve as Jefferson Davis's military adviser. In that capacity he directed General Thomas J. Jackson to undertake a diversionary campaign in the Shenandoah Valley of Virginia in order to relieve some of the pressure on Richmond by drawing off Union troops to deal with Jackson.

Having earned his sobriquet "Stonewall" by his defensive stand on Henry House Hill at the battle of Manassas (Bull Run) in July, 1861, Jackson now demonstrated the offensive strategy that made him one of the Civil War's most famous generals. During a period of one month (May 8 – June 9), he showed what could be accomplished by deception, daring, and mobility. With only 17,000 men, Jackson moved by forced marches so swift that his infantry earned the nickname "Jackson's foot cavalry." Darting here and there through the Valley, they marched 350 miles in the course of the month, and won four battles against three separate Union armies whose combined numbers were more than twice their own – but which Jackson's force always out-numbered at the point of contact. This campaign forced Lincoln to divert to the Valley some of the reinforcements McClellan needed for his offensive on the Peninsula.

Even without those reinforcements, McClellan's army substantially outnumbered the Confederate force defending Richmond, commanded by General Joseph E. Johnston. Nevertheless, Johnston took the initiative at the end of May in an attack that produced the two-day battle of Seven Pines. The most significant result of this battle was the wounding of Johnston and his replacement by Robert E. Lee.

Lee's appointment marked a turning point in the war in the East. His qualities as a commander immediately manifested themselves: boldness, a willingness to take risks, an almost uncanny ability to read the enemy commander's mind, and a quiet charisma that won the devotion of his men. While McClellan continued to dawdle and feud with Lincoln over reinforcements, Lee planned an offensive. In the last week of June, he brought Jackson's army from the Valley, and with the combined forces launched an attack on McClellan's right flank on June 26 that precipitated what became known as the Seven Days battles – the heaviest fighting of the war so far. Constantly attacking, Lee's army of 88,000 drove McClellan's 100,000 away from Richmond, and forced them to retreat to a new fortified base on the James River. These assaults cost the Confederacy heavily in casualties, but they temporarily reversed the course of the war.

Northern sentiment plunged from the heights of euphoria in May to the depths of despair in July. "The feeling of despondency here is very great," wrote a New Yorker, while a Southerner exulted that "Lee has turned the tide, and I shall not be surprised if we have a long career of successes." The tide turned in the Western theater as well. Union success there in the spring had required the detachment of several Federal divisions to occupy and administer the large area they had conquered. These forces extended deep into enemy territory, at the end of long supply lines which were vulnerable to cavalry raids. Confederate horsemen were quick to take advantage of the opportunity. During the summer and autumn of 1862, the cavalry commands of Tennesseean

Nathan Bedford Forrest and Kentuckian John Hunt Morgan, staged repeated raids in which they burned bridges, blew up tunnels, tore up tracks, and captured supply depots and the Union garrisons defending them. By August, the once-formidable Union war machine in the West seemed to have broken down.

Confederate armies commanded by Earl Van Dorn and Braxton Bragg took advantage of these opportunities to launch counteroffensives. Initially successful, these campaigns in turn recoiled after Confederate attacks were repulsed at Corinth, Mississippi (October 3–4) and Perryville, Kentucky (October 8). Even after these defeats, however, Confederate forces in the Western theater were in better shape than they had been four months previously.

Most attention at home and abroad, however, focused on the Virginia theater. After the Seven Days Battles, the scene of action again shifted to northern Virginia, where several scattered Union divisions had been combined into a new army under General John Pope. Lincoln ordered a sulking McClellan to transfer his troops from the Peninsula to reinforce Pope. In response, Lee also shifted his army, precipitating several weeks of maneuvers and skirmishes that culminated in the second battle of Manassas (Bull Run) on August 29–30. It was another smashing Confederate victory that encouraged Lee, despite the exhaustion of his troops and a shortage of supplies, to keep up the pressure by invading Maryland.

Great prospects rode with Confederate troops as they began splashing across a Potomac ford 40 miles upriver from Washington on September 4. Simultaneously, Braxton Bragg's army invaded Kentucky 500 miles to the west. Maryland might be won for the Confederacy. Another victory by Lee could influence the Northern congressional elections in November, and help Democrats who opposed Lincoln's war policies to gain control of the House and paralyze the Northern war effort. A Confederate victory in Maryland might persuade Britain and France to recognize the Confederacy and intervene to end the war – especially since the long-expected cotton famine had finally materialized. In September 1862, the British and French governments were indeed considering recognition of the Confederacy, and were awaiting the outcome of Lee's invasion to decide whether to proceed.

In Washington, too, momentous consequences awaited the outcome of this military campaign. After the humiliating defeat at Bull Run, Lincoln had combined Pope's army with McClellan's and given McClellan command of both. Lincoln ordered him to pursue Lee and "destroy the rebel army, if possible." While he awaited news from the front, Lincoln reflected on a crucial policy decision he had made two

*Rosie O'Neal Greenhow, the Confederate spy who was arrested shortly after the battle of First Manassas*

months earlier – to issue an emancipation proclamation. The controversy over slavery had brought on the war in the first place, and it remained the crux of the conflict. From the outset, many slaves regarded Union soldiers as an army of liberation, and flocked to Union lines seeking freedom. One Union general, Benjamin F. Butler – who also happened to be a clever lawyer – accepted these fugitives in his camps in Virginia, labeling them "contraband of war" – property owned by the enemy and therefore subject to confiscation. This phrase caught on. It became the legal basis for confiscation acts passed by Congress in August 1861 and July 1862, providing for a limited emancipation of slaves owned by Confederates.

Lincoln at first had resisted making the war for Union a war against slavery, for he feared that doing so would drive the border slave states into the arms of the Confederacy, and would alienate Northern Democrats from support of the Union war effort. But as time went on, and the war escalated in scope and fury, Lincoln's own moral hatred of slavery combined with his growing conviction that he must utilize every resource to win the war – including the enemy's slave population. The war took on new dimensions after the collapse of Northern hopes for imminent victory in the summer of 1862. It became a "total war", requiring mobilization of

every resource and the destruction of every enemy resource. To strike at slavery would weaken the Confederacy by undermining its labor force, and would mobilize that resource for the Union along with the moral power of fighting for freedom. To make emancipation a Northern war aim would also undercut Confederate efforts to gain British recognition and support, for anti-slavery sentiment in Britain would never countenance aid to a nation fighting for slavery against one fighting for freedom.

All of these considerations shaped Lincoln's decision in July 1862 to issue an emancipation proclamation, using his powers as Commander-in-Chief to seize enemy property being used to wage war against the United States. Lincoln announced his decision to his cabinet on July 22. According to notes kept by Secretary of the Navy, Gideon Welles, Lincoln presented emancipation as a "military necessity, absolutely essential to the preservation of the Union. We must free the slaves or be ourselves subdued. The slaves [are] undeniably an element of strength to those who [have] their service, and we must decide whether that element should be with us or against us... We wanted the army to strike more vigorous blows. The Administration must set an example, and strike at the heart of the rebellion" – slavery.

Secretary of State, William H. Seward, persuaded Lincoln to withhold announcement of the proclamation until a significant military victory could give it credibility and force. Lincoln slipped his proclamation into a desk drawer and waited for such a victory. It would prove to be a long wait, as Union military fortunes slid deeper into the slough of despond, and Lee invaded Maryland. But in the third week of September came news from the banks of Antietam Creek, near the village of Sharpsburg, that Lee had retreated back to Virginia after the bloodiest single day of the war, the battle of Antietam (called Sharpsburg by the Confederates) on September 17. McClellan missed an opportunity to inflict an even more crushing defeat on the enemy. Although Lincoln was disappointed with that failure, he considered Antietam enough of a victory to issue a preliminary Emancipation Proclamation on September 22, declaring that on January 1, 1863, slaves in all states still in rebellion "shall be then, thenceforward, and forever free." After January 1, when Lincoln issued the final Proclamation, the Union army became officially an army of liberation. The North fought to create a new Union, not to restore an old one.

*The battle of Antietam, from a lithograph by Kurz and Allison*

The news from Antietam put a brake on the British and French momentum toward recognition of the Confederacy; the Emancipation Proclamation tightened that brake. Bragg's retreat from Kentucky after the battle of Perryville further chilled Confederate prospects. Republicans retained control of the Union Congress in the 1862 elections. But the North could not win the war by issuing proclamations, turning back Confederate invasions, and winning elections. Union armies had to invade, conquer, and destroy Confederate resistance.

Lincoln urged McClellan to seize the initiative, cross the Potomac, and attack Lee's army in Virginia before it could recover from its setback. McClellan found one excuse after another for delay, until Lincoln finally gave up on him and on November 7, removed him from command. His successor was Ambrose E. Burnside, who maneuvered the lumbering Army of the Potomac into position to attack the Confederates in the hills behind the Rappahannock River at Fredericksburg on December 13. But Burnside's repeated assaults resulted in the most dispiriting defeat ever suffered by the Army of the Potomac. When Lincoln heard the news from Fredericksburg, he said: "If there is a worse place than hell, I am in it."

Reports from the West did little to dispel the renewed gloom in Washington. The Confederates had fortified the city of Vicksburg on bluffs commanding the Mississippi River, enabling them to maintain control of an important stretch of the river between Vicksburg and Port Hudson in Louisiana, which they also fortified. These strongholds denied the Federals domination of the entire Mississippi Valley, and preserved transport links between the western and eastern halves of the Confederacy. To sever those links was the goal of Ulysses S. Grant, who in November, 1862, launched a two-pronged drive against Vicksburg. With one army, Grant moved overland from the north, while with another his principal subordinate, William Tecumseh Sherman, sailed down the Mississippi to attack Vicksburg defenses from the flank while Grant came up on their rear.

But raids by Confederate cavalry frustrated Grant's progress by destroying the railroads and supply depots in his rear, forcing him to retreat. Meanwhile, Sherman attacked the Confederates at Chickasaw Bluffs on December 29, with no more success than Burnside had enjoyed at Fredericksburg.

The only gleam of light for the North in that dark winter of discontent came from central Tennessee. In that theater, Lincoln had removed General Don Carlos Buell from command of the Army of the Cumberland for the same reason he had removed McClellan – lack of vigor and aggressiveness. Buell's successor was William S. Rosecrans, who had proved himself a fighter in subordinate commands. On the day after Christmas, 1862, Rosecrans moved out from his base at Nashville to attack Braxton Bragg's Army of Tennessee at Murfreesboro. The ensuing three-day battle (called Stones River by the Union and Murfreesboro by the Confederacy) resulted in Confederate success on the first day (December 31), but defeat on the last. Both armies suffered devastating casualties equal to a third of their strength, leaving them crippled for months. The Confederate retreat to a new base 40 miles farther south enabled the North to call Stones River a victory. Lincoln expressed his gratitude to Rosecrans: "I can never forget, whilst I remember anything, that you gave us a hard-earned victory which, had there been a defeat instead, the nation could scarcely have lived over."

As it was, the nation scarcely lived over the winter of 1862–1863, for Union military prospects and morale declined further during the early months of 1863 before they finally experienced a dramatic improvement in midsummer.

*The battle of New Orleans, painted by Jo Davidson*

# Eastern Kentucky and Tennessee JANUARY 10–JUNE 18 1862

As Northern and Southern armies took the field, Kentucky sought to remain neutral. For 5 months the state was center stage in the struggle for America's heartland.

The Confederacy made the first move: on September 3, troops crossed into Kentucky and seized Columbus on the Mississippi. Federal forces, better prepared and enjoying local support, took possession of Paducah and Louisville on the Ohio River.

The Confederates, led by General Albert S. Johnston, anchored their center at Bow ling Green. The Federals confronted John ston, and considered how best to achieve one of Lincoln's cherished goals – to free East Tennessee from oppression.

In mid-December, Brigadier General Don Carlos Buell sent a little-known colonel – James A. Garfield – to clear the Confed erates, led by Brigadier General Humphrey Marshall, out of the mountains of eastern Kentucky. Using steamboats, by January 6 Garfield was within seven miles of Paintsville, where Marshall was camped.

Marshall evacuated Paintsville and posted his 1,500 Confederates on ridges bordering Middle Creek. On the 10th Garfield advanced, and by noon the fighting was intense, the 1,700 Federals carrying the fight to the Confederates by working their way up the wooded slopes. Both sides withdrew at nightfall, the Con federates to Martin's Mill, the Federals to Prestonsburg. Unable to sustain his army at Martin's Mill, Marshall returned to Virg inia, leaving the Federals in possession of eastern Kentucky.

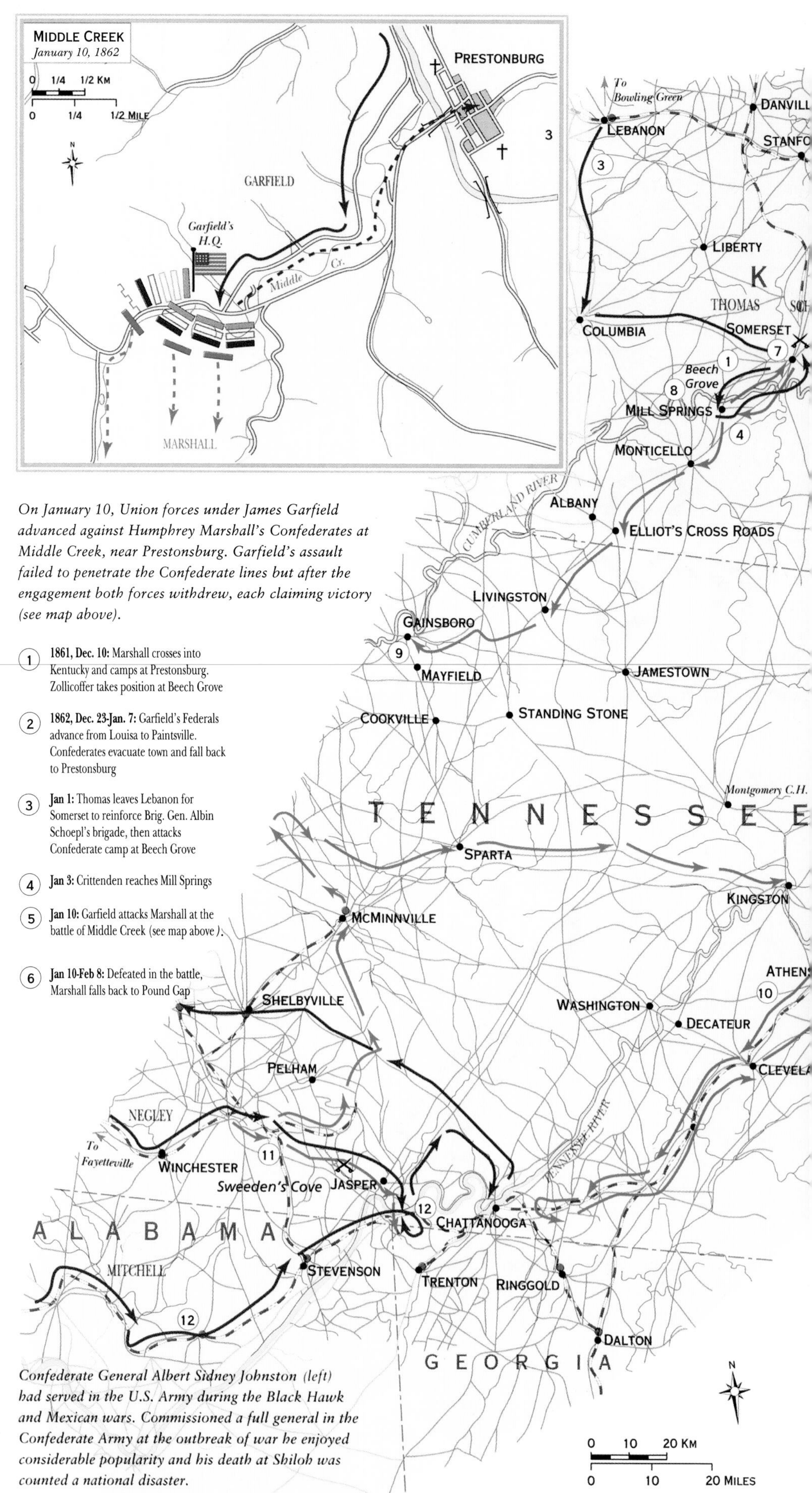
MIDDLE CREEK
January 10, 1862
0 1/4 1/2 KM
0 1/4 1/2 MILE
N
PRESTONBURG
GARFIELD
Garfield's H.Q.
Middle Cr.
MARSHALL
3
To Bowling Green
DANVILL
LEBANON
STANFO
LIBERTY
K
THOMAS
COLUMBIA
SOMERSET
Beech Grove
MILL SPRINGS
MONTICELLO
CUMBERLAND RIVER
ALBANY
ELLIOT'S CROSS ROADS
LIVINGSTON
GAINSBORO
MAYFIELD
JAMESTOWN
COOKVILLE
STANDING STONE
Montgomery C.H.
T E N N E S S E E
SPARTA
KINGSTON
MCMINNVILLE
ATHEN
SHELBYVILLE
WASHINGTON
DECATEUR
CLEVEL
PELHAM
NEGLEY
To Fayetteville
WINCHESTER
Sweeden's Cove
JASPER
TENNESSEE RIVER
CHATTANOOGA
A L A B A M A
MITCHELL
STEVENSON
TRENTON
RINGGOLD
DALTON
G E O R G I A
0 10 20 KM
0 10 20 MILES
1 1861, Dec. 10: Marshall crosses into Kentucky and camps at Prestonsburg. Zollicoffer takes position at Beech Grove
2 1862, Dec. 23-Jan. 7: Garfield's Federals advance from Louisa to Paintsville. Confederates evacuate town and fall back to Prestonsburg
3 Jan 1: Thomas leaves Lebanon for Somerset to reinforce Brig. Gen. Albin Schoepl's brigade, then attacks Confederate camp at Beech Grove
4 Jan 3: Crittenden reaches Mill Springs
5 Jan 10: Garfield attacks Marshall at the battle of Middle Creek (see map above).
6 Jan 10-Feb 8: Defeated in the battle, Marshall falls back to Pound Gap

*On January 10, Union forces under James Garfield advanced against Humphrey Marshall's Confederates at Middle Creek, near Prestonsburg. Garfield's assault failed to penetrate the Confederate lines but after the engagement both forces withdrew, each claiming victory (see map above).*

*Confederate General Albert Sidney Johnston (left) had served in the U.S. Army during the Black Hawk and Mexican wars. Commissioned a full general in the Confederate Army at the outbreak of war he enjoyed considerable popularity and his death at Shiloh was counted a national disaster.*

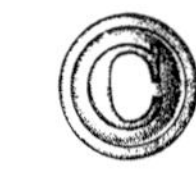

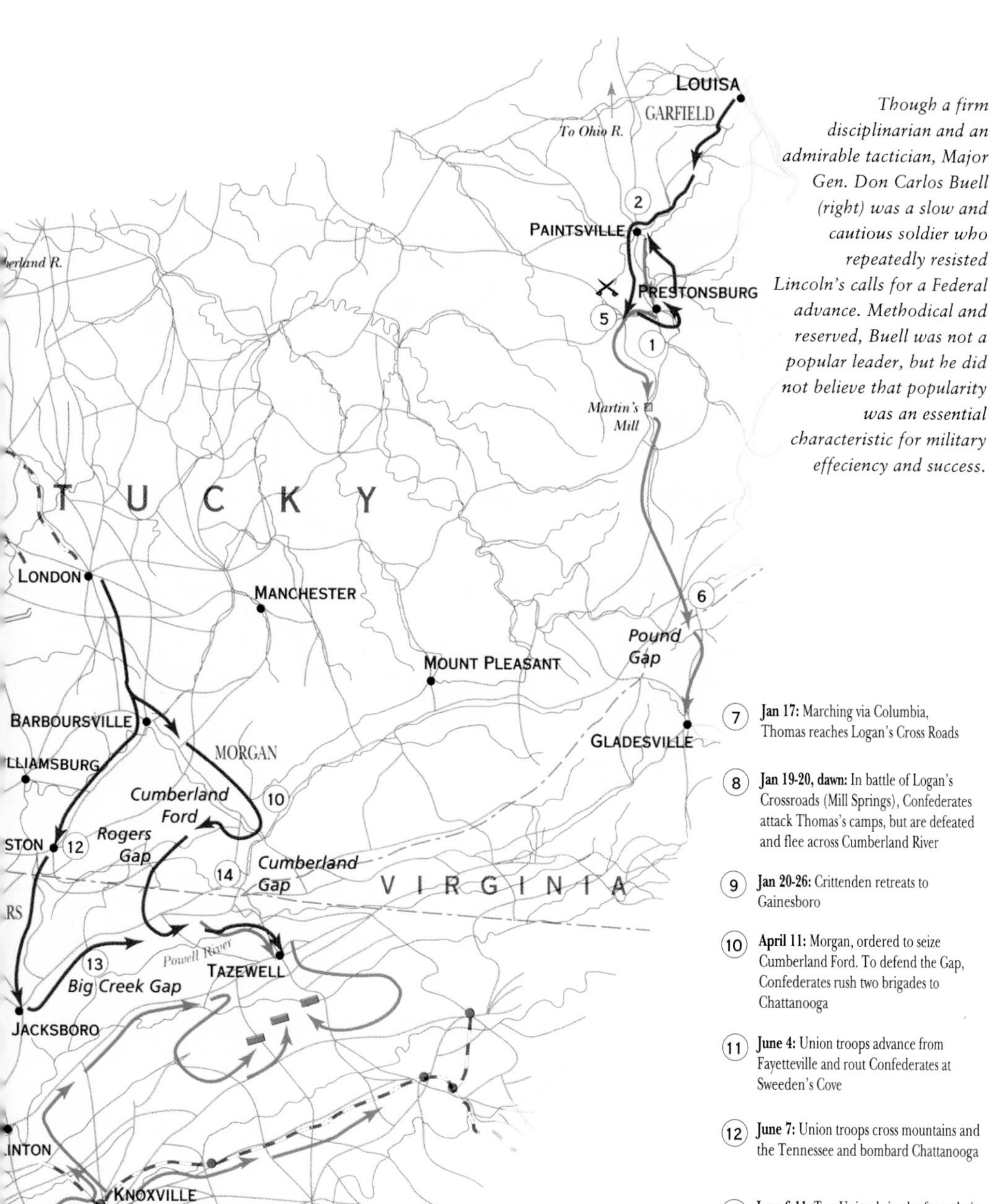

LOUISA
GARFIELD
To Ohio R.
PAINTSVILLE
PRESTONSBURG
Martin's Mill
Pound Gap
GLADESVILLE
MOUNT PLEASANT
MANCHESTER
LONDON
BARBOURSVILLE
MORGAN
Cumberland Ford
Rogers Gap
Cumberland Gap
VIRGINIA
Powell River
TAZEWELL
Big Creek Gap
JACKSBORO
KNOXVILLE
7 Jan 17: Marching via Columbia, Thomas reaches Logan's Cross Roads
8 Jan 19-20, dawn: In battle of Logan's Crossroads (Mill Springs), Confederates attack Thomas's camps, but are defeated and flee across Cumberland River
9 Jan 20-26: Crittenden retreats to Gainesboro
10 April 11: Morgan, ordered to seize Cumberland Ford. To defend the Gap, Confederates rush two brigades to Chattanooga
11 June 4: Union troops advance from Fayetteville and rout Confederates at Sweeden's Cove
12 June 7: Union troops cross mountains and the Tennessee and bombard Chattanooga
13 June 6-11: Two Union brigades force their way through Big Creek and Rogers gaps and enter Powell River Valley
14 June 17-18: Confederates evacuate Cumberland Gap; Morgan occupies it


*Though a firm disciplinarian and an admirable tactician, Major Gen. Don Carlos Buell (right) was a slow and cautious soldier who repeatedly resisted Lincoln's calls for a Federal advance. Methodical and reserved, Buell was not a popular leader, but he did not believe that popularity was an essential characteristic for military effeciency and success.*

On January 19, a second battle was fought in the region; a Confederate army led by Major General George B. Crittenden had established a fortified camp at Beech Grove on the Cumberland, opposite Mill Springs. On January 1st, Federal Brigadier General George H. Thomas marched his division through heavy rain from Lebanon reaching Logan's Cross Roads on the 17th.

With the Cumberland in flood, Crittenden ordered a night march and a dawn attack on the Federal camps, hoping to beat Thomas's forces before they concentrated. The Confederates surprised the Federals, but faltered when popular Brigadier General Felix K. Zollicoffer was killed. This gained the Federals a respite, and Thomas brought up reinforcements. After a desperate fight, the Federals turned Crittenden's left flank and the Confederates fled across the river. Middle Creek had prised the Confederates out of eastern Kentucky, and the battle of Mill Springs had breached Johnston's Kentucky line.

It was April 11 before the Union followed up their victory. Brigadier General George W. Morgan, leading his Seventh Division, was sent to capture Cumberland Gap and free East Tennessee. The Confederates had fortified the Wilderness Road route to the Gap. Favoring an indirect approach, Morgan ordered one brigade to advance via Big Creek Gap while other forces threatened Chattanooga. Confederates were rushed to Chattanooga, but realizing they were outflanked by the Federal column, they evacuated Cumberland Gap on June 17. The next day Morgan occupied the strategic gateway to East Tennessee.

*Recognising that a Union attack was imminent, Crittenden preempted the strike with an assault at Logan's Crossroads (right) during the early hours of January 19. Though initially successful, the Confederate attack wavered when Zollicoffer was killed in the confusion. Reinforced, Thomas's Federals were able to break the Confederate left. The Confederates hurriedly retreated across the Cumberland, abandoning their camp and supplies. Union casualties numbered 261, Confederate 533.*

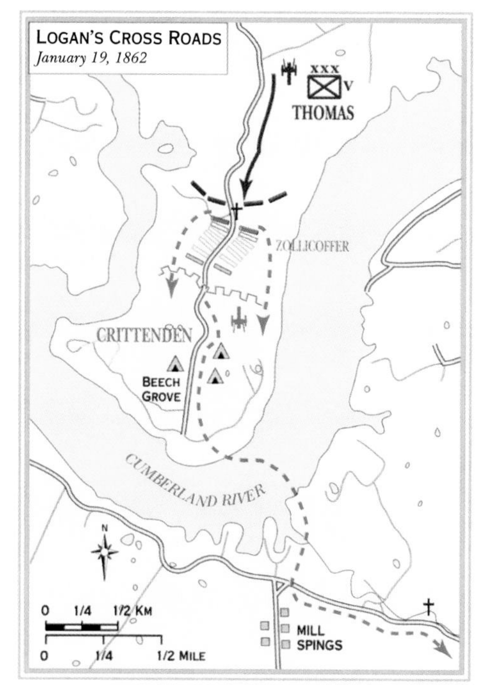

LOGAN'S CROSS ROADS
January 19, 1862
XXX
THOMAS
ZOLLICOFFER
CRITTENDEN
BEECH GROVE
CUMBERLAND RIVER
N
0 1/4 1/2 KM
0 1/4 1/2 MILE
MILL SPINGS


"The country must now be roused to make the greatest effort that it will be called upon to make during the war. No matter what the sacrifice may be, it must be made, and without loss of time... All the resources of the Confederacy are now needed for the defense of Tennessee."

*Gen. Albert Sidney Johnston.*

# Fort Henry and Donelson Campaign FEBUARY 1862

Prodded by Abraham Lincoln's call for a general movement of the army and navy against the "insurgent forces," General Halleck ordered General Grant to take Fort Henry, the sole Confederate bastion defending the Tennessee River.

Mounting only 17 cannon and constructed on low ground that was partially flooded, Fort Henry was unexpectedly weak. In view of these limitations, the fort's commander, Brigadier General Lloyd Tilghman, decided to save the major part of his Confederate garrison by sending all but some artillerists, and a few sick, overland to the stronger Fort Donelson on the Cumberland River.

On February 6, as Grant's 15,000 troops slogged south to position themselves for the attack on Fort Henry, a flotilla of seven Federal gunboats, led by Flag Officer Andrew H. Foote, steamed forward and pounded the fort into submission, thus opening the lower Tennessee River to Union control.

Fort Henry's collapse forced General A.S. Johnston, supreme Confederate commander in the West, to evacuate middle Kentucky and fall back on Nashville. Johnston sent 15,000 men under Brigadier General John B. Floyd, to reinforce and defend Fort Donelson. On

KENTUCKY
Camp Hallet
Bailey's Ford
Panther Cr. Island
C.F. SMITH
GRANT
McCLERNAND
Boyd's
ROAD TO DOVER
TELEGRAPH LINE
Fort Henry
Fort Heiman
TENNESSEE RIVER
Panther Cr.
Piney Cr.

*The commander of the naval squadron in the attacks on Forts Henry and Donelson, Andrew Hull Foote (below), was a strict disciplinarian who, nevertheless, enjoyed the respect of the men who served under him. Foote was injured when a Confederate shell from Fort Donelson crashed into the pilot-house of his flagship, the* St. Louis, *disabling the ship and making her an easy target for the rebel gunners.*

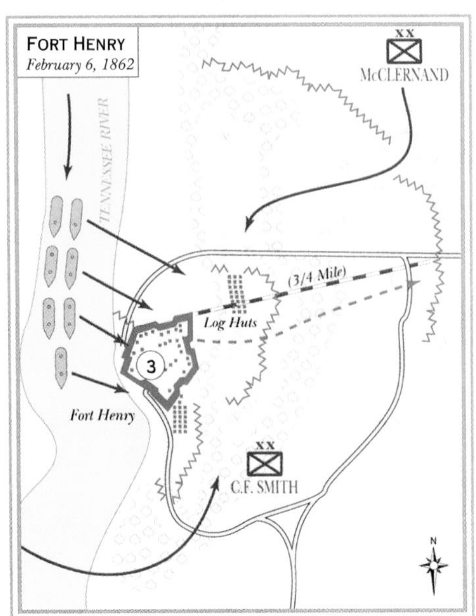
FORT HENRY
February 6, 1862
McCLERNAND
TENNESSEE RIVER
(3/4 Mile)
Log Huts
Fort Henry
C.F. SMITH
N

*As Grant's forces advanced south towards Fort Henry, and Foote's seven gunboats steamed upstream to pour fire into the partially flooded fort, the Confederate commander, Brig. Gen. Lloyd Tilghman, decided to surrender, and evacuate his garrison east to Fort Donelson. Slowed by rain-soaked terrain, McClernand's Federal division was unable to block the Confederate garrison's retreat (see map left).*

> "Fill your cartridge boxes, quick, and get into line. The enemy is trying to escape and he must not be permitted to do so."
>
> *Grant to McClernand during the attempted Confederate breakout.*

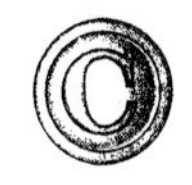

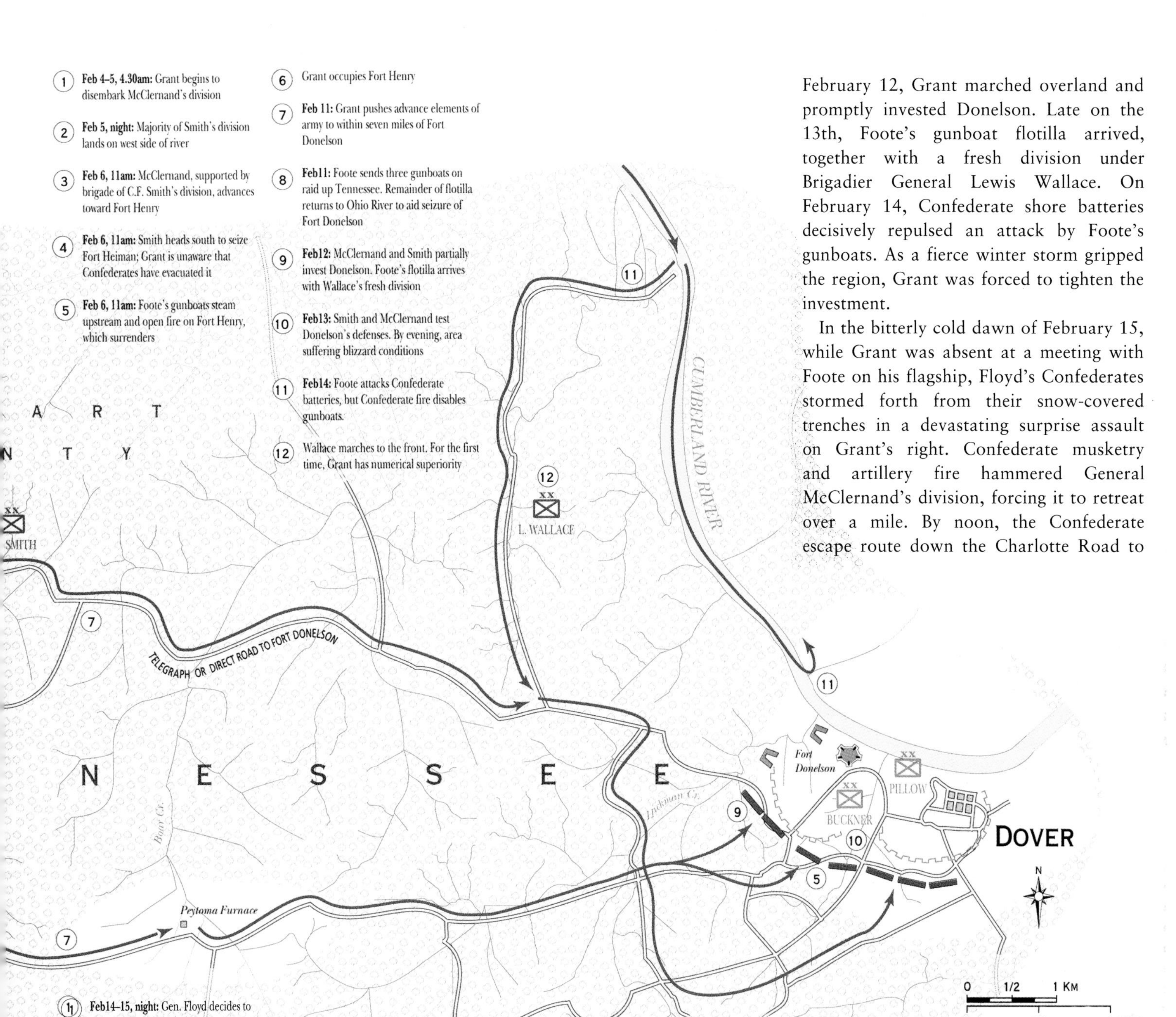

1 Feb 4–5, 4.30am: Grant begins to disembark McClernand's division
2 Feb 5, night: Majority of Smith's division lands on west side of river
3 Feb 6, 11am: McClernand, supported by brigade of C.F. Smith's division, advances toward Fort Henry
4 Feb 6, 11am: Smith heads south to seize Fort Heiman; Grant is unaware that Confederates have evacuated it
5 Feb 6, 11am: Foote's gunboats steam upstream and open fire on Fort Henry, which surrenders
6 Grant occupies Fort Henry
7 Feb 11: Grant pushes advance elements of army to within seven miles of Fort Donelson
8 Feb11: Foote sends three gunboats on raid up Tennessee. Remainder of flotilla returns to Ohio River to aid seizure of Fort Donelson
9 Feb12: McClernand and Smith partially invest Donelson. Foote's flotilla arrives with Wallace's fresh division
10 Feb13: Smith and McClernand test Donelson's defenses. By evening, area suffering blizzard conditions
11 Feb14: Foote attacks Confederate batteries, but Confederate fire disables gunboats.
12 Wallace marches to the front. For the first time, Grant has numerical superiority
CUMBERLAND RIVER
L. WALLACE
SMITH
TELEGRAPH OR DIRECT ROAD TO FORT DONELSON
T E N N E S S E E
Fort Donelson
PILLOW
BUCKNER
DOVER
Hickman Cr.
Boar Cr.
Peytoma Furnace
N
0 1/2 1 KM
0 1/2 1 MILE


February 12, Grant marched overland and promptly invested Donelson. Late on the 13th, Foote's gunboat flotilla arrived, together with a fresh division under Brigadier General Lewis Wallace. On February 14, Confederate shore batteries decisively repulsed an attack by Foote's gunboats. As a fierce winter storm gripped the region, Grant was forced to tighten the investment.

In the bitterly cold dawn of February 15, while Grant was absent at a meeting with Foote on his flagship, Floyd's Confederates stormed forth from their snow-covered trenches in a devastating surprise assault on Grant's right. Confederate musketry and artillery fire hammered General McClernand's division, forcing it to retreat over a mile. By noon, the Confederate escape route down the Charlotte Road to Nashville was opened. A great victory had been won. However, indecision now gripped Floyd and his subordinate, General Pillow, and the Confederate troops were ordered back into their fortifications. Grant now counterattacked, closing up the gap on the right with Wallace's division, while an assault by Brigadier General Charles F. Smith's division seized a portion of the outer line of Confederate trenches on the left, before night ended combat.

Continued animosity and confusion among the Confederate generals, together with the arrival of strong Federal reinforcements, led to Donelson's unconditional surrender on February 16. The entire Confederate heartland was now vulnerable to Federal invasion.

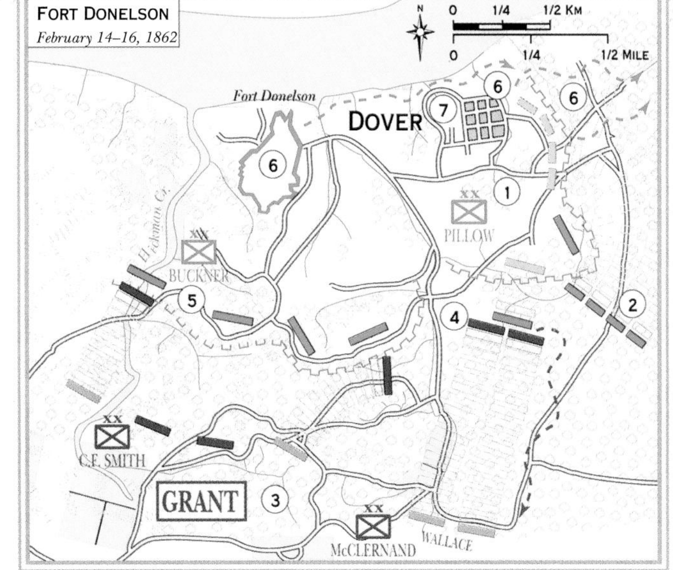

1 Feb14–15, night: Gen. Floyd decides to abandon Donelson and escapes to Nashville. Pillow and Buckner maneuver to attack Grant's right flank
2 Feb15, dawn–noon: Pillow and Buckner attack McClernand, break Federal line and secure Charlotte Road escape route, but Floyd weakens and orders troops back to entrenchments
3 Feb15, 1pm: Grant arrives; realizing Confederates are trying to escape, orders attack on entire Confederate front
4 Feb15, pm: As Confederates retire into Donelson, Wallace and McClernand attack and regain lost ground
5 Feb15, pm: Smith advances along Hickman Creek: seizes outer Confederate entrenchments, but his advance halted. Fighting ends
6 Feb15–16, night: Confederates surrender. Floyd and Pillow escape with about 2,500 men. Forrest leads cavalry to safety across Lick Creek
7 Feb16, dawn: In Dover, Grant accepts "unconditional surrender" of Donelson
FORT DONELSON
February 14–16, 1862
N
0 1/4 1/2 KM
0 1/4 1/2 MILE
Fort Donelson
DOVER
PILLOW
BUCKNER
Hickman Cr.
C.F. SMITH
GRANT
McCLERNAND
WALLACE

# Northwest Arkansas MARCH 7 – DECEMBER 8 1862

During the fall of 1861 and winter of 1862, Federal forces vied for control of southwestern Missouri with the pro-secessionist State Guard, commanded by Major General Sterling Price, and Confederate forces from Arkansas under General Ben McCulloch. The region changed hands several times.

In February 1862 a Federal army of over 10,000 men, under General Samuel R. Curtis, captured Springfield, driving the Missouri State Guard into northwestern Arkansas. There, Price united with McCulloch under the overall command of Confederate general Earl Van Dorn. With 16,500 men, Van Dorn enjoyed a rare superiority of numbers. He advanced immediately, marching his men through winter ice storms to reach the Federal army's rear on March 7. However, the men's exhaustion, poor coordination, and a shortage of ammunition robbed the attack of much of its effectiveness.

Faced with a surprise attack on his right flank and rear, Curtis performed masterfully. He gradually shifted his army to the north, parrying the uncoordinated Confederate assaults throughout the day. On March 8 he counterattacked and Van Dorn, decisively defeated, withdrew. The Union lost 1,270; the Confederates about 2,000.

The dramatic Federal successes, which began with General Grant's capture of forts Henry and Donelson in February 1862, shifted attention and resources away from the Trans-Mississippi theater. After Van Dorn led reinforcements across the Mississippi, Confederate Major General Thomas C. Hindman raised an army of 11,000 men, and in early December moved against Brigadier General James G. Blunt's 7,000 men near Fayetteville, hoping to defeat them before 3,000 men arrived from Springfield under Brigadier General Francis J. Herron.

When his own movement proved slower than Herron's, Hindman left a small force to hold Blunt, and moved instead against Herron. But rather than attacking, he assumed a defensive position at Prairie Grove. Herron's assaults on December 7 proved ineffective, nor did Blunt's arrival later give the Union victory. Athough the battle ended in a draw, Hindman withdrew, leaving Arkansas vulnerable to subsequent Federal advances.

> "Silent and sad. The vulture and the wolf have now communion, and the dead, friends and foes, sleep in the same lonely grave."
>
> *Gen. Samuel Ryan Curtis after Pea Ridge.*

1. **Early March:** Federal army under Gen. Samuel Curtis deploys along Little Sugar Creek
2. Van Dorn's combined forces bypass the strong Union position by following the Bentonville detour. The march exhausts his men and hundreds drop out en route
3. **March 7:** Van Dorn reaches Federal army rear, but is detected by Union scouts
4. **March 7:** Curtis gradually shifts his troops from their entrenched defensive positions to meet the uncoordinated Confederate attacks to his right flank and rear

BENTONVILLE & KEETSVILLE ROAD (BENTONVILLE DETO

VAN DORN

C

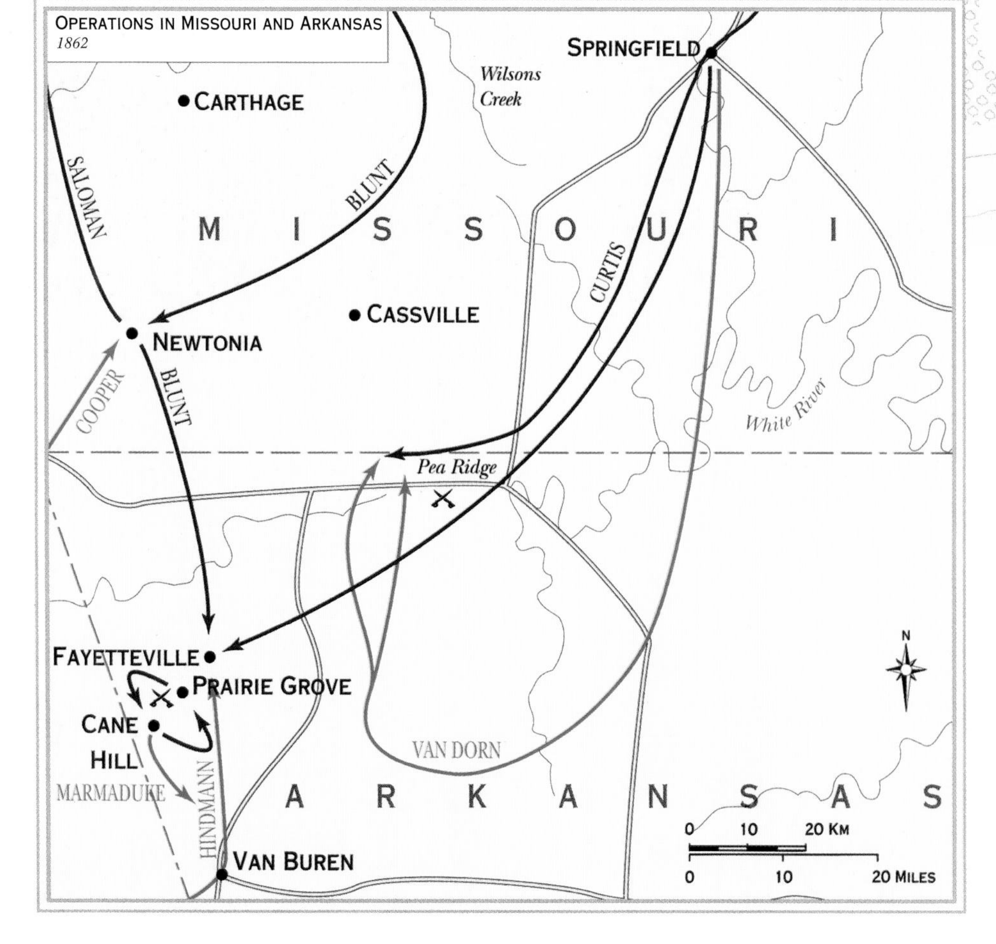
OPERATIONS IN MISSOURI AND ARKANSAS
1862
SPRINGFIELD
Wilsons Creek
CARTHAGE
SALOMAN
BLUNT
M I S S O U R I
CURTIS
CASSVILLE
NEWTONIA
COOPER
BLUNT
White River
Pea Ridge
FAYETTEVILLE
PRAIRIE GROVE
CANE HILL
MARMADUKE
HINDMANN
VAN DORN
A R K A N S A S
VAN BUREN
N
0 10 20 KM
0 10 20 MILES

*After defeating Gen. Nathaniel Lyon and occupying Lexington in 1861, Sterling Price wintered in Southern Missouri. To revenge these defeats, a Union army under Gen. Samuel Ryan Curtis was sent to the region early in 1862. Outnumbered, Price retreated until reinforced by overall Confederate commander Earl Van Dorn. Van Dorn led his forces against those of Curtis and the two converged at Pea Ridge (see map left).*

*Ben McCulloch (below) was so popular that when he was felled by a shot through the heart the fighting effectiveness of his troops evaporated, many of them simply wandering from the field, stupefied with grief.*

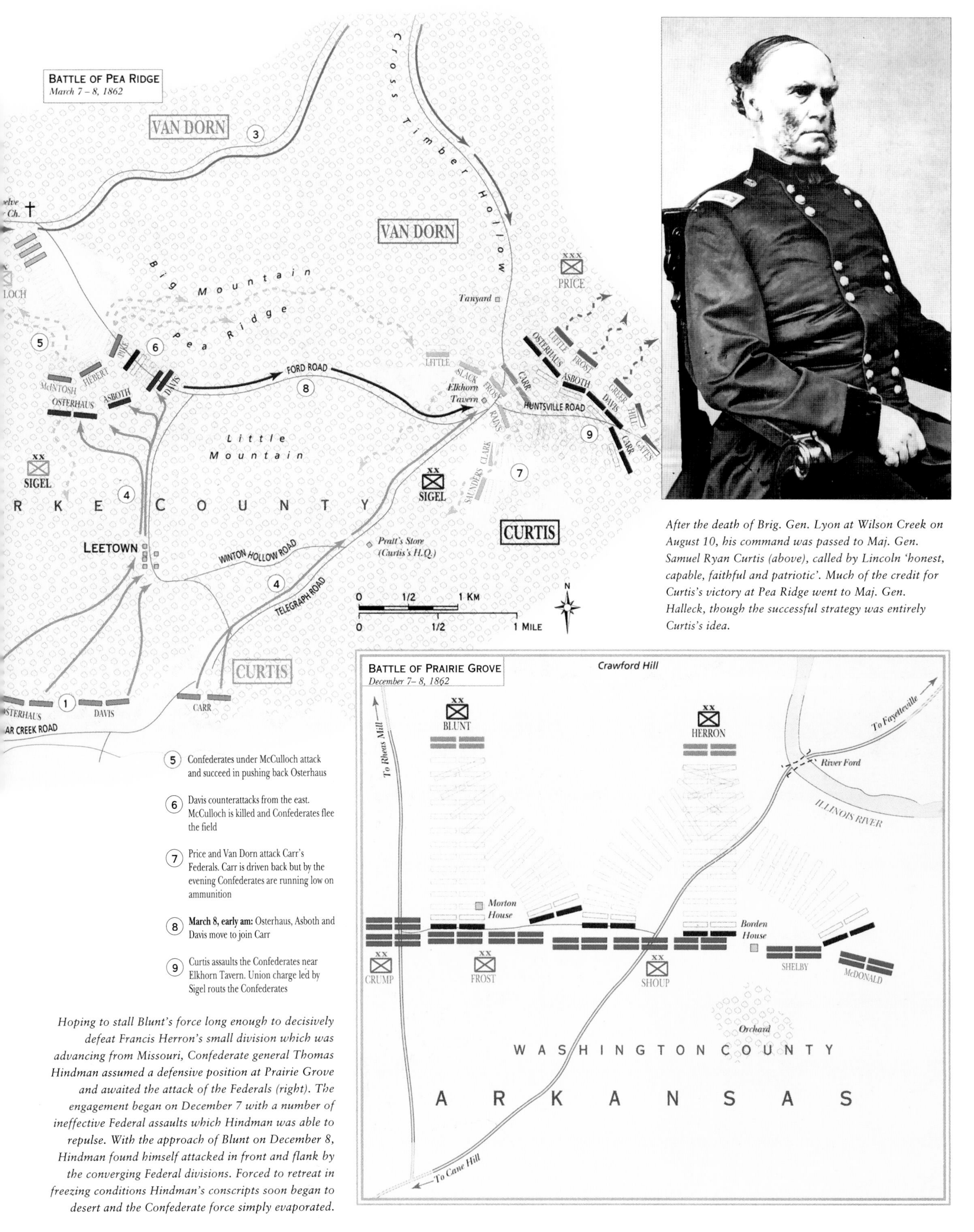
BATTLE OF PEA RIDGE
March 7 – 8, 1862
VAN DORN
VAN DORN
Cross Timber Hollow
Big Mountain
Pea Ridge
PRICE
Tanyard
LITTLE
SLACK
Elkhorn Tavern
FROST
CARR
OSTERHAUS
ASBOTH
GREER
DAVIS
HILL
GATES
HUNTSVILLE ROAD
FORD ROAD
PIKE
HEBERT
McINTOSH
OSTERHAUS
ASBOTH
DAVIS
LOCH
Little Mountain
SIGEL
SIGEL
RAINS
CLARK
SAUNDERS
CURTIS
CURTIS
LEETOWN
WINTON HOLLOW ROAD
TELEGRAPH ROAD
Pratt's Store (Curtis's H.Q.)
0 1/2 1 KM
0 1/2 1 MILE
N
CARR
DAVIS
CREEK ROAD
5 Confederates under McCulloch attack and succeed in pushing back Osterhaus
6 Davis counterattacks from the east. McCulloch is killed and Confederates flee the field
7 Price and Van Dorn attack Carr's Federals. Carr is driven back but by the evening Confederates are running low on ammunition
8 March 8, early am: Osterhaus, Asboth and Davis move to join Carr
9 Curtis assaults the Confederates near Elkhorn Tavern. Union charge led by Sigel routs the Confederates
BATTLE OF PRAIRIE GROVE
December 7– 8, 1862
Crawford Hill
BLUNT
HERRON
To Rheas Mill
To Fayetteville
River Ford
ILLINOIS RIVER
Morton House
Borden House
CRUMP
FROST
SHOUP
SHELBY
McDONALD
Orchard
WASHINGTON COUNTY
ARKANSAS
To Cane Hill

*After the death of Brig. Gen. Lyon at Wilson Creek on August 10, his command was passed to Maj. Gen. Samuel Ryan Curtis (above), called by Lincoln 'honest, capable, faithful and patriotic'. Much of the credit for Curtis's victory at Pea Ridge went to Maj. Gen. Halleck, though the successful strategy was entirely Curtis's idea.*

*Hoping to stall Blunt's force long enough to decisively defeat Francis Herron's small division which was advancing from Missouri, Confederate general Thomas Hindman assumed a defensive position at Prairie Grove and awaited the attack of the Federals (right). The engagement began on December 7 with a number of ineffective Federal assaults which Hindman was able to repulse. With the approach of Blunt on December 8, Hindman found himself attacked in front and flank by the converging Federal divisions. Forced to retreat in freezing conditions Hindman's conscripts soon began to desert and the Confederate force simply evaporated.*

# Shiloh APRIL 6 1862

With the loss of Forts Henry and Donelson in February, General Johnston withdrew his disheartened Confederate forces into west Tennessee, northern Mississippi and Alabama to reorganize. In early March, General Halleck responded by ordering General Grant to advance his Union Army of West Tennessee on an invasion up the Tennessee River.

Occupying Pittsburg Landing, Grant entertained no thought of a Confederate attack. Halleck's instructions were that following the arrival of General Buell's Army of the Ohio from Nashville, Grant would advance south in a joint offensive to seize the Memphis & Charleston Railroad, the Confederacy's only east-west all-weather supply route that linked the lower Mississippi Valley to cities on the Confederacy's east coast.

Assisted by his second-in-command, General Beauregard, Johnston shifted his scattered forces and concentrated almost 55,000 men around Corinth. Strategically located where the Memphis & Charleston crossed the Mobile & Ohio Railroad, Corinth was the western Confederacy's most important rail junction.

On April 3, realizing Buell would soon reinforce Grant, Johnston launched an offensive with his newly christened Army of the Mississippi. Advancing upon Pittsburg Landing with 43,938 men, Johnston planned to surprise Grant, cut his army off from retreat to the Tennessee River, and drive the Federals west into the swamps of Owl Creek.

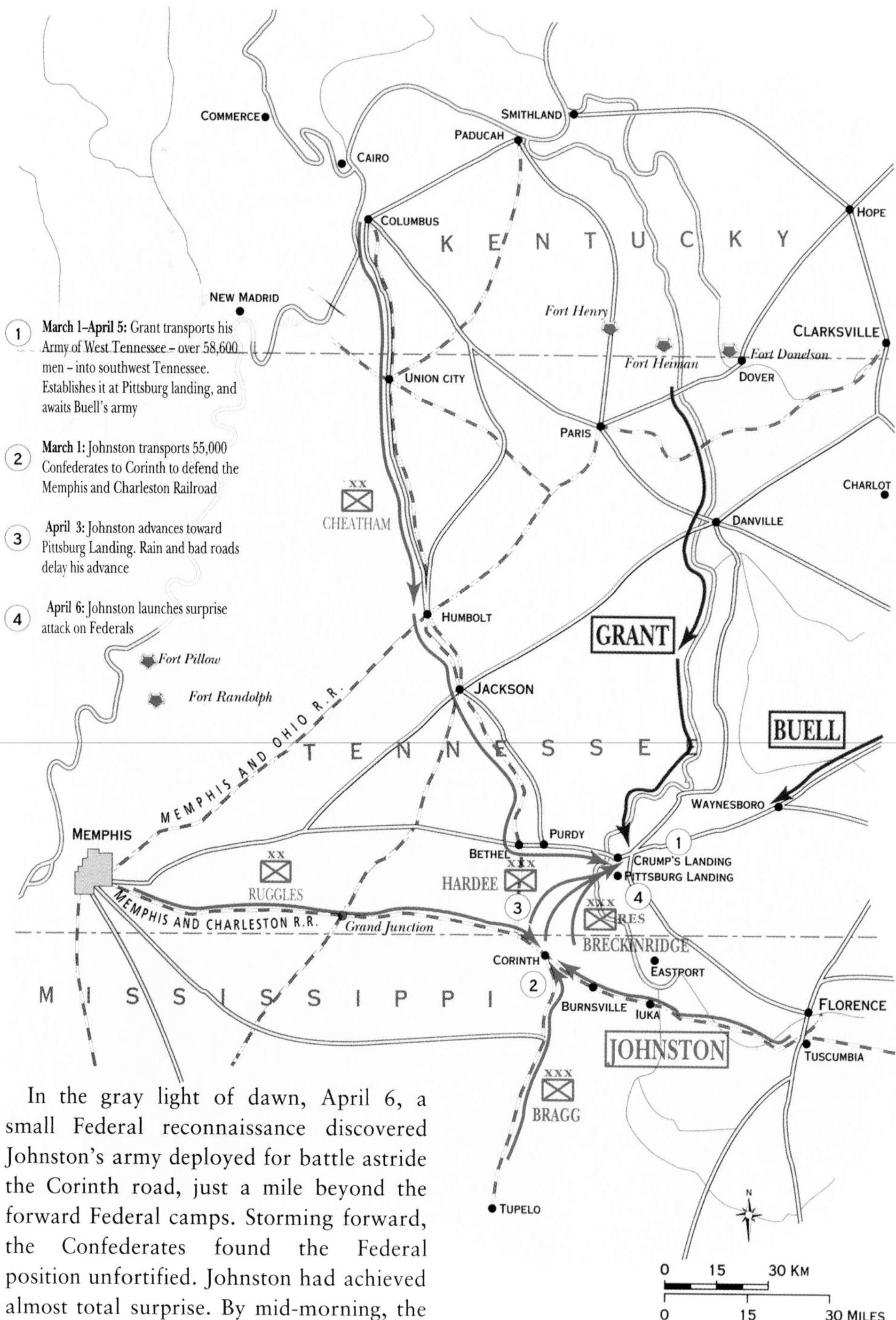
1 March 1–April 5: Grant transports his Army of West Tennessee – over 58,600 men – into southwest Tennessee. Establishes it at Pittsburg landing, and awaits Buell's army
2 March 1: Johnston transports 55,000 Confederates to Corinth to defend the Memphis and Charleston Railroad
3 April 3: Johnston advances toward Pittsburg Landing. Rain and bad roads delay his advance
4 April 6: Johnston launches surprise attack on Federals
COMMERCE
CAIRO
SMITHLAND
PADUCAH
COLUMBUS
K E N T U C K Y
HOPE
NEW MADRID
Fort Henry
Fort Heiman
Fort Donelson
CLARKSVILLE
DOVER
UNION CITY
PARIS
CHARLOT
DANVILLE
XX
CHEATHAM
HUMBOLT
GRANT
Fort Pillow
Fort Randolph
JACKSON
MEMPHIS AND OHIO R.R.
T E N N E S S E E
BUELL
WAYNESBORO
MEMPHIS
PURDY
BETHEL
CRUMP'S LANDING
PITTSBURG LANDING
XX
RUGGLES
XXX
HARDEE
XXX
RES
BRECKINRIDGE
MEMPHIS AND CHARLESTON R.R.
Grand Junction
CORINTH
EASTPORT
M I S S I S S I P P I
BURNSVILLE
IUKA
FLORENCE
JOHNSTON
TUSCUMBIA
XXX
BRAGG
TUPELO
N
0 15 30 KM
0 15 30 MILES

*Union troops flee toward the Tennessee River during the Confederate assault (below). The shock to the unprepared Federals was great and though many units offered fierce resistance, others quickly buckled and sought to escape on board the army transports.*

In the gray light of dawn, April 6, a small Federal reconnaissance discovered Johnston's army deployed for battle astride the Corinth road, just a mile beyond the forward Federal camps. Storming forward, the Confederates found the Federal position unfortified. Johnston had achieved almost total surprise. By mid-morning, the Confederates seemed within easy reach of victory, overrunning one frontline Union division and capturing its camp. However, stiff resistance on the Federal right entangled Johnston's brigades in a savage fight around Shiloh Church. Throughout the day, Johnston's army hammered the Federal right, which gave ground but did not break. Casualties upon this brutal killing ground were immense.

Meanwhile, Johnston's flanking attack stalled in front of Sarah Bell's peach orchard and the dense oak thicket labeled the "hornet's nest" by the Confederates. Grant's left flank withstood Confederate assaults for seven crucial hours before being forced to yield ground in the late afternoon. Despite inflicting heavy casualties and seizing ground, the Confederates only drove Grant towards the river, instead of away from it. The Federal survivors established a solid front before Pittsburg Landing and repulsed the last Confederate charge as dusk ended the first day of fighting.

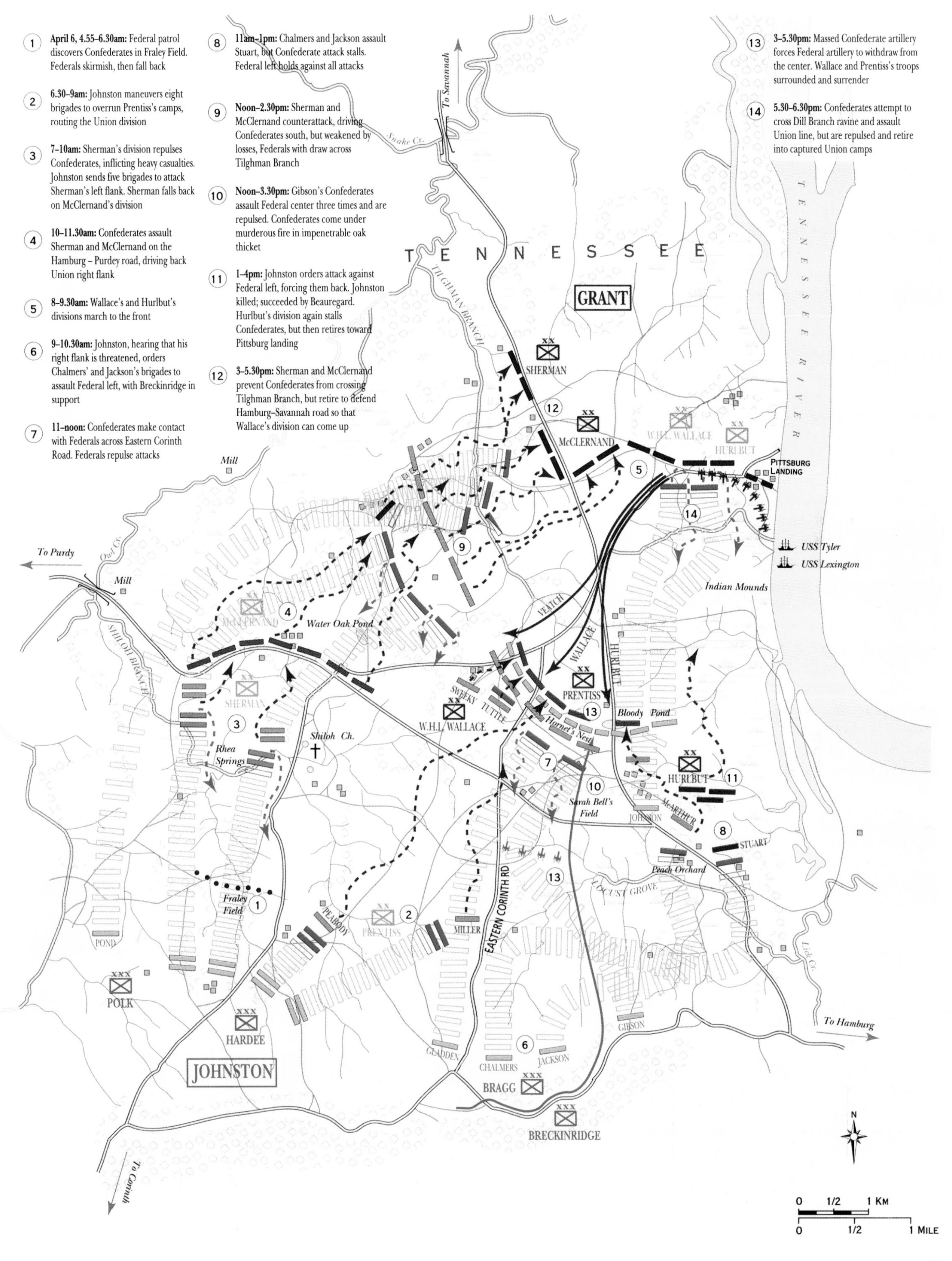

1 April 6, 4.55–6.30am: Federal patrol discovers Confederates in Fraley Field. Federals skirmish, then fall back
2 6.30–9am: Johnston maneuvers eight brigades to overrun Prentiss's camps, routing the Union division
3 7–10am: Sherman's division repulses Confederates, inflicting heavy casualties. Johnston sends five brigades to attack Sherman's left flank. Sherman falls back on McClernand's division
4 10–11.30am: Confederates assault Sherman and McClernand on the Hamburg – Purdey road, driving back Union right flank
5 8–9.30am: Wallace's and Hurlbut's divisions march to the front
6 9–10.30am: Johnston, hearing that his right flank is threatened, orders Chalmers' and Jackson's brigades to assault Federal left, with Breckinridge in support
7 11–noon: Confederates make contact with Federals across Eastern Corinth Road. Federals repulse attacks
8 11am–1pm: Chalmers and Jackson assault Stuart, but Confederate attack stalls. Federal left holds against all attacks
9 Noon–2.30pm: Sherman and McClernand counterattack, driving Confederates south, but weakened by losses, Federals with draw across Tilghman Branch
10 Noon–3.30pm: Gibson's Confederates assault Federal center three times and are repulsed. Confederates come under murderous fire in impenetrable oak thicket
11 1–4pm: Johnston orders attack against Federal left, forcing them back. Johnston killed; succeeded by Beauregard. Hurlbut's division again stalls Confederates, but then retires toward Pittsburg landing
12 3–5.30pm: Sherman and McClernand prevent Confederates from crossing Tilghman Branch, but retire to defend Hamburg–Savannah road so that Wallace's division can come up
13 3–5.30pm: Massed Confederate artillery forces Federal artillery to withdraw from the center. Wallace and Prentiss's troops surrounded and surrender
14 5.30–6.30pm: Confederates attempt to cross Dill Branch ravine and assault Union line, but are repulsed and retire into captured Union camps
TENNESSEE
TENNESSEE RIVER
To Savannah
Snake Cr.
Tilghman Branch
GRANT
SHERMAN
McCLERNAND
W.H.L. WALLACE
HURLBUT
PITTSBURG LANDING
USS Tyler
USS Lexington
Indian Mounds
To Purdy
Owl Cr.
Mill
Shiloh Branch
Water Oak Pond
VEATCH
WALLACE
PRENTISS
Sweeny
Tuttle
Hornet's Nest
Bloody Pond
Rhea Springs
Shiloh Ch.
Sarah Bell's Field
McARTHUR
JOHNSON
STUART
Peach Orchard
Locust Grove
Fraley Field
PEABODY
MILLER
EASTERN CORINTH RD
POND
POLK
HARDEE
JOHNSTON
GLADDEN
CHALMERS
JACKSON
BRAGG
BRECKINRIDGE
GIBSON
Lick Cr.
To Hamburg
To Corinth
N
0 1/2 1 KM
0 1/2 1 MILE

# Shiloh APRIL 7 1862

Shiloh's first day of slaughter also witnessed the death of the Confederate leader, General Johnston, who fell at mid-afternoon, struck down by a stray bullet while directing the action on the Confederate right. At dusk, the advance division of General Buell's Federal Army of the Ohio reached Pittsburg Landing, and crossed the river to file into line on the Union left during the night. Buell's arrival, plus the timely appearance of a reserve division from Grant's army, led by Major General Lewis Wallace, fed over 22,500 reinforcements into the Union lines. On April 7, Grant renewed the fighting with an aggressive counterattack.

Taken by surprise, General Beauregard managed to rally 30,000 of his badly disorganized Confederates, and mounted a tenacious defense. Inflicting heavy casualties on the Federals, Beauregard's troops temporarily halted the determined Union advance. However, strength in numbers provided Grant with a decisive advantage. By mid-afternoon, as waves of fresh Federal troops swept forward, pressing the exhausted Confederates back to Shiloh Church, Beauregard realized his armies' peril and ordered a retreat. During the night, the Confederates withdrew, greatly disorganized, to their fortified stronghold at Corinth. Possession of the grisly battlefield passed to the victorious Federals, who were satisfied to simply reclaim Grant's camps and make an exhausted bivouac among the dead.

General Johnston's massive and rapid concentration at Corinth, and surprise attack on Grant at Pittsburg Landing, had presented the Confederacy with an opportunity to reverse the course of the war. The aftermath, however, left the invading Union forces still poised to carry out the capture of the Corinth rail junction. Shiloh's awesome toll of 23,746 men killed, wounded, or missing brought a shocking realization to both sides that the war would not end quickly.

Gen Grant present at this charge

*The Union counterattack on the morning of April 7 (above). Although elated with their success of the previous day, Confederates were exhausted and demoralized by the arrival of Union reinforcements. Furthermore, their line was weakened by the withdrawal of Polk's division the previous evening, when the retirement order had been misinterpreted.*

*Although this photograph (below) was not taken until a few days after the fighting at Shiloh had come to an end, it gives a very accurate indication of the position of the large 24-pound siege guns used by Grant and Buell during the battle. The guns are directed toward the location of the final Confederate assault which took place during the afternoon of April 7.*

"You gentlemen have had your way today, but it will be very different tomorrow. You'll see. Buell will effect a junction with Grant tonight and we'll turn the tables on you in the morning. You'll see."

*Gen. Prentiss prophesying the events of April 7 to his Confederate captors, April 6.*

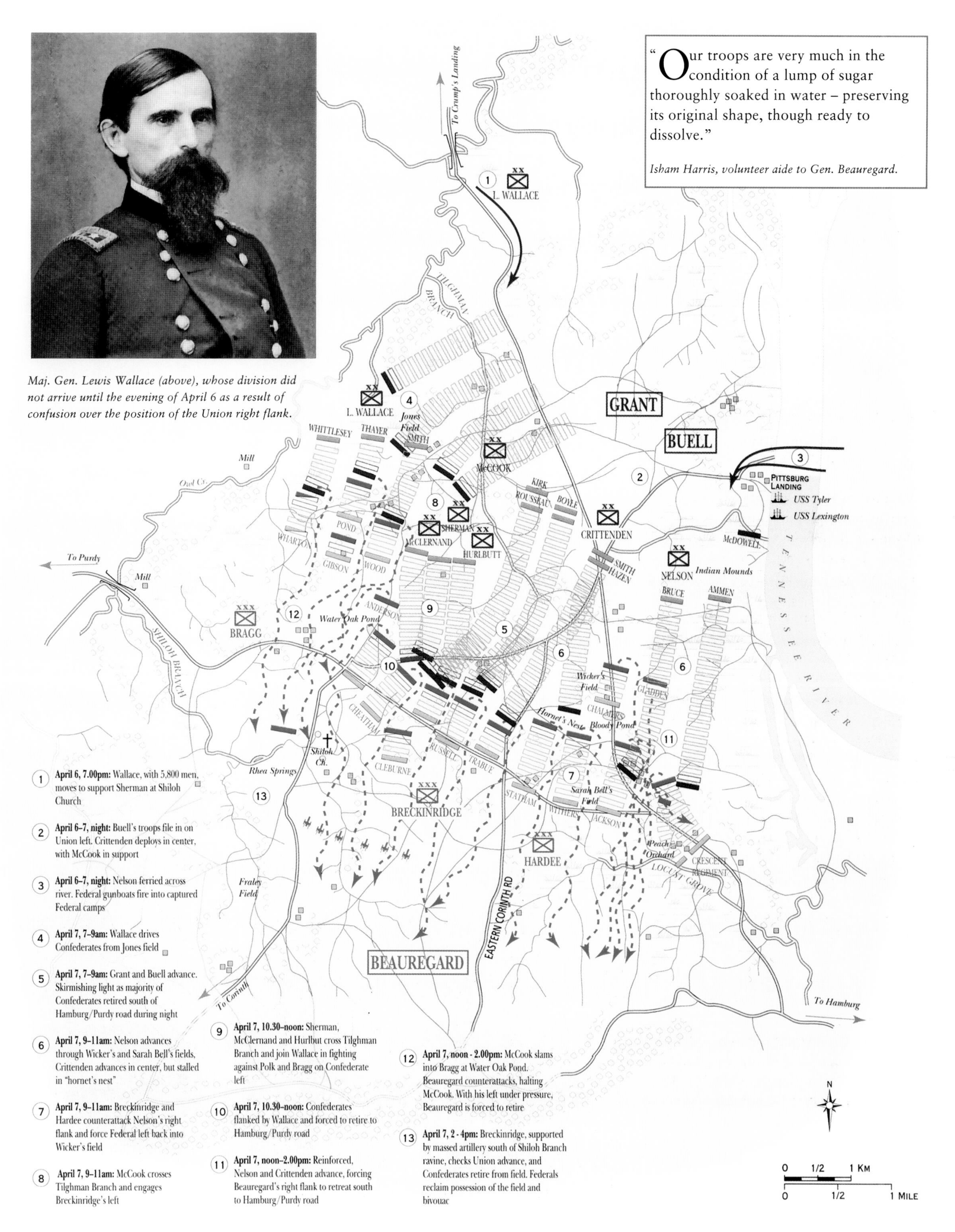

To Crump's Landing
L. WALLACE
TILGHMAN BRANCH
GRANT
BUELL
PITTSBURG LANDING
USS Tyler
USS Lexington
L. WALLACE
Jones Field
WHITTLESEY
THAYER
SMITH
McCOOK
Mill
Owl Cr.
KIRK
ROUSSEAU
BOYLE
CRITTENDEN
McDOWELL
SHERMAN
McCLERNAND
HURLBUTT
WHARTON
POND
GIBSON
WOOD
SMITH
HAZEN
NELSON
Indian Mounds
BRUCE
AMMEN
To Purdy
Mill
BRAGG
Water Oak Pond
ANDERSON
SHILOH BRANCH
TENNESSEE RIVER
Wicker's Field
GLADDEN
CHALMERS
Hornet's Nest
Bloody Pond
CHEATHAM
Shiloh Ch.
RUSSELL
TRABUE
CLEBURNE
Rhea Springs
BRECKINRIDGE
STATHAM
WITHERS
Sarah Bell's Field
JACKSON
Peach Orchard
CRESCENT REGIMENT
LOCUST GROVE
HARDEE
Fraley Field
EASTERN CORINTH RD
BEAUREGARD
To Corinth
To Hamburg
N
0 1/2 1 KM
0 1/2 1 MILE
1 April 6, 7.00pm: Wallace, with 5,800 men, moves to support Sherman at Shiloh Church
2 April 6–7, night: Buell's troops file in on Union left. Crittenden deploys in center, with McCook in support
3 April 6–7, night: Nelson ferried across river. Federal gunboats fire into captured Federal camps
4 April 7, 7–9am: Wallace drives Confederates from Jones field
5 April 7, 7–9am: Grant and Buell advance. Skirmishing light as majority of Confederates retired south of Hamburg/Purdy road during night
6 April 7, 9–11am: Nelson advances through Wicker's and Sarah Bell's fields, Crittenden advances in center, but stalled in "hornet's nest"
7 April 7, 9–11am: Breckinridge and Hardee counterattack Nelson's right flank and force Federal left back into Wicker's field
8 April 7, 9–11am: McCook crosses Tilghman Branch and engages Breckinridge's left
9 April 7, 10.30–noon: Sherman, McClernand and Hurlbut cross Tilghman Branch and join Wallace in fighting against Polk and Bragg on Confederate left
10 April 7, 10.30–noon: Confederates flanked by Wallace and forced to retire to Hamburg/Purdy road
11 April 7, noon–2.00pm: Reinforced, Nelson and Crittenden advance, forcing Beauregard's right flank to retreat south to Hamburg/Purdy road
12 April 7, noon - 2.00pm: McCook slams into Bragg at Water Oak Pond. Beauregard counterattacks, halting McCook. With his left under pressure, Beauregard is forced to retire
13 April 7, 2 - 4pm: Breckinridge, supported by massed artillery south of Shiloh Branch ravine, checks Union advance, and Confederates retire from field. Federals reclaim possession of the field and bivouac


Maj. Gen. Lewis Wallace (above), whose division did not arrive until the evening of April 6 as a result of confusion over the position of the Union right flank.

"Our troops are very much in the condition of a lump of sugar thoroughly soaked in water – preserving its original shape, though ready to dissolve."

*Isham Harris, volunteer aide to Gen. Beauregard.*

# Upper Mississippi Valley MARCH 3 – JUNE 6 1862

With General Grant advancing up the Tennessee River, General Halleck sent Major General John Pope with 18,000 men overland against New Madrid, Missouri and Island Number 10, where 7,000 Confederate troops and strong shore batteries barred passage of the Mississippi River.

Supported on the river by Flag Officer Foote's gunboat and mortar boat flotilla, Pope coordinated a month-long amphibious operation, capturing both southern bastions, nearly 7,000 Confederate troops, heavy artillery, and considerable quantities of ammunition and supplies. The Union victory cleared the way for Federal naval operations to proceed downstream to Fort Pillow, Tennessee.

On April 11, Halleck arrived at Pittsburg Landing, superseding Grant (placed second-in-command) in direct command of the Union invasion. On April 29, Halleck's force, totaling 123,000 men in three Union armies (under Generals Buell, Pope and George H. Thomas), headed inland to seize the heavily fortified rail junction at Corinth.

Fresh from the tragic and bitter Shiloh experience, both sides displayed a reluctance to engage in pitched battle. Waging a classic campaign of offensive entrenchment, Halleck advanced cautiously against General Beauregard, who managed to stall the Union juggernaut for a month. Finally, with half his reported 112,000 men listed sick or absent, low on fresh water and supplies, and facing annihilation, Beauregard evacuated Corinth and withdrew into central Mississippi.

On May 30, Halleck took possession of Corinth, securing Union control of the Confederacy's sole east-west link railroad, the Memphis & Charleston, and furthering the Union objective of reclaiming the lower Mississippi Valley.

Meanwhile, upstream from Fort Pillow, the Federal flotilla lay anchored off Plum Run Bend. On May 10, Captain James E. Montgomery's makeshift Confederate River Defense Fleet unexpectedly appeared and boldly attacked the Federal fleet. In a brief action, the Confederate boats – lacking heavy armament and armor and outfitted instead with wood and cotton-bale armor – rammed and sank two of the big Union ironclads. During the affair, four of Montgomery's eight boats were damaged, forcing him to withdraw to Fort Pillow and then Memphis.

After enduring a month-long bombardment, followed by Beauregard's abandonment of Corinth, the Confederates

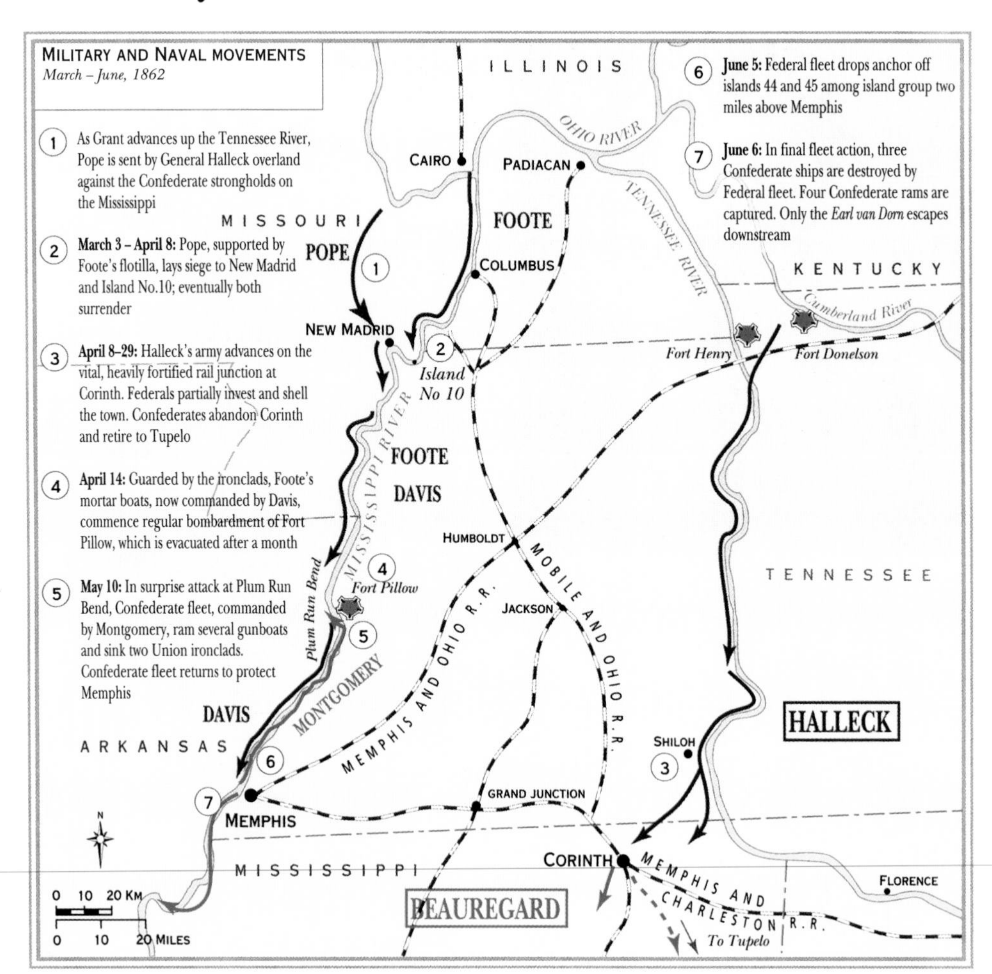
MILITARY AND NAVAL MOVEMENTS
March – June, 1862
1 As Grant advances up the Tennessee River, Pope is sent by General Halleck overland against the Confederate strongholds on the Mississippi
2 March 3 – April 8: Pope, supported by Foote's flotilla, lays siege to New Madrid and Island No.10; eventually both surrender
3 April 8-29: Halleck's army advances on the vital, heavily fortified rail junction at Corinth. Federals partially invest and shell the town. Confederates abandon Corinth and retire to Tupelo
4 April 14: Guarded by the ironclads, Foote's mortar boats, now commanded by Davis, commence regular bombardment of Fort Pillow, which is evacuated after a month
5 May 10: In surprise attack at Plum Run Bend, Confederate fleet, commanded by Montgomery, ram several gunboats and sink two Union ironclads. Confederate fleet returns to protect Memphis
6 June 5: Federal fleet drops anchor off islands 44 and 45 among island group two miles above Memphis
7 June 6: In final fleet action, three Confederate ships are destroyed by Federal fleet. Four Confederate rams are captured. Only the Earl van Dorn escapes downstream
ILLINOIS
MISSOURI
KENTUCKY
TENNESSEE
ARKANSAS
MISSISSIPPI
CAIRO
PADIACAN
OHIO RIVER
TENNESSEE RIVER
Cumberland River
MISSISSIPPI RIVER
POPE
FOOTE
DAVIS
MONTGOMERY
HALLECK
BEAUREGARD
COLUMBUS
NEW MADRID
Island No 10
Fort Henry
Fort Donelson
HUMBOLDT
JACKSON
Fort Pillow
Plum Run Bend
MOBILE AND OHIO R.R.
MEMPHIS AND OHIO R.R.
SHILOH
GRAND JUNCTION
MEMPHIS
CORINTH
MEMPHIS AND CHARLESTON R.R.
FLORENCE
To Tupelo
0 10 20 KM
0 10 20 MILES

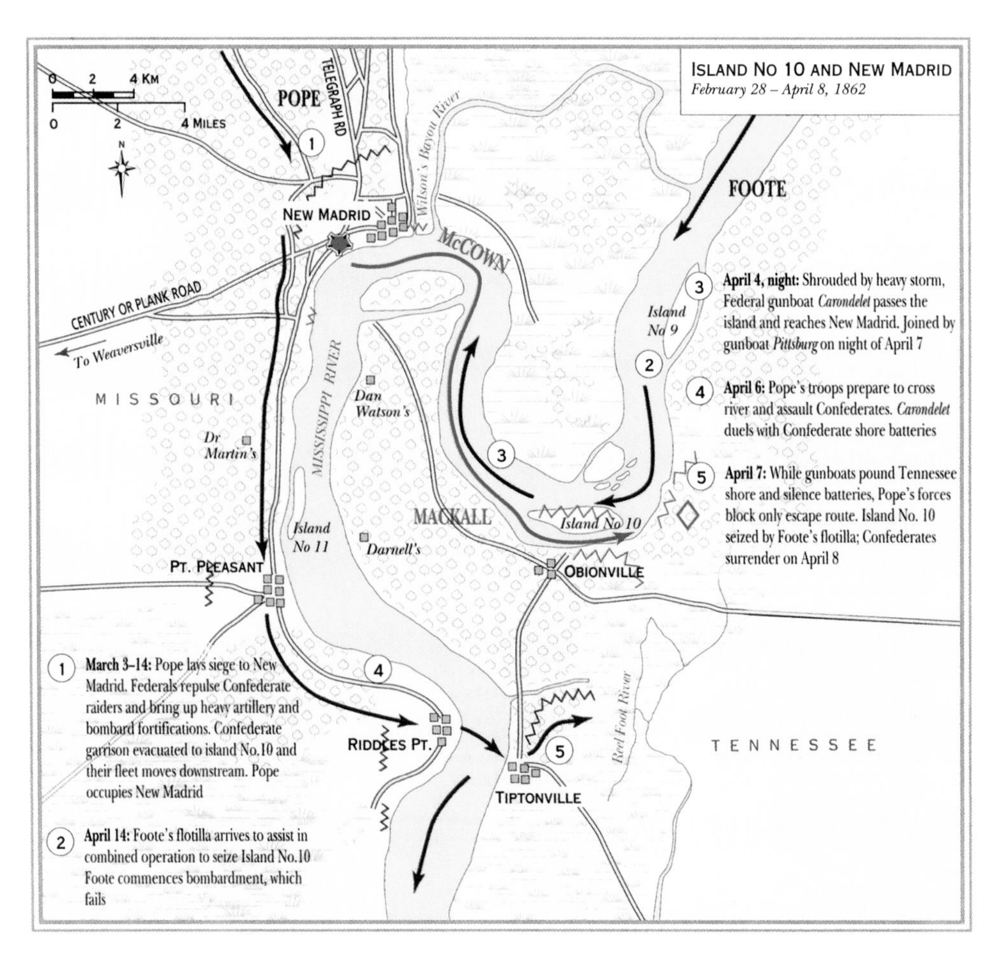
ISLAND NO 10 AND NEW MADRID
February 28 – April 8, 1862
1 March 3-14: Pope lays siege to New Madrid. Federals repulse Confederate raiders and bring up heavy artillery and bombard fortifications. Confederate garrison evacuated to island No.10 and their fleet moves downstream. Pope occupies New Madrid
2 April 14: Foote's flotilla arrives to assist in combined operation to seize Island No.10 Foote commences bombardment, which fails
3 April 4, night: Shrouded by heavy storm, Federal gunboat Carondelet passes the island and reaches New Madrid. Joined by gunboat Pittsburg on night of April 7
4 April 6: Pope's troops prepare to cross river and assault Confederates. Carondelet duels with Confederate shore batteries
5 April 7: While gunboats pound Tennessee shore and silence batteries, Pope's forces block only escape route. Island No. 10 seized by Foote's flotilla; Confederates surrender on April 8
0 2 4 KM
0 2 4 MILES
POPE
TELEGRAPH RD
FOOTE
NEW MADRID
McCOWN
Wilson's Bayou River
CENTURY OR PLANK ROAD
To Weaversville
MISSOURI
MISSISSIPPI RIVER
Island No 9
Dan Watson's
Dr Martin's
Island No 11
Darnell's
MACKALL
Island No 10
PT. PLEASANT
OBIONVILLE
RIDDLES PT.
TIPTONVILLE
Reel Foot River
TENNESSEE

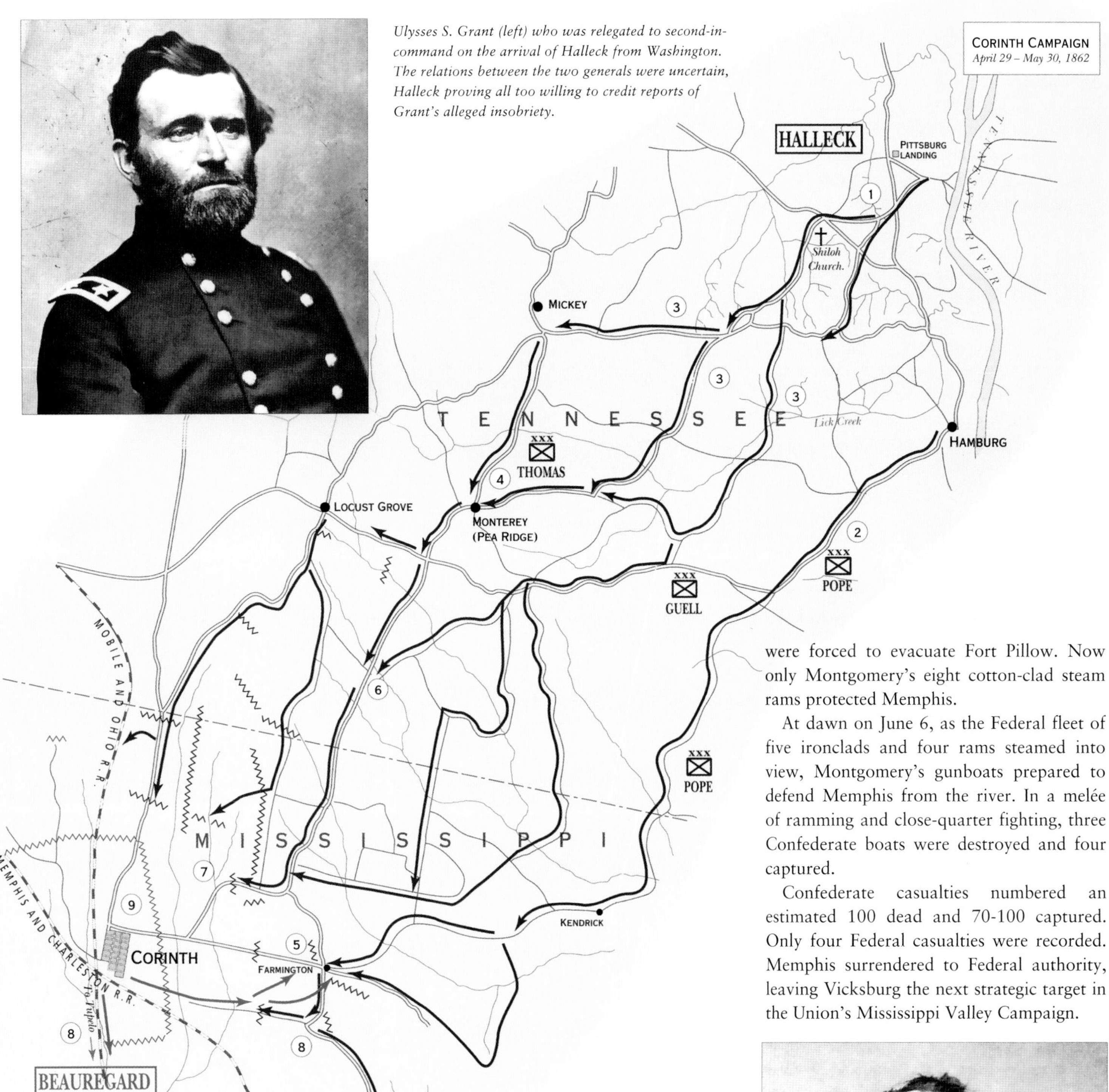
CORINTH CAMPAIGN
April 29 – May 30, 1862
HALLECK
PITTSBURG LANDING
TENNESSEE RIVER
Shiloh Church.
MICKEY
TENNESSEE
Lick Creek
HAMBURG
THOMAS
LOCUST GROVE
MONTEREY (PEA RIDGE)
POPE
GUELL
MOBILE AND OHIO R.R.
MISSISSIPPI
MEMPHIS AND CHARLESTON R.R.
To Tupelo
CORINTH
FARMINGTON
KENDRICK
BEAUREGARD

*Ulysses S. Grant (left) who was relegated to second-in-command on the arrival of Halleck from Washington. The relations between the two generals were uncertain, Halleck proving all too willing to credit reports of Grant's alleged insobriety.*

were forced to evacuate Fort Pillow. Now only Montgomery's eight cotton-clad steam rams protected Memphis.

At dawn on June 6, as the Federal fleet of five ironclads and four rams steamed into view, Montgomery's gunboats prepared to defend Memphis from the river. In a melée of ramming and close-quarter fighting, three Confederate boats were destroyed and four captured.

Confederate casualties numbered an estimated 100 dead and 70-100 captured. Only four Federal casualties were recorded. Memphis surrendered to Federal authority, leaving Vicksburg the next strategic target in the Union's Mississippi Valley Campaign.

1. **April 8–29:** Halleck assembles three separate armies as one massive army group
2. **April 22:** Pope's army begins to arrive in preparation for offensive. Davis's division arrives to reinforce Halleck
3. **April 29:** Halleck's army group advances inland along a ten mile front
4. **May 4:** Sherman's division entrenches. Concerned about his exposed right flank, Halleck entrenches Thomas's army upon each stage of the advance
5. **May 9:** Beauregard attacks elements of Pope's army in Farmington, but disorganized assault enables Pope to extract his forces
6. Hampered by spring storms and sickness, Halleck's Federals dig rifle pits and redoubts, and improve roads to bring up army and its supplies
7. **May 28–29:** Federals partially invest Corinth, and shell the town
8. **May 29:** Beauregard, unable to defeat Federal army, evacuates Corinth during the night and moves to Tupelo
9. **May 30:** Halleck takes possession of Corinth

*Author of a number of standard works on tactics, Maj. Gen. Henry Wager Halleck (right) was known by the nickname "Old Brains". Many who had heard of the general's reputation experienced "a distinct feeling of disappointment" when they met him in person.*

# Lower Mississippi Valley APRIL 24 - JULY 1862

Militarily for the Union, only the avoidance of European intervention and the retention of the border states exceeded the importance of controlling the Mississippi River. After efforts to descend it failed, the Lincoln administration ordered an attack on New Orleans. While Federal forces under Captain David G. Farragut and General Butler organized at Ship Island, Confederate officials reduced the garrison at New Orleans. When Farragut's ships entered the Mississippi in early April, Confederate opposition consisted of 500 men and 80 cannon in Forts Jackson and St. Philip, a chain floated on barges to barricade the river, and an ineffective fleet.

After a six-day mortar bombardment failed to silence the forts, Farragut decided to run the gauntlet. Early on April 24, his fleet penetrated the barricade, passed the forts, and destroyed the Confederate flotilla. Continuing upstream, Farragut captured New Orleans, the Confederates having abandoned the city to avoid its destruction by bombardment.

Going on to Vicksburg, Farragut discovered that its high bluffs rendered naval bombardment ineffective. Regardless, his superiors ordered him to continue the shelling. In late June, Farragut passed the Vicksburg batteries to join a Federal flotilla upstream. On July 15, the Confederate ironclad *Arkansas* passed through both Federal fleets before anchoring at Vicksburg. That night, Farragut ran the gauntlet again in an unsuccessful attempt to destroy the *Arkansas*.

Within two weeks, increasing sickness and the falling river compelled Farragut to abandon Vicksburg. After landing General Thomas Williams and 3,200 soldiers at Baton Rouge, the fleet continued to New Orleans. Farragut's departure from Vicksburg prompted Confederate General Van Dorn to secure the Mississippi between Vicksburg and Port Hudson.

On August 5, Confederate forces attacked Baton Rouge. Fog, friendly fire, and excessive troop movements hampered their advance, but once the first Federal regiment broke and Williams was killed, a rout quickly ensued. The Confederates chased the Federals to the river, where shells from Federal gunboats halted their pursuit. Although *Arkansas*'s engine failure robbed the Confederates of a tactical victory, the threat they posed caused the Federals to evacuate Baton Rouge 16 days later, enabling Van Dorn to fortify Port Hudson.

*The Union mortar flotilla was commanded by Farragut's adoptive brother, David D. Porter. For six days forts Jackson and St. Philip were bombarded by the 20 mortar schooners, each of which mounted one heavy 13-inch mortar and two or more long 32-pounders (above).*

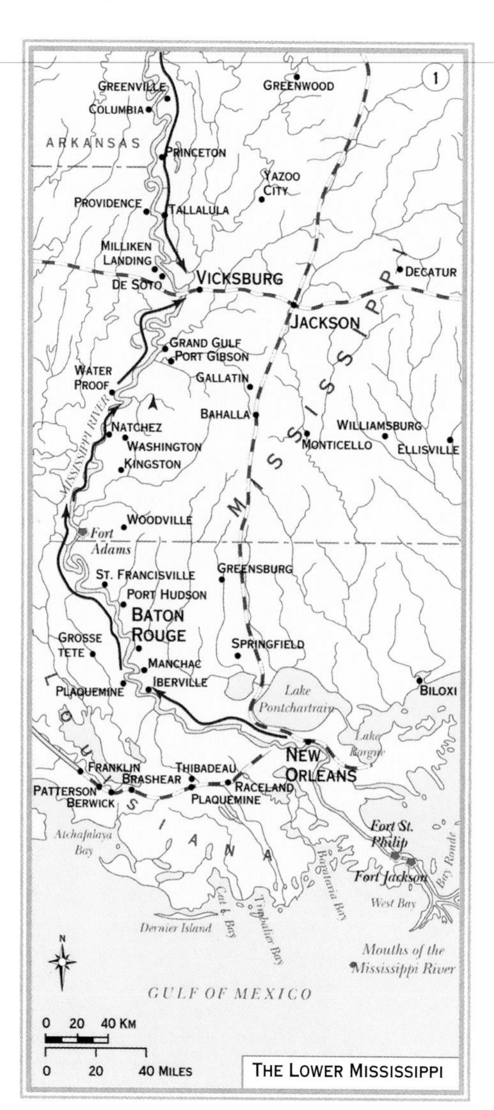

1
GREENVILLE
COLUMBIA
GREENWOOD
ARKANSAS
PRINCETON
YAZOO CITY
PROVIDENCE
TALLALULA
MILLIKEN LANDING
DE SOTO
VICKSBURG
DECATUR
JACKSON
GRAND GULF
PORT GIBSON
WATER PROOF
GALLATIN
BAHALLA
MISSISSIPPI
NATCHEZ
WASHINGTON
KINGSTON
WILLIAMSBURG
MONTICELLO
ELLISVILLE
WOODVILLE
Fort Adams
ST. FRANCISVILLE
GREENSBURG
PORT HUDSON
BATON ROUGE
GROSSE TETE
SPRINGFIELD
MANCHAC
IBERVILLE
PLAQUEMINE
Lake Pontchartrain
BILOXI
NEW ORLEANS
FRANKLIN
BRASHEAR
THIBADEAU
PATTERSON
BERWICK
RACELAND
PLAQUEMINE
LOUISIANA
Atchafalaya Bay
Fort St. Philip
Fort Jackson
West Bay
Dernier Island
Mouths of the Mississippi River
GULF OF MEXICO
N
0 20 40 KM
0 20 40 MILES
THE LOWER MISSISSIPPI


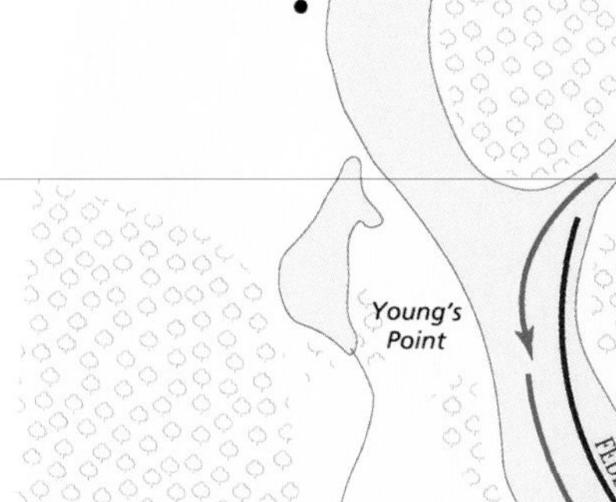

Young's Point
FEDERAL FLOTILLA
C.S.S. ARKANSAS


MISSISSIPPI RIVER

DUCKPORT

VICKSBURG, SHREVEPORT & TEXAS R. R.

1 **April 28 – May 17:** After the surrender of New Orleans, Farragut's fleet continues northward, briefly occupying Natchez (see map left)

2 **May 18:** Reaching Vicksburg, Farragut demands the city's surrender. Request refused by Confederate military governor

3 **May 19 – June 27:** Federal fleet bombards Vicksburg but fails to force surrender

4 **June 28:** Federal fleet passes the Vicksburg batteries (see map right)

5 **July 1:** Farragut's fleet meets with Western Flotilla which has sailed south from Memphis

6 **July 15:** *C.S.S. Arkansas*, commanded by Isaac Newton Brown, attacks the Union fleet crippling *U.S.S. Carondelet*

7 **July 15:** *Arkansas* suffers 60 casualties and severe damage but reaches safety under Vicksburg batteries

8 **July 24:** Combined Union fleets abandon attempts to capture Vicksburg and steam south toward New Orleans

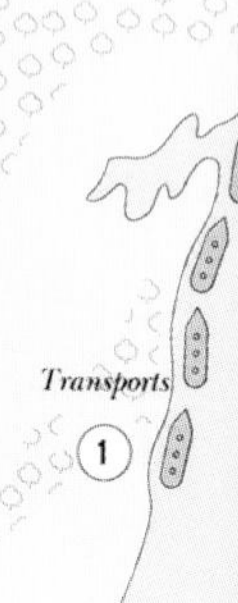

Transports
1

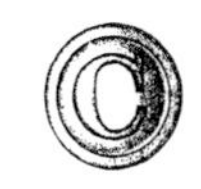

*Captain Theodorus Bailey and party rowing ashore to demand the surrender of New Orleans (above). The mayor refused to lower the Confederate flag and this action was performed by two Union officers in the face of armed and hostile crowds of the city's inhabitants.*

*The certain destruction of New Orleans in the face of overwhelming numbers and fire-power, forced the Confederate commander, Maj. Gen. Mansfield Lovell (above), to abandon the city's defenses. The mayor offered surrender on April 29.*

> "Though it was a starlight night we were not discovered until we were well under the forts; then they opened a tremendous fire on us... the air was filled with shells and explosions which almost blinded me as I stood on the forecastle trying to see my way... Although our masts and rigging got badly shot through, our hull was but little damaged."
>
> *G. Perkins, a lieutenant in Farragut's fleet.*

> "The people of New Orleans, while unable to resist your force, do not... transfer allegiance from the government of their choice... they yield the obedience which the conqueror is entitled to extort from the conquered."
>
> *The mayor of New Orleans to Farragut, April 29.*

Davis's Flotilla
De Soto Point
6
Maynadiers's Mortars
Water Battery
Sciota
4
Hartford
Wyman's Battery
5
Supply Boats
7
Arkansas
3
2
VICKSBURG
8
Sydney C. Jones
Marine Hospital Battery
HALLS FERRY ROAD
Bluff Batteries
Renshaw's "Bombers"
Smedes' Point
WARRENTON ROAD
N
0 1 2 KM
0 1 2 MILES


*At 2am Farragut's fleet steamed northward in an attempt to bypass the city's formidable batteries. By 4am the Confederates were alerted to Farragut's intentions and the batteries opened fire. By 6am the entire fleet, with the exception of three vessels, had run the gauntlet. The action was important in that it revealed how a flotilla could escape serious damage while passing land batteries, but it had also become clear that Vicksburg could not be captured by a fleet unaided by land-based troops. Federal casualties numbered 45.*

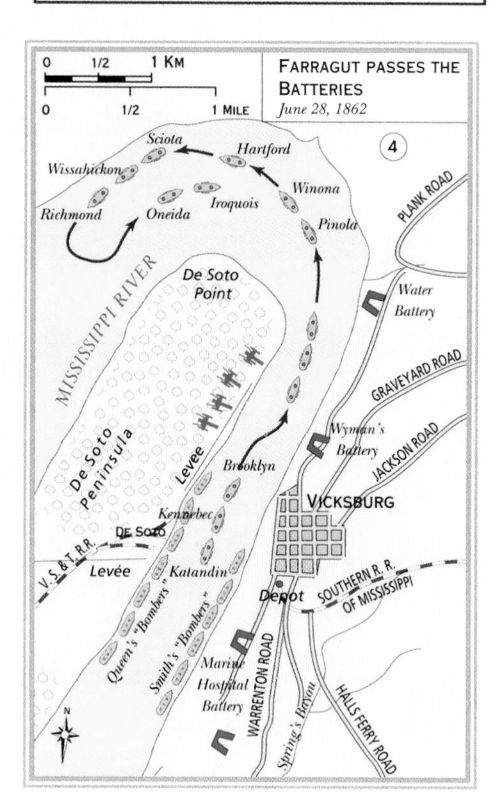

FARRAGUT PASSES THE BATTERIES
June 28, 1862
0 1/2 1 KM
0 1/2 1 MILE
4
Sciota
Hartford
Wissahickon
Winona
Iroquois
Richmond
Oneida
Pinola
PLANK ROAD
MISSISSIPPI RIVER
De Soto Point
Water Battery
GRAVEYARD ROAD
De Soto Peninsula
Levee
Wyman's Battery
JACKSON ROAD
Brooklyn
Kennebec
VICKSBURG
DE SOTO
V.S.&T.R.R.
Levee
Katandin
Depot
SOUTHERN R.R. OF MISSISSIPPI
Queen's "Bombers"
Smith's "Bombers"
Marine Hospital Battery
WARRENTON ROAD
Spring's Bayou
HALLS FERRY ROAD
N

*Having moved from his Virginia home at the state's secession, Farragut was at first suspected by the Washington hierarchy, but his victory at New Orleans proved both his loyalty and his worth and secured his promotion to the rank of rear-admiral (above).*

# New Mexico Campaign JULY 1861 – JULY 1862

Texans had long coveted New Mexico. In July 1861 a small force of Texans entered New Mexico Territory and occupied Fort Fillmore. Later, at San Augustin Spring, they captured a much larger force of U.S. regulars. A Territory of Arizona was set up, and a delegate sent to Richmond.

This was the beginning of a dream. To make it a reality, Confederate Brigadier General Henry H. Sibley raised a 3,500-strong brigade, concentrating it at Fort Thorn. On February 7, 1862, Sibley's force started north. Posted at Fort Craig was Colonel Edward R. S. Canby and 3,800 Federal soldiers.

The Confederates closed on Fort Craig on February 16, but found they were outnumbered, and the Federals prepared. Sibley bypassed Fort Craig by fording the Rio Grande south of the fort, and continuing north. On February 21, Canby rushed a third of his command to hold Valverde Ford. Seizing the initiative, the Federals crossed the Rio Grande, and compelled the Confederates to fall back to the river. The fighting escalated as both sides committed reinforcements. The Federals held their own until a Confederate charge captured a battery on Canby's left. This turned the tide and Canby's force recrossed the Rio Grande and withdrew into Fort Craig. The Confederates had opened the door to Santa Fé and Colorado.

Leaving Canby's troops to "wither on the vine," the Confederates occupied Albuquerque and Santa Fé. Their next objective was Fort Union, long the major Federal supply depot in New Mexico. But, by the time the Confederates resumed the advance, reinforcements from the Denver mining camps, led by Colonel John Slough, had reached Fort Union. On March 22, Slough led 1,342 men from the fort toward Confederate-held Santa Fé.

On hearing of the Federal advance, some 400 Confederates took to the field. On

> "Except for its geographical position, the Territory of New Mexico is not worth a quarter of the blood and treasure expended in its conquest. As a field for military operations it possesses not a single element, except in the multiplicity of its defensible positions. The indispensable element, food, cannot be relied on."
>
> *Brig. Gen. Sibley after his return from New Mexico.*

*Col. Edward Canby (above) commanded a force of approximately 1,000 territorial militia, inadequately trained and equipped. In preparation for Sibley's anticipated assault, Canby was able to raise five regiments with the assistance of prominent New Mexicans. As a result of his having won the battle of Glorieta Pass, Canby was promoted and transferred to Washington where he was named Assistant Adjutant General. Having survived the war, in which his last act was to accept the surrender of Confederate generals Edmund Kirby Smith and Richard Taylor, Canby met his death at the hands of the Madoc Indians in 1873.*

> "General Canby was an officer of great merit. He was naturally studious, and inclined to the law. There have been in the army but very few, if any, officers who took as much interest in reading and digesting every act of Congress... His knowledge gained in this way made him a most valuable staff officer."
>
> *Ulysses S. Grant, in his "Memoirs."*

1. **July 3-4 1861:** Small force of Confederates, led by Lt. Col. John Baylor, occupies Fort Bliss
2. **July 27:** Baylor compels Major Isaac Lynde to surrender his force at San Augustin Spring
3. **Feb 7–20:** Sibley marches north up Rio Grande
4. **Feb 21:** Federals defeated at Valverde
5. **Feb 23:** Sibley bypasses Canby at Fort Craig and marches north
6. **March 1:** Federals evacuate Albuquerque
7. **March 11:** Slough reaches Fort Union
8. **March 22:** Slough leaves Fort Union to recapture Santa Fé from Confederates
9. **March 25-26:** Confederates march east from Santa Fé, but are defeated by Chivington's Federals in Apache Canyon
10. **March 28:** Scurry defeats Slough's Federals at Glorieta Pass, but after raid on their supplies at Johnson's Ranch, Confederates withdraw to Santa Fé
11. **April 11–12:** Sibley evacuates Sante Fé, then Albuquerque
12. **April 17–25:** Sibley crosses Rio Grande, makes 100-mile detour to bypass Fort Craig, returning to the river 40 miles south of the fort
13. **July 23:** Confederates withdraw from Mesilla and retreat toTexas

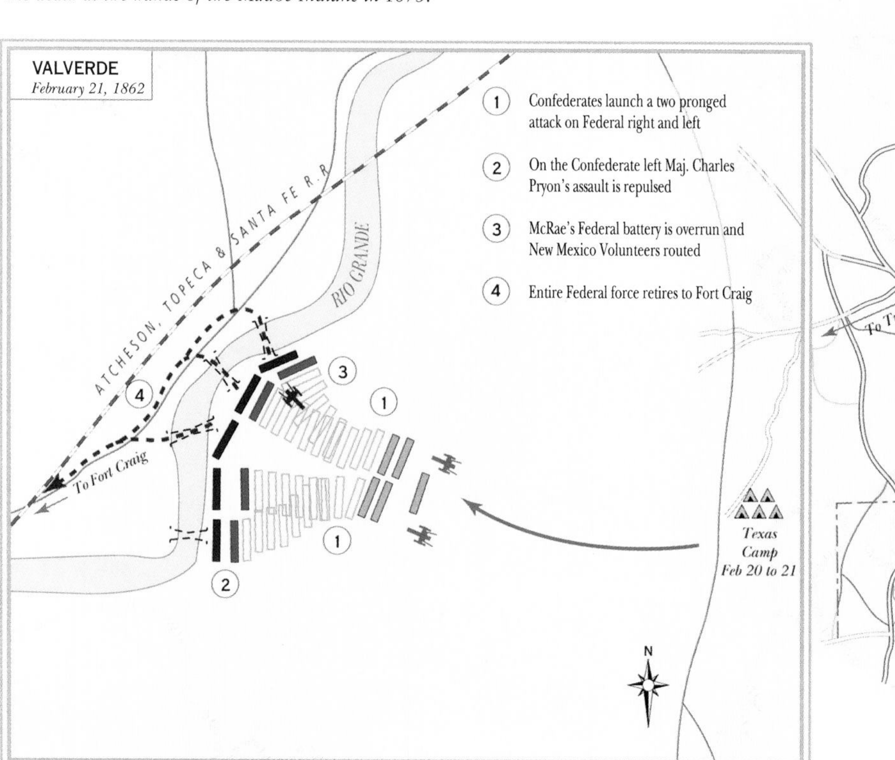
VALVERDE
February 21, 1862
ATCHESON, TOPECA & SANTA FE R.R
RIO GRANDE
To Fort Craig
Texas
Camp
Feb 20 to 21
N
1 Confederates launch a two pronged attack on Federal right and left
2 On the Confederate left Maj. Charles Pryon's assault is repulsed
3 McRae's Federal battery is overrun and New Mexico Volunteers routed
4 Entire Federal force retires to Fort Craig

A graduate of West Point, Brig. Gen. Henry Hopkins Sibley (right) had gained considerable experience of the Southwest when a major in the U.S. army before the war. Though in favor of the expedition, Jefferson Davis was unable to offer any material support and Sibley was responsible for raising and equipping his force.

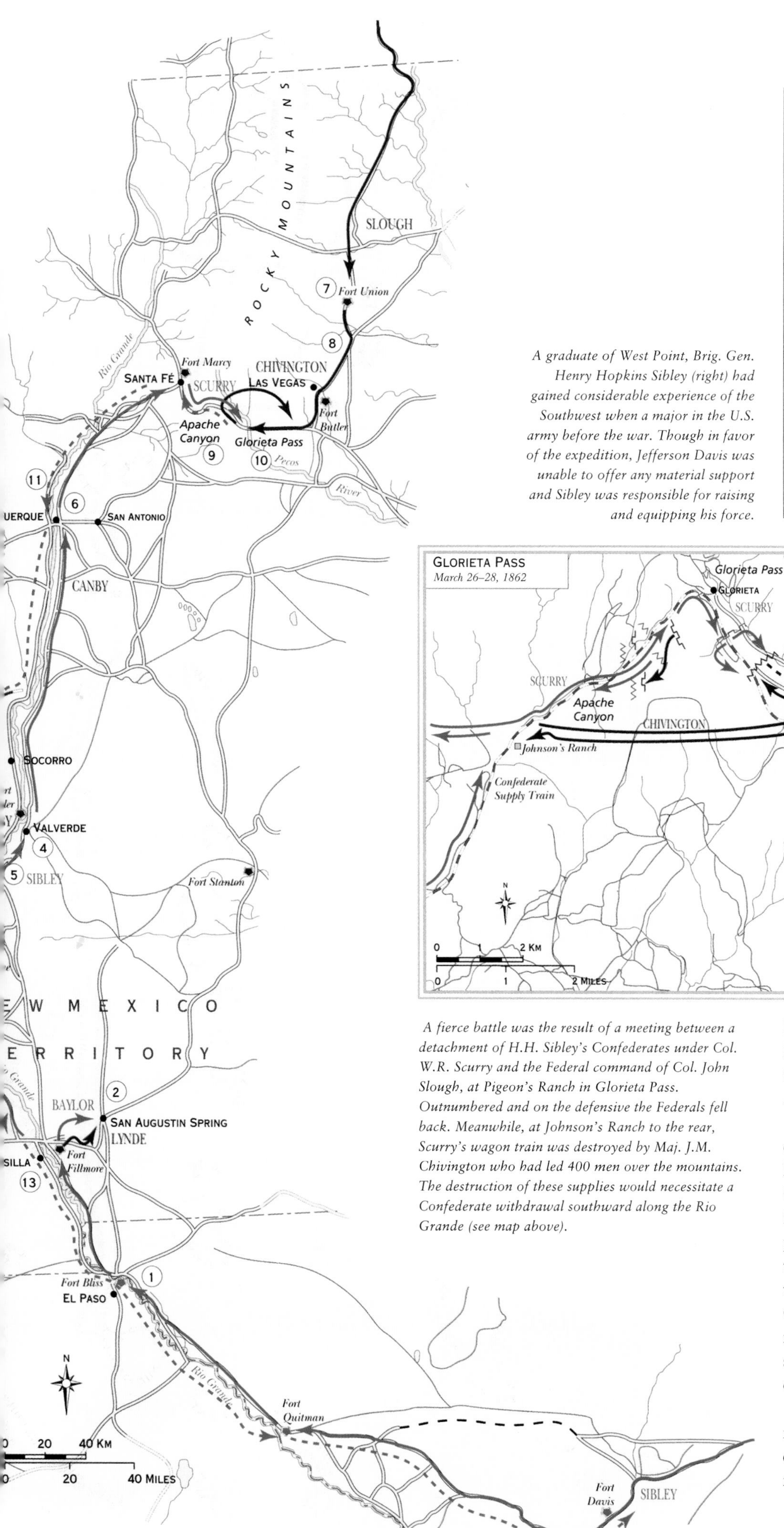
ROCKY MOUNTAINS
SLOUGH
Fort Union
Fort Marcy
SANTA FÉ
SCURRY
CHIVINGTON
LAS VEGAS
Fort Butler
Apache Canyon
Glorieta Pass
Pecos
River
Rio Grande
SAN ANTONIO
CANBY
SOCORRO
VALVERDE
SIBLEY
Fort Stanton
NEW MEXICO
TERRITORY
BAYLOR
SAN AUGUSTIN SPRING
LYNDE
Fort Fillmore
Fort Bliss
EL PASO
Rio Grande
Fort Quitman
Fort Davis
SIBLEY
N
0 20 40 KM
0 20 40 MILES
GLORIETA PASS
March 26–28, 1862
Glorieta Pass
GLORIETA
SCURRY
PECOS
SLOUGH
Apache Canyon
CHIVINGTON
Johnson's Ranch
Confederate Supply Train
Pecos River
Kozlowski's Ranch
Union H.Q.
0 1 2 KM
0 1 2 MILES

A fierce battle was the result of a meeting between a detachment of H.H. Sibley's Confederates under Col. W.R. Scurry and the Federal command of Col. John Slough, at Pigeon's Ranch in Glorieta Pass. Outnumbered and on the defensive the Federals fell back. Meanwhile, at Johnson's Ranch to the rear, Scurry's wagon train was destroyed by Maj. J.M. Chivington who had led 400 men over the mountains. The destruction of these supplies would necessitate a Confederate withdrawal southward along the Rio Grande (see map above).

March 26, the Federal vanguard, under Major John Chivington, met and defeated the Confederates in a running fight in Apache Canyon.

March 28 was a decisive day for the campaign: the reinforced Confederates headed east, but as they debouched from Glorieta Pass, they encountered Slough's force. A desperate battle ensued. About 5 pm, the Federals disengaged and returned to their camp at Kozlowski's Ranch. But that was not the end. Earlier in the day Chivington with 490 men had crossed the mountain and destroyed 80 Confederate supply waggons at Johnson's Ranch. This daring Federal raid destroyed a dream.

As a result of the raid, the Confederates retreated to Santa Fe. On July 23 they evacuated Mesilla and began the 700-mile retreat to San Antonio, Texas.

# North Carolina Campaign FEBRUARY 8 – APRIL 25 1862

In late August, 1861, the Union captured Forts Clark and Hatteras, which commanded the principal inlet to Pamlico Sound, but it was four months before the Lincoln administration took advantage of this success. On January 11, 1862, an armada of 63 ships sailed from Hampton Roads for the outer banks of North Carolina. Federal Brigadier General Ambrose E. Burnside commanded the 15,000 soldiers.

On January 13, the expedition hove to off Hatteras Inlet, but it took until February 5 to maneuver the various vessels over the bar and into Pamlico Sound. On the 7th, 17 Federal gunboats engaged eight Confederate gunboats and compelled them to withdraw up Croatan Sound. Burnside's troops landed at Ashby's Harbor, drove the Confederates from their defenses, and pursued them to the north end of Roanoke Island, where they surrendered. This Federal victory could – if exploited – have opened the back door to Richmond.

On February 10, 14 Federal gunboats entered Albermarle Sound, ascended the Pasquotank River and overwhelmed the seven vessels of the Confederate "Mosquito" fleet that had escaped the Roanoke Island debacle. The Union now controlled Albermarle Sound.

On March 11 Burnside, with 11,000 soldiers, sailed from Roanoke Island and was joined by 13 Federal gunboats at Hatteras Inlet, crossed Pamlico Sound and ascended the Neuse River. As the gunboats shelled the woods, Burnside landed his troops at Slocum Creek. The Federals pushed ahead, and on March 14, engaged and defeated 4,000 Confederates, who retreated to Kinston. The Federals occupied New Bern.

Federal Brigadier General John G. Parke moved his command from New Bern to Carolina City, intent on forcing the surrender of Fort Macon which was denying access to Burnside's army of the Morehead City wharves and railroad. When, on March 23, the fort commander refused to surrender, Parke began siege operations. These were followed on April 25 by a bombardment, and on April 26, Fort Macon surrendered.

A week earlier, Burnside's Federal army had suffered its first defeat. On April 18, two Federal brigades landed at Elizabeth City on the Pasquotank River. Their aim was to destroy locks on the Dismal Swamp Canal thus preventing two Confederate

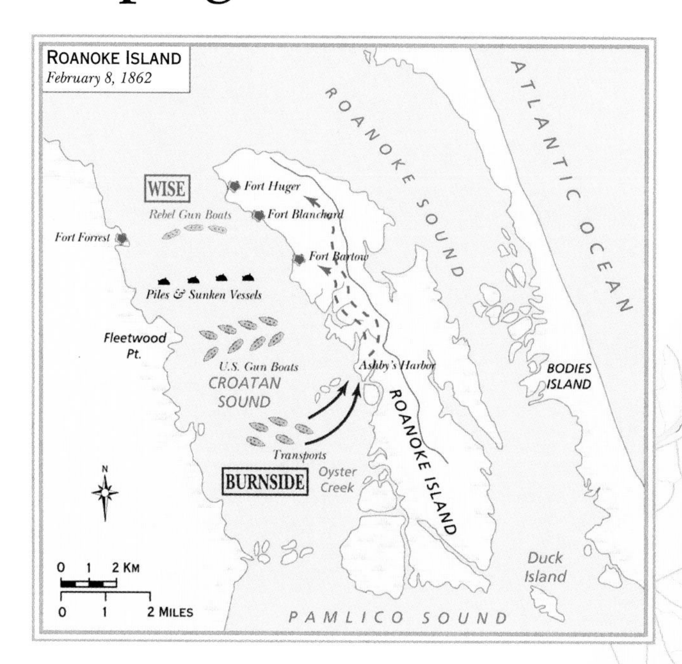
ROANOKE ISLAND
February 8, 1862
ATLANTIC OCEAN
ROANOKE SOUND
WISE
Fort Huger
Rebel Gun Boats
Fort Blanchard
Fort Forrest
Fort Bartow
Piles & Sunken Vessels
Fleetwood Pt.
U.S. Gun Boats
CROATAN SOUND
Ashby's Harbor
ROANOKE ISLAND
BODIES ISLAND
Transports
BURNSIDE
Oyster Creek
N
0 1 2 KM
0 1 2 MILES
Duck Island
PAMLICO SOUND

"I need not say how unfavorable an influence these defeats, following in such quick succession, have produced in public sentiment. If not soon counterbalanced by some decisive success of our arms, we may... bid adieu to all hopes of reasonable recognition."

*John Slidell, Confederate Envoy to Paris, reporting on French public opinion after the defeats in North Carolina.*

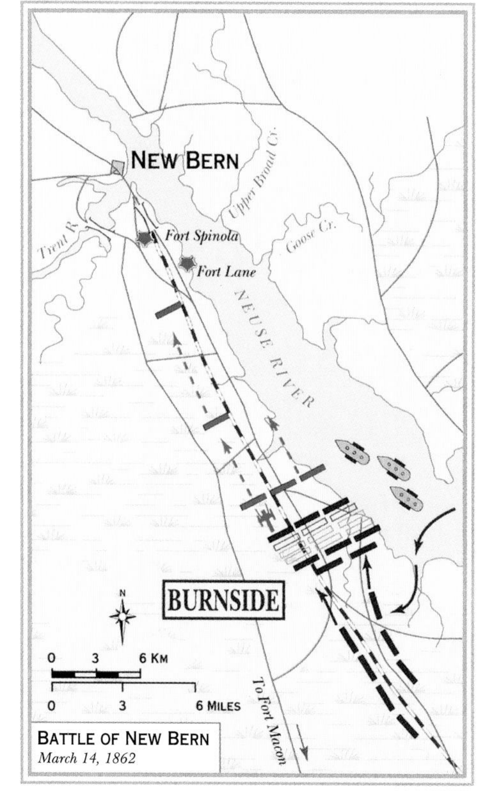
NEW BERN
Upper Broad Cr.
Trent R.
Fort Spinola
Goose Cr.
Fort Lane
NEUSE RIVER
BURNSIDE
N
0 3 6 KM
0 3 6 MILES
To Fort Macon
BATTLE OF NEW BERN
March 14, 1862

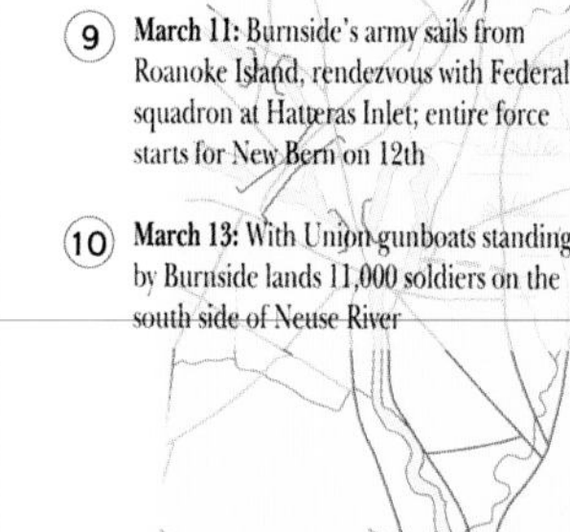
1 Jan 11: Fleet of over 60 vessels with some 15,000 troops sails from Hampton Roads to the North Carolina Sounds
2 Jan 13: Fleet arrives in Hatteras Inlet
3 Jan 22: Brig. Gen. Henry A. Wise named to command Confederate forces defending Roanoke Island
4 Jan 24-Feb 5: Union fleet crosses Hatteras Inlet Bar and enters Pamlico Sound
5 Feb 7: Burnside's troops land on Roanoke Island
6 Feb 8: Burnside overwhelms Confederates, and the island and its 2,000 defenders surrender (see map left)
7 Feb 10: Union squadron destroys rest of Confederate flotilla; Burnside ocupies Elizabeth City
8 Feb 12-20: Union amphibious forces occupy Edenton and Winston and reconnoiter Curratuck Sound
9 March 11: Burnside's army sails from Roanoke Island, rendezvous with Federal squadron at Hatteras Inlet; entire force starts for New Bern on 12th
10 March 13: With Union gunboats standing by Burnside lands 11,000 soldiers on the south side of Neuse River
11 March 14: Burnside advances on New Bern, defeats 4,000 Confederates and captures town (see map bottom left)
12 March 20-21: Union forces sail from New Bern and briefly occupy Washington
13 March 19-22: Federal force moved by boat and rail from New Bern to Carolina City
14 March 23: Federals demand surrender of Confederate fort guarding entrance to Beaufort Inlet, demand refused
15 March 29-April 26: Union troops land on Bogue Banks. Siege operations begin. Federals open fire on Fort Macon. Four gunboats join in bombardment of Confederates (see map bottom right)
WINS
WINDSOR
ROANOKE
WILLIAMSTON
TAR RIVER
Grindle Cr.
MAYSVILLE
GREENVILLE
PACLOTUS
WASHING
MARLBOROUGH
PLEASANT MT.
COXVILLE
BELLE VIEW X RDS.
TRINITY
BLOUNT HALL
BLOUNT'S
NEUSE RIVER
To Goldsboro
KINSTON
TRENT RIVER
POLLOCKSVILLE
ONSLOW
To Wilmington

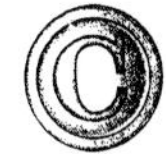

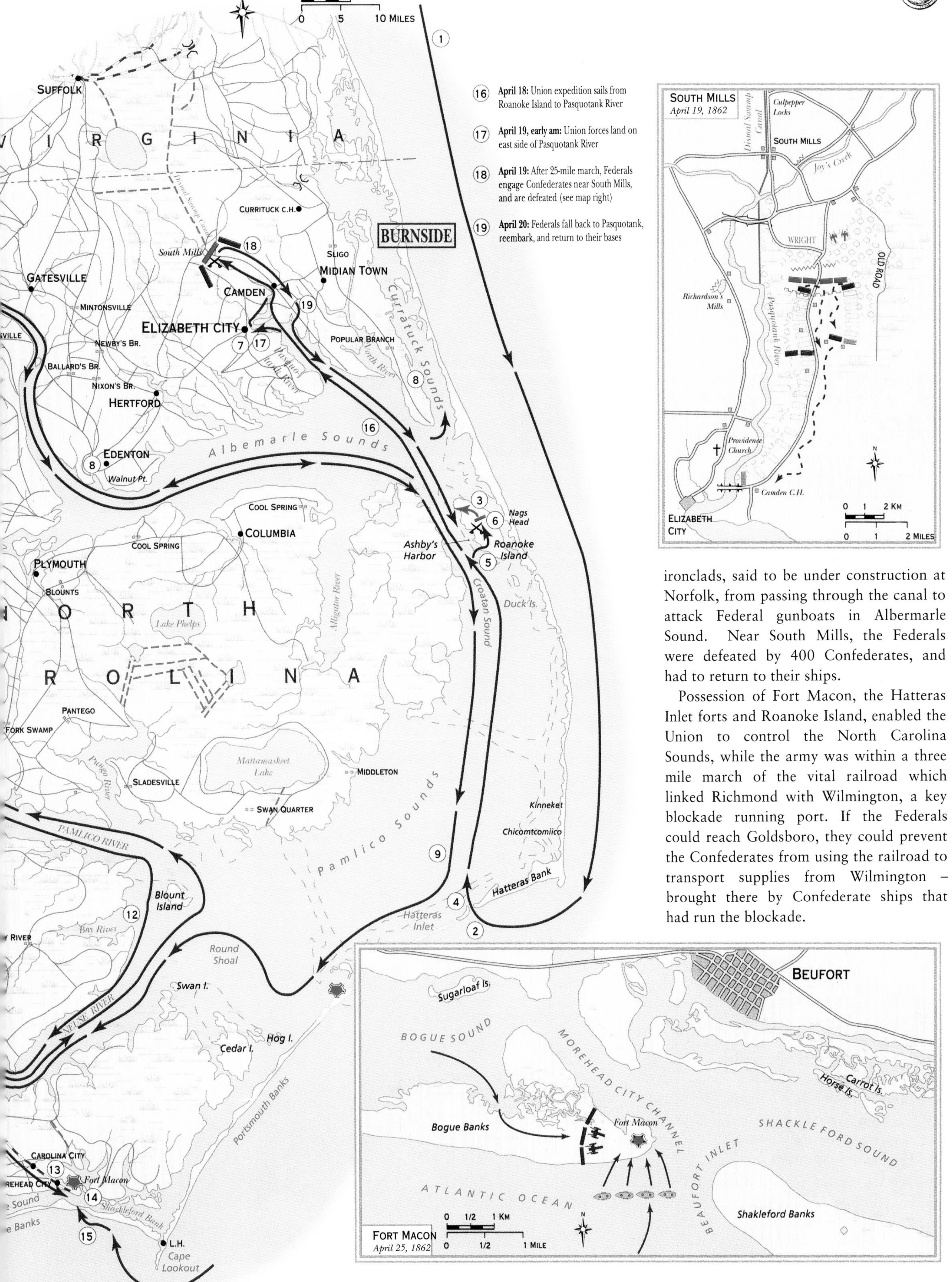
N
0 5 10 KM
0 5 10 MILES
16 April 18: Union expedition sails from Roanoke Island to Pasquotank River
17 April 19, early am: Union forces land on east side of Pasquotank River
18 April 19: After 25-mile march, Federals engage Confederates near South Mills, and are defeated (see map right)
19 April 20: Federals fall back to Pasquotank, reembark, and return to their bases
BURNSIDE
SUFFOLK
V I R G I N I A
CURRITUCK C.H.
South Mills
SLIGO
MIDIAN TOWN
GATESVILLE
CAMDEN
MINTONSVILLE
ELIZABETH CITY
NEWBY'S BR.
BALLARD'S BR.
NIXON'S BR.
HERTFORD
POPULAR BRANCH
Currituck Sounds
North River
Pasquotank River
Albemarle Sounds
EDENTON
Walnut Pt.
COOL SPRING
COLUMBIA
COOL SPRING
PLYMOUTH
BLOUNTS
Nags Head
Ashby's Harbor
Roanoke Island
Croatan Sound
Duck Is.
Alligator River
Lake Phelps
N O R T H
C A R O L I N A
PANTEGO
FORK SWAMP
Pungo River
Mattamuskeet Lake
MIDDLETON
SLADESVILLE
SWAN QUARTER
Kinneket
Chicomtcomiico
PAMLICO RIVER
Pamlico Sounds
Hatteras Bank
Blount Island
Hatteras Inlet
Bay River
RIVER
Round Shoal
Swan I.
NEUSE RIVER
Hog I.
Cedar I.
Portsmouth Banks
CAROLINA CITY
REHEAD CITY
Fort Macon
Sound
Banks
Shackleford Bank
L.H.
Cape Lookout
SOUTH MILLS
April 19, 1862
Dismal Swamp Canal
Culpepper Locks
SOUTH MILLS
Joy's Creek
WRIGHT
OLD ROAD
Richardson's Mills
Pasquotank River
Providence Church
Camden C.H.
ELIZABETH CITY
0 1 2 KM
0 1 2 MILES
BEUFORT
Sugarloaf Is.
BOGUE SOUND
MOREHEAD CITY CHANNEL
Carrot Is.
Horse Is.
Bogue Banks
Fort Macon
SHACKLE FORD SOUND
BEAUFORT INLET
ATLANTIC OCEAN
Shakleford Banks
0 1/2 1 KM
0 1/2 1 MILE
FORT MACON
April 25, 1862

ironclads, said to be under construction at Norfolk, from passing through the canal to attack Federal gunboats in Albermarle Sound. Near South Mills, the Federals were defeated by 400 Confederates, and had to return to their ships.

Possession of Fort Macon, the Hatteras Inlet forts and Roanoke Island, enabled the Union to control the North Carolina Sounds, while the army was within a three mile march of the vital railroad which linked Richmond with Wilmington, a key blockade running port. If the Federals could reach Goldsboro, they could prevent the Confederates from using the railroad to transport supplies from Wilmington – brought there by Confederate ships that had run the blockade.

# Showdown at Hampton Roads MARCH 8 – 9 1862

Before abandoning the Norfolk Navy Yard in April, 1861, the Federals scuttled the vessels they left behind. However, the Confederates managed to raise the steam frigate *Merrimac*, which they converted to an ironclad, renaming her *Virginia*.

On March 8, 1862, *Virginia*, under the command of Captain Franklin Buchanan, left Norfolk. Her rapid conversion had left no time for a trial cruise, and she proved capable of barely six knots. Adding to her difficulties was her deep draft, which gave her a limited capacity to maneuver in the shoal water of Hampton Roads.

Among the Federal vessels in the Roads were two sailing frigates, *Cumberland* and *Congress*, both anchored near the mouth of the James River. *Virginia* steamed directly for *Cumberland*, opening fire at a distance of less than a mile. *Cumberland* returned fire, joined by Federal shore batteries, but they failed to stop *Virginia's* advance. *Virginia* fired a broadside at *Congress*, then rammed *Cumberland* beneath the waterline. Drawing off, she fired another broadside at *Congress*, then rammed *Cumberland* again. The Federal frigate sank almost immediately.

Still under fire from *Virginia*, *Congress* tried to get underway, but ran aground. *Congress* struck her flag, but in the confusion, Federal shore batteries continued firing. *Virginia* withdrew, leaving *Congress* a burning wreck. She blew up around midnight.

The following day *Virginia* left Norfolk

*Franklin Buchanan, commander of the* C.S.S. Virginia, *photographed in his U.S. Navy uniform before the war (above). Believing that his native state would secede, Buchanan resigned his commission in 1861; when Maryland did not join the Confederacy, the Secretary of the Navy refused to allow him to withdraw his resignation and Buchanan instead joined the Confederate navy. Wounded by a Union sharpshooter at Hampton Roads, he was not present at the duel between the* Virginia *and the* Monitor.

"I laid the *Monitor* close alongside the *Merrimac*, and gave her a shot. She returned our compliment by a shell weighing one hundred and fifty pounds, fired when we were close together, which struck the turret so squarely that it received the whole force... it did not start a rivet-head or a nut!"

Capt. Warden when showing Lincoln around the Monitor.

1. *USS Merrimac* is converted to ironclad *CSS Virginia* at Norfolk navy yard
2. **March 8, c. 1pm:** *Virginia* enters Hampton Roads and steams toward *Cumberland* and *Congress*
3. *Virginia* attacks and rams *Cumberland* twice. *Cumberland* sinks
4. *Virginia* engages *Congress* and sets her on fire. At 6.06pm *Virginia* returns to Norfolk
5. **4.10pm:** *Monitor* arrives from New York and takes up position near *Minnesota*
6. **Midnight:** *Congress* explodes
7. **March 9, am:** *Virginia* appears off Craney Island steaming toward *Minnesota*
8. *Monitor* gets upsteam and places herself between *Virginia* and *Minnesota*. *Monitor* opens fire against *Virginia*
9. In trying to maneuver *Minnesota* runs aground
10. Action between ironclads continues until 12.15pm. *Virginia* withdraws

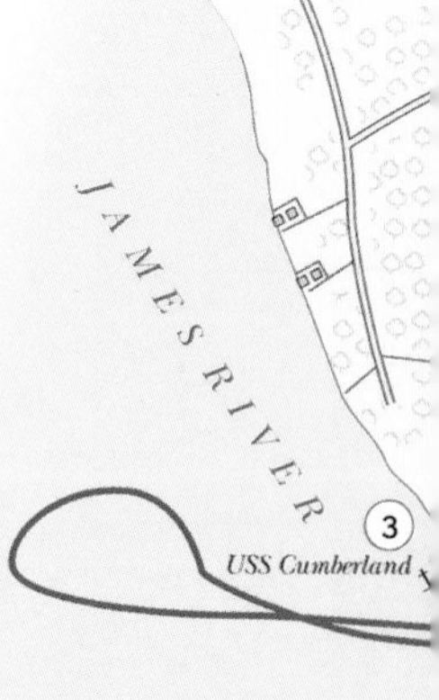
JAMES RIVER
3
USS Cumberland

N

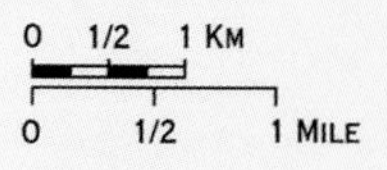
0 1/2 1 KM
0 1/2 1 MILE

*The crew of the* Monitor *relax on the ship's deck (below). Though enjoying the safety afforded them by armor-plating, in action the men fought in hot, crowded, and airless conditions.*

*Captain of the* Monitor, *John Warden, in the uniform of a rear admiral (below). When a rifle-pointed shell exploded in the ship's conning-tower, Warden was temporarily blinded. Having recovered from his injuries, Warden was later able to accompany Lincoln on a tour of inspection.*

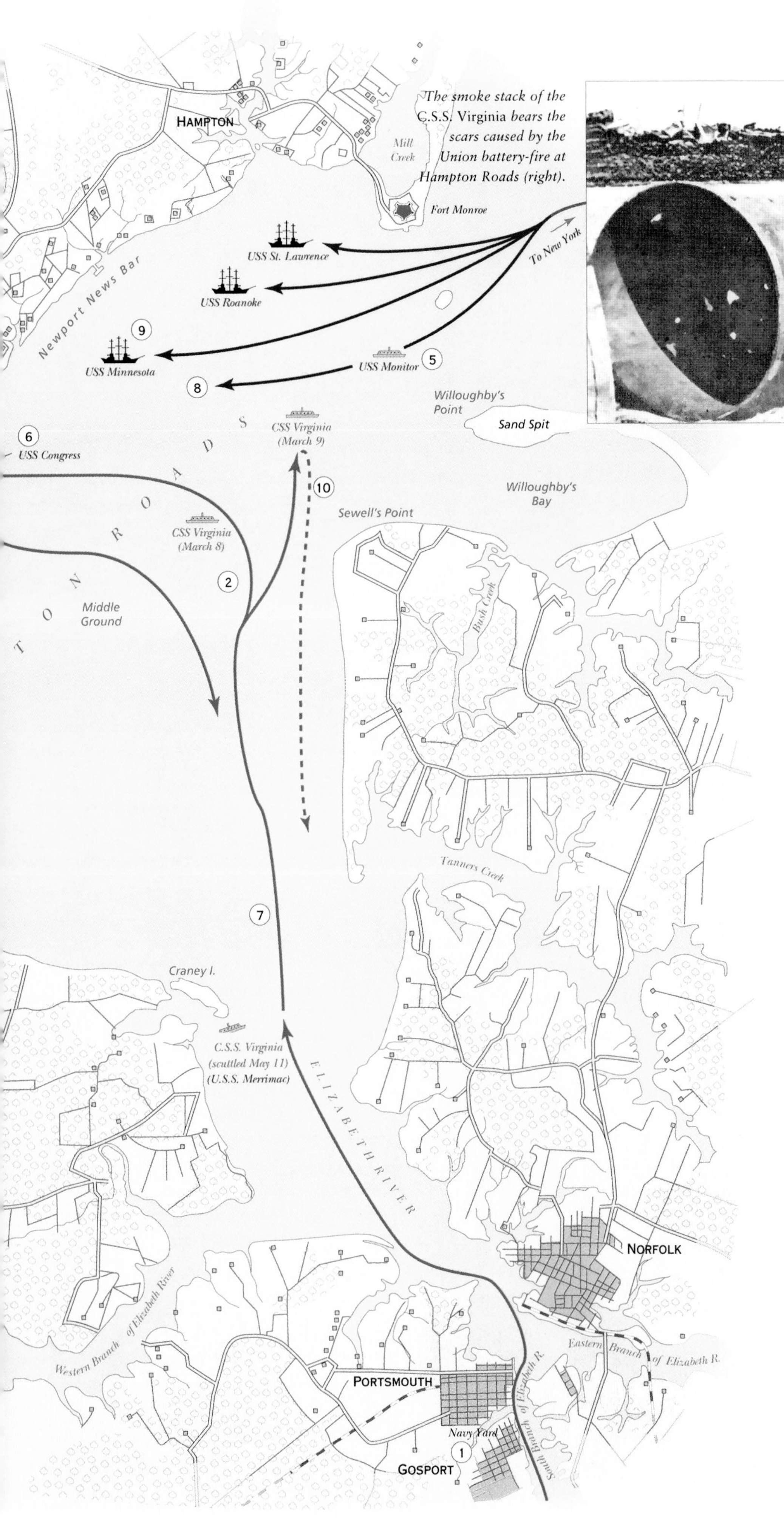
HAMPTON
Mill Creek
Fort Monroe
To New York
USS St. Lawrence
USS Roanoke
Newport News Bar
9
USS Minnesota
USS Monitor
5
8
Willoughby's Point
Sand Spit
CSS Virginia (March 9)
6
USS Congress
HAMPTON ROADS
10
Willoughby's Bay
Sewell's Point
CSS Virginia (March 8)
2
Middle Ground
Bush Creek
Tanners Creek
7
Craney I.
C.S.S. Virginia (scuttled May 11) (U.S.S. Merrimac)
ELIZABETH RIVER
NORFOLK
Western Branch of Elizabeth River
Eastern Branch of Elizabeth R.
PORTSMOUTH
Navy Yard
1
GOSPORT
South Branch of Elizabeth R.

*The smoke stack of the C.S.S.* Virginia *bears the scars caused by the Union battery-fire at Hampton Roads (right).*

again, and made for the frigate *Minnesota*, which was lying close to the Federal ironclad *Monitor*. *Minnesota* had gone aground in Hampton's shallow waters; prevented by her deep draft from approaching, *Virginia* opened fire on her from a distance. *Monitor* steamed ahead to place herself between *Minnesota* and *Virginia*. *Virginia's* captain hoped to range close enough to deliver a full broadside against the approaching *Monitor*, but about half a mile from the frigate, *Virginia* struck bottom, and it took 15 minutes to free her. Finally, the ironclad was close enough to deliver a broadside. *Virginia's* shot ricocheted off *Monitor*, with little effect, while the Federal shot did no more than rattle a few of *Virginia's* plates.

For four hours the two ironclads slugged it out. Never more than 200 yards apart, often with barely 30 yards between them, neither vessel was able to gain the advantage.

When *Monitor's* captain was hit by a flying fragment and taken below, *Monitor* briefly withdrew. In that time, *Virginia* came about and steamed toward Norfolk. The battle ended with both sides simultaneously withdrawing, and both claiming victory. In reality, history's first battle between ironclads was a draw.

"Our vessel, *The Virginia*, has invented a new way of destroying the blockade. Instead of raising it she sinks it or I believe she is good at both, for the one she burned was raised to a pretty considerable height when the magazine exploded."

*A Confederate soldier reporting on the destruction of the* Cumberland *and* Congress.

# McClellan's Peninsula Campaign APRIL 5 – JUNE 16 1862

General McClellan believed the shortest and safest route to Richmond was up the Peninsula flanked by the York and James rivers. Advancing on April 4, McClellan halted abruptly before Major General John B. Magruder's fortifications around Yorktown. Believing Magruder's scant 10,000 men actually outnumbered his 105,000, McClellan elected to besiege Yorktown. This gave General Johnston's Confederate army a chance to join Magruder.

Fearing that McClellan could flank the Yorktown defenses, Johnston abandoned the line on May 4. On the following day, McClellan clashed with Johnston's rearguard at Williamsburg, but could not force the Confederate army to engage. McClellan tried to flank the retreating Confederates by disembarking Franklin's division at Eltham's Landing on the York River. On May 7, Hood's Confederate brigade attacked Franklin, halting him in his tracks.

With Johnston's retreat, the Confederates abandoned Norfolk and scuttled their ironclad, *Virginia* (originally the Federal steam frigate *Merrimac*). On May 15, Federal ships sailed up the James River for Richmond, but were checked by the strong fortifications and obstructions at Drewry's Bluff. Johnston established a defensive line near the capital. Unable to flank Johnston, and lacking the confidence to attack him, McClellan prepared for another siege.

The threat of Federal reinforcements from Fredericksburg spurred Johnston's 65,000 Confederates into action. The sudden reversal of the Federal march from Fredericksburg – and a powerful squall – focused Johnston's attack on McClellan's weak left. Striking south of the Chickahominy River on May 31, Johnston's offensive crumbled at Seven Pines (called Fair Oaks by the Union). The Confederates lacked direction or coordination, and when Federal reinforcements poured across the raging Chickahominy, the Confederates retreated on June 1. The Federals suffered 5,000 casualties; the Confederates 6,134. Among the wounded Confederates was General Johnston. President Jefferson Davis replaced the fallen officer with General Robert E. Lee.

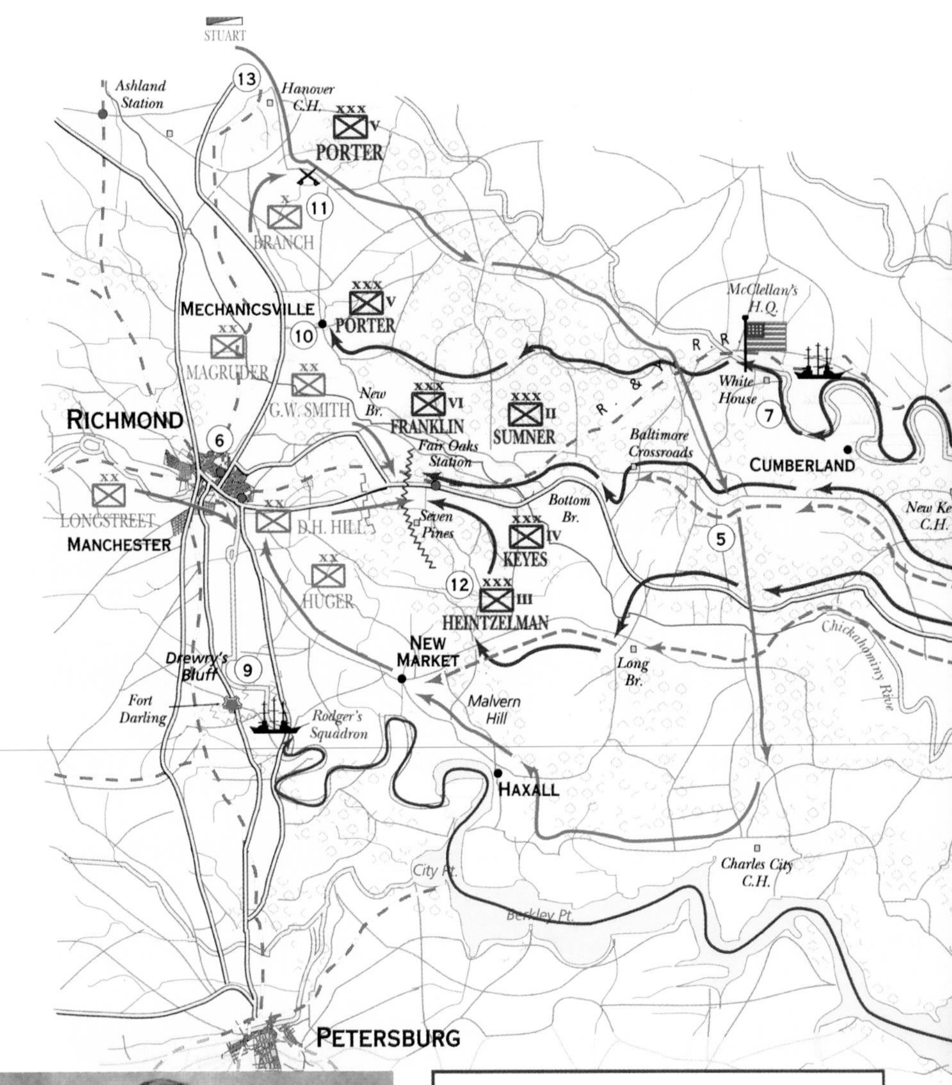
STUART
Ashland Station
Hanover C.H.
PORTER
BRANCH
MECHANICSVILLE
PORTER
McClellan's H.Q.
MAGRUDER
G.W. SMITH
New Br.
FRANKLIN
SUMNER
White House
RICHMOND
Fair Oaks Station
Baltimore Crossroads
CUMBERLAND
LONGSTREET
D.H. HILL
Seven Pines
Bottom Br.
KEYES
MANCHESTER
HUGER
HEINTZELMAN
NEW MARKET
Long Br.
Chickahominy River
Drewry's Bluff
Fort Darling
Rodger's Squadron
Malvern Hill
HAXALL
City Pt.
Charles City C.H.
Berkley Pt.
PETERSBURG

*Though supremely confident when given command of the Union forces, Gen. McClellan (right) proved to be over-cautious. When commenting on his performance Secretary Stanton stated that "If he had a million men, he would swear the enemy had two millions, and then he would sit down in the mud and yell for three."*

"Tomorrow night I hope to have twelve new guns and five to ten heavy mortars in battery... I will not open fire unless the enemy annoys us, hoping to get all the guns in battery and the trenches well advanced before meeting with serious opposition."

*Gen. McClellan before Yorktown.*

"I suppose ere this you have heard that we are having a fight at Yorktown at last... Though the enemy have shelled us for eleven days, they have injured us hardly any; I think we have lost 4 or 5 killed and very few wounded. I have had bombs and Minnie Balls to whistle all around me, but fortunately they did me no harm."

*A Georgia infantryman writing home from the besieged Yorktown.*

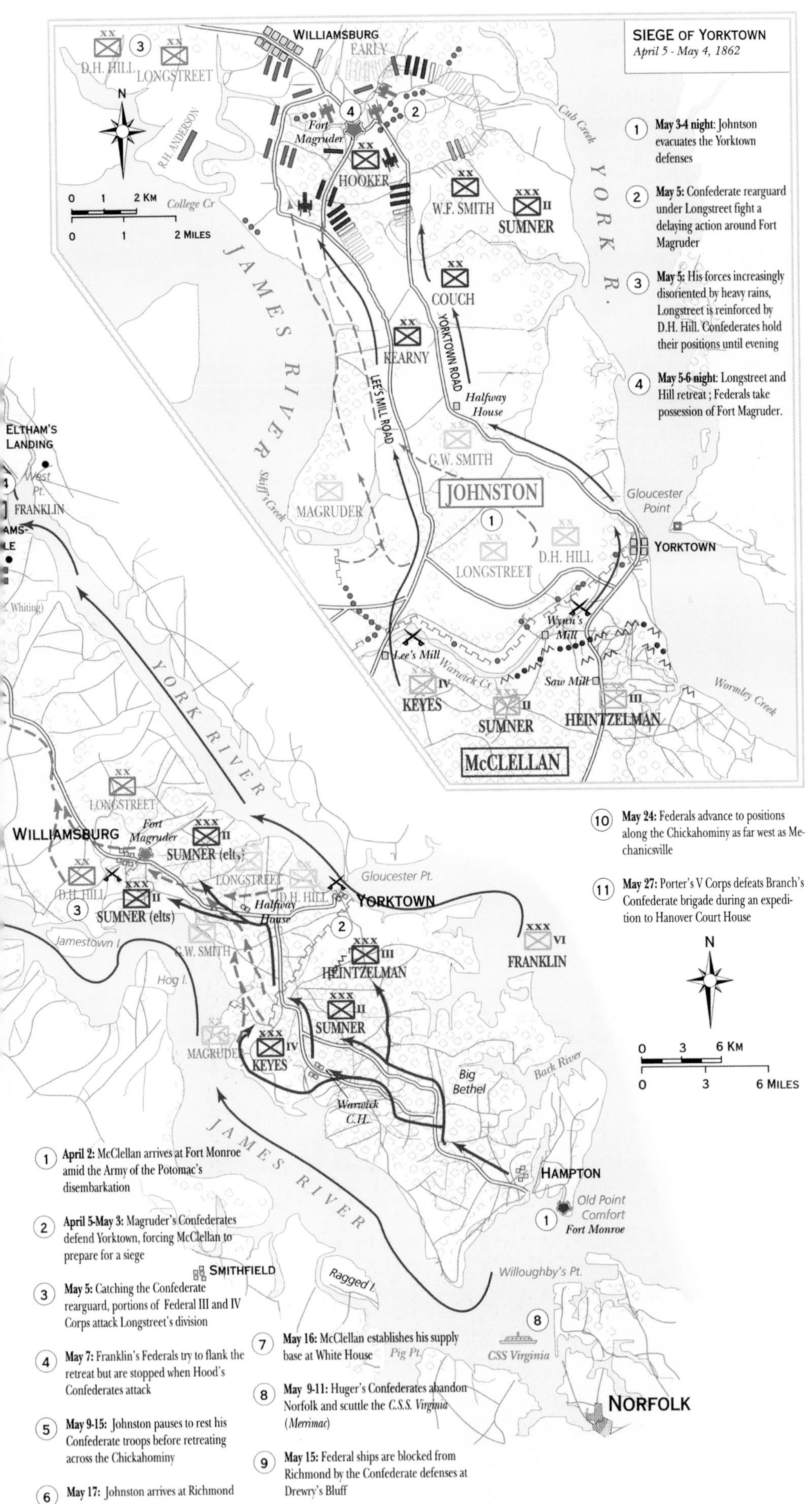
SIEGE OF YORKTOWN
April 5 - May 4, 1862
1 May 3-4 night: Johnston evacuates the Yorktown defenses
2 May 5: Confederate rearguard under Longstreet fight a delaying action around Fort Magruder
3 May 5: His forces increasingly disoriented by heavy rains, Longstreet is reinforced by D.H. Hill. Confederates hold their positions until evening
4 May 5-6 night: Longstreet and Hill retreat; Federals take possession of Fort Magruder.
WILLIAMSBURG
EARLY
D.H. HILL
LONGSTREET
R.H. ANDERSON
Fort Magruder
HOOKER
W.F. SMITH
SUMNER
COUCH
KEARNY
YORKTOWN ROAD
LEE'S MILL ROAD
Halfway House
G.W. SMITH
JOHNSTON
MAGRUDER
LONGSTREET
D.H. HILL
Gloucester Point
YORKTOWN
Lee's Mill
Warwick Cr
Wynn's Mill
Saw Mill
KEYES
SUMNER
HEINTZELMAN
McCLELLAN
Cub Creek
College Cr
Skiff's Creek
Wormley Creek
YORK R.
JAMES RIVER
0 1 2 KM
0 1 2 MILES
ELTHAM'S LANDING
West Pt.
FRANKLIN
YORK RIVER
WILLIAMSBURG
Fort Magruder
SUMNER (elts)
LONGSTREET
D.H. HILL
SUMNER (elts)
Halfway House
YORKTOWN
Gloucester Pt.
Jamestown I.
G.W. SMITH
HEINTZELMAN
FRANKLIN
Hog I.
SUMNER
MAGRUDER
KEYES
Big Bethel
Back River
Warwick C.H.
JAMES RIVER
HAMPTON
Old Point Comfort
Fort Monroe
SMITHFIELD
Ragged I.
Willoughby's Pt.
CSS Virginia
Pig Pt.
NORFOLK
0 3 6 KM
0 3 6 MILES
1 April 2: McClellan arrives at Fort Monroe amid the Army of the Potomac's disembarkation
2 April 5-May 3: Magruder's Confederates defend Yorktown, forcing McClellan to prepare for a siege
3 May 5: Catching the Confederate rearguard, portions of Federal III and IV Corps attack Longstreet's division
4 May 7: Franklin's Federals try to flank the retreat but are stopped when Hood's Confederates attack
5 May 9-15: Johnston pauses to rest his Confederate troops before retreating across the Chickahominy
6 May 17: Johnston arrives at Richmond
7 May 16: McClellan establishes his supply base at White House
8 May 9-11: Huger's Confederates abandon Norfolk and scuttle the C.S.S. Virginia (Merrimac)
9 May 15: Federal ships are blocked from Richmond by the Confederate defenses at Drewry's Bluff
10 May 24: Federals advance to positions along the Chickahominy as far west as Mechanicsville
11 May 27: Porter's V Corps defeats Branch's Confederate brigade during an expedition to Hanover Court House

*On a rainy May 31, 6,000 Confederate soldiers fell during the battle of Seven Pines; one of them was Johnston (above) who was struck by two musket balls and thrown from his horse, breaking a number of ribs. Although he recovered from his wounds, Johnston was temporarily incapacitated and was replaced by the untried Robert E. Lee.*

(12) **May 31-June 1:** Johnston attacks the Federals at Seven Pines but is repulsed and Johnston wounded; Robert E. Lee assumes command after the battle

(13) **June 12-15:** Stuart's Confederate cavalry rides around McClellan's army, gathering intelligence for Lee's proposed counter-offensive

*The Confederates retreating from Yorktown offered fierce resistance to the advancing Federals – necessitating reinforcement by a division commanded by Philip Kearny (below), a gallant and courageous officer who was subsequently killed at Second Manassas.*

# Jackson's Shenandoah Valley Campaign: Phase 1 MARCH 23 - MAY 25 1862

The dawn of 1862 was nearly the sunset of the Confederacy. As spring brought new life to the landscape, the obscure and eccentric Stonewall Jackson brought new hope to a besieged South. Jackson, the Confederate commander in the Shenandoah Valley, had a meager 6,000 troops with which to contest the 38,000 Federals under General Nathaniel P. Banks.

After forcing Jackson to retreat to Mount Jackson, Banks confidently began sending his troops east to reinforce McClellan's army. Learning this, Jackson boldly struck the Federals at Kernstown on March 23. The Federals defeated Jackson – the only defeat of his career – and inflicted 718 casualties, with the loss of 590. However, Jackson gained a strategic victory when Lincoln, believing that Jackson would not have attacked Banks unless he had a sizable force, cancelled the transfer of Banks's division to McClellan.

Cooperating with Major General Richard S. Ewell's 8,500-man division, Jackson spirited his troops to the mountain hamlet of McDowell, where he defeated a Federal force on May 8. Forcing the Federals to retreat, Jackson's Valley Army drove a wedge between the converging Federal forces of Banks and Major General John C. Frémont.

Turning his attention back to Banks, Jackson advanced down the Valley to New Market. Crossing the Massanutten Mountain into the Luray Valley, the Confederates surprised and overran a Federal outpost at Front Royal. Capturing the town on May 23, the Confederates assaulted Banks's rear. Banks retreated to Winchester, arriving just ahead of the Confederates. Banks attempted to defend the strategic city, but Jackson flanked his position with ease on May 25, and routed the Federals from their defenses. Jackson pursued the fleeing Federal army to the outskirts of Harper's Ferry.

Stonewall Jackson's small Valley Army had turned the tables on Banks and the Washington government, and now held control of the entire Shenandoah Valley.

> "He classed all who were weak and weary, who fainted by the wayside, as men wanting in patriotism. If a man's face was as white as cotton and his pulse so low you could hardly feel it, he looked upon him merely as an inefficient soldier and rode off impatiently."
>
> *One of Jackson's officers describing his commander during the Shenandoah Valley campaign.*

*Gen. Robert Huston Milroy (above) whose division was defeated by Jackson at McDowell, on May 8. Although he lacked sufficient numbers to halt the Confederate advance, Milroy's defense was spirited. When forced to withdraw he hampered his pursuers by setting fire to the woods along the road.*

*Though lacking co-ordination, and bewildered by the lightning advance of the Confederates, many Federal units exhibited a courageous and pugnacious spirit. Lieut. Col. Thomas Kane (below) led his 13th Pennsylvania Reserves against Jackson at Harrisonburg on June 6. His small command was decimated and Kane was captured after being wounded.*

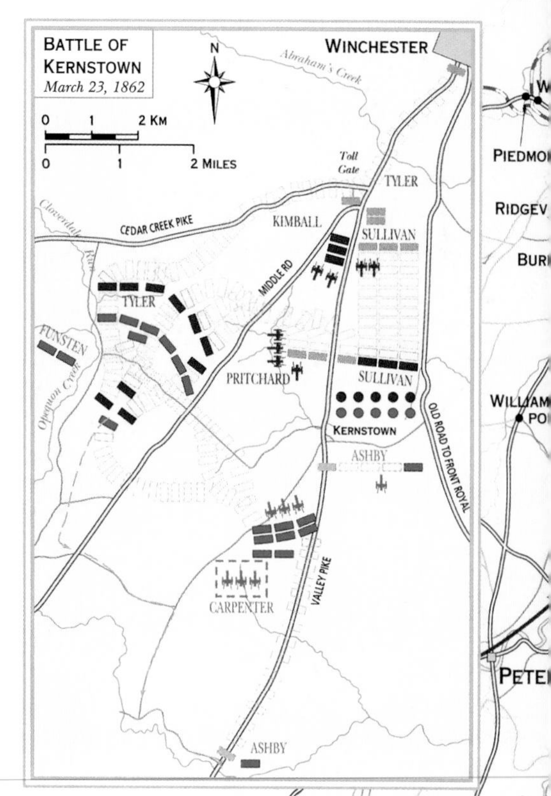
BATTLE OF KERNSTOWN
March 23, 1862
N
0 1 2 KM
0 1 2 MILES
WINCHESTER
Abraham's Creek
Toll Gate
TYLER
CEDAR CREEK PIKE
KIMBALL
SULLIVAN
TYLER
MIDDLE RD
FUNSTEN
Opequon Creek
PRITCHARD
SULLIVAN
KERNSTOWN
ASHBY
OLD ROAD TO FRONT ROYAL
VALLEY PIKE
CARPENTER
ASHBY

*Unaware that Union Gen. Shields commanded a force more than double the size of his own, Jackson attacked Kernstown on March 23. Though heavily outnumbered, Jackson's force fought well before executing an orderly withdrawal. Confederate casualties 718, Union 590.*

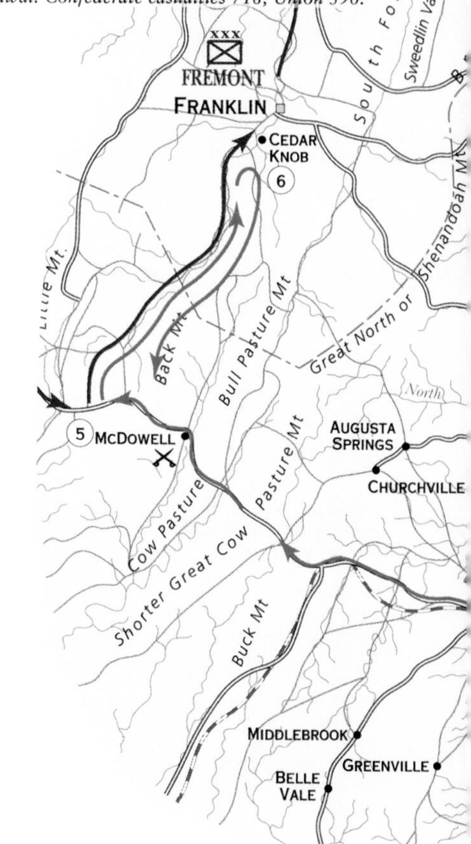
FREMONT
FRANKLIN
CEDAR KNOB
6
South Fork Mountain
Sweedlin Valley
Little Mt
Back Mt
Bull Pasture Mt
Great North or Shenandoah Mt
North
5
McDOWELL
AUGUSTA SPRINGS
CHURCHVILLE
Cow Pasture
Shorter Great Cow Pasture
Pasture Mt
Buck Mt
MIDDLEBROOK
GREENVILLE
BELLE VALE

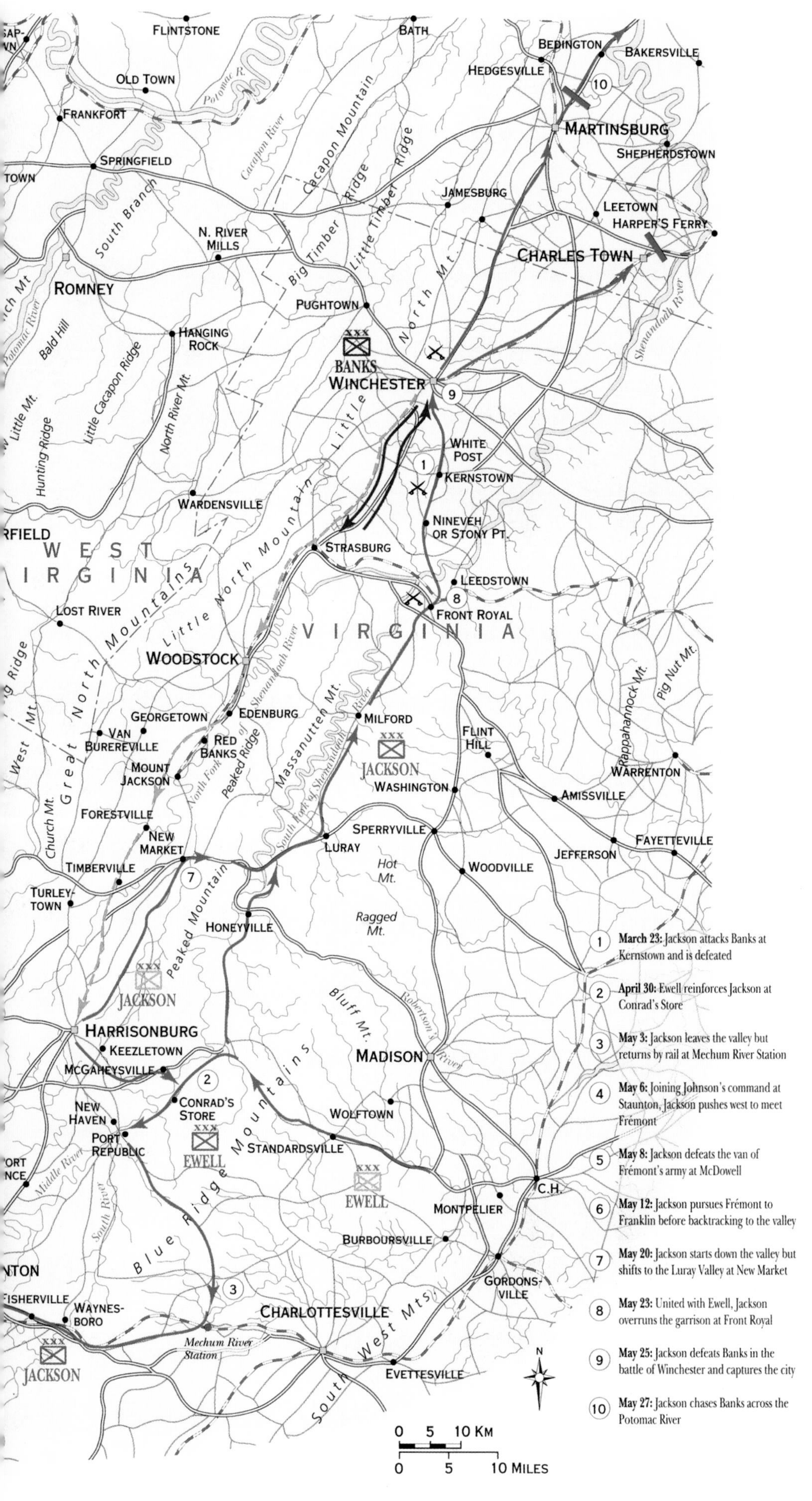

FLINTSTONE
BATH
BEDINGTON
BAKERSVILLE
HEDGESVILLE
OLD TOWN
FRANKFORT
Potomac R.
MARTINSBURG
SHEPHERDSTOWN
SPRINGFIELD
Cacapon Mountain
Cacapon River
Little Timber Ridge
Big Timber Ridge
South Branch
JAMESBURG
LEETOWN
HARPER'S FERRY
N. RIVER MILLS
CHARLES TOWN
ROMNEY
North Mt.
PUGHTOWN
Shenandoah River
HANGING ROCK
Bald Hill
Little Cacapon Ridge
North River Mt.
Hunting Ridge
BANKS
WINCHESTER
WHITE POST
KERNSTOWN
WARDENSVILLE
NINEVEH OR STONY PT.
STRASBURG
WEST VIRGINIA
LEEDSTOWN
FRONT ROYAL
LOST RIVER
Little North Mountains
VIRGINIA
WOODSTOCK
Great North Mountain
Rappahannock Mt.
Pig Nut Mt.
GEORGETOWN
EDENBURG
MILFORD
VAN BUREREVILLE
RED BANKS
FLINT HILL
MOUNT JACKSON
Massanutten Mt.
JACKSON
WASHINGTON
WARRENTON
AMISSVILLE
FORESTVILLE
NEW MARKET
SPERRYVILLE
LURAY
FAYETTEVILLE
JEFFERSON
Church Mt.
TIMBERVILLE
Hot Mt.
WOODVILLE
TURLEYTOWN
HONEYVILLE
Ragged Mt.
Peaked Mountain
JACKSON
HARRISONBURG
Bluff Mt.
KEEZLETOWN
MADISON
McGAHEYSVILLE
CONRAD'S STORE
NEW HAVEN
PORT REPUBLIC
EWELL
WOLFTOWN
STANDARDSVILLE
Blue Ridge Mountains
EWELL
C.H.
MONTPELIER
BURBOURSVILLE
GORDONSVILLE
FISHERVILLE
WAYNESBORO
CHARLOTTESVILLE
South West Mts.
Mechum River Station
JACKSON
EVETTESVILLE
0 5 10 KM
0 5 10 MILES
1 March 23: Jackson attacks Banks at Kernstown and is defeated
2 April 30: Ewell reinforces Jackson at Conrad's Store
3 May 3: Jackson leaves the valley but returns by rail at Mechum River Station
4 May 6: Joining Johnson's command at Staunton, Jackson pushes west to meet Frémont
5 May 8: Jackson defeats the van of Frémont's army at McDowell
6 May 12: Jackson pursues Frémont to Franklin before backtracking to the valley
7 May 20: Jackson starts down the valley but shifts to the Luray Valley at New Market
8 May 23: United with Ewell, Jackson overruns the garrison at Front Royal
9 May 25: Jackson defeats Banks in the battle of Winchester and captures the city
10 May 27: Jackson chases Banks across the Potomac River


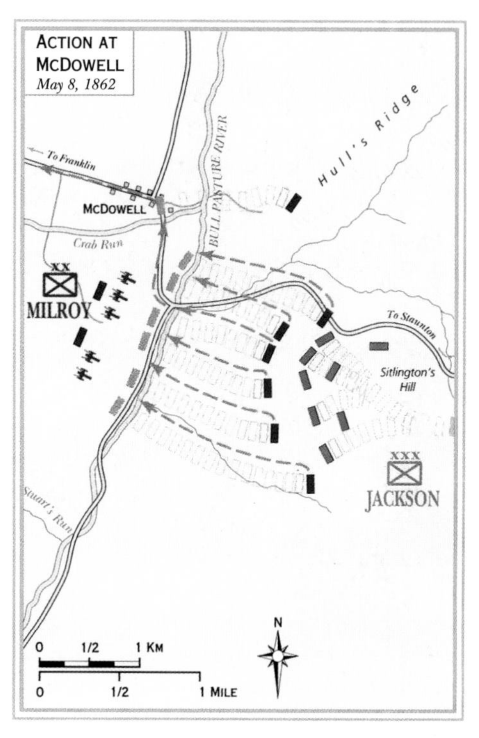

ACTION AT McDOWELL
May 8, 1862
To Franklin
McDOWELL
BULL PASTURE RIVER
Hull's Ridge
Crab Run
MILROY
To Staunton
Sitlington's Hill
JACKSON
0 1/2 1 KM
0 1/2 1 MILE
N


*Having concentrated his forces on the village of McDowell (above), Gen. Milroy, reinforced by Shenk's brigade from Franklin, pre-empted the Confederate attack and caught Jackson unawares. Lacking sufficient strength to follow-up his assault, Milroy retreated. Casualties were 256 Union and 498 Confederate.*

*At Winchester, Federal forces were initially able to withstand the Confederate assaults (below) but eventually the Union line broke, precipitating a confused retreat toward Harper's Ferry. A considerable stockpile of Union supplies were lost. Union casualties totaled 1,714, Confederate 400.*

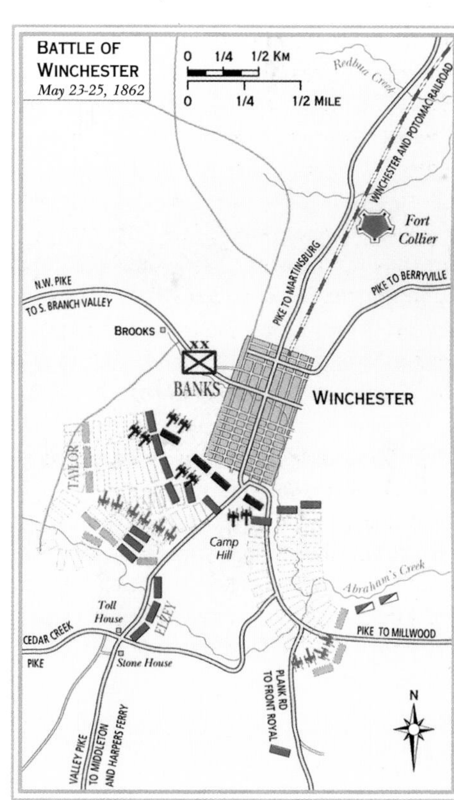

BATTLE OF WINCHESTER
May 23-25, 1862
0 1/4 1/2 KM
0 1/4 1/2 MILE
Redbud Creek
WINCHESTER AND POTOMAC RAILROAD
Fort Collier
N.W. PIKE
TO S. BRANCH VALLEY
PIKE TO MARTINSBURG
PIKE TO BERRYVILLE
BROOKS
BANKS
WINCHESTER
TAYLOR
Camp Hill
Abraham's Creek
Toll House
EWELL
CEDAR CREEK PIKE
Stone House
PIKE TO MILLWOOD
PLANK RD TO FRONT ROYAL
VALLEY PIKE TO MIDDLETON AND HARPERS FERRY
N

# Jackson's Shenandoah Valley Campaign: Phase 2

## MAY 30 – JUNE 9 1862

Threatened by Stonewall Jackson's close proximity to Washington, Lincoln diverted Federal troops to surround the Confederates in the Shenandoah Valley. Jackson watched three Federal columns converge to destroy him, then neatly eluded them by falling back from Harpers Ferry on May 3l. Part of Jackson's army had to march 50 miles in two days, narrowly escaping the closing Federal trap by only a few hours.

The Confederates continued to withdraw up the Valley to Harrisonburg, with the Federals in hot pursuit. Federals, under General Frémont and Brigadier General James Shields, raced southward on either side of Massanutten Mountain. Jackson realized that the two Federal columns would converge on the sleepy village of Port Republic. The Confederates con-centrated in that area to keep Frémont and Shields separated. Jackson's cavalry, under Turner Ashby, delayed the Federal advance, but Ashby was killed in a small battle at Harrisonburg on June 6.

On June 8, Frémont's column encountered General Ewell's Confederates who were positioned on a series of ridges near the Cross Keys tavern. Though Frémont outnumbered Ewell – 10,500 Federals to 5,000 Confederates – Frémont haltingly tested the Confederate lines. Holding Frémont at Cross Keys throughout the day, Ewell provided Jackson with time to plan a strike four miles to the east at Port Republic. Jackson's troops crossed the South Fork of the Shenandoah River, and attacked Shields' Federal division on June 9. After a vicious battle, the Federals were routed and retreated northward. Frémont also prudently withdrew to the north, leaving Stonewall Jackson master of the Valley.

Jackson had won a spectacular string of victories and thwarted 64,000 Federal troops with an army of no more than 16,000. In the course of three months, he had inflicted 7,000 Federal casualties, with the loss of only 2,500 men, and had become the most famous soldier of the day.

*Edwin Forbes's sketch of the battle of Cross Keys (above) shows a reserve cavalry unit and field ambulances in the foreground. Forbes identifies the small house in the woods as the Union field hospital.*

*The brigade of Maj. Gen. Richard Stoddart Ewell, nicknamed "Old Baldy," (below) arrived at Cross Keys just when it was most needed. The presence of Ewell's command meant that the Confederates outnumbered the Federals by two to one.*

> "Always mystify, mislead, and suprise the enemy, if possible. And when you strike and overcome him, never let up in the pursuit so long as your men have strength to follow... Such tactics will win every time, and a small army may thus destroy a large one in detail, and repeated victory will make it invincible"
>
> *Jackson to one of his officers after the battle of Cross Keys*

1. **May 30:** McDowell's troops recapture Front Royal in an attempt to cut off Jackson's retreat.
2. **May 30:** Frémont attempts to close Jackson's retreat route from the west.
3. **May 31:** Jackson abandons his positions above Winchester and slips through the Union vice.
4. **June 6:** Ashby ambushes the Union vanguard at Harrisonburg but is killed in the battle.
5. **June 8:** Ewell delays Frémont in the battle of Cross Keys
6. **June 8:** Union cavalry dashes through Port Republic and almost captures Jackson
7. **June 9:** Jackson defeats McDowell's advance at Port Republic

FRANKLIN
CEDAR KNOB
Bull Pasture Mountain
Great North or Shenandoah Mt
South Fork Mountain
North Fork Mountain
Middle Mountain
Sweedlin Valley

N

0 5 10 KM
0 5 10 MILES

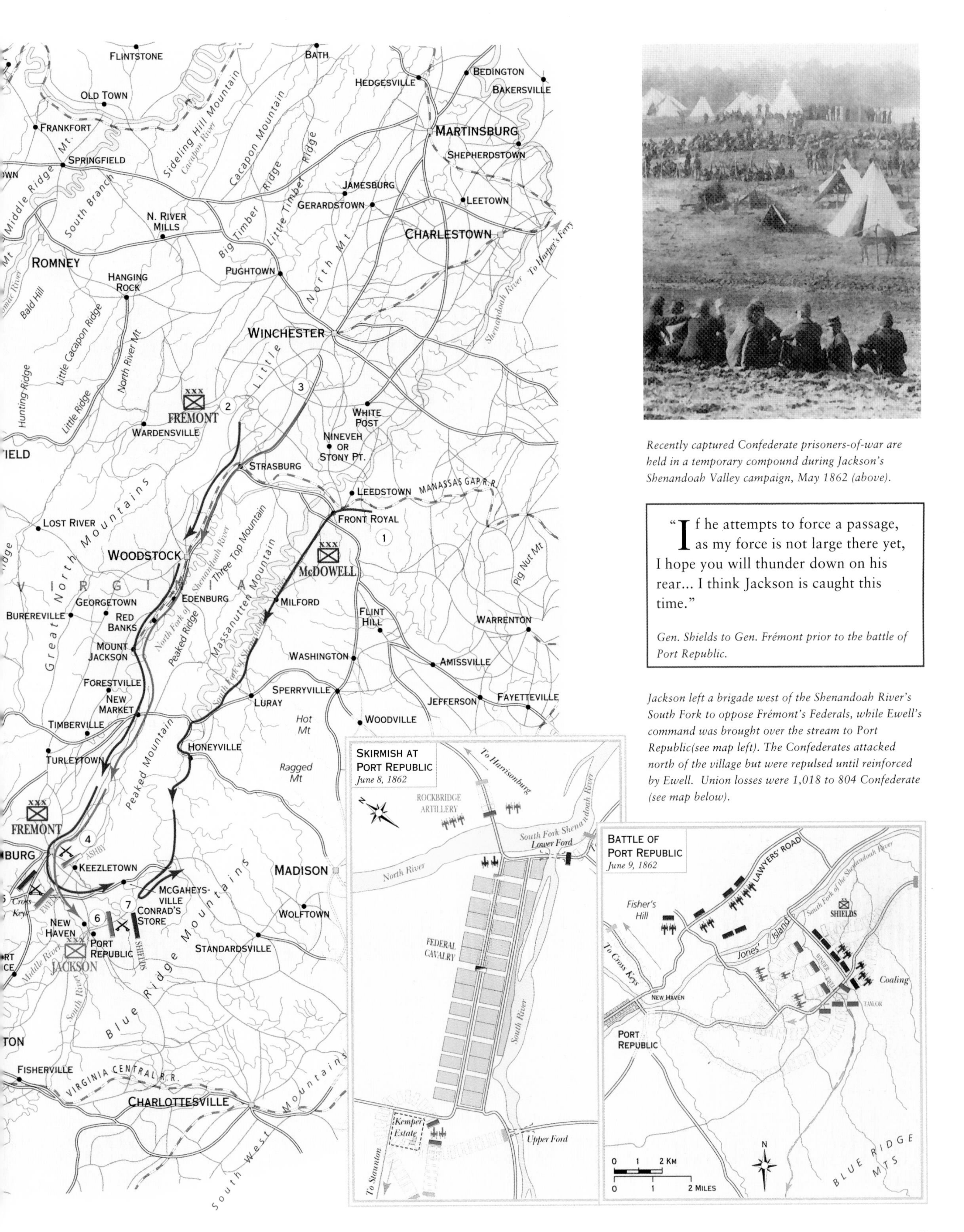

FLINTSTONE
BATH
BEDINGTON
HEDGESVILLE
BAKERSVILLE
OLD TOWN
FRANKFORT
MARTINSBURG
SHEPHERDSTOWN
SPRINGFIELD
Sideling Hill Mountain
Cacapon River
Cacapon Mountain
Middle Ridge Mt.
South Branch
Timber Ridge
Little Timber Ridge
JAMESBURG
GERARDSTOWN
LEETOWN
N. RIVER MILLS
Big Timber
CHARLESTOWN
To Harper's Ferry
ROMNEY
HANGING ROCK
PUGHTOWN
North Mt.
Bald Hill
Little Cacapon Ridge
Hunting Ridge
Little Ridge
North River Mt
Shenandoah River
WINCHESTER
Little
FREMONT
WARDENSVILLE
WHITE POST
NINEVEH OR STONY PT.
STRASBURG
LEEDSTOWN
MANASSAS GAP R.R.
FRONT ROYAL
LOST RIVER
Great North Mountains
Three Top Mountain
Massanutten Mountain
McDOWELL
WOODSTOCK
VIRGINIA
Pig Nut Mt
EDENBURG
MILFORD
GEORGETOWN
BUREREVILLE
RED BANKS
FLINT HILL
WARRENTON
North Fork of Shenandoah River
Peaked Ridge
MOUNT JACKSON
WASHINGTON
AMISSVILLE
FORESTVILLE
NEW MARKET
SPERRYVILLE
LURAY
JEFFERSON
FAYETTEVILLE
TIMBERVILLE
Hot Mt
WOODVILLE
HONEYVILLE
TURLEYTOWN
Ragged Mt
Peaked Mountain
KEEZLETOWN
ASHBY
MADISON
Cross Keys
McGAHEYSVILLE
CONRAD'S STORE
NEW HAVEN
PORT REPUBLIC
SHIELDS
JACKSON
WOLFTOWN
STANDARDSVILLE
Blue Ridge Mountains
Middle River
South River
FISHERVILLE
VIRGINIA CENTRAL R.R.
CHARLOTTESVILLE
South West Mountains
1
2
3
4
6
7
SKIRMISH AT PORT REPUBLIC
June 8, 1862
To Harrisonburg
ROCKBRIDGE ARTILLERY
South Fork Shenandoah River
Lower Ford
North River
FEDERAL CAVALRY
South River
Kemper Estate
Upper Ford
To Staunton
BATTLE OF PORT REPUBLIC
June 9, 1862
LAWYERS' ROAD
South Fork of the Shenandoah River
Fisher's Hill
SHIELDS
Jones' Island
To Cross Keys
WINDER
Coaling
NEW HAVEN
TAYLOR
PORT REPUBLIC
BLUE RIDGE MTS
0 1 2 KM
0 1 2 MILES
N


*Recently captured Confederate prisoners-of-war are held in a temporary compound during Jackson's Shenandoah Valley campaign, May 1862 (above).*

> "If he attempts to force a passage, as my force is not large there yet, I hope you will thunder down on his rear... I think Jackson is caught this time."
>
> *Gen. Shields to Gen. Frémont prior to the battle of Port Republic.*

*Jackson left a brigade west of the Shenandoah River's South Fork to oppose Frémont's Federals, while Ewell's command was brought over the stream to Port Republic(see map left). The Confederates attacked north of the village but were repulsed until reinforced by Ewell. Union losses were 1,018 to 804 Confederate (see map below).*

# Seven Days Battles: Phase 1 JUNE 1862

Throughout June, General Lee fortified Richmond and made preparations to attack McClellan's Federal army. J.E.B. Stuart's Confederate cavalry confirmed McClellan's exact positions in a daring and celebrated ride around the Federal army.

Lee kept Magruder's and Huger's 25,000 Confederates near Richmond to guard the capital, while the rest of the army united with Stonewall Jackson to turn McClellan's right flank at Mechanicsville. Lee wanted to force Fitz-John Porter's V Corps out of a strong line, and then attack during its retreat when the Federals would be at their most vulnerable. McClellan almost upset the plan on June 25 with a probing attack at Oak Grove. When the Federals stalled, Lee pressed ahead with his offensive.

Venturing into unfamiliar country, and uncharacteristically lethargic, Stonewall Jackson lagged behind as Lee's lieutenants attacked the Federal line at Beaver Dam Creek late in the afternoon of June 26. Lee's forces resorted to a series of bloody frontal assaults – the hottest action centering on Ellerson's Mill – but without success. The Confederates lost 1,484 men, while the Federals suffered only 361 casualities. The Federals had to abandon their defenses after dark when Jackson took position at Hundley's Corner and threatened their rear. Porter's V Corps retreated five miles to a commanding position behind Boatswain's Swamp, near Gaines' Mill.

Lee followed on June 27, and attacked the Federals near Gaines' Mill. In five hours of repeated assaults, the Confederates hammered at the Federal defenses. At dusk, two Confederate brigades pierced the Federal line, forcing Porter to retreat in some confusion. As the Federals withdrew south of the Chickahominy River, Lee had won his first clear victory. McClellan prepared to abandon his supply base at White House. Confederate losses amounted to 8,750; the Federals to 6,837.

"The timber between us and the enemy hid them from view, but we pulled our triggers nevertheless, and rushed down the hill... at the Yankees in the first line of breastworks. They waited not for the onset, but fled like a flock of sheep, carrying with them their supports in the second and third lines."

*A veteran of John B. Hood's Texas Brigade at Gaines' Mill.*

US

*As the Confederates pursued Porter's retreating forces, they were subjected to murderous artillery fire from the Federal batteries. Capt. John C. Tidball and his staff (above) served at Mechanicsville and contributed to the severe Confederate casualties.*

*(Below) Shorn of its timber stories, the skeleton of Gaines' Mill marks the passage of war. Beneath the mill's shadow , Porter's line had broken in the face of a bayonet charge led by John B. Hood. The Confederates lost 1,000 men, but captured two regiments.*

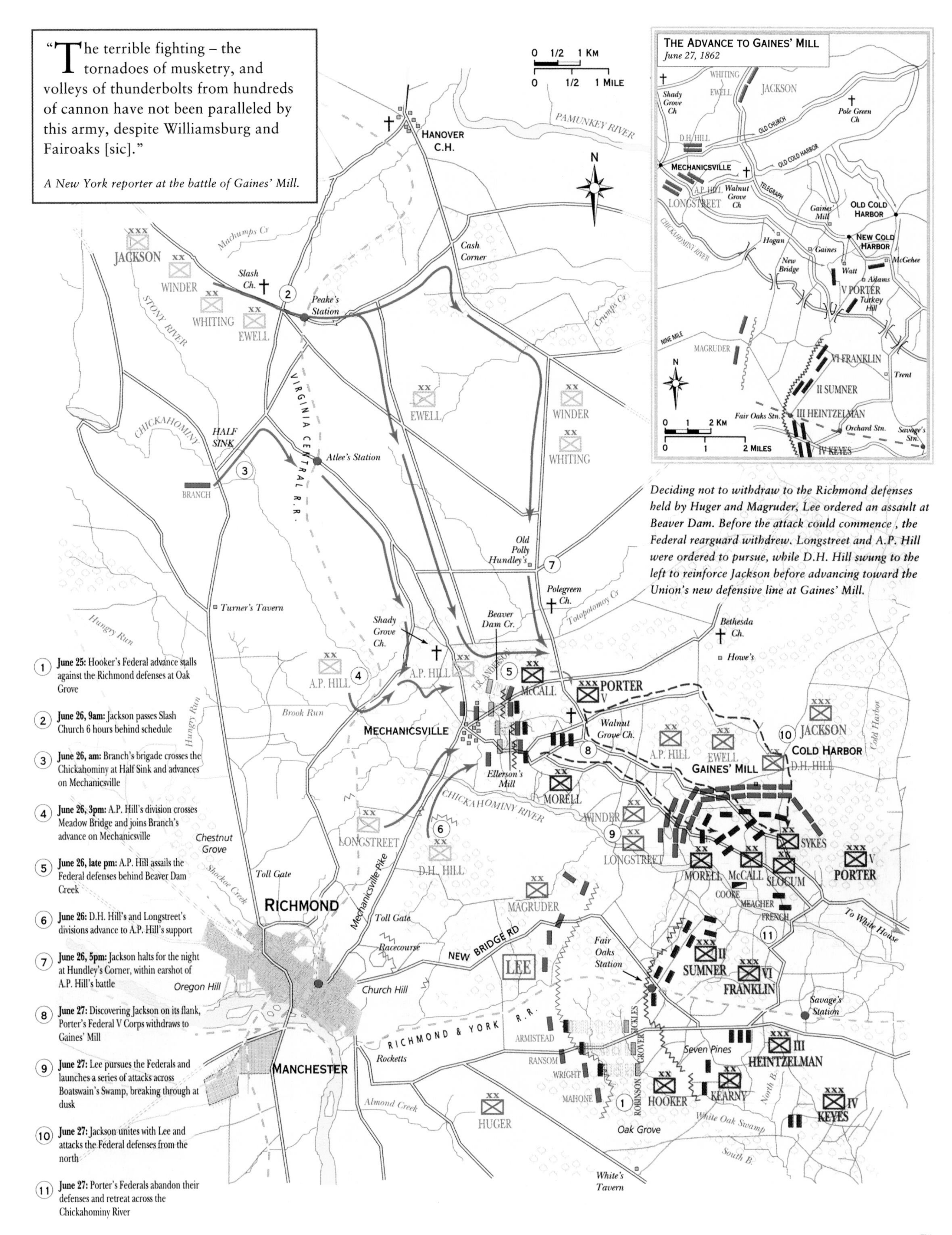
"The terrible fighting – the tornadoes of musketry, and volleys of thunderbolts from hundreds of cannon have not been paralleled by this army, despite Williamsburg and Fairoaks [sic]."
A New York reporter at the battle of Gaines' Mill.
0 1/2 1 KM
0 1/2 1 MILE
N
THE ADVANCE TO GAINES' MILL
June 27, 1862
Deciding not to withdraw to the Richmond defenses held by Huger and Magruder, Lee ordered an assault at Beaver Dam. Before the attack could commence, the Federal rearguard withdrew. Longstreet and A.P. Hill were ordered to pursue, while D.H. Hill swung to the left to reinforce Jackson before advancing toward the Union's new defensive line at Gaines' Mill.
1 June 25: Hooker's Federal advance stalls against the Richmond defenses at Oak Grove
2 June 26, 9am: Jackson passes Slash Church 6 hours behind schedule
3 June 26, am: Branch's brigade crosses the Chickahominy at Half Sink and advances on Mechanicsville
4 June 26, 3pm: A.P. Hill's division crosses Meadow Bridge and joins Branch's advance on Mechanicsville
5 June 26, late pm: A.P. Hill assails the Federal defenses behind Beaver Dam Creek
6 June 26: D.H. Hill's and Longstreet's divisions advance to A.P. Hill's support
7 June 26, 5pm: Jackson halts for the night at Hundley's Corner, within earshot of A.P. Hill's battle
8 June 27: Discovering Jackson on its flank, Porter's Federal V Corps withdraws to Gaines' Mill
9 June 27: Lee pursues the Federals and launches a series of attacks across Boatswain's Swamp, breaking through at dusk
10 June 27: Jackson unites with Lee and attacks the Federal defenses from the north
11 June 27: Porter's Federals abandon their defenses and retreat across the Chickahominy River
HANOVER C.H.
PAMUNKEY RIVER
JACKSON
WINDER
WHITING
EWELL
Slash Ch.
Peake's Station
Cash Corner
STONY RIVER
VIRGINIA CENTRAL R.R.
HALF SINK
BRANCH
Atlee's Station
Old Polly Hundley's
Polegreen Ch.
Turner's Tavern
Shady Grove Ch.
Beaver Dam Cr.
A.P. HILL
McCALL
PORTER
MECHANICSVILLE
Walnut Grove Ch.
Ellerson's Mill
MORELL
CHICKAHOMINY RIVER
LONGSTREET
D.H. HILL
GAINES' MILL
COLD HARBOR
SYKES
SLOCUM
MAGRUDER
RICHMOND
NEW BRIDGE RD
LEE
Fair Oaks Station
SUMNER
FRANKLIN
Savage's Station
Seven Pines
HEINTZELMAN
HOOKER
KEARNY
KEYES
HUGER
Oak Grove
White's Tavern
MANCHESTER
Oregon Hill
Church Hill
Rocketts
RICHMOND & YORK R.R.
To White House

# Seven Days' Battles: Phase 2 JUNE 29 – JULY 1 1862

As he changed his base of supplies from White House on the Pamunkey River to Harrison's Landing on the James River, General McClellan began moving his army south across the Virginia peninsula. From June 28 to July 1, Lee tried repeatedly to destroy the retreating Federal columns while they were at their most vulnerable.

On June 29, Magruder's Confederates advanced through McClellan's abandoned works, and then east to Savage's Station. The Federal rearguard checked the Confederates before withdrawing across White Oak Swamp. Lee hurried to race ahead of McClellan's columns, while Stonewall Jackson crowded the Federal rear, but a defiant Federal rearguard easily stalled Jackson at White Oak Swamp on June 30. At the same time, General Longstreet and A.P. Hill pitched into McClellan's infantry at Glendale, but were unable to sever the Federal column. Reunited on July 1, Lee's army launched a number of headlong assaults against the strong Federal battle line at Malvern Hill. McClellan's artillery tore the Confederates apart until sunset finally ended the futile attacks. After dark, McClellan fell back to the protection of the navy's gunboats at Harrison's Landing.

In the course of the Seven Days battles, the losses had been sobering: Lee incurred 20,141 in his continuous attacks, while the Federal defenders lost 15,849. Lee could not destroy the Federal army with his elaborate plans; his commands were too spread out; his general staff was still too unseasoned for so intricate a scheme of attack. The result was an exasperating lack of coordination between the Confederate leaders. In the end, however, McClellan opted to retreat from the gates of Richmond, and with the Confederate capital no longer threatened, the initiative passed to Lee. The Confederate general now focused on northern Virginia.

> "Charge after charge is made on our artillery, with a demoniac will to take it, if it costs them half their army. Down it mows their charging ranks, till they lie in heaps and rows, from behind which our men fight as securely as if in rifle pits... The slaughter is terrible, and to add to the carnage, our gunboats are throwing their murderous missiles with furious effect into the ranks of our enemy."
>
> *A Wisconsin infantryman present at Malvern Hill.*

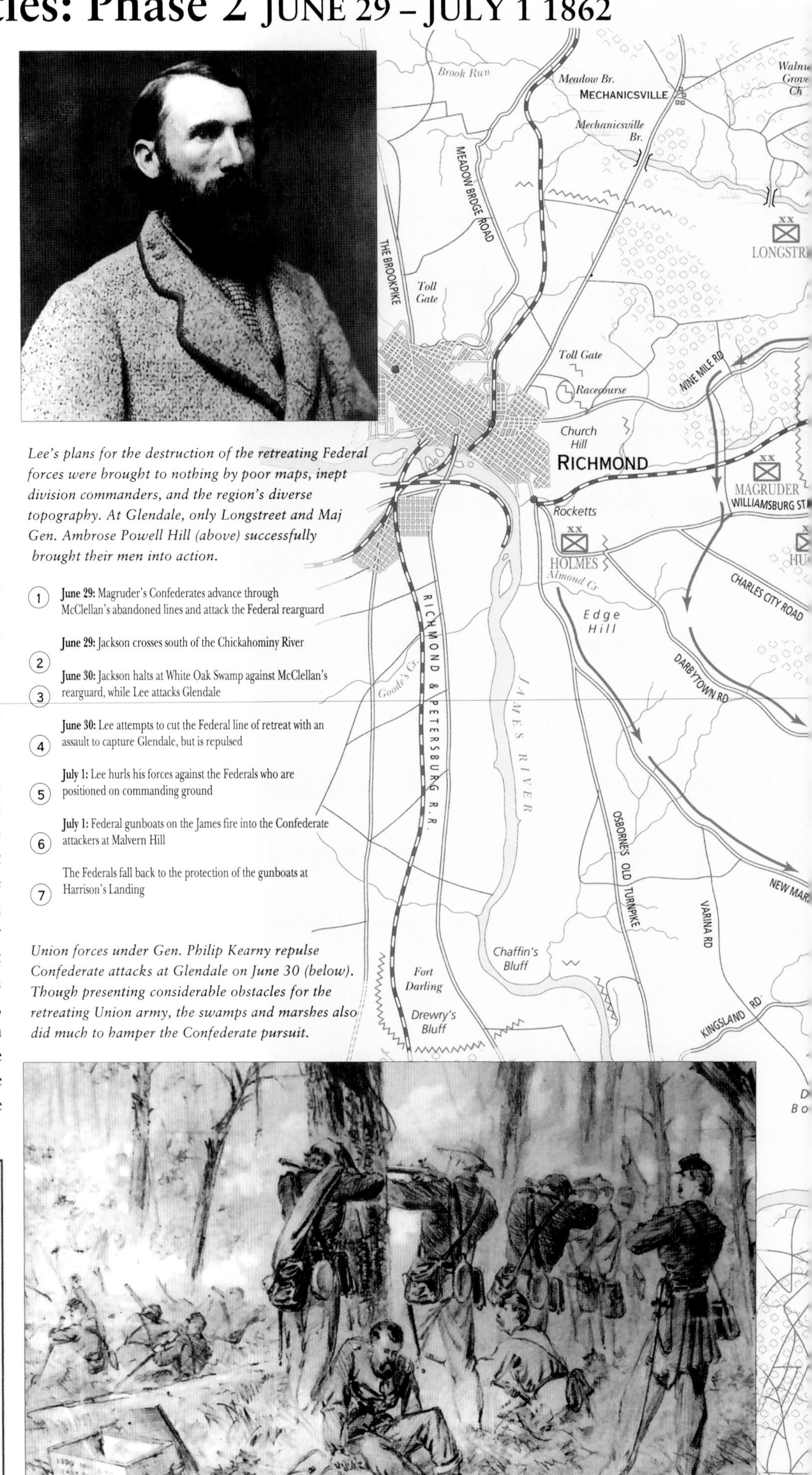

Brook Run
Meadow Br.
MECHANICSVILLE
Mechanicsville Br.
MEADOW BRIDGE ROAD
THE BROOKPIKE
Toll Gate
Toll Gate
Racecourse
LONGSTR
NINE MILE RD
Church Hill
RICHMOND
MAGRUDER
WILLIAMSBURG ST
Rocketts
HOLMES
Almond Cr.
Edge Hill
CHARLES CITY ROAD
DARBYTOWN RD
RICHMOND & PETERSBURG R.R.
Goode's Cr.
JAMES RIVER
OSBORNE'S OLD TURNPIKE
VARINA RD
NEW MAR
Chaffin's Bluff
Fort Darling
Drewry's Bluff
KINGSLAND RD
1 June 29: Magruder's Confederates advance through McClellan's abandoned lines and attack the Federal rearguard
2 June 29: Jackson crosses south of the Chickahominy River
3 June 30: Jackson halts at White Oak Swamp against McClellan's rearguard, while Lee attacks Glendale
4 June 30: Lee attempts to cut the Federal line of retreat with an assault to capture Glendale, but is repulsed
5 July 1: Lee hurls his forces against the Federals who are positioned on commanding ground
6 July 1: Federal gunboats on the James fire into the Confederate attackers at Malvern Hill
7 The Federals fall back to the protection of the gunboats at Harrison's Landing


*Lee's plans for the destruction of the retreating Federal forces were brought to nothing by poor maps, inept division commanders, and the region's diverse topography. At Glendale, only Longstreet and Maj Gen. Ambrose Powell Hill (above) successfully brought their men into action.*

*Union forces under Gen. Philip Kearny repulse Confederate attacks at Glendale on June 30 (below). Though presenting considerable obstacles for the retreating Union army, the swamps and marshes also did much to hamper the Confederate pursuit.*

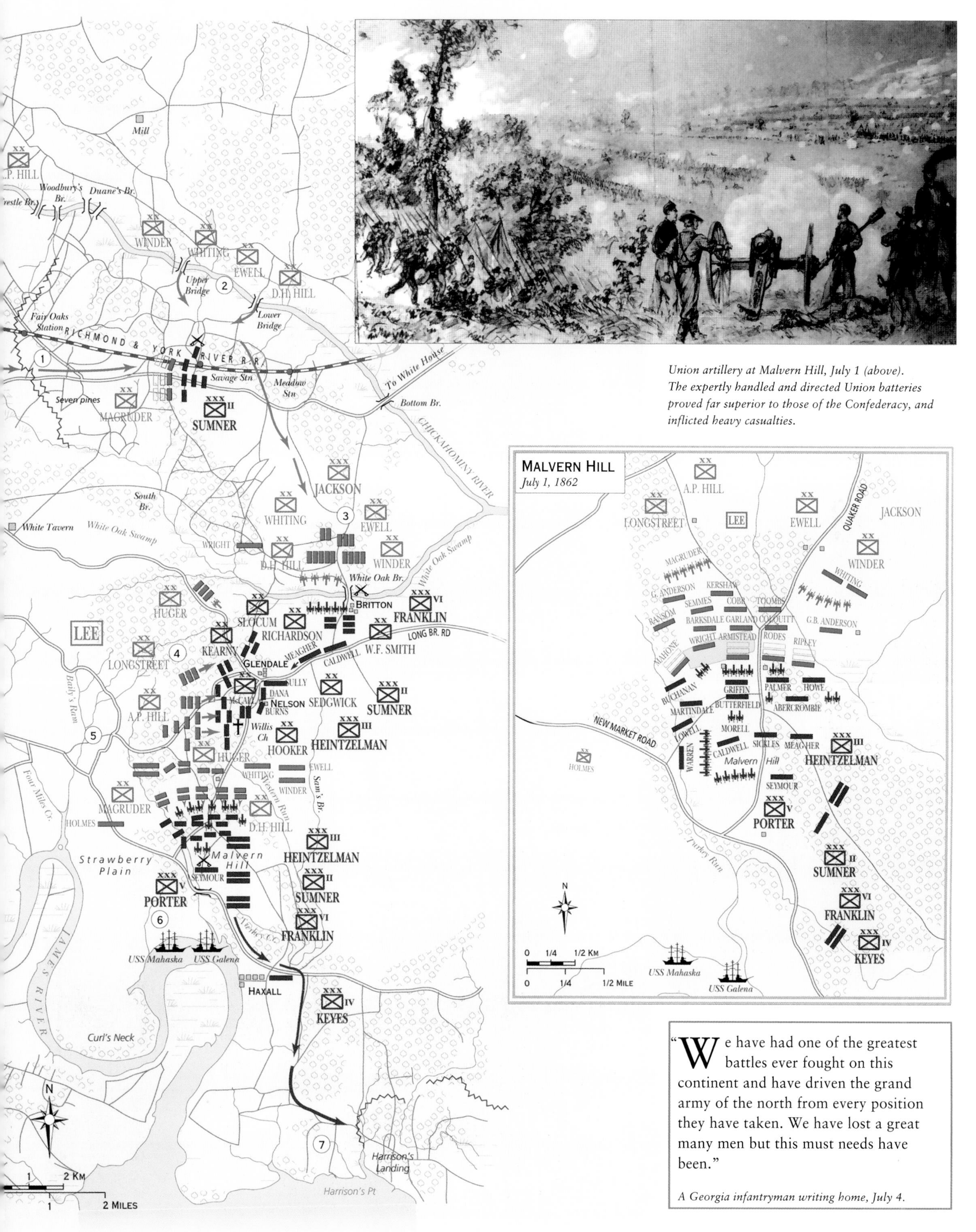
Mill
Woodbury's Br.
Duane's Br.
Upper Bridge
Lower Bridge
Fair Oaks Station
RICHMOND & YORK RIVER R.R.
Savage Stn
Meadow Stn
Seven pines
To White House
Bottom Br.
CHICKAHOMINY RIVER
South Br.
White Tavern
White Oak Swamp
White Oak Br.
BRITTON
LONG BR. RD
GLENDALE
NELSON
Willis Ch
Baily's Run
Four Miles Cr.
Western Run
Sam's Br.
Strawberry Plain
Malvern Hill
Turkey Cr.
USS Mahaska
USS Galena
HAXALL
JAMES RIVER
Curl's Neck
Harrison's Landing
Harrison's Pt
1 2 KM
1 2 MILES
N
LEE
LONGSTREET
A.P. HILL
HUGER
MAGRUDER
HOLMES
JACKSON
WHITING
EWELL
WINDER
D.H. HILL
SUMNER
FRANKLIN
SLOCUM
RICHARDSON
KEARNY
W.F. SMITH
SEDGWICK
HOOKER
HEINTZELMAN
PORTER
KEYES
MALVERN HILL
July 1, 1862
QUAKER ROAD
NEW MARKET ROAD
Turkey Run
0 1/4 1/2 KM
0 1/4 1/2 MILE

*Union artillery at Malvern Hill, July 1 (above). The expertly handled and directed Union batteries proved far superior to those of the Confederacy, and inflicted heavy casualties.*

"We have had one of the greatest battles ever fought on this continent and have driven the grand army of the north from every position they have taken. We have lost a great many men but this must needs have been."

*A Georgia infantryman writing home, July 4.*

# Second Manassas (Bull Run) Campaign: Phase 1

## 14 JULY – 27 AUGUST 1862

In the sweltering heat of late July, 1862, a new threat to the Confederacy loomed out of Northern Virginia. Federal General John Pope massed the beaten elements of the 1862 Valley Campaign and started south toward Culpeper Court House. Lee detested Pope's bombastic proclamations against Southern civilians, and dispatched Stonewall Jackson to supress Pope's Army of Virginia.

In early August, Jackson gathered his 22,000 troops at Orange and Gordonsville, but soon advanced to strike at an isolated Federal corps led by General Nathaniel Banks. Crossing the Rapidan River, Jackson attacked Banks at Cedar Mountain on August 9. The battle hung in the balance, but Jackson's superior numbers eventually drove Banks from the field. Pope reinforced Banks, forcing the Confederates back across the Rapidan. Jackson had 1,400 casualties at Cedar Mountain; Banks incurred 2,500.

When Lee learned that McClellan would join forces with Pope, the Confederate general united Longstreet's command with Jackson's to strike at Pope. Lee and Pope sparred for two weeks along the Rapidan, and then the Rappahannock rivers. At one point Lee's cavalry, under J.E.B. Stuart, raided Pope's line of communications and supplies, attacking Catlett's Station on a stormy August night. Unable to burn the rain-soaked railroad bridges, Stuart consoled himself with the capture of Pope's headquarters' equipment, including the Federal commander's dress uniform.

On August 25, Stonewall Jackson marched 54 miles in 36 hours in a daring move against Pope's rear. Turning Pope's right flank, Jackson captured and burned Pope's supply depot at Manassas Junction. Pope moved to isolate Jackson, but the Confederates eluded his trap by marching five miles north, taking cover in the woods and behind an unfinished railroad embankment north of the Warrenton Turnpike. Jackson had chosen his ground well; the stage was set for battle as the opposing armies converged on Manassas.

> "Jackson ordered forward the "Stonewall Brigade"... He waved his hat and told them remember they were the Stonewall Brigade, and with a shout they rushed forward and hurled back the insolent foe."
>
> *A Confederate captain at Cedar Mountain.*

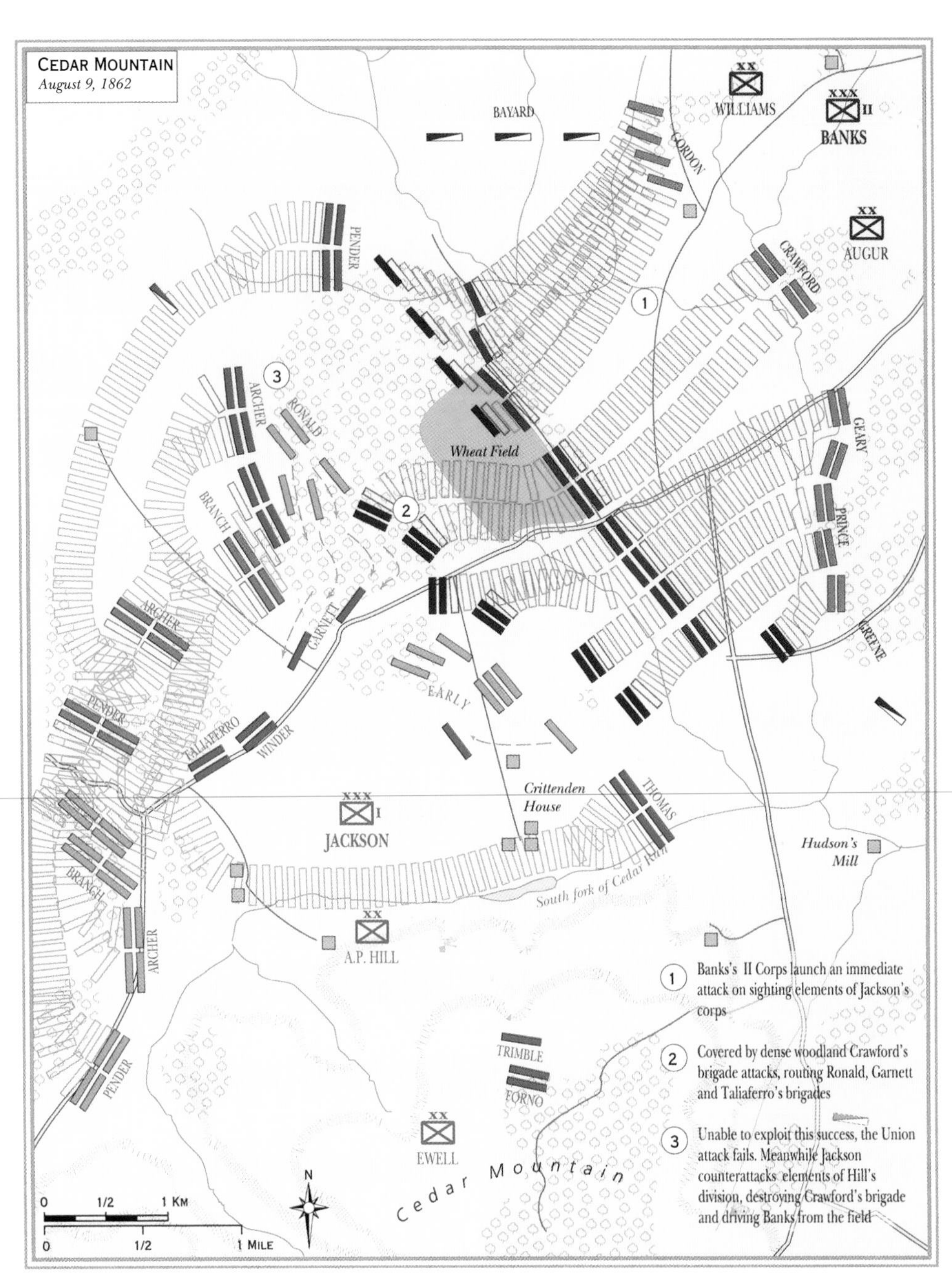
CEDAR MOUNTAIN
August 9, 1862
BAYARD
WILLIAMS
BANKS
GORDON
AUGUR
CRAWFORD
PENDER
ARCHER
RONALD
Wheat Field
BRANCH
GEARY
PRINCE
GARNETT
GREENE
ARCHER
EARLY
PENDER
TALIAFERRO
WINDER
Crittenden House
THOMAS
JACKSON
Hudson's Mill
South fork of Cedar Run
BRANCH
A.P. HILL
ARCHER
TRIMBLE
FORNO
PENDER
EWELL
Cedar Mountain
N
0 1/2 1 KM
0 1/2 1 MILE
1 Banks's II Corps launch an immediate attack on sighting elements of Jackson's corps
2 Covered by dense woodland Crawford's brigade attacks, routing Ronald, Garnett and Taliaferro's brigades
3 Unable to exploit this success, the Union attack fails. Meanwhile Jackson counterattacks elements of Hill's division, destroying Crawford's brigade and driving Banks from the field

*Union troops charging the left flank of Jackson's army at Cedar Mountain (below). Successive Union attacks had so weakened Jackson's left that collapse seemed imminent. Confederate hopes were revived by the timely arrival of Ewell's artillery and the commencement of a devastating bombardment.*

"Our boys fought like heroes or devils; and although met by an immense force of the enemy, they succeeded in driving him back through one piece of woods into the open field beyond. The fighting in this woods [sic] was most terrible; men fought bayonet to bayonet."

*Edwin Forbes, Union artist present at Cedar Mountain.*

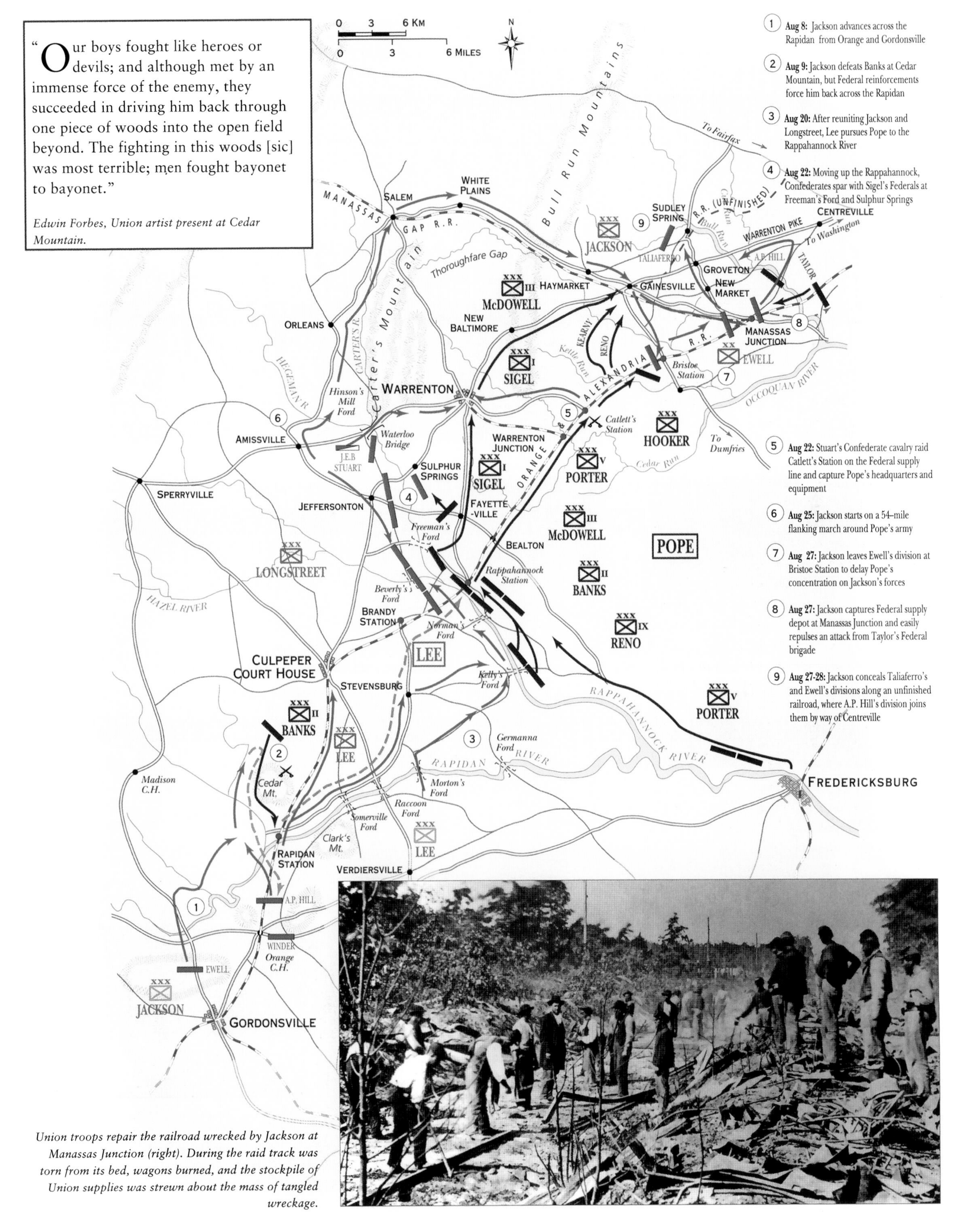

0 3 6 KM
0 3 6 MILES
N
1 Aug 8: Jackson advances across the Rapidan from Orange and Gordonsville
2 Aug 9: Jackson defeats Banks at Cedar Mountain, but Federal reinforcements force him back across the Rapidan
3 Aug 20: After reuniting Jackson and Longstreet, Lee pursues Pope to the Rappahannock River
4 Aug 22: Moving up the Rappahannock, Confederates spar with Sigel's Federals at Freeman's Ford and Sulphur Springs
5 Aug 22: Stuart's Confederate cavalry raid Catlett's Station on the Federal supply line and capture Pope's headquarters and equipment
6 Aug 25: Jackson starts on a 54-mile flanking march around Pope's army
7 Aug 27: Jackson leaves Ewell's division at Bristoe Station to delay Pope's concentration on Jackson's forces
8 Aug 27: Jackson captures Federal supply depot at Manassas Junction and easily repulses an attack from Taylor's Federal brigade
9 Aug 27-28: Jackson conceals Taliaferro's and Ewell's divisions along an unfinished railroad, where A.P. Hill's division joins them by way of Centreville
Bull Run Mountains
To Fairfax
WHITE PLAINS
SALEM
MANASSAS GAP R.R.
SUDLEY SPRING
R.R. (UNFINISHED)
CENTREVILLE
WARRENTON PIKE
To Washington
JACKSON
TALIAFERRO
A.P. HILL
GROVETON
TAYLOR
Thoroughfare Gap
McDOWELL
HAYMARKET
GAINESVILLE
NEW MARKET
Carter's Mountain
NEW BALTIMORE
ORLEANS
MANASSAS JUNCTION
EWELL
KEARNY
RENO
SIGEL
Bristoe Station
OCCOQUAN RIVER
HEDGEMAN R.
CARTER'S R.
Hinson's Mill Ford
WARRENTON
ORANGE & ALEXANDRIA
Catlett's Station
HOOKER
To Dumfries
AMISSVILLE
Waterloo Bridge
J.E.B STUART
WARRENTON JUNCTION
SULPHUR SPRINGS
PORTER
Cedar Run
SPERRYVILLE
JEFFERSONTON
FAYETTE-VILLE
Freeman's Ford
BEALTON
POPE
LONGSTREET
HAZEL RIVER
Rappahannock Station
BANKS
Beverly's Ford
BRANDY STATION
Norman's Ford
RENO
LEE
CULPEPER COURT HOUSE
STEVENSBURG
Kelly's Ford
RAPPAHANNOCK RIVER
PORTER
Germanna Ford
RAPIDAN RIVER
FREDERICKSBURG
Madison C.H.
Cedar Mt.
Morton's Ford
Raccoon Ford
Somerville Ford
Clark's Mt.
RAPIDAN STATION
VERDIERSVILLE
A.P. HILL
WINDER
Orange C.H.
EWELL
JACKSON
GORDONSVILLE


Union troops repair the railroad wrecked by Jackson at Manassas Junction (right). During the raid track was torn from its bed, wagons burned, and the stockpile of Union supplies was strewn about the mass of tangled wreckage.

# Second Manassas: Phase 2 AUGUST 28 – 30 1862

From August 27 to the evening of the 28th, Pope's forces scoured the area looking for Jackson's elusive raiders. At 6 p.m. on the 28th, Jackson attacked a detached Federal brigade at Brawner's Farm. This fierce engagement of the Federal brigade – which later became known as the Iron Brigade – gave Pope a point on which to concentrate his forces and destroy Jackson. The Confederates prepared a strong defensive line, with their left at Sudley Springs and their right at Brawner's Farm. The infantry waited behind an unfinished railroad embankment, while the artillery took up a position behind them on Stony Ridge.

Pope attacked Jackson's line throughout August 29. Most of the assaults were small and uncoordinated. At 5 p.m. a Federal division captured a portion of the railroad, but was soon expelled by a vicious counter-attack from A.P. Hill's Confederates. Un-

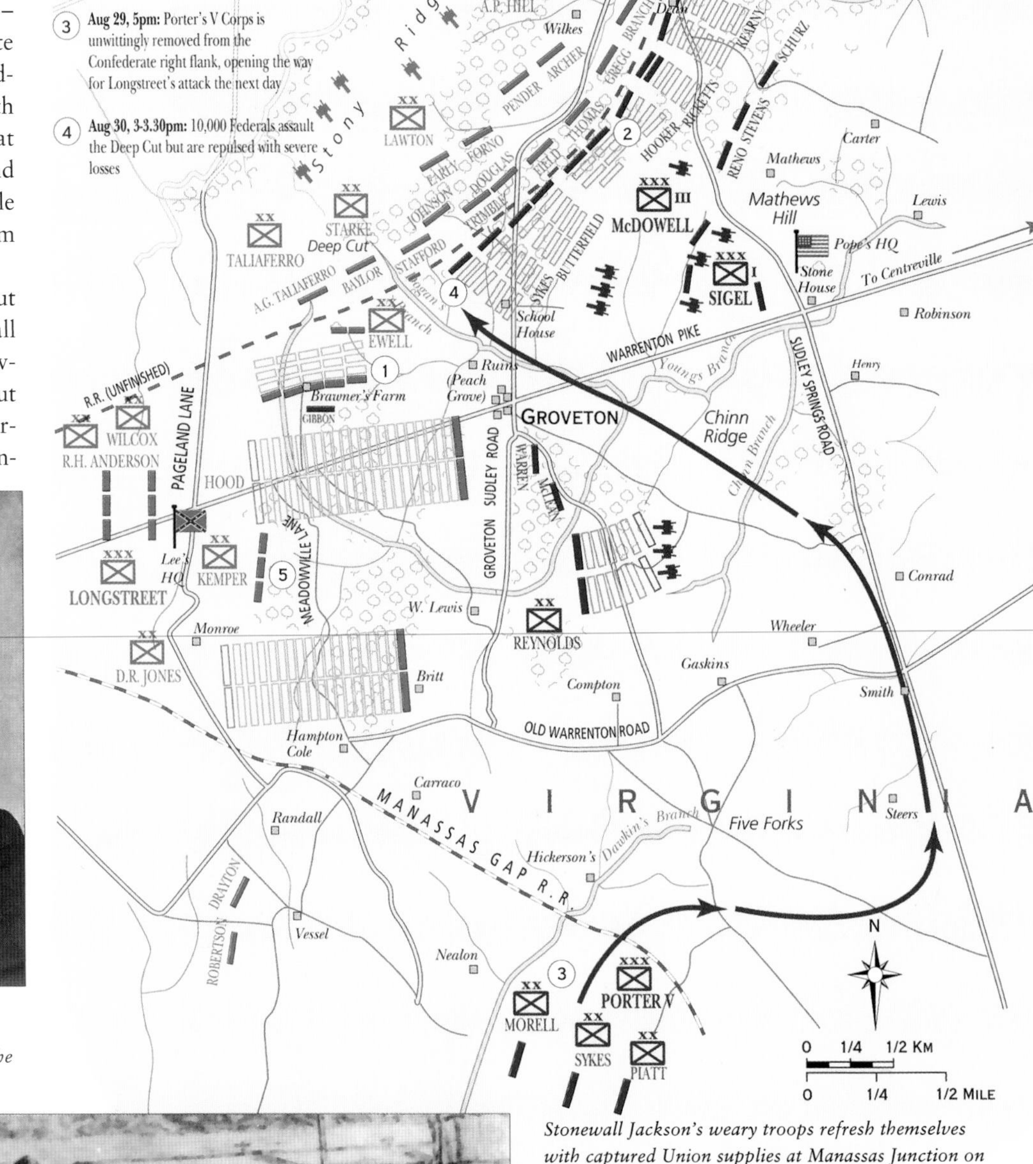
1 Aug 28: Jackson attacks Gibbon's lone Federal brigade; Confederate Generals Taliaferro and Ewell are both wounded in the action
2 Aug 29: Pope assaults different parts of Jackson's lines without success. Kearny's Federals break A.P. Hill's front momentarily at 5pm
3 Aug 29, 5pm: Porter's V Corps is unwittingly removed from the Confederate right flank, opening the way for Longstreet's attack the next day
4 Aug 30, 3-3.30pm: 10,000 Federals assault the Deep Cut but are repulsed with severe losses
5 Aug 30, 3.30pm: Longstreet strikes Pope's exposed left flank. As the Federal line collapses, Jackson joins Longstreet's attack
SUDLEY SPRINGS
Sudley Mansion
Sudley Mill
Benson
Sudley Ch.
Bull Run
JACKSON
A.P. HILL
Cushing
Wilkes
Stony Ridge
LAWTON
PENDER
ARCHER
GREGG
BRANCH
THOMAS
FIELD
FORNO
EARLY
DOUGLAS
JOHNSON
TRIMBLE
STAFFORD
BAYLOR
A.G. TALIAFERRO
KEARNY
SCHURZ
RENO
STEVENS
HOOKER
RICKETTS
BUTTERFIELD
SYKES
FITZ LEE
STARKE
Deep Cut
TALIAFERRO
McDOWELL
SIGEL
Mathews
Mathews Hill
Carter
Lewis
Pope's HQ
Stone House
To Centreville
Robinson
Henry
School House
EWELL
WARRENTON PIKE
Youngs Branch
R.R. (UNFINISHED)
Brawner's Farm
GIBBON
Ruins
(Peach Grove)
GROVETON
Chinn Ridge
Chinn Branch
SUDLEY SPRINGS ROAD
WILCOX
R.H. ANDERSON
PAGELAND LANE
HOOD
Lee's HQ
KEMPER
LONGSTREET
MEADOWVILLE LANE
GROVETON SUDLEY ROAD
WARREN
McLEAN
W. Lewis
REYNOLDS
Conrad
Monroe
D.R. JONES
Britt
Wheeler
Gaskins
Compton
Smith
Hampton Cole
OLD WARRENTON ROAD
Carraco
VIRGINIA
Randall
MANASSAS GAP R.R.
Dawkin's Branch
Five Forks
Steers
Hickerson's
DRAYTON
ROBERTSON
Vessel
Nealon
MORELL
PORTER V
SYKES
PIATT
N
0 1/4 1/2 KM
0 1/4 1/2 MILE

Gen. Longstreet (above), whose unexpected and devastating attack almost succeeded in annihilating the Federal army on August 30.

*Stonewall Jackson's weary troops refresh themselves with captured Union supplies at Manassas Junction on August 27 (left). Many of the Confederate soldiers were bootless and reduced to rags but their physical wretchedness did nothing to quell their optimism.*

> "Oh, the horrid scenes around us! Brains, fractured skulls, broken arms and legs, and the human form mangled in every conceivable and inconceivable manner... The fugitives will not be rallied, but, broken and dismayed, are pursued by our victorious troops until darkness closes around."
>
> *A Confederate private on the disorganised retreat of the Union troops on August 30.*

> "So long as the interests of our country are entrusted to a lying braggart like Pope, we have little reason to hope successfully to compete with an army led by Lee, Johnston and old 'Stonewall' Jackson."
>
> *A Union officer after Second Manassas.*

*The Union retreat became a disorganised flight, with the beaten troops ignoring the attempts of their officers to instil order and discipline (right). Recriminations were rife and Lincoln quickly decided to give command to Gen. George B. McClellan .*

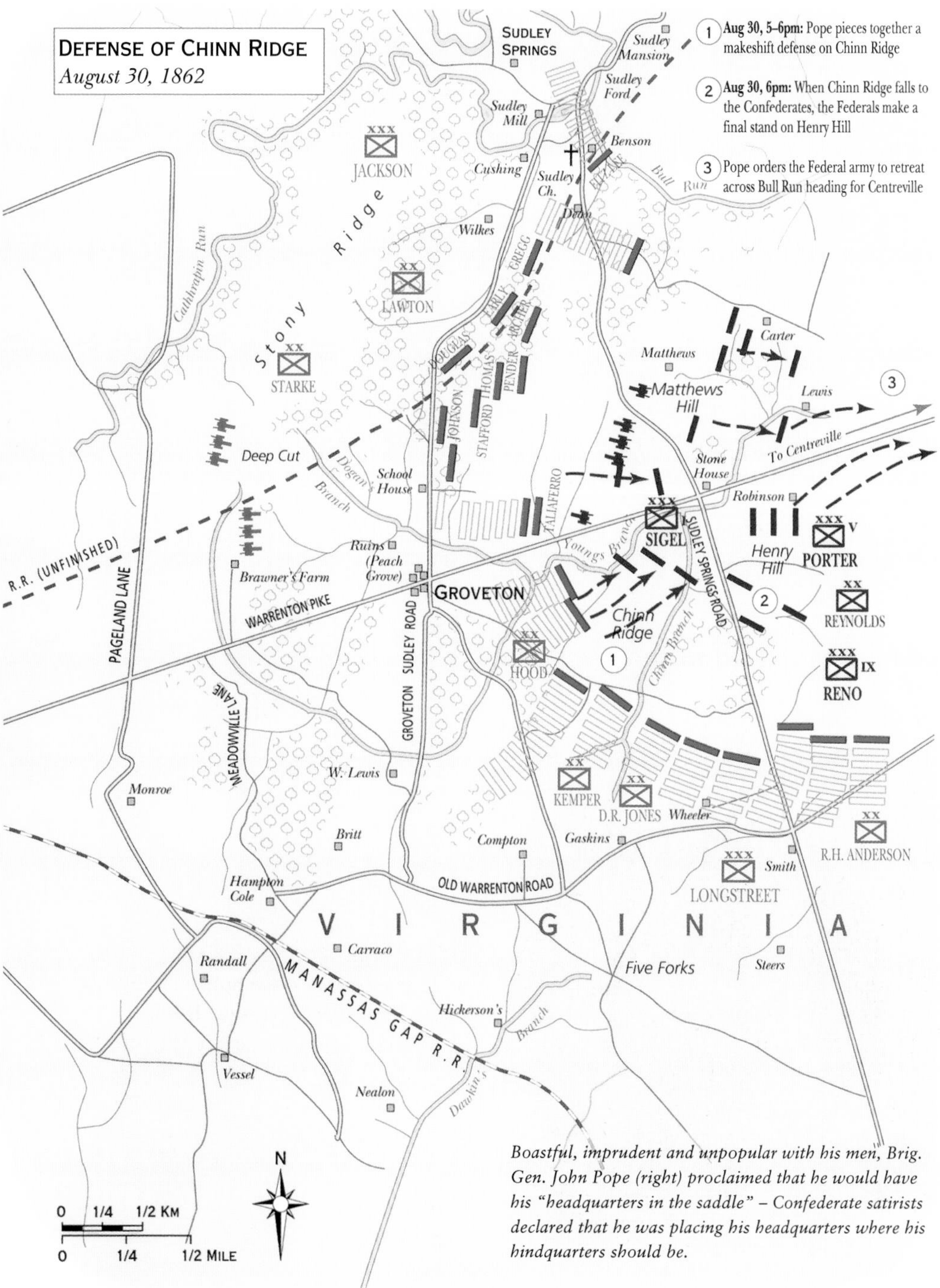
DEFENSE OF CHINN RIDGE
August 30, 1862
1 Aug 30, 5–6pm: Pope pieces together a makeshift defense on Chinn Ridge
2 Aug 30, 6pm: When Chinn Ridge falls to the Confederates, the Federals make a final stand on Henry Hill
3 Pope orders the Federal army to retreat across Bull Run heading for Centreville
SUDLEY SPRINGS
Sudley Mansion
Sudley Ford
Sudley Mill
Benson
Cushing
Sudley Ch.
Dean
Bull Run
JACKSON
Wilkes
Stony Ridge
LAWTON
STARKE
GREGG
FIELD
BRANCH
THOMAS
PENDER
ARCHER
DOUGLASS
JOHNSON
STAFFORD
Catharpin Run
Matthews
Carter
Matthews Hill
Lewis
To Centreville
Stone House
Robinson
Deep Cut
Dogan Branch
School House
TALIAFERRO
SIGEL
Youngs Branch
SUDLEY SPRINGS ROAD
Henry Hill
PORTER
V
REYNOLDS
RENO
IX
R.R. (UNFINISHED)
Ruins (Peach Grove)
Brawner's Farm
GROVETON
WARRENTON PIKE
PAGELAND LANE
GROVETON SUDLEY ROAD
Chinn Ridge
Chinn Branch
HOOD
MEADOWVILLE LANE
W. Lewis
Monroe
KEMPER
D.R. JONES
Wheeler
Britt
Compton
Gaskins
Smith
LONGSTREET
R.H. ANDERSON
Hampton Cole
OLD WARRENTON ROAD
VIRGINIA
Carraco
Randall
Five Forks
Steers
MANASSAS GAP R.R.
Hickerson's
Branch
Vessel
Nealon
Dawkin's Branch
N
0 1/4 1/2 KM
0 1/4 1/2 MILE

*Boastful, imprudent and unpopular with his men, Brig. Gen. John Pope (right) proclaimed that he would have his "headquarters in the saddle" – Confederate satirists declared that he was placing his headquarters where his hindquarters should be.*

known to Pope, Lee and Longstreet – with 30,000 reinforcements – had marched through Thoroughfare Gap and joined Jackson during the day, extending the Confederate right.

On August 30, at 3 pm, 10, 000 Federals assaulted Jackson's position at the Deep Cut, and were decimated. As the surviving Federals fell back, Lee unleashed Longstreet on Pope's left. Pope tried to stave off disaster with a makeshift line at Chinn Ridge. When it collapsed under relentless Confederate pressure, the Federals pieced together one final defense on Henry Hill. These lines gave Pope the time he needed to evacuate his army across Bull Run creek.

For the next two days, Lee's Confederates pursued the retreating Federals. On September 1, amid a torrential thunderstorm, Jackson harried a portion of Pope's column at Chantilly. The Federals escaped into the Washington defenses, but Lee saw the way open for a move into Maryland. The Confederates lost 9,197 men at Second Manassas and the Federals 16,054.

# Antietam Campaign: Phase 1 SEPTEMBER 4 – 20 1862

General Lee's Confederate army of 45,000 crossed the Potomac River into Maryland, riding a tide of victory. Plunging across White's Ferry on September 4, the Confederates entered Frederick three days later. Finding the Federal garrison standing defiantly across the Confederate line of communications at Harper's Ferry, Lee decided to divide his forces to capture them, and issued an order to that effect to his commanders. Jackson was to take three columns and converge on Harper's Ferry, while Longstreet would move to Boonsboro to guard against McClellan's cautious Federal approach from Washington.

By an unlucky chance, Lee's order to his commanders, Order 191, had been found by Federal soldiers at Frederick, wrapped around three cigars. It was sent to McClellan who jubilantly studied the divided position of Lee's forces. Surging ahead on September 14, McClellan's advance clashed with Longstreet's command at South Mountain. Federal I and IX Corps pushed the Confederates out of their stronghold at Turner's Gap. At the same time, VI Corps secured Crampton's Gap. As Lee's veterans fell back from South Mountain, Lee began retreating toward the Potomac. Hearing that Harper's Ferry would probably fall the next day, Lee boldly turned Longstreet's forces, and took a stand along Antietam Creek at Sharpsburg.

When Jackson's columns had closed around Harper's Ferry on September 13, they had bombarded the Federal defenses throughout the following day. The Federals finally surrendered on September 15. Jackson had captured 12,500 prisoners and 73 cannon at the cost of 286 casualties, making the battle of Harper's Ferry the largest Federal surrender of the Civil War. Jackson's success had not come a moment too soon. Following the surrender of Harper's Ferry, Jackson reunited his troops with Lee, leaving only A.P. Hill's division behind to organize the capitulation.

> "The rebels are wretchedly clad... The cavalry men are mostly barefooted, and the feet of the infantry are bound up in rags and pieces of rawhide."
>
> *A resident of Harper's Ferry observing the arrival of Jackson's army.*

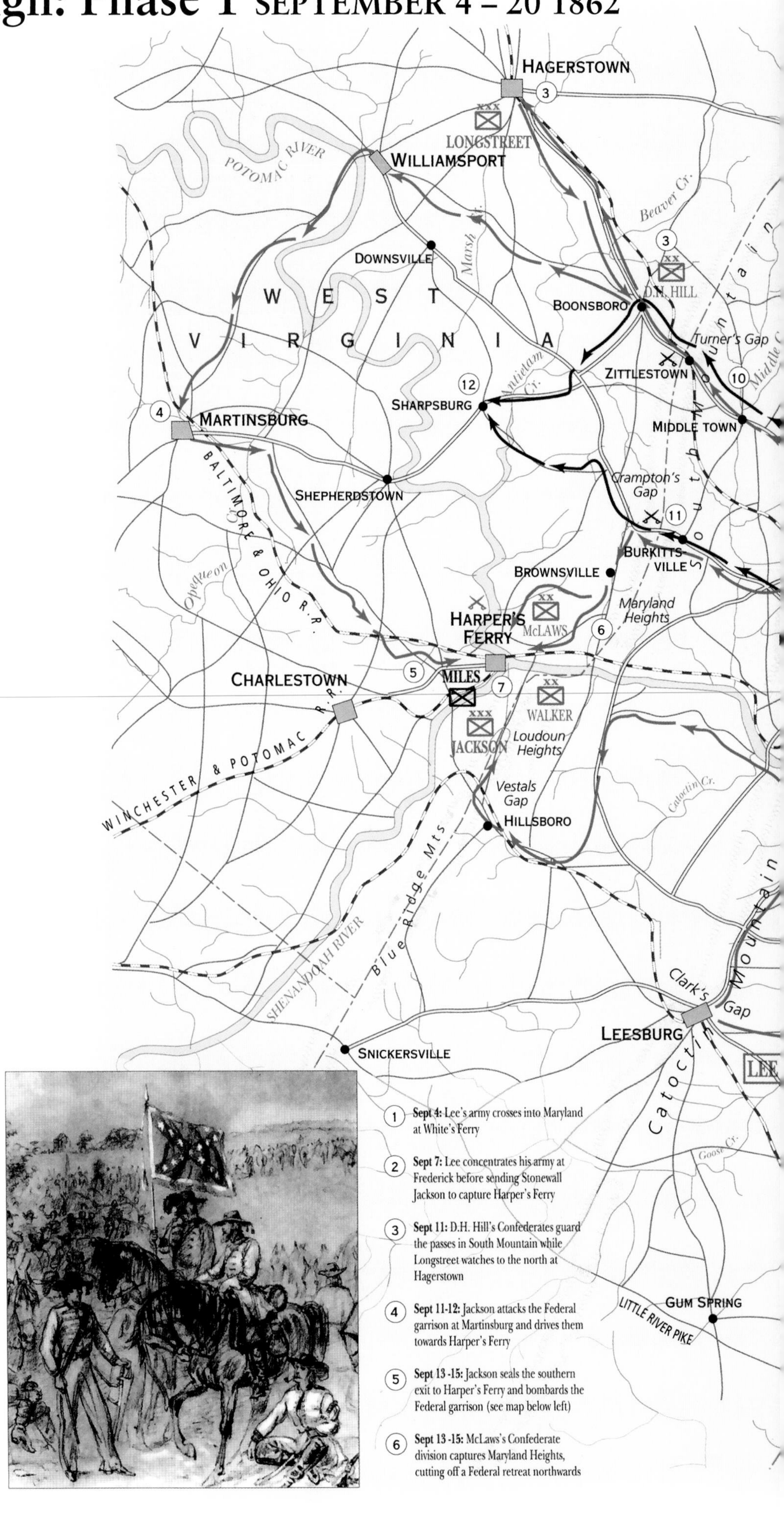
HAGERSTOWN
LONGSTREET
WILLIAMSPORT
POTOMAC RIVER
Beaver Cr.
DOWNSVILLE
Marsh Cr.
D.H. HILL
WEST VIRGINIA
BOONSBORO
Turner's Gap
ZITTLESTOWN
Antietam Cr.
SHARPSBURG
MARTINSBURG
MIDDLE TOWN
BALTIMORE & OHIO R.R.
SHEPHERDSTOWN
Crampton's Gap
Opequon Cr.
BURKITTSVILLE
BROWNSVILLE
South Mountain
Maryland Heights
HARPER'S FERRY
McLAWS
CHARLESTOWN
MILES
WALKER
JACKSON
Loudoun Heights
WINCHESTER & POTOMAC R.R.
Vestals Gap
HILLSBORO
Catoctin Cr.
Blue Ridge Mts
SHENANDOAH RIVER
Clark's Gap
Catoctin Mountain
LEESBURG
SNICKERSVILLE
LEE
Goose Cr.
GUM SPRING
LITTLE RIVER PIKE
1 Sept 4: Lee's army crosses into Maryland at White's Ferry
2 Sept 7: Lee concentrates his army at Frederick before sending Stonewall Jackson to capture Harper's Ferry
3 Sept 11: D.H. Hill's Confederates guard the passes in South Mountain while Longstreet watches to the north at Hagerstown
4 Sept 11-12: Jackson attacks the Federal garrison at Martinsburg and drives them towards Harper's Ferry
5 Sept 13-15: Jackson seals the southern exit to Harper's Ferry and bombards the Federal garrison (see map below left)
6 Sept 13-15: McLaws's Confederate division captures Maryland Heights, cutting off a Federal retreat northwards

*The 1st Virginia Cavalry halt during Lee's invasion of Maryland (right). Lee had hoped that the appearance of his troops would inspire enthusiasm in the state's residents but the reception was, in fact, only lukewarm.*

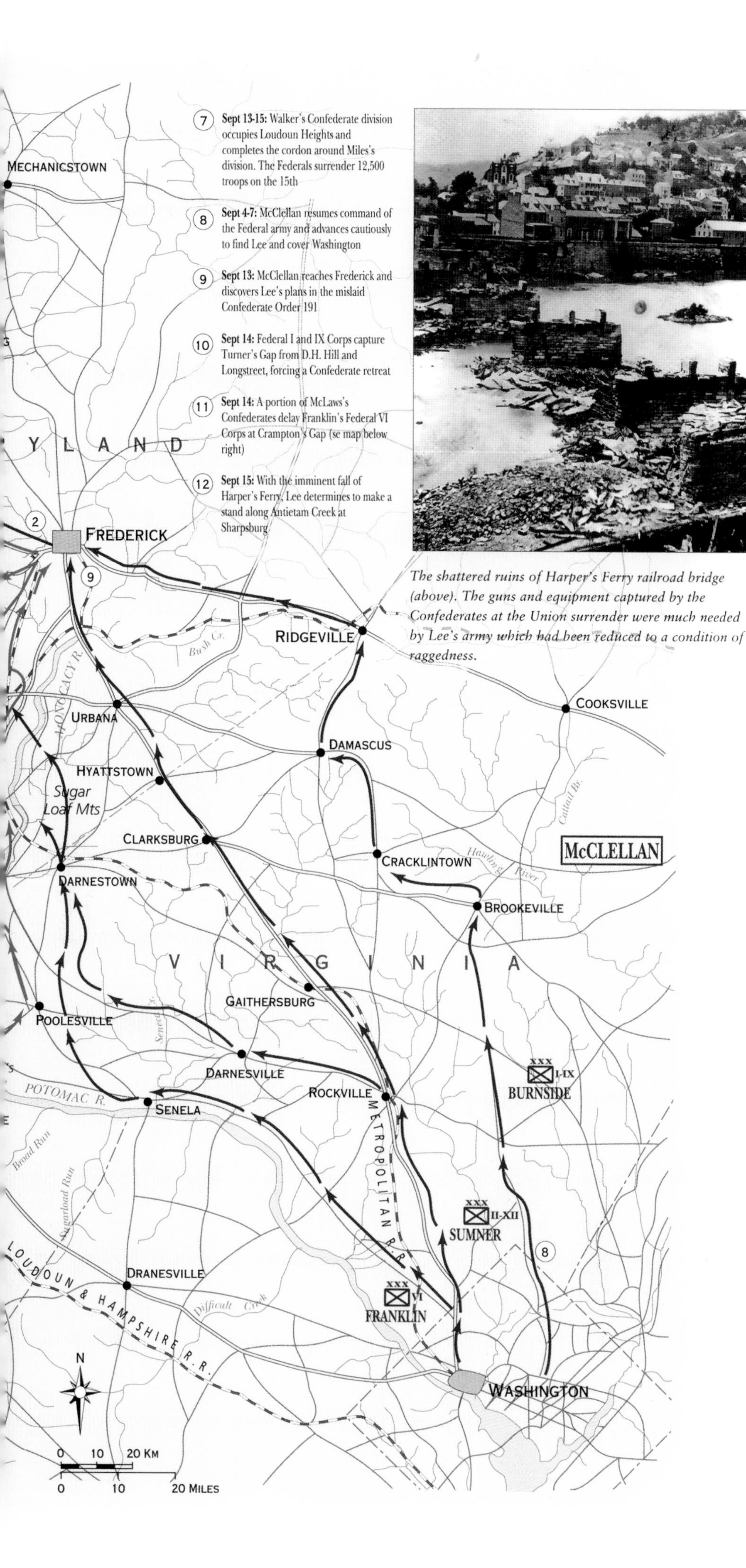

7 Sept 13-15: Walker's Confederate division occupies Loudoun Heights and completes the cordon around Miles's division. The Federals surrender 12,500 troops on the 15th
8 Sept 4-7: McClellan resumes command of the Federal army and advances cautiously to find Lee and cover Washington
9 Sept 13: McClellan reaches Frederick and discovers Lee's plans in the mislaid Confederate Order 191
10 Sept 14: Federal I and IX Corps capture Turner's Gap from D.H. Hill and Longstreet, forcing a Confederate retreat
11 Sept 14: A portion of McLaws's Confederates delay Franklin's Federal VI Corps at Crampton's Gap (se map below right)
12 Sept 15: With the imminent fall of Harper's Ferry, Lee determines to make a stand along Antietam Creek at Sharpsburg
MECHANICSTOWN
Y L A N D
FREDERICK
RIDGEVILLE
Bush Cr.
MONOCACY R.
COOKSVILLE
URBANA
DAMASCUS
HYATTSTOWN
Sugar Loaf Mts
CLARKSBURG
CRACKLINTOWN
Hawling River
Cattail Br.
McCLELLAN
DARNESTOWN
BROOKEVILLE
V I R G I N I A
GAITHERSBURG
POOLESVILLE
DARNESVILLE
ROCKVILLE
POTOMAC R.
SENELA
BURNSIDE
I-IX
METROPOLITAN R.R.
Broad Run
Sugarland Run
SUMNER
II-XII
LOUDOUN & HAMPSHIRE R.R.
DRANESVILLE
Difficult Creek
FRANKLIN
VI
WASHINGTON
N
0 10 20 KM
0 10 20 MILES


*The shattered ruins of Harper's Ferry railroad bridge (above). The guns and equipment captured by the Confederates at the Union surrender were much needed by Lee's army which had been reduced to a condition of raggedness.*

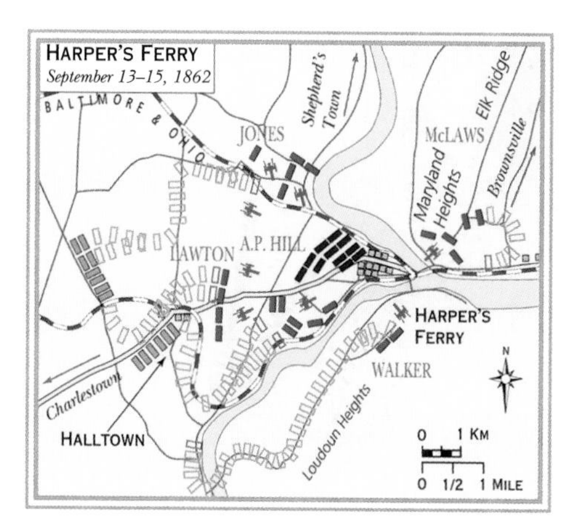

HARPER'S FERRY
September 13–15, 1862
BALTIMORE & OHIO
Shepherd's Town
Elk Ridge
JONES
McLAWS
Maryland Heights
Brownsville
LAWTON
A.P. HILL
HARPER'S FERRY
WALKER
Charlestown
HALLTOWN
Loudoun Heights
0 1 KM
0 1/2 1 MILE


*The Confederates under Stonewall Jackson, Walker and McLaws besieged the 12,500-strong Federal garrison at Harper's Ferry (above), bombarding the defenses during the 14th. The garrison surrendered the following day, thus opening a Confederate supply line through the Shenandoah Valley.*

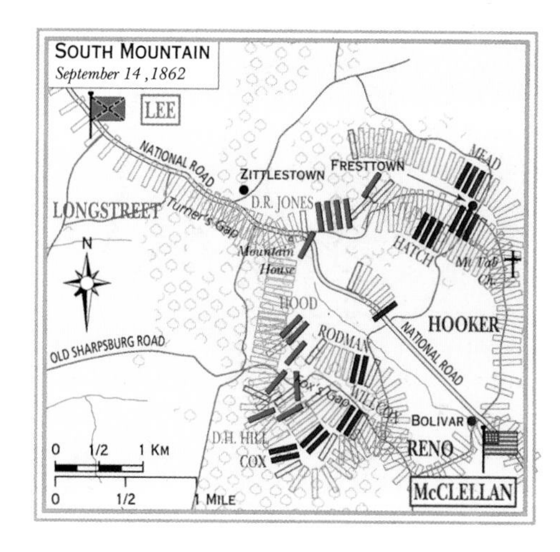

SOUTH MOUNTAIN
September 14 ,1862
LEE
NATIONAL ROAD
ZITTLESTOWN
FRESTTOWN
MEAD
LONGSTREET
Turner's Gap
D.R. JONES
Mountain House
HATCH
Mt Tab Ch.
HOOD
HOOKER
OLD SHARPSBURG ROAD
RODMAN
NATIONAL ROAD
WILLCOX
BOLIVAR
D.H. HILL
RENO
COX
0 1/2 1 KM
0 1/2 1 MILE
McCLELLAN


*Lee, his forces scattered, ordered his remaining troops to defend the passes through South Mountain (above). Begining before dawn, the Federal attack was advancing toward Turner's Gap by midday, but was stopped by the Confederates. McClellan now threw further forces against them, which in enveloping the Confederate flanks. Darkness ended the fighting and the Confederates withdrew.*

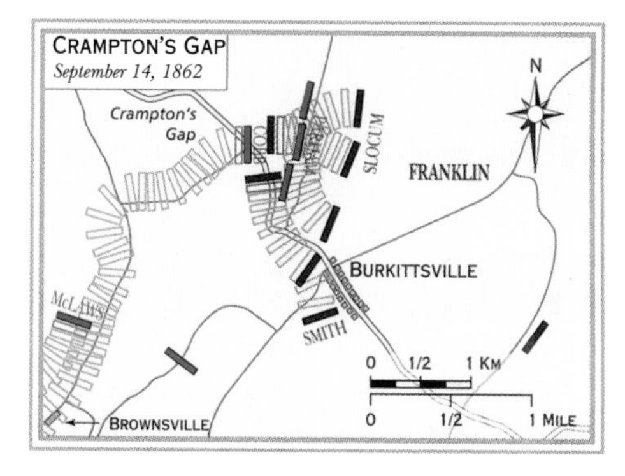

CRAMPTON'S GAP
September 14, 1862
Crampton's Gap
SLOCUM
FRANKLIN
BURKITTSVILLE
McLAWS
SMITH
0 1/2 1 KM
0 1/2 1 MILE
BROWNSVILLE


*In a well-conceived attempt to relieve the besieged Federal garrison at Harper's Ferry, the left wing of McClellan's army, commanded by Maj. Gen. William B. Franklin, easily defeated a small Confederate force at Crampton's Gap, 6 miles to the south of South Mountain (above).*

# Battle of Antietam SEPTEMBER 17 1862

With the majority of Stonewall Jackson's command arriving from Harper's Ferry, Lee posted his 38,000 troops on a four mile line along Antietam Creek, near Sharpsburg. McClellan massed his army of 75,000 to attack on September 17, a day destined to be the bloodiest of the war.

McClellan unleashed his attack at dawn, striking Jackson's position on the Confederate left. Fighting raged back and forth around the West Woods, the Dunker Church, and David Miller's cornfield. Wave after wave of Federals hammered the Confederates, only to be driven back by desperate counterattacks. As the fighting diminished on the Confederate left, it renewed in intensity along the center.

Repeated Federal assaults eventually captured the Confederate main line of defense, which ran along a sunken road. The carnage was so terrible that the sunken road became known as Bloody Lane. McClellan had broken the Confederate line, but he refused to commit his reserves to exploit it. Instead, the action shifted south to the area of Burnside's Bridge across Antietam Creek. Major General Ambrose E. Burnside's Federal IX Corps struggled for several hours against Confederate fire to capture the bridge and cross the creek before it could advance on Sharpsburg.

Pushing Longstreet's Confederates back to the edge of town, the Federals were suddenly assailed on their left by A.P. Hill. Marching 17 miles from Harper's Ferry, Hill's division arrived just in time to paralyze Burnside's drive and end the fighting for the day.

Throughout September 18, the two armies glared at each other in restless stalemate, but after dark the Confederate army recrossed the Potomac into Virginia. Belatedly, McClellan followed, but on September 19 his vanguard was roughy checked at Shepherdstown.

The whirlwind of battle along the Antietam claimed 12,401 dead and wounded Federals, and 10,318 Confederates. The Confederacy had lost its best chance to rally Maryland to its banner and gain European recognition. Upon Lee's retreat, Abraham Lincoln issued the preliminary Emancipation Proclamation, changing the original aims of the war – to preserve the Union – into a crusade to free the slaves.

*Confederate dead near David Miller's cornfield on September 17. Initially pushed back by Gen. Hooker's assault, the rebels were rallied by Jackson and successfully counterattacked. Hooker later commented that "it was never my fortune to witness a more bloody, dismal battlefield."*

"We are in the midst of the most terrible battle of the war – perhaps of history... it will be either a great defeat or a most glorious victory. I think and hope that God will give us a glorious victory."

*Gen. McClellan to Gen. Halleck, September 17, 1862*

1 **Sept 17, 6 am:** Hooker's Federal I Corps begins the attack but his left bogs down under artillery fire from Nicodemus Hill

2 **7 am:** Hood's Confederates counterattack and stop I Corps' advance at the Miller cornfield

3 **7.30–9 am:** Mansfield's XII Corps attacks to the Dunker Church but fresh Confederate reinforcements drive them back

4 **10 am:** Sedgwick's division of Sumner's II Corps attacks into the West Woods but is flanked and repulsed with heavy losses

Cox
LEE
STUART
WALKER
West
JACKSON
SEMMES
RANSOM
MANNING
ARMISTEAD
COOKE
LANDING ROAD
S.D. Piper
SHARPSBURG
Lee's H.Q.
To Sheperdstown
TOOMBS
9
A.P. HILL


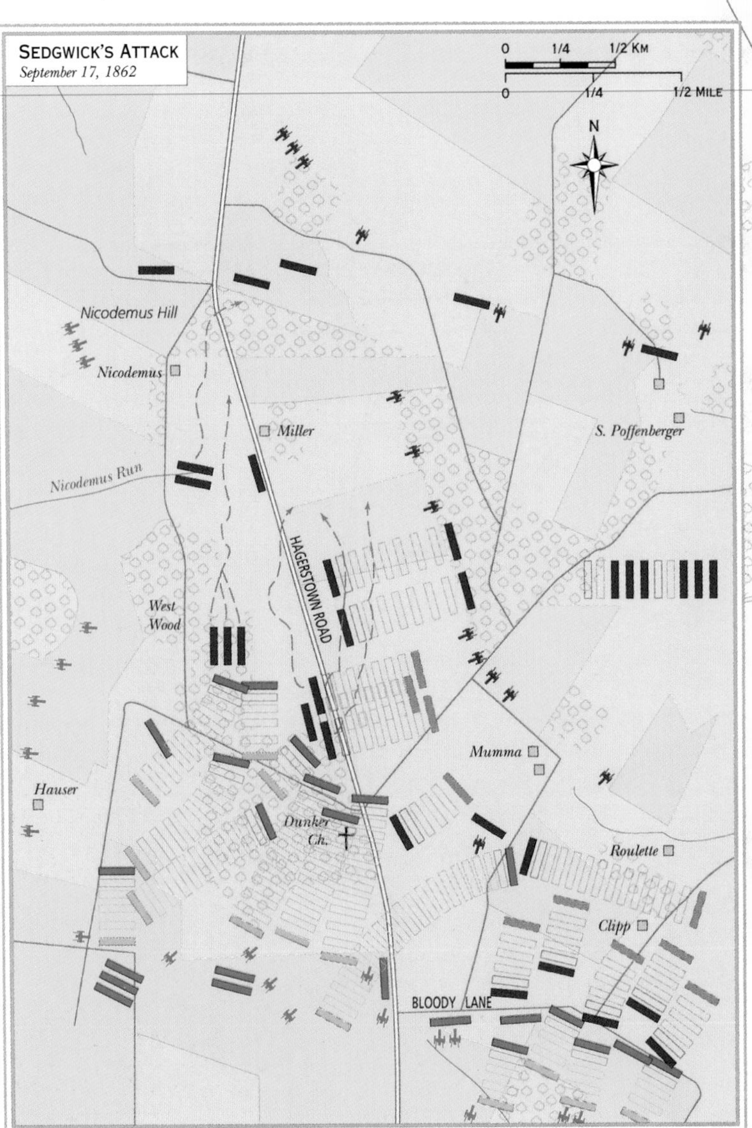

SEDGWICK'S ATTACK
September 17, 1862
0 1/4 1/2 KM
0 1/4 1/2 MILE
N
Nicodemus Hill
Nicodemus
Miller
S. Poffenberger
Nicodemus Run
HAGERSTOWN ROAD
West Wood
Mumma
Hauser
Dunker Ch.
Roulette
Clipp
BLOODY LANE


*Although initially successful in their attack on the Confederate line at the West Woods, Sedgwick's division was all but annihilated in a suprise counter-attack launched by two Confederate divisions (right).*

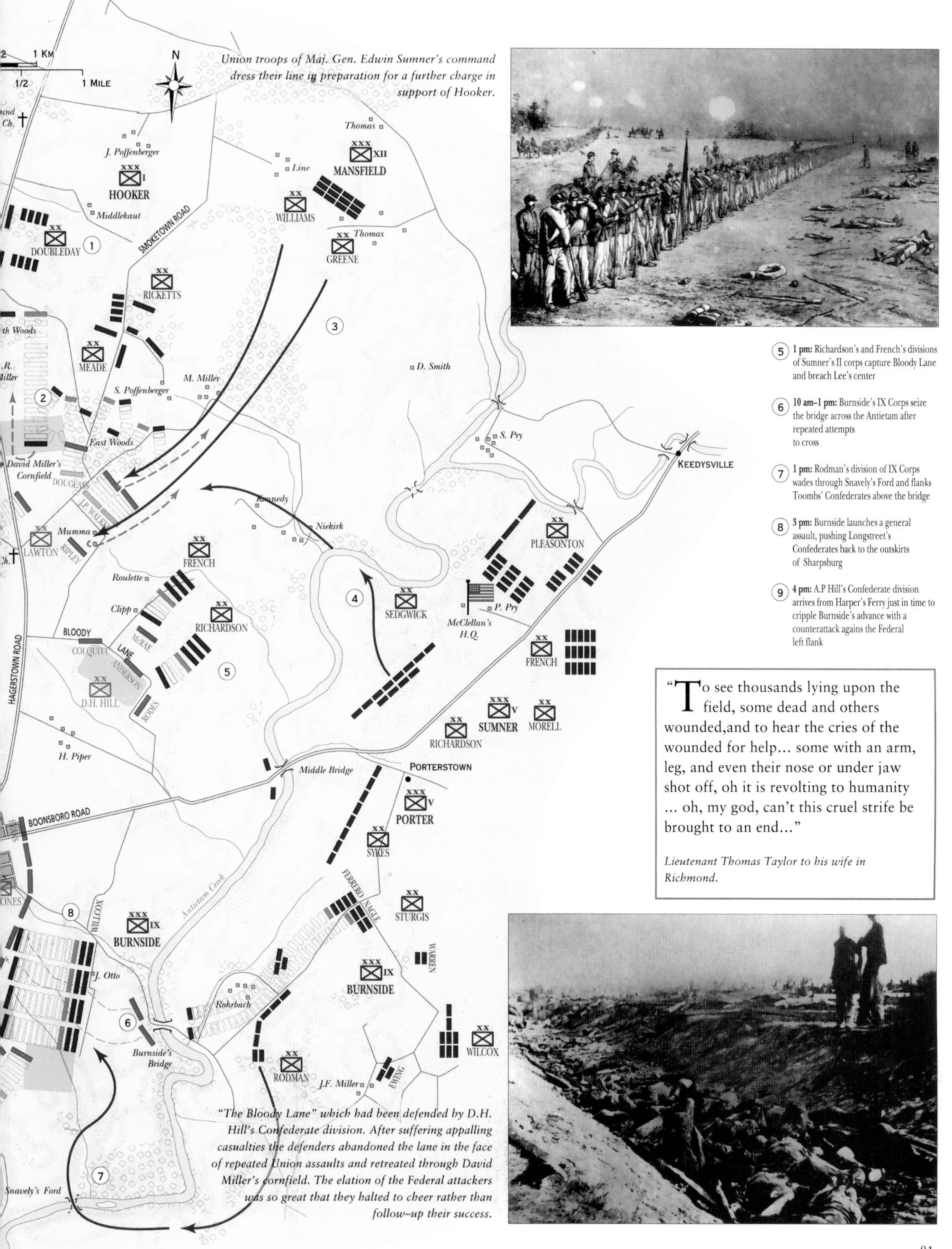

1 KM
1/2
1 MILE
N
HOOKER
MANSFIELD
WILLIAMS
GREENE
DOUBLEDAY
RICKETTS
MEADE
J. Poffenberger
Middlekauf
Line
Thomas
SMOKETOWN ROAD
S. Poffenberger
M. Miller
East Woods
David Miller's Cornfield
DOUGLASS
J.R. WALKER
Mumma
LAWTON
RIPLEY
Kennedy
Niekirk
D. Smith
S. Pry
KEEDYSVILLE
PLEASONTON
FRENCH
RICHARDSON
SEDGWICK
Roulette
Clipp
BLOODY LANE
COLQUITT
McRAE
ANDERSON
RODES
D.H. HILL
HAGERSTOWN ROAD
H. Piper
P. Pry
McClellan's H.Q.
SUMNER
MORELL
Middle Bridge
PORTERSTOWN
PORTER
SYKES
BOONSBORO ROAD
Antietam Creek
WILCOX
BURNSIDE
FERRERO
NAGLE
STURGIS
WARREN
J. Otto
Rohrbach
Burnside's Bridge
RODMAN
J.F. Miller
EWING
Snavely's Ford


*Union troops of Maj. Gen. Edwin Sumner's command dress their line in preparation for a further charge in support of Hooker.*

*"The Bloody Lane" which had been defended by D.H. Hill's Confederate division. After suffering appalling casualties the defenders abandoned the lane in the face of repeated Union assaults and retreated through David Miller's cornfield. The elation of the Federal attackers was so great that they halted to cheer rather than follow-up their success.*

5. **1 pm:** Richardson's and French's divisions of Sumner's II corps capture Bloody Lane and breach Lee's center
6. **10 am–1 pm:** Burnside's IX Corps seize the bridge across the Antietam after repeated attempts to cross
7. **1 pm:** Rodman's division of IX Corps wades through Snavely's Ford and flanks Toombs' Confederates above the bridge
8. **3 pm:** Burnside launches a general assault, pushing Longstreet's Confederates back to the outskirts of Sharpsburg
9. **4 pm:** A.P Hill's Confederate division arrives from Harper's Ferry just in time to cripple Burnside's advance with a counterattack agains the Federal left flank

"To see thousands lying upon the field, some dead and others wounded,and to hear the cries of the wounded for help... some with an arm, leg, and even their nose or under jaw shot off, oh it is revolting to humanity ... oh, my god, can't this cruel strife be brought to an end..."

*Lieutenant Thomas Taylor to his wife in Richmond.*

# Fredericksburg DECEMBER 13 1862

Embarrassed by General McClellan's repeated defeats and apparent lack of commitment in prosecuting the war, Lincoln replaced him on November 7 with General Ambrose Burnside. Burnside launched a winter campaign against the Confederate capital, Richmond, by way of Fredericksburg, a strategically important town on the Rappahannock River. The Federal Army of the Potomac, 115,000-strong, raced to Fredericksburg, arriving on November 17. There were only a few thousand Confederates on hand to challenge them, yet the Federal advance ground to a halt on the eastern bank of the Rappahannock, opposite the city. Burnside's campaign was delayed for over a week when material he had ordered for pontoon bridges failed to arrive. Disappointed by the delay, Burnside marked time for a further two weeks. Meanwhile, Lee took advantage of the stalled Federal drive to concentrate and entrench his Army of Northern Virginia, some 78,000-strong, on the high ground behind Fredericksburg.

With the arrival of the pontoons, Burnside crossed the river on December 11, despite fierce fire from Confederate snipers concealed in buildings along the city's river front. When the Confederates withdrew, Federal soldiers looted the town, from which the inhabitants had been evacuated. By December 13, Burnside was prepared to launch a two-pronged attack to drive Lee's forces from an imposing set of hills just outside Fredericksburg.

The main assault struck south of the city. Misunderstandings and bungled leadership on the part of the commander of the Federal left, Major General William B. Franklin, limited the attacking force to two small divisions – Major General George G. Meade to lead; Major General John Gibbon in support. Meade's troops broke through an unguarded gap in the Confederate lines, but Jackson's men expelled the unsupported Federals, inflicting heavy losses. Burnside launched his second attack from Fredericksburg against the Confederate left on Marye's Heights. Wave after wave of Federal attackers were mown down by Confederate troops firing from an unas-sailable position in a sunken road protected by a stone wall. Over the course of the afternoon, no fewer than fourteen successive Federal brigades charged the wall of Confederate fire. Not a single Federal soldier reached Longstreet's line.

On December 15, Burnside ordered his beaten army back across the Rappahannock.

"Once more unsuccessful, and only a bloody record to show our men were brave... [The army] has strong limbs to march and meet the foe, stout arms to strike heavy blows, brave hearts to dare – but the brains, the brains! Have we no brains to use the arms and limbs and eager hearts with cunning? Perhaps Old Abe has some funny story to tell appropriate to the occasion."

*Newspaper reporter describing the Union attacks on Gen. Longstreet's corps.*

*Unable to ford the Rappahannock River near the town, Burnside was forced – despite fierce sniper fire from the opposite bank – to construct pontoon bridges (above).*

*Three bridges originally straddled the Rappahanhannock; by the time Burnside advanced, however, all had been destroyed. In an attempt to aid their reconstruction, Union artillery began a barrage of the town (below).*

1. **Dec 11, am:** Confederate sniper fire delays completion of Federal pontoon bridges
2. **Dec 11:** Federal artillery attempt to drive away Confederate sharpshooters, but without success
3. **Dec 11, pm:** Federal infantry cross river in boats and force Confederates out of Fredericksburg
4. **Dec 13, 10-11am:** Pelham prevents main Federal attack with only one gun on Federals' flank
5. **Dec 13, noon:** Following a Federal bombardment, Meade attacks and is repulsed

RAPPAHANNOCK RIVER
R.H. ANDERS

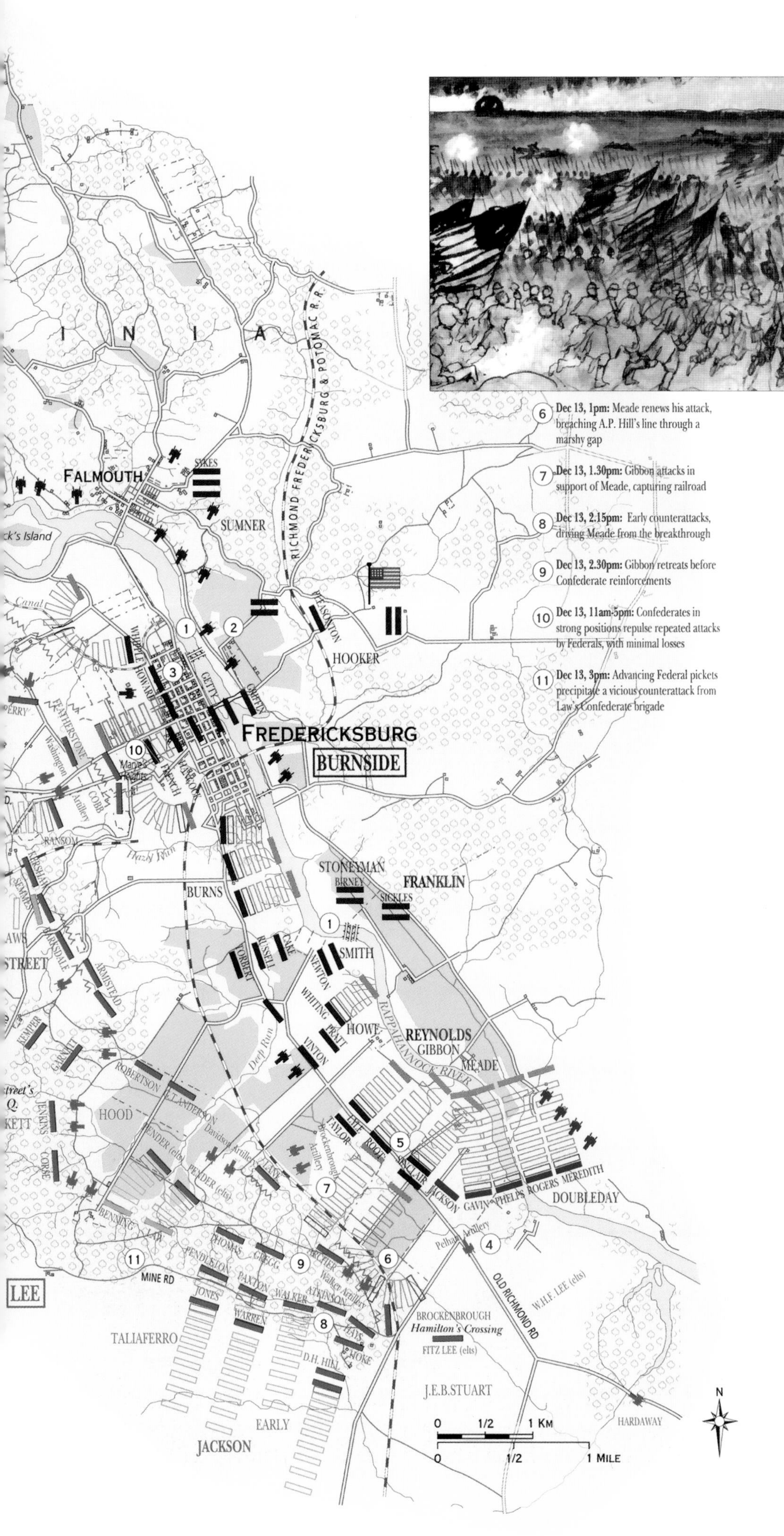

Dec 13, 1pm: Meade renews his attack, breaching A.P. Hill's line through a marshy gap
Dec 13, 1.30pm: Gibbon attacks in support of Meade, capturing railroad
Dec 13, 2.15pm: Early counterattacks, driving Meade from the breakthrough
Dec 13, 2.30pm: Gibbon retreats before Confederate reinforcements
Dec 13, 11am-5pm: Confederates in strong positions repulse repeated attacks by Federals, with minimal losses
Dec 13, 3pm: Advancing Federal pickets precipitate a vicious counterattack from Law's Confederate brigade
FALMOUTH
FREDERICKSBURG
BURNSIDE
SUMNER
SYKES
HOOKER
PLEASONTON
RICHMOND FREDERICKSBURG & POTOMAC R.R.
FRANKLIN
STONEMAN
BIRNEY
SICKLES
SMITH
HOWE
REYNOLDS
GIBBON
MEADE
DOUBLEDAY
RAPPAHANNOCK RIVER
OLD RICHMOND RD
MINE RD
Hamilton's Crossing
BROCKENBROUGH
FITZ LEE (elts)
W.H.F. LEE (elts)
J.E.B.STUART
HARDAWAY
TALIAFERRO
EARLY
JACKSON
D.H. HILL
HOOD
LEE
Deep Run
Hazel Run
Canal
0 1/2 1 KM
0 1/2 1 MILE
N


*Obstinately refusing to learn from the lessons of previous frontal assaults, in the late afternoon of December 13 Burnside ordered a further attack with exactly the same results (above). The Union troops were eventually forced to withdraw, having sustained 13,000 casualties.*

The Union had lost 13,000 soldiers in a battle in which the dreadful carnage was matched only by its futility. Federal morale plummeted, and Burnside was swiftly relieved of his command. By contrast, the morale of the Confederacy reached a peak. Their casualties had been considerably lighter than the Union's, totalling only 5,000. Lee's substantial victory at Fredericksburg, won with relative ease, increased the already buoyant confidence of the Army of Northern Virginia, which led subsequently to the invasion of the North the following summer.

*The delayed Union attack enabled Gen. Longstreet (above) to take up a strong defensive position on Marye's Heights behind Fredericksburg. The stone wall at the base of the Heights proved an insurmountable obstacle, and Burnside's troops faced devastating volleys from the Confederate artillery and infantry.*

# Confederate Cavalry Raids JULY 4 – JULY 27 1862

To disrupt General Buell's offensive to seize Chattanooga, Confederate Colonels John Hunt Morgan and Nathan Bedford Forrest led slashing cavalry raids into central Kentucky and middle Tennessee to cut Federal supply lines, and throw Buell's operation into confusion.

Morgan left Knoxville on July 4, with 876 officers and men in the newly formed Second Kentucky Cavalry. Led by a vanguard of 60 scouts, the column rode north for the bluegrass state of Kentucky. For 24 days, Morgan's Kentuckians raided throughout their home state, winning engagements at Tomkinsville (July 9); Lebanon (July 12), and Cynthiana (July 17), before returning to Livingston, Tennessee on July 28. The raid covered over 1,000 miles, during which Morgan captured – and paroled – 1,200 prisoners, at a loss of fewer than 90 raiders. He also destroyed railroad and telegraph lines, bridges and supply depots. Morgan's success damaged northern morale and resulted in sharp criticism of Buell.

On August 12, Morgan and his Kentuckians struck again, this time at Gallatin above Nashville, where they completely destroyed an 800-foot tunnel on the Louisville & Nashville Railroad. This action successfully severed Buell's invading army from its main supply base at Louisville, Kentucky.

Concurrent with Morgan's Kentucky raid, Forrest left Chattanooga on July 9, and proceeded to devastate Federal occupation forces in middle Tennessee. At 4.30 am on July 13 – his 41st birthday – having ridden 50 miles in just over fourteen hours, Forrest thundered into Murfreesboro ahead of his 1,400 screaming troopers and surprised and captured General Thomas T. Crittenden and his 1,040-man Federal garrison. After securing Union supplies estimated at a million dollars, Forrest withdrew to McMinnville. On the 18th, "old Bedford" had his men back in the saddle and marching to Lebanon where they forced the Union garrison to abandon the town. Following the retreating Federals northward to Nashville, Forrest destroyed two bridges below the city on the Nashville & Chattanooga Railroad, putting the line out of operation for a week; forced the moblization of two Federal divisions to protect the railroad and destroy the Confederate raiders; and panicked Governor Andrew Johnson into thinking Nashville faced imminent capture.

The combined raids of Morgan and Forrest immobilized Buell's 40,000-man Union army; stalled the campaign to seize Chattanooga; and illustrated the Confederacy's strategic advantage of fighting an aggressive defensive in their own territory.

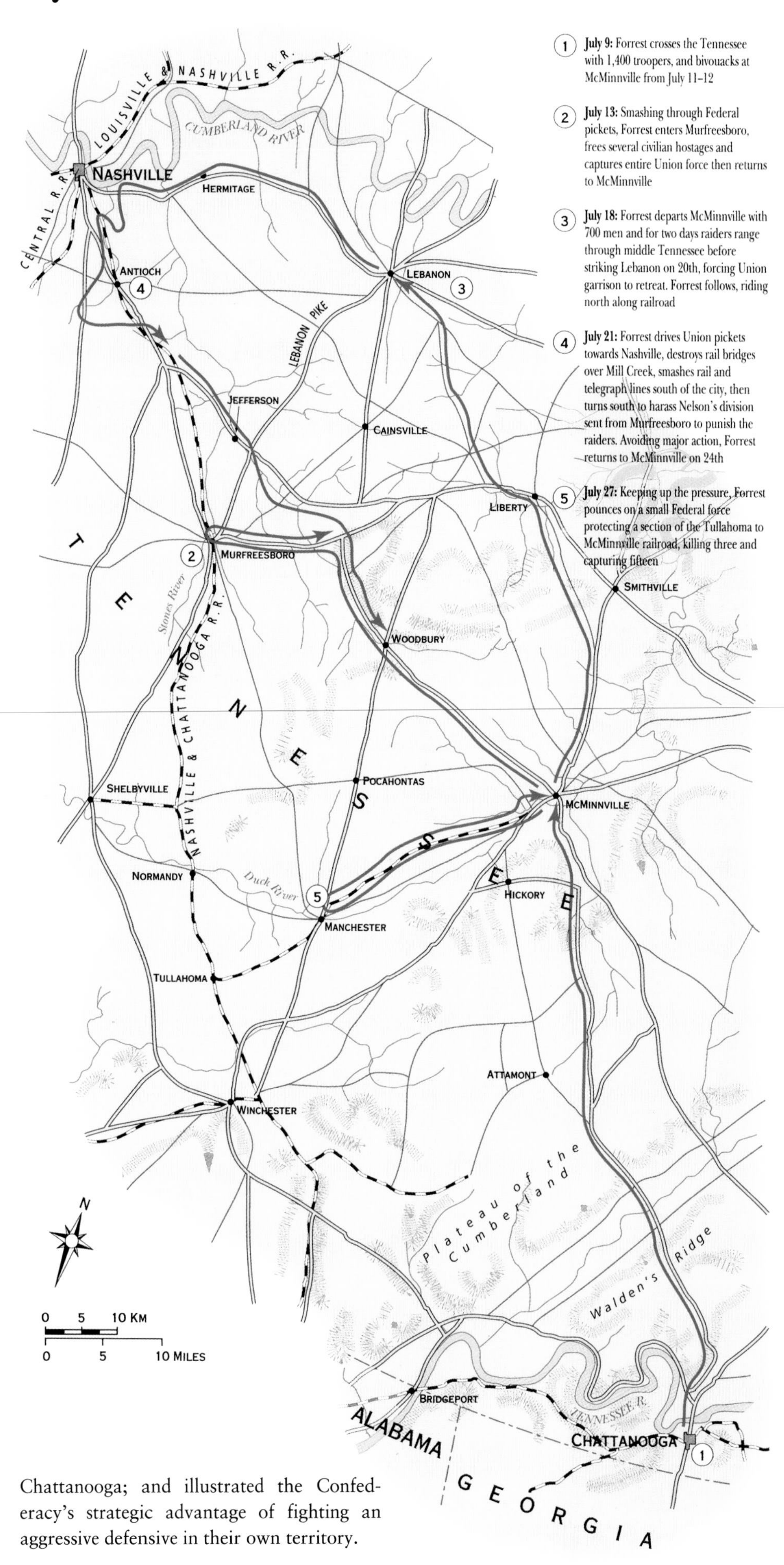
1 July 9: Forrest crosses the Tennessee with 1,400 troopers, and bivouacks at McMinnville from July 11–12
2 July 13: Smashing through Federal pickets, Forrest enters Murfreesboro, frees several civilian hostages and captures entire Union force then returns to McMinnville
3 July 18: Forrest departs McMinnville with 700 men and for two days raiders range through middle Tennessee before striking Lebanon on 20th, forcing Union garrison to retreat. Forrest follows, riding north along railroad
4 July 21: Forrest drives Union pickets towards Nashville, destroys rail bridges over Mill Creek, smashes rail and telegraph lines south of the city, then turns south to harass Nelson's division sent from Murfreesboro to punish the raiders. Avoiding major action, Forrest returns to McMinnville on 24th
5 July 27: Keeping up the pressure, Forrest pounces on a small Federal force protecting a section of the Tullahoma to McMinnville railroad, killing three and capturing fifteen
LOUISVILLE & NASHVILLE R.R.
CUMBERLAND RIVER
NASHVILLE
HERMITAGE
CENTRAL R.R.
ANTIOCH
LEBANON
LEBANON PIKE
JEFFERSON
CAINSVILLE
LIBERTY
MURFREESBORO
SMITHVILLE
Stones River
NASHVILLE & CHATTANOOGA R.R.
WOODBURY
TENNESSEE
POCAHONTAS
SHELBYVILLE
MCMINNVILLE
NORMANDY
Duck River
HICKORY
MANCHESTER
TULLAHOMA
ATTAMONT
WINCHESTER
Plateau of the Cumberland
N
Walden's Ridge
0 5 10 KM
0 5 10 MILES
BRIDGEPORT
TENNESSEE R.
ALABAMA
CHATTANOOGA
GEORGIA

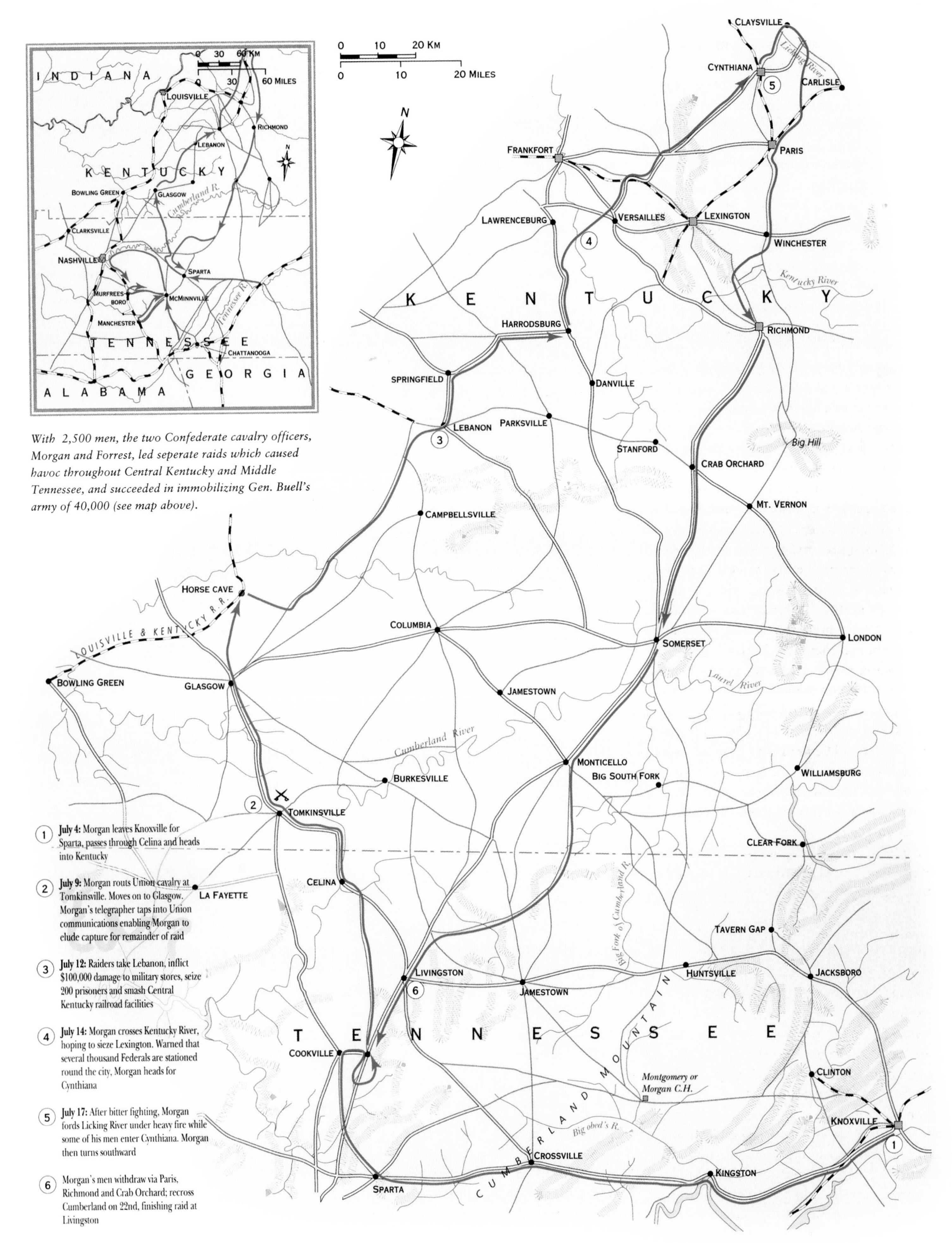

0 10 20 KM
0 10 20 MILES
N
INDIANA
KENTUCKY
TENNESSEE
GEORGIA
ALABAMA
LOUISVILLE
RICHMOND
LEBANON
BOWLING GREEN
GLASGOW
Cumberland R.
CLARKSVILLE
NASHVILLE
SPARTA
MURFREESBORO
MCMINNVILLE
MANCHESTER
Tennessee R.
CHATTANOOGA
0 30 60 KM
0 30 60 MILES
CLAYSVILLE
Licking River
CYNTHIANA
CARLISLE
PARIS
FRANKFORT
LAWRENCEBURG
VERSAILLES
LEXINGTON
WINCHESTER
Kentucky River
K E N T U C K Y
HARRODSBURG
RICHMOND
SPRINGFIELD
DANVILLE
LEBANON
PARKSVILLE
STANFORD
Big Hill
CRAB ORCHARD
MT. VERNON
CAMPBELLSVILLE
HORSE CAVE
LOUISVILLE & KENTUCKY R.R.
COLUMBIA
SOMERSET
LONDON
BOWLING GREEN
GLASGOW
Laurel River
JAMESTOWN
Cumberland River
MONTICELLO
BIG SOUTH FORK
WILLIAMSBURG
BURKESVILLE
TOMKINSVILLE
CLEAR FORK
CELINA
LA FAYETTE
Big Fork of Cumberland R.
TAVERN GAP
LIVINGSTON
HUNTSVILLE
JACKSBORO
JAMESTOWN
T E N N E S S E E
COOKVILLE
CUMBERLAND MOUNTAIN
CLINTON
Montgomery or Morgan C.H.
KNOXVILLE
Big obed's R.
CROSSVILLE
KINGSTON
SPARTA
1 July 4: Morgan leaves Knoxville for Sparta, passes through Celina and heads into Kentucky
2 July 9: Morgan routs Union cavalry at Tomkinsville. Moves on to Glasgow. Morgan's telegrapher taps into Union communications enabling Morgan to elude capture for remainder of raid
3 July 12: Raiders take Lebanon, inflict $100,000 damage to military stores, seize 200 prisoners and smash Central Kentucky railroad facilities
4 July 14: Morgan crosses Kentucky River, hoping to sieze Lexington. Warned that several thousand Federals are stationed round the city, Morgan heads for Cynthiana
5 July 17: After bitter fighting, Morgan fords Licking River under heavy fire while some of his men enter Cynthiana. Morgan then turns southward
6 Morgan's men withdraw via Paris, Richmond and Crab Orchard; recross Cumberland on 22nd, finishing raid at Livingston


*With 2,500 men, the two Confederate cavalry officers, Morgan and Forrest, led seperate raids which caused havoc throughout Central Kentucky and Middle Tennessee, and succeeded in immobilizing Gen. Buell's army of 40,000 (see map above).*

# Confederate Invasion of Kentucky AUGUST 14 – OCT 9 1862

In the summer of 1862, with General Buell's Army of the Ohio threatening Chattanooga, General Bragg shifted the Confederate Army of the Mississippi to southeast Tennessee to cooperate with General Edward Kirby Smith's army in an invasion of Kentucky. Bragg's plan was to draw Buell away from Chattanooga, defeat his army, and rally Kentuckians to join the Confederacy. Leaving Knoxville on August 14, Smith's 10,000-man command forced the Federals to evacuate Cumberland Gap. Entering Kentucky, Smith overwhelmed a small force of green, hastily assembled Federal troops at Richmond, before advancing to occupy Lexington.

Bragg left Chattanooga with 30,000 men on August 28, and moved into Kentucky. Capturing 4,000 Federals at Munfordsville, Bragg sat squarely across Buell's supply line to Louisville. However, Bragg now decided to move northeast to collect supplies and unite with Smith. Sidestepping Louisville, he occupied Bardstown and installed a Confederate state government at Frankfort.

Buell responded to the massive Confederate invasion by backtracking his forces into Kentucky. Reaching Louisville on September 25, Buell advanced against the Confederates, now scattered across a 60-mile front. Feinting against Frankfort with two divisions, Buell advanced the main portion of his 58,000-man army in three columns so as to converge on Bragg's left near Harrodsburg. Deceived by these tactics, Bragg and Smith concentrated over half their forces to defend Frankfort.

On October 7, alerted by heavy skirmishing on Major General William J. Hardee's front, Bragg instructed General Polk to join Hardee at Perryville and attack what he believed to be only a detached part of Buell's army. The next morning, Buell's left wing – probing forward through the drought-plagued countryside in search of water – met and engaged the Confederates occupying the hills along Doctor's Creek. Even with Polk's troops present, Bragg could only muster 16,000 men to fight what became the major

"I deem it useless and inexpedient to continue the pursuit... [The army] has not accomplished all that I had hoped or all that faction might demand; yet, composed as it is, one half of perfectly new troops, it has defeated a powerful and thoroughly disciplined army... and has driven it away baffled and dispirited."

*Gen. Buell to Gen. Halleck after Perryville.*

"With the whole southwest thus in the enemy's possession, my crime would have been unpardonable had I kept my noble little army to be ice-bound in a northern clime, without tents or shoes, and obliged to forage daily for bread etc."

*Gen. Bragg's justification of his decision to withdraw from Kentucky.*

1. **Aug 14, 1862:** Smith leaves Knoxville and bypasses Cumberland Gap. Gen. Morgan's Federal force, defending the gap, retreats
2. **Aug 30:** Smith attacks and overwhelms Federals defending Richmond. Confederates occupy Lexington
3. **Aug 28:** As Smith closes on Lexington, Bragg leaves Chattanooga and advances rapidly north
4. In order to strike Bragg' s column near Sparta, Buell withdraws to Nashville. Leaving a garrison there, Buell continues north to Bowling Green
5. **Sep 14:** Bragg reaches Glasgow. Continues on to Munfordsville.
6. Bragg forces surrender of Munfordsville garrison on 17th. Continues to Bardstown, arriving on 30th.
7. **Sep 25:** With Bragg abandoning his ho on road to Louisville, Buell concentrat his forces on Ohio River
8. **Oct 1:** Buell, now resupplied and reinforced, sends force on a feint towa Frankfort, and another force toward Bardstown
9. Bragg and Smith defend Frankfort wit half of their forces. Alerted by heavy skirmishing on Hardee's front near Perryville, Bragg sends Polk from Harrodsburg to attack Federals (see ma right)
10. **Oct 10:** Following action at Perryville, Bragg and Smith reunite at Harrodsbu decide to abandon campaign and withdraw to Tennessee

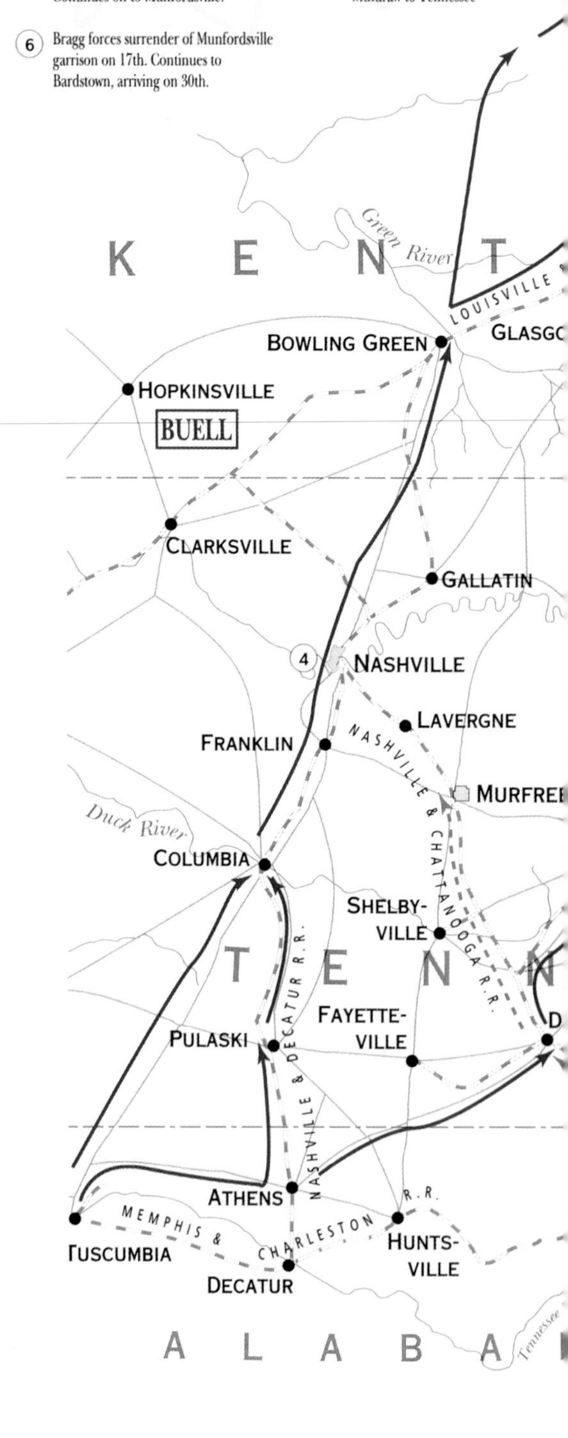

K E N T
Green River
LOUISVILLE
BOWLING GREEN
GLASGC
HOPKINSVILLE
BUELL
CLARKSVILLE
GALLATIN
4
NASHVILLE
LAVERGNE
MURFREI
FRANKLIN
NASHVILLE & CHATTANOOGA R.R.
Duck River
COLUMBIA
SHELBY-VILLE
T E N N
DECATUR R.R.
FAYETTE-VILLE
PULASKI
NASHVILLE &
ATHENS
MEMPHIS & CHARLESTON R.R.
TUSCUMBIA
HUNTS-VILLE
DECATUR
D
A L A B A
Tennessee


*A Federal brigade under Col. John Starkweather successfully resists an attack by elements of Gen. Benjamin Cheatham's Confederate division, on the Union left during the battle of Perryville (below).*

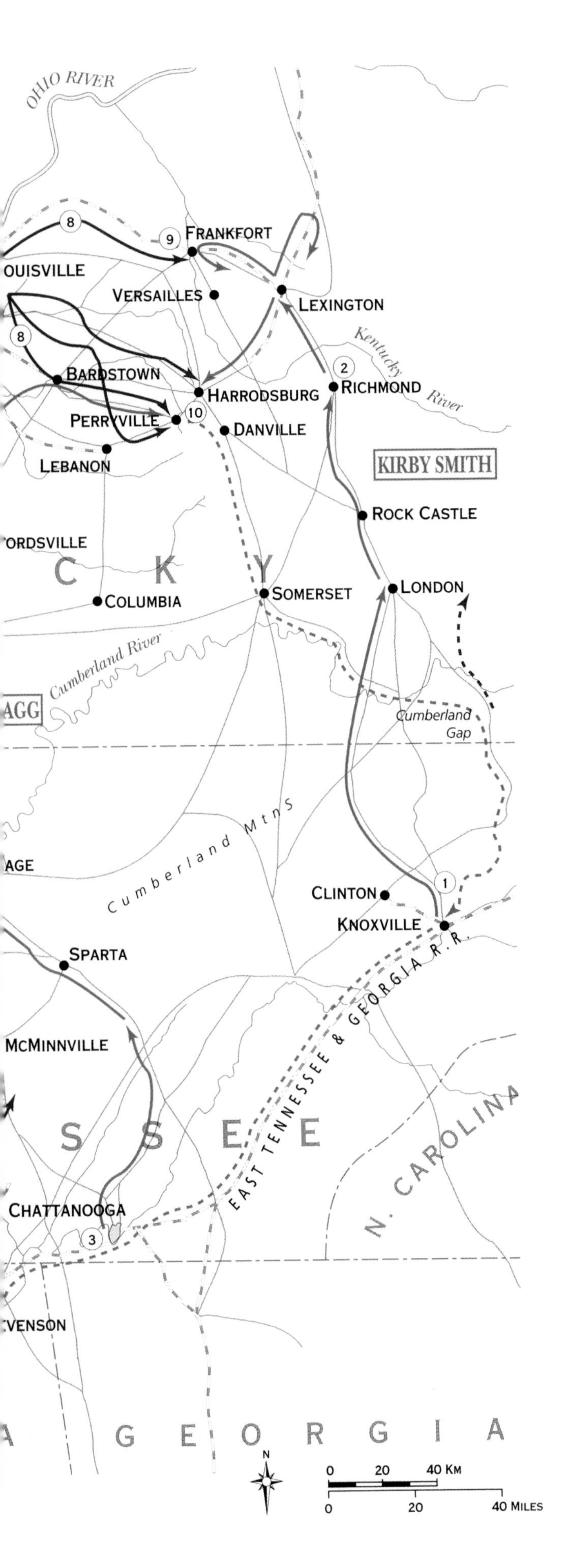
OHIO RIVER
FRANKFORT
VERSAILLES
LEXINGTON
Kentucky River
BARDSTOWN
HARRODSBURG
RICHMOND
PERRYVILLE
DANVILLE
LEBANON
KIRBY SMITH
ROCK CASTLE
SOMERSET
LONDON
COLUMBIA
Cumberland River
Cumberland Gap
Cumberland Mtns
CLINTON
KNOXVILLE
SPARTA
MCMINNVILLE
EAST TENNESSEE & GEORGIA R.R.
N. CAROLINA
CHATTANOOGA
GEORGIA
0 20 40 KM
0 20 40 MILES

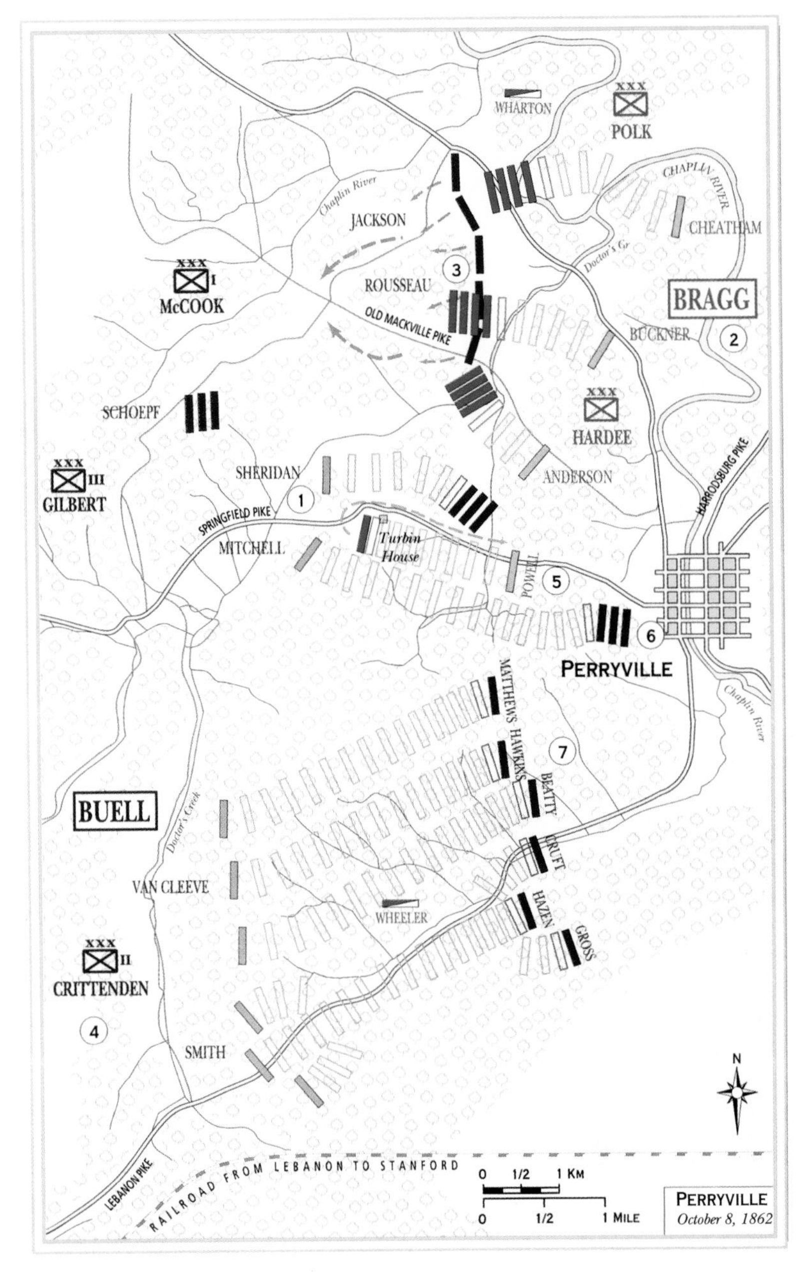
WHARTON
POLK
CHAPLIN RIVER
CHEATHAM
JACKSON
Chaplin River
McCOOK
ROUSSEAU
OLD MACKVILLE PIKE
BRAGG
BUCKNER
SCHOEPF
HARDEE
GILBERT
SHERIDAN
ANDERSON
HARRODSBURG PIKE
SPRINGFIELD PIKE
MITCHELL
Turbin House
POWELL
PERRYVILLE
MATTHEWS
HAWKINS
BEATTY
HAZEN
BUELL
Doctor's Creek
VAN CLEEVE
WHEELER
CRITTENDEN
SMITH
LEBANON PIKE
RAILROAD FROM LEBANON TO STANFORD
0 1/2 1 KM
0 1/2 1 MILE
PERRYVILLE
October 8, 1862

1 **Oct 8, 1862, dawn:** Buell's left wing, probing toward Doctor's Creek in search of water, meets and engages Confederates. Sheridan's division drives them past Turbin House and digs in

2 **10 am:** Bragg arrives, orders Polk to attack Buell's left flank with Cheatham's and Buckner's Confederate divisions

3 **2 pm:** Confederates slam into McCook's Federal corps; McCook's front collapses. Across Doctor's Creek, Sheridan – under orders not to engage – watches disaster befall McCook

4 Crittenden is unaware of the battle; bluffed by Wheeler's cavalry, his corps is not engaged

5 **4.15pm:** Advancing along Springfield Pike, Powell's brigade attacks Sheridan , but is repulsed. Sheridan pushes toward Perryville. Buell finally learns that a major battle is in progress

6 **Late pm:** Federals attack Confedererate flank and chase Powell's men into Perryville. Finding themselves isolated , Federals withdraw

7 **Night:** Buell brings up the rest of his forces to the battlefield. Outnumbered, Bragg retreats to Harrodsburg

battle of the war in Kentucky.

From his headquarters, hearing only sporadic cannon fire, Buell did not realise until late afternoon that a major battle was in progress, and only nine out of his twenty-four brigades managed to engage. The Confederates fought stubbornly, pushing the Union left back a mile before darkness ended the fighting. Bragg, however, deciding that he was too heavily outnumbered to continue, retreated during the night and joined Smith near Harrodsburg. The Confederates now abandoned the campaign, withdrawing through the Cumberland Gap into Tennessee. Buell failed to pursue, and was relieved of his command. Kentucky, however, remained securely in Union hands for the rest of the war.

# Iuka and Corinth SEPTEMBER 19 – OCTOBER 4 1862

While national attention focused on Gen-eral Bragg's Kentucky invasion, 14,000 Confederates under General Sterling Price occupied Iuka on September 14. Price planned to move to the Tennessee River and prevent General Grant from sending reinforcements to General Buell, whose Federal army was pursuing Bragg into Kentucky. In response, Grant set a trap to destroy Price, ordering General Rosecrans with 10,000 troops to advance on Iuka from the south and attack Price. Meanwhile, Grant himself advanced from Corinth with 8,000 men to block Price's movement north.

On September 19, Rosecrans's vanguard surprised Price southwest of Iuka. Price struck back with Brigadier General Lewis Henry Little's division, and the tough Confederate veterans badly mauled Rosecrans before night halted the battle.

Realizing that he was caught between two Federal columns, Price retreated south during the night. The following morning, Rosecrans occupied Iuka and ordered a pursuit. Rough terrain, poor roads, and a stubborn Confederate rearguard ended the pointless chase by late afternoon. Price slipped away safely to Baldwyn.

In late September, using his seniority to control Price's movements, General Van Dorn – in command of the Department of Mississippi and East Louisiana – ordered Price to join him at Ripley for an advance on the Corinth rail junction, key to Federal defenses in northern Mississippi. Van Dorn gambled that victory at Corinth would force Grant to evacuate west Tennessee.

On October 3, Van Dorn and Price hurled 22,000 men against Corinth's outer defenses, held by Rosecrans's 23,000 troops. During the long, hot day, the tenacious Confederates forced Rosecrans to fall back two miles. Night halted the fighting, with Rosecrans holding the city's inner line of defensive redoubts.

The next morning, Van Dorn threw a massive frontal assault against the Union fortifications, penetrating them at several points. But artillery and rifle fire from the redoubts cut the exposed Confederate brigades to pieces. Repulsed, and having sustained devastating losses, Van Dorn retreated west towards the Hatchie River, slowly pursued by Rosecrans.

On October 7, 7,000 Union troops, sent by Grant from Bolivar, Tennessee to intercept Van Dorn, caught the Confederates at Davis' Bridge. Forcing Van Dorn's vanguard back across the river, the Federals seized the

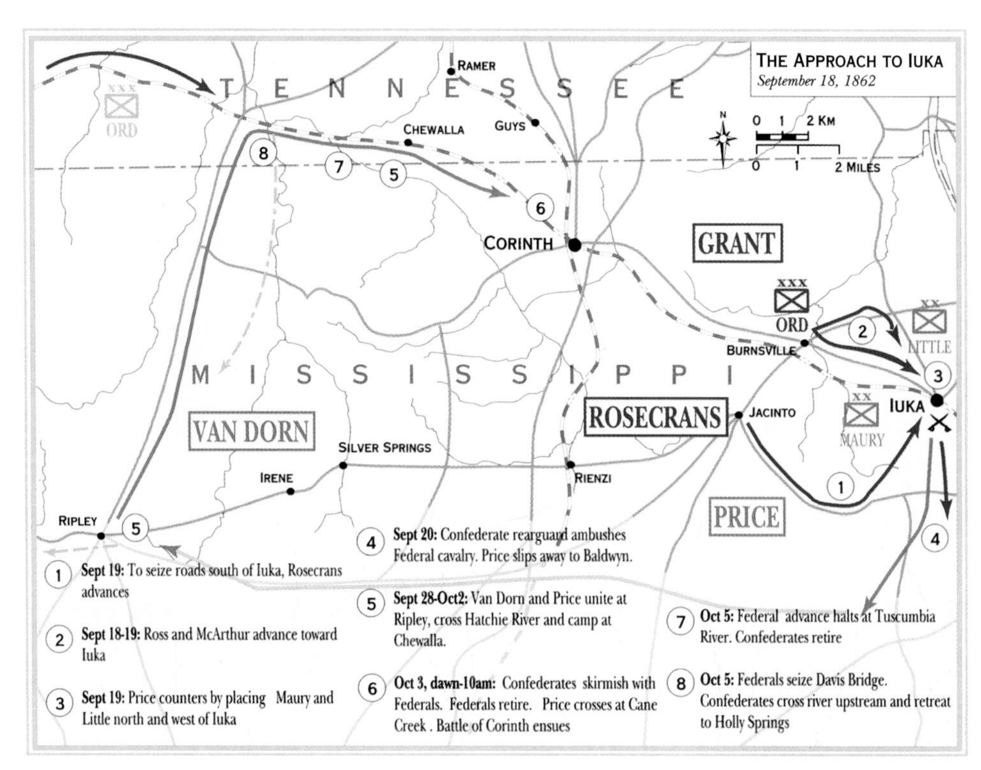
THE APPROACH TO IUKA
September 18, 1862
TENNESSEE
MISSISSIPPI
RAMER
CHEWALLA
GUYS
CORINTH
GRANT
ORD
LITTLE
BURNSVILLE
IUKA
JACINTO
MAURY
ROSECRANS
VAN DORN
SILVER SPRINGS
IRENE
RIENZI
PRICE
RIPLEY
0 1 2 KM
0 1 2 MILES
1 Sept 19: To seize roads south of Iuka, Rosecrans advances
2 Sept 18-19: Ross and McArthur advance toward Iuka
3 Sept 19: Price counters by placing Maury and Little north and west of Iuka
4 Sept 20: Confederate rearguard ambushes Federal cavalry. Price slips away to Baldwyn.
5 Sept 28-Oct2: Van Dorn and Price unite at Ripley, cross Hatchie River and camp at Chewalla.
6 Oct 3, dawn-10am: Confederates skirmish with Federals. Federals retire. Price crosses at Cane Creek. Battle of Corinth ensues
7 Oct 5: Federal advance halts at Tuscumbia River. Confederates retire
8 Oct 5: Federals seize Davis Bridge. Confederates cross river upstream and retreat to Holly Springs

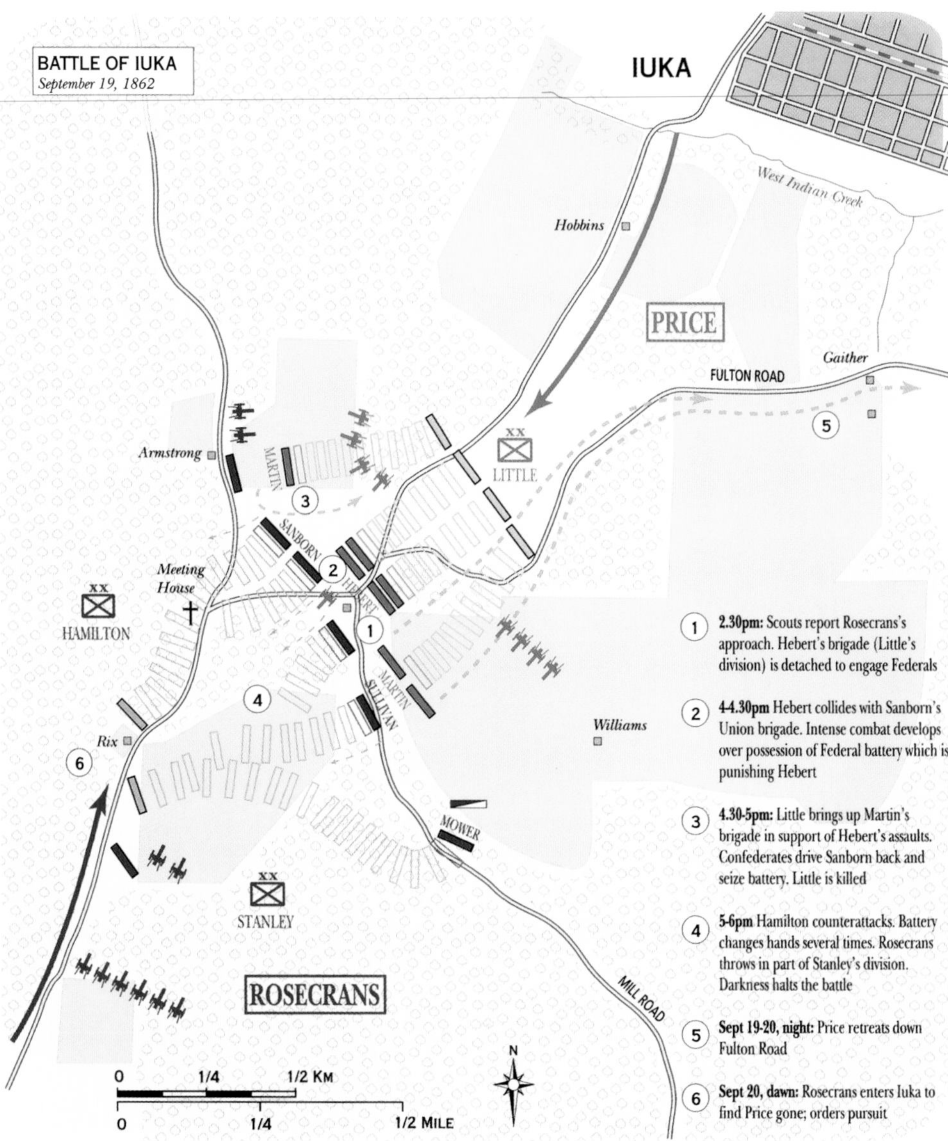
BATTLE OF IUKA
September 19, 1862
IUKA
West Indian Creek
Hobbins
PRICE
Gaither
FULTON ROAD
LITTLE
Armstrong
MARTIN
SANBORN
HEBERT
Meeting House
HAMILTON
SULLIVAN
Williams
Rix
MOWER
STANLEY
ROSECRANS
MILL ROAD
0 1/4 1/2 KM
0 1/4 1/2 MILE
1 2.30pm: Scouts report Rosecrans's approach. Hebert's brigade (Little's division) is detached to engage Federals
2 4-4.30pm Hebert collides with Sanborn's Union brigade. Intense combat develops over possession of Federal battery which is punishing Hebert
3 4.30-5pm: Little brings up Martin's brigade in support of Hebert's assaults. Confederates drive Sanborn back and seize battery. Little is killed
4 5-6pm Hamilton counterattacks. Battery changes hands several times. Rosecrans throws in part of Stanley's division. Darkness halts the battle
5 Sept 19-20, night: Price retreats down Fulton Road
6 Sept 20, dawn: Rosecrans enters Iuka to find Price gone; orders pursuit

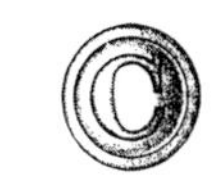

bridge. Van Dorn, however, managed to extricate his army from between the two converging Union columns, and crossing the river upstream, turned west to Ripley. The ill-fated Corinth attack achieved nothing but heavy losses. A complete failure, the autumn campaign was the last Confederate offensive in the Mississippi theater.

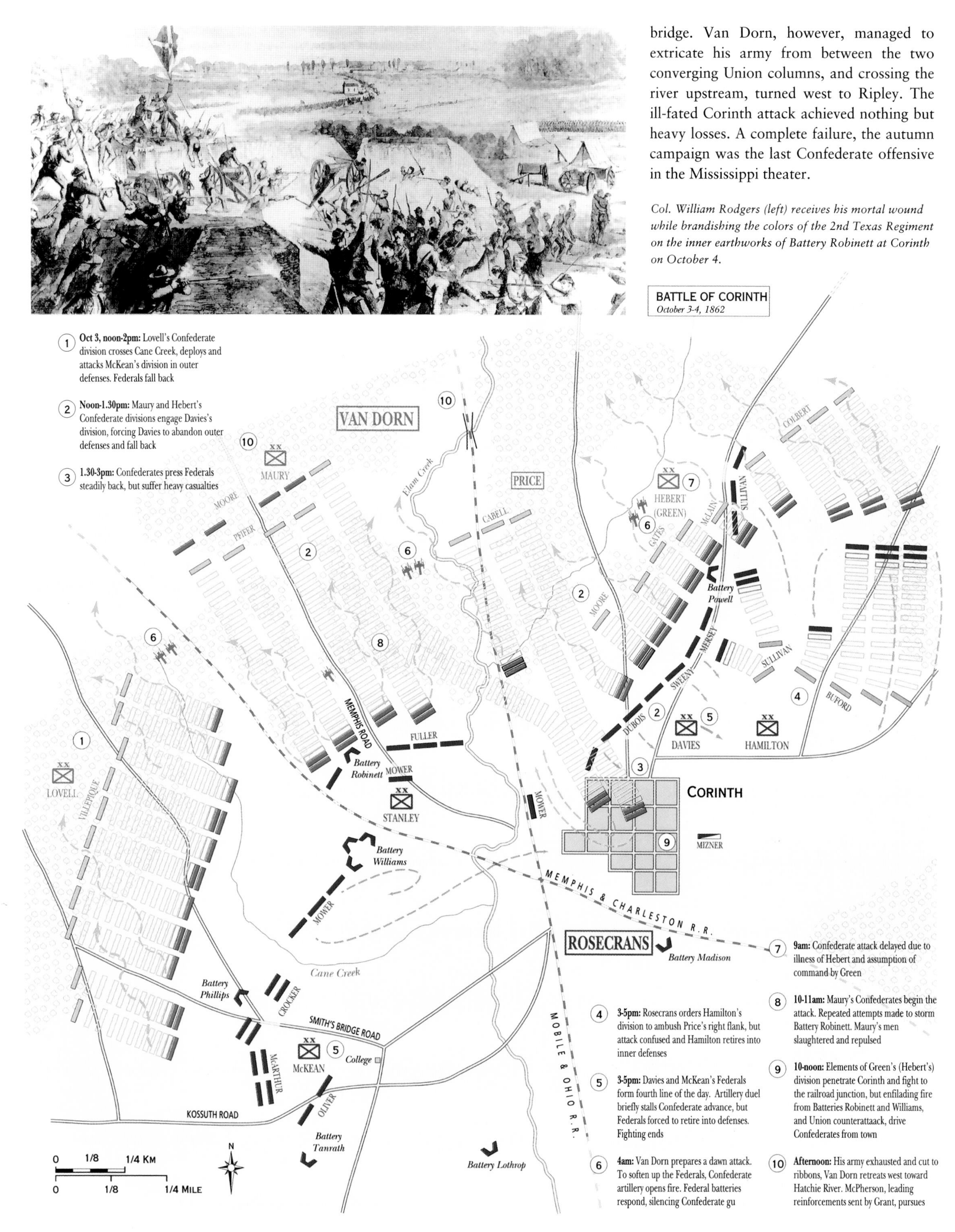
BATTLE OF CORINTH
October 3-4, 1862
1 **Oct 3, noon-2pm:** Lovell's Confederate division crosses Cane Creek, deploys and attacks McKean's division in outer defenses. Federals fall back
2 **Noon-1.30pm:** Maury and Hebert's Confederate divisions engage Davies's division, forcing Davies to abandon outer defenses and fall back
3 **1.30-3pm:** Confederates press Federals steadily back, but suffer heavy casualties
4 **3-5pm:** Rosecrans orders Hamilton's division to ambush Price's right flank, but attack confused and Hamilton retires into inner defenses
5 **3-5pm:** Davies and McKean's Federals form fourth line of the day. Artillery duel briefly stalls Confederate advance, but Federals forced to retire into defenses. Fighting ends
6 **4am:** Van Dorn prepares a dawn attack. To soften up the Federals, Confederate artillery opens fire. Federal batteries respond, silencing Confederate gu
7 **9am:** Confederate attack delayed due to illness of Hebert and assumption of command by Green
8 **10-11am:** Maury's Confederates begin the attack. Repeated attempts made to storm Battery Robinett. Maury's men slaughtered and repulsed
9 **10-noon:** Elements of Green's (Hebert's) division penetrate Corinth and fight to the railroad junction, but enfilading fire from Batteries Robinett and Williams, and Union counterattaack, drive Confederates from town
10 **Afternoon:** His army exhausted and cut to ribbons, Van Dorn retreats west toward Hatchie River. McPherson, leading reinforcements sent by Grant, pursues
VAN DORN
MAURY
MOORE
PFIFER
Elam Creek
PRICE
CABELL
HEBERT (GREEN)
GATES
McLAIN
SULLIVAN
COLBERT
Battery Powell
MOORE
MERSEY
SWEENY
SULLIVAN
BUFORD
DUBOIS
DAVIES
HAMILTON
CORINTH
MIZNER
MEMPHIS ROAD
FULLER
Battery Robinett
MOWER
STANLEY
MOWER
LOVELL
VILLEPIQUE
Battery Williams
MOWER
MEMPHIS & CHARLESTON R.R.
ROSECRANS
Battery Madison
Cane Creek
Battery Phillips
CROCKER
SMITH'S BRIDGE ROAD
McKEAN
College
McARTHUR
OLIVER
KOSSUTH ROAD
Battery Tanrath
Battery Lothrop
MOBILE & OHIO R.R.
0 1/8 1/4 KM
0 1/8 1/4 MILE
N

*Col. William Rodgers (left) receives his mortal wound while brandishing the colors of the 2nd Texas Regiment on the inner earthworks of Battery Robinett at Corinth on October 4.*

# Grant's First Vicksburg Campaign DECEMBER 1862

Since the occupation of Corinth, Mississippi, in May, 1862, the Federal Army of the Tennessee had operated in West Tennessee and North Mississippi. Grant had concentrated 36,000 troops at Grand Junction, Tennessee, from where he proposed marching south to repair the Mississippi Central Railroad. The capture of Jackson would enable the Federals to approach Vicksburg from the east. The Confederate forces, now led by Lieutenant General John C. Pemberton who had replaced Van Dorn, were based at Jackson.

Van Dorn remained in charge of the army camp at Holly Springs, but satisfied that the camp was untenable, he retired behind the Tallahatchie River and threw up earthworks covering the river crossings. Crossing the river would now be too risky, so Grant determined to outflank Van Dorn. On November 27, about 7,000 Federal troops landed near Friar's Point, and crossed the upper Delta.

Van Dorn was unaware of this threat until November 28, by which time the Federals were crossing the Tallahatchie. With the enemy across the river and threatening the railroads 50 miles to Van Dorn's rear, Pemberton ordered the army to retire behind the Yalobusha River. Torrential rain flooded the rivers, slowing the Federal pursuit as the Confederates withdrew southward to Grenada.

Grant's pursuing columns, now reinforced by Sherman, halted to rebuild the railroad and stockpile the depot at Holly Springs. Intending to employ naval power to flank the Confederates, Grant decided to deploy the divisions being assembled by Major General John A. McClernand for an amphibious assault on Vicksburg. Sherman embarked McClernand's troops and proceeded with them down the Mississippi to Helena, where he was joined by a fourth division. With this force, Sherman planned to ascend the Yazoo River, disembark, and seize the high ground northeast of Vicksburg. Before Grant was ready, the Confederates turned the tables on the Federals: Van Dorn, now reassigned to the cavalry, moved against Grant's supply depots.

Meanwhile, General Bragg at Murfreesboro was confronted by the Army of the Cumberland. Bragg ordered Nathan Bedford Forrest to attack the Federal supply lines. Forrest duly crossed the Tennessee River at Clifton on December 15–17 and wrecked the railroad from Jackson to the Kentucky line, causing Grant to suspend his advance.

Unknown to Grant, Van Dorn with 3,500 horsemen, left the Grenada area on December 17, swinging east of the Federals. Early on December 20, Van Dorn swept into Holly Springs, surprised the Federals, and destroyed their main supply depot. To avoid pursuit, he headed northward and returned to Grenada.

The Holly Springs raid had far-reaching repercussions on Grant's plans; Van Dorn's raiders had destroyed his most important depot, Forrest had wrecked his vital railroad link, and the country had been exhausted by the armies. On December 21 Grant began to withdraw. Apprised that steamboats crowded with Federal troops were descending the Mississippi, Pemberton rushed reinforcements by rail from Grenada to Vicksburg.

The day after Christmas, three of Sherman's divisions landed at Johnson's plantation and marched toward Walnut Hills. But Sherman was too late: Confederate brigades had already reinforced the Vicksburg defenders.

On December 28, Major General Frederick Steele's Federal division landed near Thompson Lake, and thrust toward Blake's Levee, but his advance was checked by Confederate cannon. The following day – Steele having failed to turn the Confederate right – Sherman assaulted Pemberton's center, but his troops were slaughtered as they crossed the Chickasaw Bayou causeway, and Sherman suspended his attack. On January 2, 1863, he re-embarked his 30,000 men and left the Yazoo River. Paraphrasing Julius Caesar, he described his Chickasaw Bayou campaign: "I reached Vicksburg at the time appointed, landed, assaulted, and failed."

McClernand, now in command of Federal troops in the area, sailed up the Arkansas River and captured Arkansas Post and its 5,000 defenders on January 11.

*The defender of Vicksburg, Gen. John C. Pemberton (above). Abrupt and domineering in manner, Pemberton was not widely liked, though he enjoyed the favor of Jefferson Davis.*

(1) **Nov 3-4:** Grant's center wing moves south; two wings of his Army of the Tennessee rendezvous at Grand Junction

(2) **Nov 8-10:** Pemberton's field army, led by Van Dorn withdraws from Holly Springs and takes position covering the Tallahatchie crossings

(3) **Nov 26:** Sherman, commanding Grant's right wing, marches southeast from Memphis

(4) **Nov 27:** Grant's center and left wings advance south, occupy Holly Springs and approach the Tallahatchie crossings

(5) **Nov 27-Dec 7:** Union troops land at Friar's Point, cross the Delta and wreck railroads near Granada

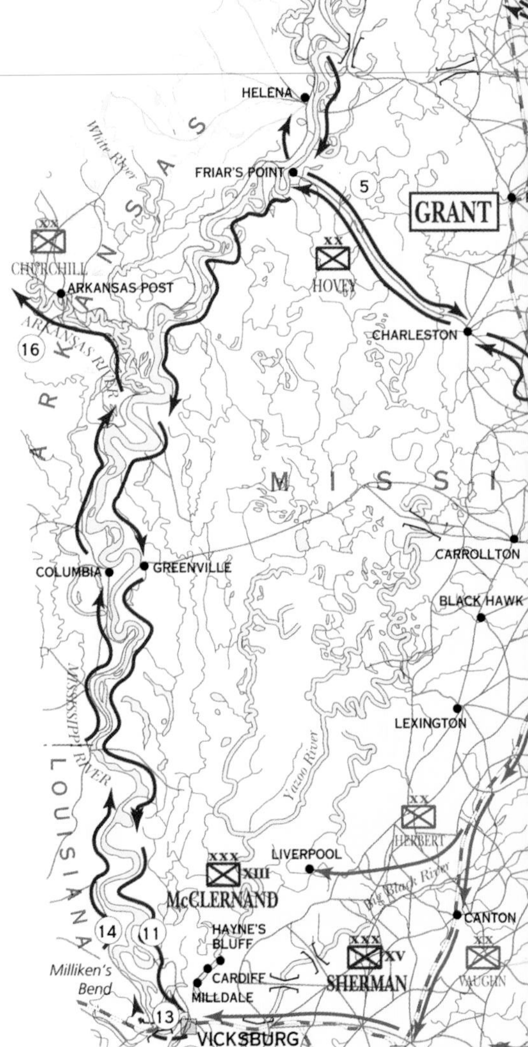
HERNANDO
HELENA
FRIAR'S POINT
5
GRANT
HOVEY
ARKANSAS POST
16
CHARLESTON
COLUMBIA
GREENVILLE
CARROLLTON
BLACK HAWK
LEXINGTON
Yazoo River
LIVERPOOL
XIII
McCLERNAND
14
11
HAYNE'S BLUFF
CARDIFF
MILLDALE
XV
SHERMAN
CANTON
Milliken's Bend
13
VICKSBURG
12
PEMBERTON
JACKSON

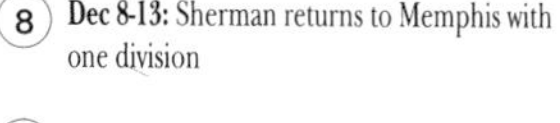

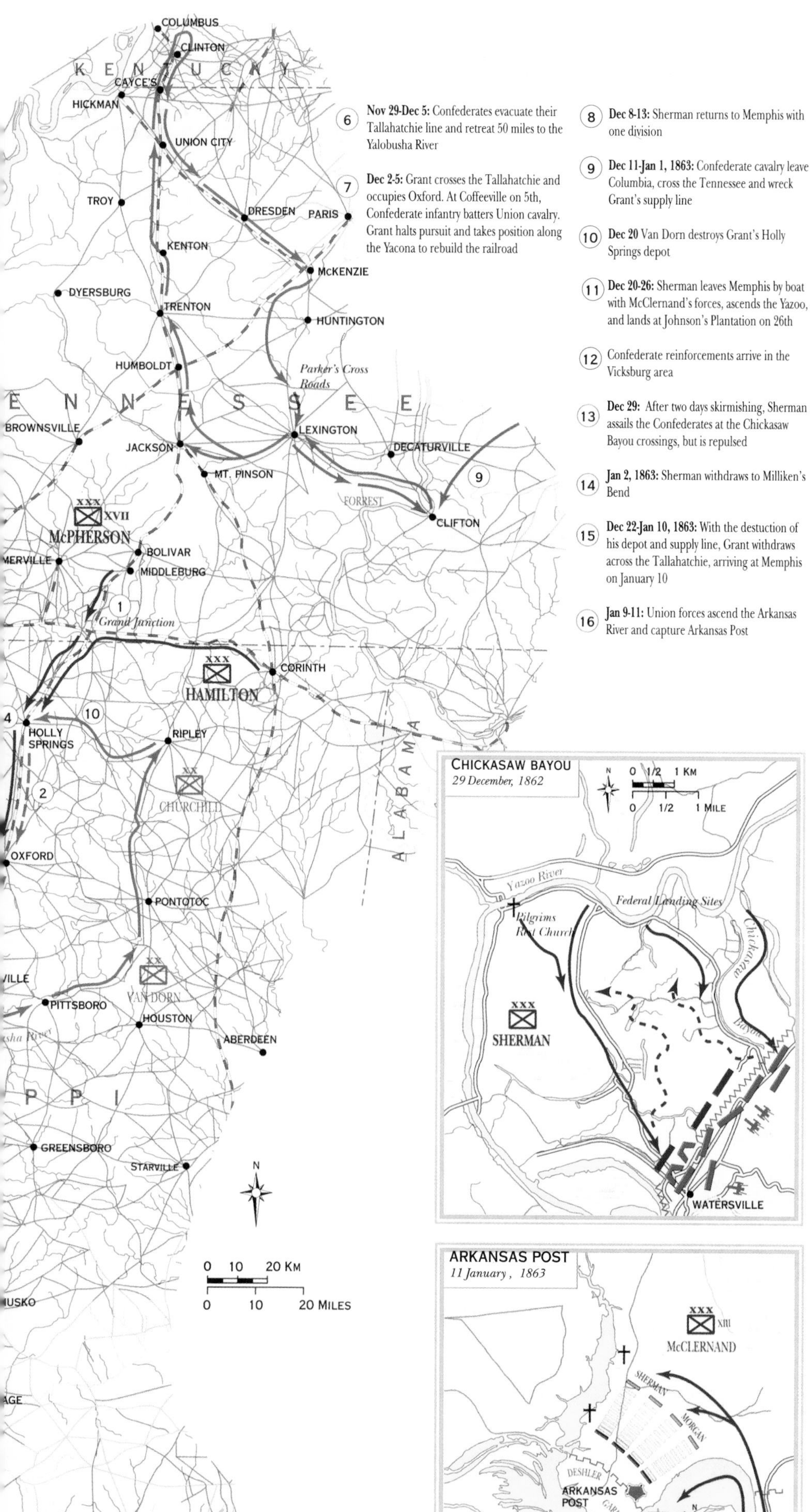

COLUMBUS
CLINTON
KENTUCKY
CAYCE'S
HICKMAN
UNION CITY
TROY
DRESDEN
PARIS
KENTON
MCKENZIE
DYERSBURG
TRENTON
HUNTINGTON
HUMBOLDT
Parker's Cross Roads
TENNESSEE
BROWNSVILLE
JACKSON
LEXINGTON
DECATURVILLE
MT. PINSON
FORREST
CLIFTON
XVII
McPHERSON
BOLIVAR
MIDDLEBURG
Grand Junction
CORINTH
HAMILTON
HOLLY SPRINGS
RIPLEY
CHURCHILL
OXFORD
PONTOTOC
PITTSBORO
VAN DORN
HOUSTON
ABERDEEN
GREENSBORO
STARVILLE
ALABAMA
0 10 20 KM
0 10 20 MILES
Nov 29-Dec 5: Confederates evacuate their Tallahatchie line and retreat 50 miles to the Yalobusha River
Dec 2-5: Grant crosses the Tallahatchie and occupies Oxford. At Coffeeville on 5th, Confederate infantry batters Union cavalry. Grant halts pursuit and takes position along the Yacona to rebuild the railroad
Dec 8-13: Sherman returns to Memphis with one division
Dec 11-Jan 1, 1863: Confederate cavalry leave Columbia, cross the Tennessee and wreck Grant's supply line
Dec 20 Van Dorn destroys Grant's Holly Springs depot
Dec 20-26: Sherman leaves Memphis by boat with McClernand's forces, ascends the Yazoo, and lands at Johnson's Plantation on 26th
Confederate reinforcements arrive in the Vicksburg area
Dec 29: After two days skirmishing, Sherman assails the Confederates at the Chickasaw Bayou crossings, but is repulsed
Jan 2, 1863: Sherman withdraws to Milliken's Bend
Dec 22-Jan 10, 1863: With the destuction of his depot and supply line, Grant withdraws across the Tallahatchie, arriving at Memphis on January 10
Jan 9-11: Union forces ascend the Arkansas River and capture Arkansas Post
CHICKASAW BAYOU
29 December, 1862
0 1/2 1 KM
0 1/2 1 MILE
Yazoo River
Federal Landing Sites
Pilgrims Rest Church
Chickasaw Bayou
SHERMAN
WATERSVILLE
ARKANSAS POST
11 January, 1863
XIII
McCLERNAND
SHERMAN
MORGAN
DESHLER
ARKANSAS POST
GARLAND
CHURCHILL
ARKANSAS RIVER
0 1/2 1 KM
0 1/2 1 MILE

*Once temporarily relieved of command on the suspicion of insanity, William T. Sherman (above) still enjoyed Grant's full confidence. Though unsuccessful in his assault on Vicksburg's defenses, he had repeatedly proved his willingness and aptitude in combat.*

*Despite devastating fire from the Confederate defenses, Sherman's force advanced toward the bluffs north of Vicksburg near Chickasaw Bayou (left). The Confederates were able to hold off their more numerous assailants and the attack failed. Union casualties were 1,776, Confederate 207.*

> "Vicksburg is the key. The war can never be brought to a close until the key is in our pocket. We can take all the northern parts of the Confederacy, and they can still defy us from Vicksburg. It means... fresh troops from the States of the far South, and a cotton country where they can raise the staple without interference."
>
> *President Abraham Lincoln.*

> "Vicksburg is the nailhead that [holds] the South's two halves together."
>
> *President Jefferson Davis.*

*Confederate guns at Arkansas Post (left) were battered by the gunboats of Admiral Porter's fleet, while McClernand's force attacked over land. After some four hours the vastly outnumbered Confederates surrendered. Casualties: 1061 Union, 4791 Confederate.*

# Stones River (Murfreesboro)

## DECEMBER 30 1862 – JANUARY 2 1863

In October 1862, General Rosecrans succeeded General Buell as commander of the Department and Army of the Cumberland. Northern authorities demanded aggressive action against General Bragg's Army of Tennessee.

On December 26, Rosecrans's 44,000-man army left Nashville and moved southeast in three columns against Bragg's 34,000 men, who awaited attack astride the axis of rail and stage roads northwest of Murfreesboro. Despite heavy skirmishing, Rosecrans failed to assault, and Bragg seized the initiative, maneuvering his army to outflank the Federal right and cut their line of retreat north.

Shortly after dawn on December 3l, General Hardee's corps launched a crushing assault on Rosecrans's right wing, throwing the surprised Federals onto the defensive. When resistance stiffened, Bragg ordered General Polk's corps to sustain the offensive. Fighting desperately to halt the relentless Confederate tide, Rosecrans was forced back several miles and pinned against Stones River. At mid-afternoon, unable to break Rosecrans's line, Bragg instructed Major General John C. Breckinridge's division to reinforce Polk on the right. However, Polk sent Breckinridge's brigades in piecemeal, in separate frontal attacks against the Union left, and they were slaughtered. The Confederate assaults continued into the evening, but Rosecrans's army refused to break. That night, Bragg entrenched; Rosecrans, after considering retreat, also remained on the field.

New Year's day was quiet; each side buried their dead and awaited a renewal of the struggle. On January 2, Bragg resolved to finally destroy Rosecrans; in an ill-conceived but savage assault, Breckinridge drove a division of Crittenden's corps from a commanding height northeast of Stones River. However, Union artillery, firing across the river into the Confederate left, halted Breckinridge's charge, inflicting murderous losses on the already crippled division. Fresh Union infantry counterattacked across the river, and by nightfall had driven Breckinridge back to his original position.

For the next 24 hours the armies remained poised on the battlefield, but lack of confidence in Bragg's leadership forced his subordinates to write a memorandum counseling retreat. Deeply angered, Bragg initially refused, but on the morning of the 3rd, he ordered the army to retreat toward Tullahoma. Leaving nearly 2,000 wounded behind, the Confederates withdrew during the night. Rosecrans did not pursue, remaining content with possession of the battlefield. Although not a decisive victory, Lincoln thanked Rosecrans's army for its "hard-earned victory" and said that had the battle "been a defeat instead, the nation could scarcely have lived over [it]."

*Gen. James S. Negley's Federal division counterattacks across Stones River (above). Only the onset of darkness brought a halt to the Union pursuit of Breckinridge's shattered division.*

> "This attack is made against my judgement and by the special orders of General Bragg... if it should result in disaster and I be among the slain... tell the people that I believed this attack to be very unwise and tried to prevent it."
>
> *Gen. Breckinridge, prior to the attack on January 2.*

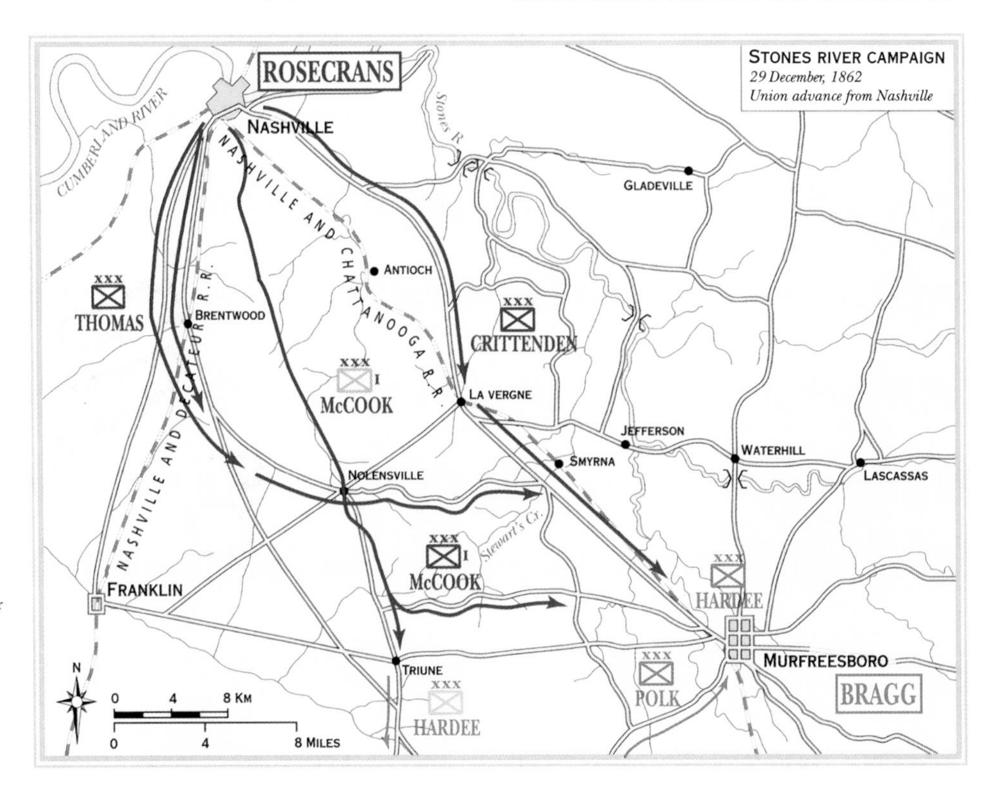
STONES RIVER CAMPAIGN
29 December, 1862
Union advance from Nashville
ROSECRANS
NASHVILLE
CUMBERLAND RIVER
Stones R.
GLADEVILLE
NASHVILLE AND CHATTANOOGA R.R.
ANTIOCH
THOMAS
BRENTWOOD
NASHVILLE AND DECATUR R.R.
CRITTENDEN
McCOOK
LA VERGNE
JEFFERSON
SMYRNA
WATERHILL
LASCASSAS
NOLENSVILLE
Stewart's Cr.
McCOOK
FRANKLIN
HARDEE
MURFREESBORO
TRIUNE
POLK
BRAGG
HARDEE
N
0 4 8 KM
0 4 8 MILES

*Rosecrans's advance (right), in a broad front, was one of constant skirmish, as Confederate cavalry under Joseph Wheeler forced Union infantry to deploy repeatedly at every piece of strong defensive terrain. By December 30, Rosecrans's three corps converged on Murfreesboro and deployed in line of battle.*

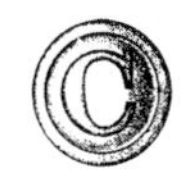

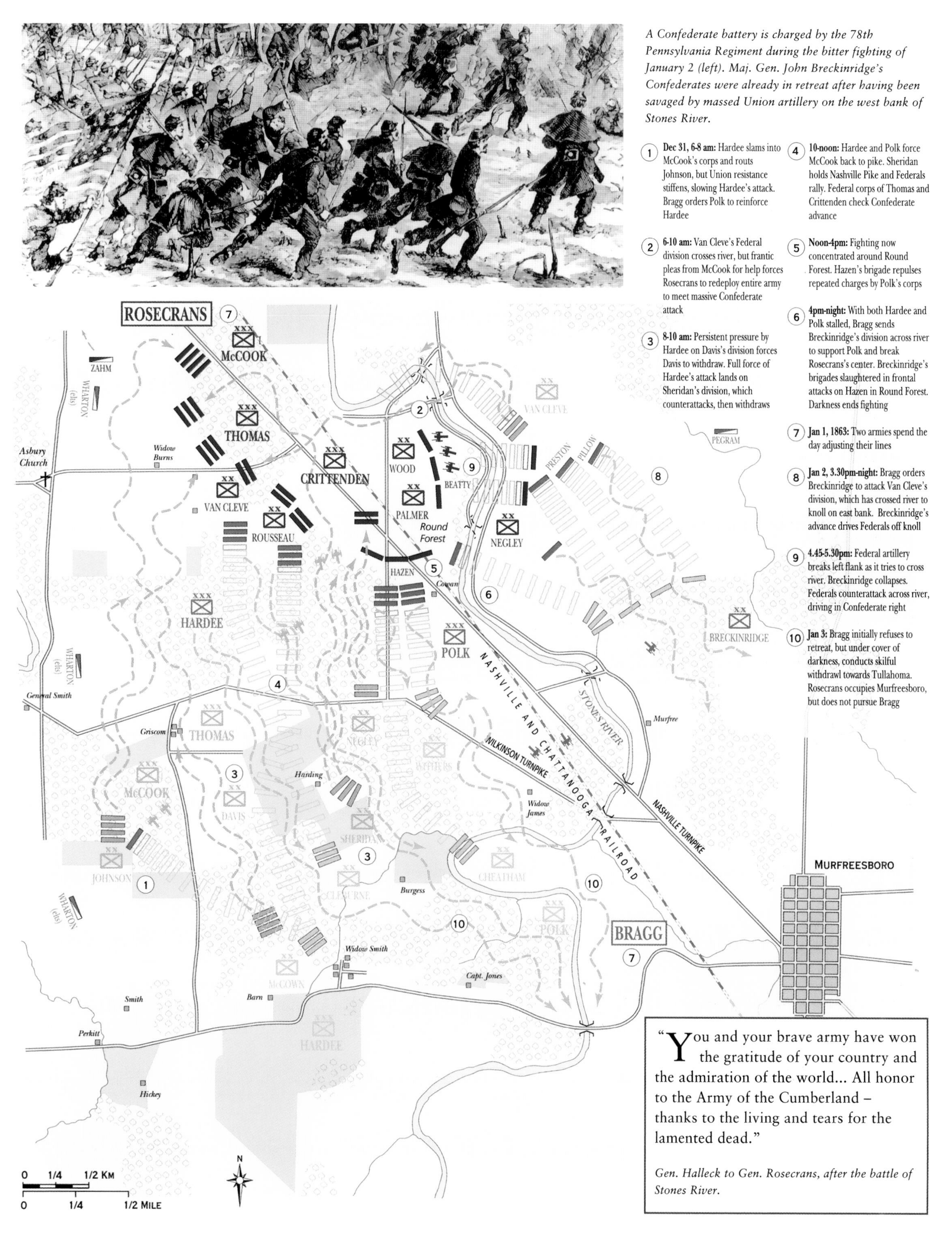
ROSECRANS
7
McCOOK
THOMAS
CRITTENDEN
WOOD
BEATTY
PALMER
Round Forest
HAZEN
Cowan
NEGLEY
VAN CLEVE
ROUSSEAU
HARDEE
POLK
PRESTON
PILLOW
PEGRAM
BRECKINRIDGE
ZAHM
WHARTON (cls)
Asbury Church
Widow Burns
General Smith
Griscom
THOMAS
NEGLEY
WITHERS
McCOOK
Harding
DAVIS
SHERIDAN
JOHNSON
CLEBURNE
Burgess
CHEATHAM
Widow James
Murfree
NASHVILLE AND CHATTANOOGA RAILROAD
WILKINSON TURNPIKE
NASHVILLE TURNPIKE
STONES RIVER
MURFREESBORO
BRAGG
POLK
Widow Smith
McCOWN
Capt. Jones
Barn
Smith
Perkitt
Hickey
HARDEE
N
0 1/4 1/2 KM
0 1/4 1/2 MILE

*A Confederate battery is charged by the 78th Pennsylvania Regiment during the bitter fighting of January 2 (left). Maj. Gen. John Breckinridge's Confederates were already in retreat after having been savaged by massed Union artillery on the west bank of Stones River.*

1. **Dec 31, 6-8 am:** Hardee slams into McCook's corps and routs Johnson, but Union resistance stiffens, slowing Hardee's attack. Bragg orders Polk to reinforce Hardee
2. **6-10 am:** Van Cleve's Federal division crosses river, but frantic pleas from McCook for help forces Rosecrans to redeploy entire army to meet massive Confederate attack
3. **8-10 am:** Persistent pressure by Hardee on Davis's division forces Davis to withdraw. Full force of Hardee's attack lands on Sheridan's division, which counterattacks, then withdraws
4. **10-noon:** Hardee and Polk force McCook back to pike. Sheridan holds Nashville Pike and Federals rally. Federal corps of Thomas and Crittenden check Confederate advance
5. **Noon-4pm:** Fighting now concentrated around Round Forest. Hazen's brigade repulses repeated charges by Polk's corps
6. **4pm-night:** With both Hardee and Polk stalled, Bragg sends Breckinridge's division across river to support Polk and break Rosecrans's center. Breckinridge's brigades slaughtered in frontal attacks on Hazen in Round Forest. Darkness ends fighting
7. **Jan 1, 1863:** Two armies spend the day adjusting their lines
8. **Jan 2, 3.30pm-night:** Bragg orders Breckinridge to attack Van Cleve's division, which has crossed river to knoll on east bank. Breckinridge's advance drives Federals off knoll
9. **4.45-5.30pm:** Federal artillery breaks left flank as it tries to cross river. Breckinridge collapses. Federals counterattack across river, driving in Confederate right
10. **Jan 3:** Bragg initially refuses to retreat, but under cover of darkness, conducts skilful withdrawl towards Tullahoma. Rosecrans occupies Murfreesboro, but does not pursue Bragg

"You and your brave army have won the gratitude of your country and the admiration of the world... All honor to the Army of the Cumberland – thanks to the living and tears for the lamented dead."

*Gen. Halleck to Gen. Rosecrans, after the battle of Stones River.*

# 1863: The Turning of the Tide

MORALE IN THE UNION armies and on the home front reached a new low in the early months of 1863. Captain Oliver Wendell Holmes, Jr., a future Supreme Court Justice who was recovering from the second of three wounds he would receive in the war, wrote that "the army is tired with its hard and terrible experience. I've pretty much made up my mind that the South have achieved their independence." The staunchly patriotic editor of the *Chicago Tribune* vented his frustration and depression in a private letter on January 14: "An armistice is bound to come during the year '63. The rebs can't be conquered by the present machinery."

Military operations in the first three months of the year seemed only to confirm this pessimism. In mid-January, Burnside began a new campaign to cross the Rappahannock River above the Confederate defenses, and come in on their rear. The movement started promisingly, but then the heavens opened, releasing several inches of sleet and rain that trapped the army helplessly in the mud. In the aftermath of this notorious "Mud March," Lincoln removed Burnside from command of the Army of the Potomac, replacing him with Joseph Hooker. A controversial general whose nickname was "Fighting Joe," Hooker at least seemed to promise an aggressive campaign once the warmer days of spring made operations possible.

A thousand miles to the west another aggressive general, Ulysses S. Grant, also seemed mired in the endless swamps and rivers that protected three sides of the Confederate "Gibraltar of the West" at Vicksburg. Only on the east, away from the river, was there high ground suitable for an assault on Vicksburg's defenses. Grant's problem was to move his army, supplies and transportation across the Mississippi to that high ground. For three months he floundered in the Mississippi-Yazoo bottomlands, while disease and exposure took a fearful toll of his troops. False rumors of excessive drinking, that had dogged Grant for years, broke out anew, but Lincoln resisted pressures to remove him from command. "What I want," Lincoln said, "is generals who will fight battles and win victories. Grant has done this, and I propose to stand by him." It was at this time that Lincoln reportedly said he would like to know Grant's brand of whiskey so that he could send some to his other generals.

Lincoln's own reputation reached a low point at this time. A visitor to Washington in February found that "the lack of respect for the President in all parties is unconcealed... If a Republican convention were to be held tomorrow, he would not get the vote of a state." In this depressed climate, the anti-war wing of the Democratic party (called Peace Democrats by themselves and Copperheads by the Republicans) found a ready audience for their message that the war was a failure and should be abandoned. Having won control of the Illinois and Indiana legislatures the preceding fall, Democrats in those states called for an armistice and a peace conference. They also demanded retraction of the "wicked, inhuman, and unholy" Emancipation Proclamation, and threatened to withdraw Illinois and Indiana troops from the war. Republican governors of the two states blocked that action.

But in Ohio the foremost Peace Democrat, Clement L. Vallandigham, was planning to run for governor. What had this wicked war accomplished? he asked Northern audiences during the winter and spring of 1863. "Let the dead at Fredericksburg and Vicksburg answer." The Confederacy could never be conquered; the only solution was to "stop the fighting. Make an armistice. Withdraw your army from the seceded states." Above all, give up the unconstitutional effort to abolish slavery. "I see more of barbarism and sin," said Vallandigham, "a thousand times more, in the continuance of this war and the enslavement of the white race by debt and taxes than in the continuation of slavery."

Vallandigham and other Copperhead spokesmen had a dire effect on Northern morale. Alarmed by the wave of desertions, the army commander in Ohio had Vallandigham arrested in May 1863. A military court convicted him of treason for aiding and abetting the enemy by uttering disloyal sentiments. The court's action raised serious questions of civil liberties. Was the conviction a violation of Vallandigham's First Amendment right of free speech? Could a military court try a civilian under martial law in a state like Ohio, where civil courts were functioning?

Lincoln was embarrassed by the swift arrest and trial of Vallandigham, which he learned about from the newspapers. To prevent Vallandigham from becoming a martyr, Lincoln commuted his sentence from imprisonment to banishment – to the Confederacy! He soon escaped from the South to Canada on a blockade runner, making his way to Windsor, Ontario, from where he ran his campaign for governor of Ohio.

Discouragement and defeatism in the North were a malaise of the spirit caused by military defeat. By contrast, Southerners were buoyed up by military success, but were suffering from hyperinflation and shortages. The tightening Union blockade, the weaknesses and imbalances of the Confederate economy, the escape of slaves to Union lines, and enemy occupation of some of the South's best agricultural areas, made it increasingly difficult for the Southern economy to produce both guns and butter. Despite the conversion of

*Lieutenant General Thomas Jonathan "Stonewall" Jackson, mortally wounded at the battle of Chancellorsville*

hundreds and thousands of acres from cotton to food crops, the deterioration of railroads and the priority given to army shipments made food scarce in some areas. A drought in the summer of 1862 had made matters worse. The price of salt – necessary to preserve meat in that pre-refrigeration age – shot out of sight. Even the middle class suffered, especially in Richmond, where the population had more than doubled since the start of the war. "The shadow of the gaunt form of famine is upon us," wrote a War Department clerk in March 1863. "I have lost twenty pounds, and my wife and children are emaciated." The rats in his kitchen were so hungry that they nibbled bread crumbs from his daughter's hand "as tame as kittens. Perhaps we shall have to eat them!"

Things were even worse for the poor, especially for the wives and children of non-slaveholding farmers and workers who had gone into the army. By the spring of 1863, food supplies were almost gone. Some women took matters into their own hands; denouncing "speculators" who allegedly hoarded goods to drive up prices, they would march in a body to stores, ask the price of bacon or cornmeal or salt, denounce such "extortion," take what they wanted, and march away. Bread riots broke out in several places. On April 2, 1863, a mob of more than a thousand women and boys looted several shops in Richmond before the militia, under the personal command of Jefferson Davis, forced them to disperse. The government subsequently released some emergency food stocks to civilians, and state and country governments provided aid to families of soldiers. Better crops in 1863 helped to alleviate the worst shortages, but many problems persisted.

In both South and North, the draft intensified social unrest. The burst of patriotic enthusiasm that had caused a million men to don blue or gray uniforms in 1861 had cooled by the spring of 1862. The Confederacy enacted a conscription law in April 1862 that made all white men (with certain occupational exemptions) aged 18 to 35 liable to the draft. The law allowed a drafted man to hire a substitute, but the price of substitutes soon rose beyond the means of the average farmer or worker. This gave rise to a bitter cry that it was a "rich man's war and a poor man's fight."

The cry grew louder in October 1862 when the Confederate Congress raised the draft age to 45, and added a clause

*The "Mud marsh," a pencil sketch by Union artist Alfred R. Waud*

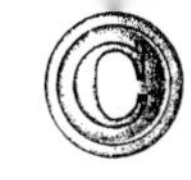

exempting from the draft one white man on every plantation with twenty or more slaves. The purpose of this "overseer exemption" was to maintain discipline over the slaves. It had been prompted by complaints from planters' wives that they had been left alone to manage the slaves after the departure of husbands, sons, and overseers for the army. But this so-called Twenty Negro Law seemed like blatant discrimination against non-slaveholding farm families whose men were also at the front. And raising the age limit to 45 meant that many fathers with large families of young children were eligible for the draft.

Similar discontent greeted the enactment of conscription in the North. In March 1863 the Union Congress decreed that all male citizens aged 20 to 45 must enroll for the draft. Not all of them would be called (married men over 35 were in effect exempted), but all would be liable. The purpose of this law was more to stimulate volunteering than to draft men directly into the army. In three of the President's four calls for troops under the law, the War Department set a quota for each congressional district, and gave it 50 days to meet the quota with volunteers before resorting to a draft lottery to make up the deficiency. Some districts never had to hold a draft, but fulfil led their quotas by offering large bounties to volunteers – the reservoir of men who had enlisted from patriotic motives had been pretty well exhausted by 1863. The bounty system produced glaring abuses, including "bounty jumpers," who enlisted and then deserted as soon as they received their money – often to enlist elsewhere under another name.

Unlike the Confederate law, the Union draft law permitted the hiring of substitutes. To keep the price of substitutes from skyrocketing as it had in the Confederacy, the Northern law allowed a drafted man the alternative of paying a "commutation fee" of $300 that exempted him from the current draft call (but not necessarily the next one). That provision raised the cry of rich man's war/poor man's fight in the North as well. Indeed, the sense of grievance was more dangerous in the Union than in the Confederacy, because it was nurtured by the Democratic party and intensified by racism. Democrats opposed conscription just as they opposed emancipation. Their newspapers told white workers, especially the large Irish-American population, that the draft would force them to fight a war to free the slaves who would then come North to take their jobs. This volatile issue helped spark widespread violence when drafting began in the summer of 1863. The worst riot occurred in New York City, where huge mobs – consisting mainly of Irish-Americans – demolished draft offices, lynched several blacks, and destroyed huge areas of the city in four days of looting and burning (July 13–16).

Draft riots in the North and bread riots in the South exposed alarming class fissures that were widened by the strains of total war. Although inflation was much less serious in the North than in the South (about 80% over four years compared with 9,000%), wages in the North lagged behind price increases. Labor unions sprang up in several industries and went on strike for higher wages. In some areas, such as the anthracite coalfields of Pennsylvania, labor organizations – dominated by Irish-Americans – combined opposition to the draft and to emancipation with violent strikes against industries owned by Protestant Republicans. Troops sent into such areas to enforce the draft sometimes suppressed strikes as well. These crosscurrents of class, ethnic, and racial hostilities produced a dangerous mixture in several Northern communities.

But the perception that it was a rich man's war and a poor man's fight was greater than the reality in both North and South. The principal forms of taxation to sustain the war were property, excise, and income taxes, most of which bore down proportionately more heavily on the wealthy than on the poor. In the South, property of the rich – including plantations and slaves – suffered greater damage and confiscation than did the property of non-slaveholders. The war liberated four million slaves, the poorest class in America. Both the Union and Confederate armies contained men from all strata of society in proportion to their percentage of the population. The overwhelming majority of Civil War soldiers were volunteers; among those who volunteered in 1861–1862 the planter class was over-represented in the Confederate army,

*The battle of Gettysburg, section of a Cyclorama by Paul Philippoteaux*

and the Northern middle class in the Union army, for these groups believed they had more at stake in the war and joined up in larger numbers during the early months of enthusiasm. It was these volunteers – especially the officers, most of whom came from the middle and upper classes – who suffered the highest casualty rates.

Even conscription did not fall much more heavily on the poor than on the rich. Those who escaped the draft by decamping to the woods, the territories, or Canada, came mainly from the poor. The Confederacy abolished substitution in December 1863, and made men who had previously sent substitutes liable to the draft. In the North several city councils, political machines, and businesses contributed funds to pay the commutation fees of drafted men who could not pay out of their own pockets. In the end, it was neither a rich man's war nor a poor man's fight; it was an American war.

*The battle of Gettysburg, painted by Thure de Thrulstrup*

General Hooker revived morale in the Army of the Potomac and instilled some of his own self-confidence into the troops. He increased unit pride by devising insignia badges for each corps. In the spring, Hooker resumed the offensive with hopes of redeeming the December disaster at Fredericksburg, and the January fiasco of the Mud March. On April 30, instead of charging straight across the Rappahannock River, Hooker crossed several miles upstream, and came in on Lee's rear. But instead of retreating, as Hooker expected him to do, Lee faced about most of his troops, and confronted the Federals in the dense woods known locally as the Wilderness – near the crossroads hostelry of Chancellorsville – where Union superiority in numbers and artillery would count for less. Nonplussed by Lee's action, Hooker lost his nerve, lost the initiative, and lost the ensuing battle.

This outcome deepened the gloom in the north. "My God!" exclaimed Lincoln when he heard the news of Chancellorsville. "What will the country say?" Copperhead opposition to the war intensified. Southern sympathizers in Britain renewed efforts for diplomatic recognition of the Confederacy. Lee decided to exploit his tactical victory at Chancellorsville by again invading the North. He hoped that another victory, this time on Northern soil, would convince Northerners and foreigners alike that the Confederacy was invincible. Lee believed his own army invincible. "There never were such men in an army before," he wrote of his troops. "They will go anywhere and do anything if properly led." So he led them into Pennsylvania during those halcyon days of June 1863 – in retrospect a move that became known as the high tide of the Confederacy.

The Army of the Potomac was not as demoralized by its defeats as Lee believed. Nor was his own army invincible. After three days of desperate battle around the Pennsylvania village of Gettysburg, the crippled army of Northern Virginia limped back to its home state, shorn of a third of its strength by the approximately 25,000 casualties it had suffered. The news of Gettysburg electrified the North. Lincoln prodded General George G. Meade, who had replaced Hooker on the eve of the battle, to pursue the retreating Confederates with vigor, but the cautious Meade allowed Lee to escape across the Potomac. Lincoln was acutely depressed. He vented his frustration in a letter to Meade that, upon reflection, he never sent. "My dear general," wrote the President, "I do believe you appreciate the magnitude of the misfortune involved in Lee's escape. He was within your easy grasp, and to have closed upon him would, in connection with our other late successes, have ended the war. As it is, the war will be prolonged indefinitely."

Among the "other late successes" Lincoln had in mind was the capture of Vicksburg. In mid-April Grant had begun a move that would enclose Vicksburg in a vice. After the Union fleet ran downriver past the Confederate batteries, and the infantry slogged down the west bank – while Benjamin Grierson's Union cavalry brigade distracted Confederate defenders with a spectacular raid through the whole state of Mississippi – Grant crossed his army to the east bank 40 miles below Vicksburg on May 1. During the next three weeks the Federals marched 180 miles, won five battles, and penned up 32,000 Confederate troops and 3,000 civilians in Vicksburg between the Union army on land and the gunboats on the river. After a siege of almost seven weeks, the starving defenders surrendered to Grant on July 4. For the second time in his career (at Fort Donelson and now at Vicksburg), Grant captured an entire enemy army. Five days later, on receiving the news from Vicksburg, the Confederate garrison at Port Hudson 200 river miles south of Vicksburg also surrendered to a besieging Union army. Northern forces thenceforth controlled the whole length of the Mississippi River. "The Father of Waters again goes unvexed to the sea," said Lincoln. The Confederacy was rent in twain. Lincoln knew who deserved the credit: "Grant is my man," he said, "and I am his the rest of the war."

The twin victories of Gettysburg and Vicksburg caused Northern spirits to soar. "Copperheads are palsied and dumb for the moment at least," wrote a jubilant New York Republican. "Government is strengthened four-fold at home and abroad." From the American legation in London the son of the American minister, Henry Adams, wrote that "the disasters of the rebels are unredeemed by even any hope of success. It is now conceded that all hope of [European] intervention is at an end."

As Northern morale rose, Southern spirits sank. "This is the darkest day of the war," wrote a Richmond diarist when he learned of the outcome at Gettysburg. Even the usually buoyant Josiah Gorgas, brilliant Chief of Ordnance for Confederate armies, wrote in his diary on July 28:

> *Events have succeeded one another with disastrous rapidity. One brief month ago we were apparently at the point of success. Lee was in Pennsylvania, threatening Harrisburg, and even Philadelphia. Vicksburg seemed to laugh all Grant's efforts to scorn. ...Now the picture is just as sombre as it was bright then. ...It seems incredible that human power could effect such a change in so brief a space. Yesterday we rode on the pinnacle of success – today absolute ruin seems to be our portion.*

Northerners had scarcely finished celebrating Gettysburg and Vicksburg when they learned of an important – and almost bloodless – triumph in Tennessee. After the battle of

Stones River, the Union Army of the Cumberland and the Confederate Army of Tennessee shadow-boxed for nearly six months as they recovered from the trauma of that battle. On June 23, Union commander William S. Rosecrans finally began an offensive to dislodge the Confederates from their defenses in the Cumberland foothills of east central Tennessee. He used cavalry and a mounted infantry brigade armed with new repeating rifles to circumvent the Confederate flanks, while his infantry threatened their front. In the first week of July, the Confederates retreated all the way to Chattanooga. After a pause for resupply, Rosecrans came on again, pushing across the Tennessee River below Chattanooga and compelling Bragg to evacuate the city on September 9. Meanwhile, another Union army captured Knoxville. Lincoln's cherished goal of liberating East Tennessee, whose inhabitants were almost as strongly Unionist as those of West Virginia, had been achieved.

But then Confederate commander Braxton Bragg counterattacked, precipitating the two-day battle of Chickamauga (September 19–20) in which the total casualties of 35,000 were second only to Gettysburg's 50,000 in the war as a whole. The Confederates' tactical victory at Chickamauga proved to be barren of strategic results. The Lincoln administration transferred two small army corps from the Army of the Potomac to Chattanooga, and gave Grant command of the combined forces there, including two corps of the Army of the Tennessee under Sherman, which also moved to Chattanooga. For the first time in the war, troops from all three of the principal Union armies fought together, under the overall command of the North's best general. The combination was too much for Braxton Bragg. On November 25, in the most successful frontal assault of the war, Union forces broke Bragg's defensive line on Missionary Ridge east of Chattanooga, driving the routed Confederates all the way to Georgia.

The Union victory at Chattanooga had important consequences: it brought to a climax a string of Federal triumphs in the second half of 1863 which made that year one of "calamity...defeat...utter ruin," in the words of a Confederate official. The Southern diarist, Mary Boykin Chesnut, found "gloom and unspoken despondency hang[ing] like a pall everywhere." Jefferson removed the discredited Bragg from command of the Army of Tennessee, replacing him

*Siege of Vicksburg, from a lithograph by Kurz and Allison*

reluctantly with Joseph E. Johnston, in whom Davis had little confidence. In March 1864, Lincoln summoned Grant to Washington and appointed him General-in-Chief of all Union armies, signifying a relentless fight to the finish.

Events in 1863 also confirmed emancipation as a Union war aim. Many Northerners had not greeted the Emancipation Proclamation with enthusiasm. Democrats and border-state Unionists denounced it, and many Union soldiers resented the idea that they would now be risking their lives for black freedom. The Democratic party hoped to capitalize on this opposition, and on Union military failures to win important off-year elections – especially for governor of Pennsylvania and governor of Ohio, where Vallandigham was the candidate, despite his exile in Canada. Northern military victories knocked one prop out from under the Democratic platform, and the performance of black soldiers fighting for the Union, and for their own freedom, knocked out another.

The enlistment of black soldiers was a logical corollary of emancipation. One purpose of declaring the freedom of slaves was to deprive the Confederacy of black labor, and to use that labor for the Union cause. Putting some of those laborers in uniform was a compelling idea, especially as white enlistments lagged and the North had to enact conscription in 1863. The Emancipation Proclamation announced an intention to enroll able-bodied male freedmen in new black regiments. They were to serve at first as labor battalions, supply troops, and garrison forces rather than as frontline combat troops. They were paid less than white soldiers, and their officers were white – in other words, black men in blue would be second-class soldiers, just as free blacks in the North were second-class citizens.

But pressure from abolitionists, as well as military necessity, eroded some of this discrimination. Congress enacted equal pay in 1864. Officers worked for better treatment of their men. Above all, they fought for their right to *fight* as combat soldiers. Even some previously hostile white soldiers came around to the notion that black men might just as well stop enemy bullets as white men. In May and June 1863, new black regiments in Louisiana fought well in an assault on Port Hudson, and in defense of a Union outpost at Milliken's Bend, near Vicksburg.

Even more significant was the action of the 54th Massachusetts Infantry, the first black regiment raised in the North. Its officers, headed by Colonel Robert Gould Shaw, came from prominent New England anti-slavery families. Two sons of the black leader Frederick Douglass were in the regiment. Shaw worked hard to win the right for the regiment to fight. On July 18, 1863, he won his point: the 54th was assigned "the post of honor" to lead an assault on Fort

*Union winter uniform, pencil sketch by Edwin Forbes*

54TH MASS

Wagner, part of the network of Confederate defenses protecting Charleston. Though the attack failed, the 54th fought courageously, suffering almost 50% casualties, including Colonel Shaw who was killed by a bullet through his heart. This battle "made Fort Wagner such a name to the colored race as Bunker Hill had been for ninety years to the white Yankees," declared the *New York Tribune*.

The battle took place just after the draft riots in New York. Abolitionists and Republican commentators drew the moral: black men who fought for the Union deserved more respect than white men who rioted against it. Lincoln made this point eloquently in a widely published letter to a Union meeting in August 1863. "Some of the commanders of our armies in the field who have given us our most important successes (he meant Grant), believe the emancipation policy, and the use of colored troops constitute the heaviest blow yet dealt to the rebellion," wrote the President. When final victory was won, he continued, "there will be some black men who can remember that, with silent tongue, and clenched teeth, and steady eye, and well-poised bayonet, they have helped mankind on to this great consummation; while, I fear, there will be some white ones, unable to forget that, with malignant heart, and deceitful speech, they have strove to hinder it."

Lincoln's letter set the tone for Republican campaigns in state elections that fall. The party swept them all – including Ohio – where they buried Vallandigham under a 100,000-vote margin swelled by the soldier vote, which went 94% against him. In effect, the elections were an endorsement of the administration's emancipation policy. If the Emancipation Proclamation had been submitted to a referendum a year earlier, observed an Illinois newspaper in December 1863, "the voice of a majority would have been against it. And yet not a year has passed before it is approved by an overwhelming majority."

But the final abolition of slavery would require Union victory in the war. Despite Northern success in the latter half of 1863, the war was far from won. Some of the heaviest fighting still lay ahead.

*The storming of Fort Wagner and the death of Col. Robert Gould Shaw, from a lithograph by Kirz and Allison*

# Grant's Second Vicksburg Campaign: Phase 1 JANUARY 8 – MAY 2 1863

Grant determined to operate against Vicksburg from a base on the Mississippi, and in the last days of January 1863 he established his headquarters at Milliken's Bend. This position would enable him to utilize Admiral David D. Porter's gunboats operating above Vicksburg, and supply his army by steam boat. Grant's Army of the Tennessee consisted of four corps: Sherman's XV Corps at Young's Point; McClernand's XIII at Milliken's Bend; McPherson's XVII at Lake Providence; and Stephen A. Hurlbut's XVI in West Tennessee.

Grant's aim was to position his army on high ground north or south of Vicksburg, and on the east side of the river. To achieve this plan with safety required bypassing Vicksburg.

Sherman's corps, therefore, began to dig a canal across De Soto Point, but his men were plagued by high water. McPherson's corps cut a canal from the Mississippi to Lake Providence. But efforts to open the circuitous waterways between the lake and the Tensas River proved futile.

Grant then sent engineers 325 river miles north of Vicksburg to open a route from the Mississippi into the Coldwater by way of Yazoo Pass. A navigable channel was opened, but it was March 2 before vessels carrying a Federal division debouched into the Coldwater, mooring north of Greenwood on March 10.

But Pemberton had sufficient warning. He despatched a small party to Yazoo Pass to obstruct the waterway, and another to Green wood to fortify the narrow neck separating the Tallahatchie and Yazoo rivers, naming the fortifications Fort Pemberton. Two Federal ironclads attacked the fort on March 11, and again on 13th, but the Confederates held their own. Discouraged, the Federals retreated up the Tallahatchie.

Grant feared disaster, but accepted Porter's proposal that his ironclads, supported by the army, could relieve the Yazoo Pass Expedition by navigating the waterways of The Delta to reach the Yazoo above Confederate-held Snyder's Bluff. On March 14 five ironclads, followed by Sherman's corps in transports, ascended the Yazoo to Steele's Bayou, but were met and turned back in Deer Creek by a Confederate force sent by Pemberton. Porter was relieved by Federal infantry which drove off the the Confederates. The gunboats, covered by

> "I attribute our failure to the strength of the enemy's position, both natural and artificial, and not to his superior fighting, but as we must all in the future have ample opportunities to test this quality, it is foolish to discuss it."
>
> *Gen. William Tecumseh Sherman.*

1. **Jan 24-March 29:** Sherman tries but fails to cut canal across De Soto Point
2. **Jan 30:** Grant establishes his H.Q. at Milliken's Bend
3. **Feb 2-March 29:** McPherson tries but fails to open route from Lake Providence to Bayou Macon

*For annotations 4 - 13 see map right*

4. **Feb 3:** Federals breach levée at Yazoo Pass and gain access to Moon Lake and Coldwater River
5. **Feb 23-March 10:** Union boats navigate Yazoo Pass and descend Coldwater and Tallahatchie rivers
6. **March 11-April 4:** Confederates at Fort Pemberton stall the Federals
7. **March 16-27:** Admiral Porter's gunboats, supported by elements from Sherman's corps, try to reach the Yazoo River above the Confederates at Snyder's Bluff
8. **March 20:** Confederate force turns back Porter's gunboats at Rolling Fork
9. **March 21-27:** Porter and Sherman abandon the Steele's Bayou route and return to their Mississippi River bases
10. **April 2-25:** Steele's division advances down Deer Creek, destroying food supplies
11. **April 5-10:** The Yazoo Pass expedition returns to Helena, Arkansas
12. **April 17-May 2:** Grierson's cavalry divert the Confederates' attention from Grant's march south by their dash from La Grange, Tennessee, to Baton Rouge, Louisiana
13. **April 29-May 1:** Sherman ascends Yazoo and threatens Snyder's Bluff distracting Confederate attention from place where Grant intends to cross the Mississippi
14. **March 31-April 28:** McClernand's corps and two divisions of McPherson's corps move from Milliken's Bend to Hard Times Landing

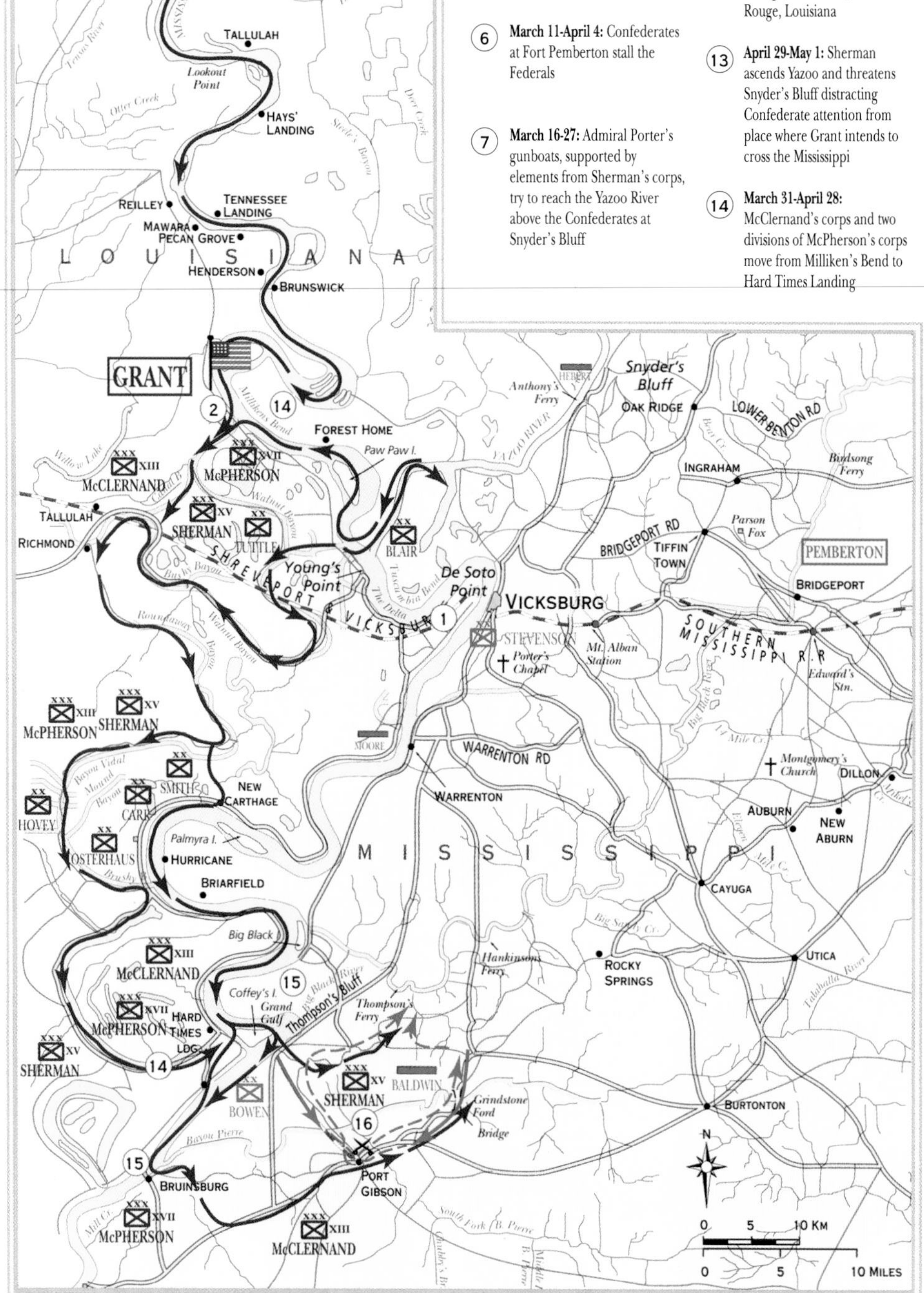
MILLIKEN'S BEND TO PORT GIBSON
January 30 - May 1, 1863
McPHERSON
TALLULAH
Lookout Point
HAYS' LANDING
REILLEY
TENNESSEE LANDING
MAWARA
PECAN GROVE
HENDERSON
BRUNSWICK
L O U I S I A N A
GRANT
FOREST HOME
Paw Paw I.
McCLERNAND
McPHERSON
SHERMAN
TUTTLE
BLAIR
TALLULAH
RICHMOND
Young's Point
De Soto Point
VICKSBURG
Snyder's Bluff
OAK RIDGE
LOWER BENTON RD
INGRAHAM
Burlsong Ferry
Anthony's Ferry
BRIDGEPORT RD
TIFFIN TOWN
Parson Fox
PEMBERTON
BRIDGEPORT
SHREVEPORT
VICKSBURG
STEVENSON
Porter's Chapel
Mt. Alban Station
SOUTHERN MISSISSIPPI R R
Edward's Stn.
McPHERSON
SHERMAN
MOORE
WARRENTON RD
WARRENTON
SMITH
CARR
HOVEY
OSTERHAUS
NEW CARTHAGE
Palmyra I.
HURRICANE
BRIARFIELD
Montgomery's Church
DILLON
AUBURN
NEW ABURN
M I S S I S S I P P I
CAYUGA
Big Black
McCLERNAND
McPHERSON
HARD TIMES LDG
Coffey's I.
Grand Gulf
Thompson's Bluff
Thompson's Ferry
Hankinsons Ferry
ROCKY SPRINGS
UTICA
SHERMAN
BOWEN
SHERMAN
BALDWIN
Grindstone Ford Bridge
BURTONTON
PORT GIBSON
BRUINSBURG
McPHERSON
McCLERNAND
N
0 5 10 KM
0 5 10 MILES

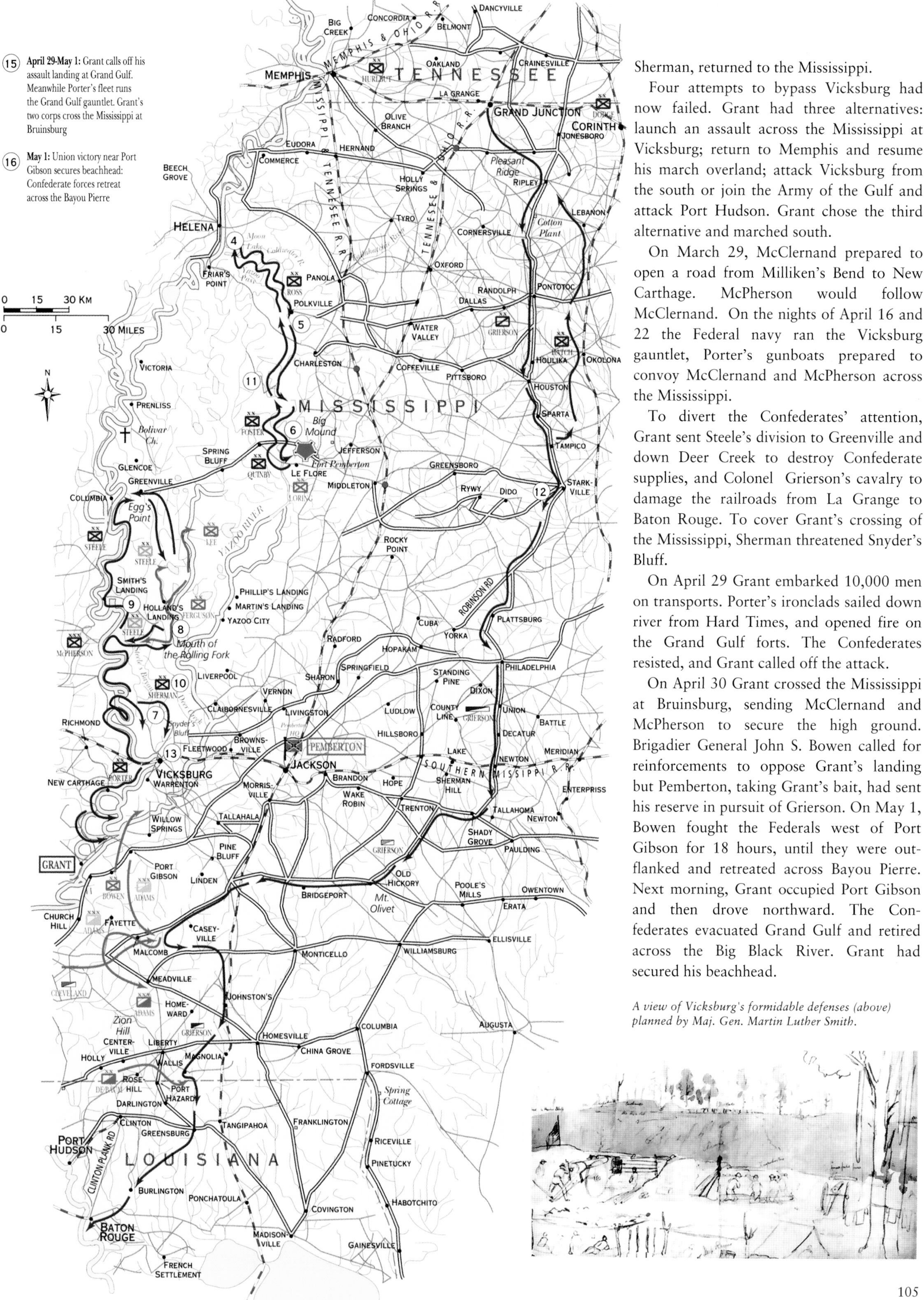

15 April 29-May 1: Grant calls off his assault landing at Grand Gulf. Meanwhile Porter's fleet runs the Grand Gulf gauntlet. Grant's two corps cross the Mississippi at Bruinsburg
16 May 1: Union victory near Port Gibson secures beachhead: Confederate forces retreat across the Bayou Pierre
0 15 30 KM
0 15 30 MILES
N
TENNESSEE
MISSISSIPPI
LOUISIANA
MEMPHIS
CORINTH
GRAND JUNCTION
HELENA
VICKSBURG
JACKSON
PEMBERTON
GRANT
PORT HUDSON
BATON ROUGE
MEMPHIS & OHIO R.R.
MISSISSIPPI & TENNESSEE R.R.
TENNESSEE & OHIO R.R.
SOUTHERN MISSISSIPPI R.R.
ROBINSON RD
CLINTON PLANK RD
YAZOO RIVER
Fort Pemberton
Big Mound
Egg's Point
Mouth of the Rolling Fork
Snyder's Bluff
GREENVILLE
GRAND GULF
PORT GIBSON
GRIERSON
SHERMAN
McPHERSON
STEELE
PORTER


Sherman, returned to the Mississippi.

Four attempts to bypass Vicksburg had now failed. Grant had three alternatives: launch an assault across the Mississippi at Vicksburg; return to Memphis and resume his march overland; attack Vicksburg from the south or join the Army of the Gulf and attack Port Hudson. Grant chose the third alternative and marched south.

On March 29, McClernand prepared to open a road from Milliken's Bend to New Carthage. McPherson would follow McClernand. On the nights of April 16 and 22 the Federal navy ran the Vicksburg gauntlet, Porter's gunboats prepared to convoy McClernand and McPherson across the Mississippi.

To divert the Confederates' attention, Grant sent Steele's division to Greenville and down Deer Creek to destroy Confederate supplies, and Colonel Grierson's cavalry to damage the railroads from La Grange to Baton Rouge. To cover Grant's crossing of the Mississippi, Sherman threatened Snyder's Bluff.

On April 29 Grant embarked 10,000 men on transports. Porter's ironclads sailed down river from Hard Times, and opened fire on the Grand Gulf forts. The Confederates resisted, and Grant called off the attack.

On April 30 Grant crossed the Mississippi at Bruinsburg, sending McClernand and McPherson to secure the high ground. Brigadier General John S. Bowen called for reinforcements to oppose Grant's landing but Pemberton, taking Grant's bait, had sent his reserve in pursuit of Grierson. On May 1, Bowen fought the Federals west of Port Gibson for 18 hours, until they were outflanked and retreated across Bayou Pierre. Next morning, Grant occupied Port Gibson and then drove northward. The Confederates evacuated Grand Gulf and retired across the Big Black River. Grant had secured his beachhead.

*A view of Vicksburg's formidable defenses (above) planned by Maj. Gen. Martin Luther Smith.*

# Grant's Second Vicksburg Campaign: Phase 2

## MAY 2 – 17 1863

Having secured his beachhead, Grant assessed his options: attacking Vicksburg from the south would capture the city, but Pemberton would escape northeast; Grant could march east and strike the Southern Railroad, pivot west, and close on Vicksburg from the east. This approach could cost Pemberton his army as well as the city. Grant, a great captain, chose the latter course.

Sherman duly crossed the Mississippi at Grand Gulf. The Army of the Tennessee resumed its advance on May 8. Grant sent McPherson through Utica toward Raymond. McClernand watched the Big Black River crossings, while Sherman closed on Auburn.

On May 11, Confederate Brigadier General John Gregg's brigade marched from Jackson to Raymond. At 10 a.m. on May 12, Gregg sighted McPherson's vanguard on the Utica road, and the fight began. Initially, Gregg's 3,000 troops held the initiative, McPherson failing to use his greater numbers to coordinate an attack. But finally, Gregg retreated toward Jackson.

The battle had important repercussions: when McPherson reported it, he doubled Gregg's true strength. This, together with rumors that General Joseph Johnston was expected at Jackson, caused Grant to alter his plan. Instead of striking for Bolton on May 13, he turned his army east. McPherson and Sherman were to capture Jackson and destroy the railroads. McClernand's corps would cover them.

Pemberton ordered his army to concentrate at Edward's Station, within easy striking distance of a Federal corps. McClernand bluffed an attack on Edward's Station. The Confederates dug in on hills south of the village, while McClernand broke contact. Stealing a march on Pemberton, he camped at Raymond on the 13th. Johnston reached Jackson late that day. After conferring with Gregg, he telegraphed Richmond: "I am too late," and prepared to evacuate the city.

On May 14, in driving rain, McPherson and Sherman converged on Jackson. After some skirmishing, the Confederates retired into the fortifications. Before the Federals could attack, word arrived that the Confederates had evacuated Jackson and were retiring up the Canton Road.

With the Federals in Jackson, Confederate brigades en route to reinforce Johnston were halted at Crystal Springs, Meridian and at Brandon. Johnston's evacuation of Jackson had scattered his army and critical days

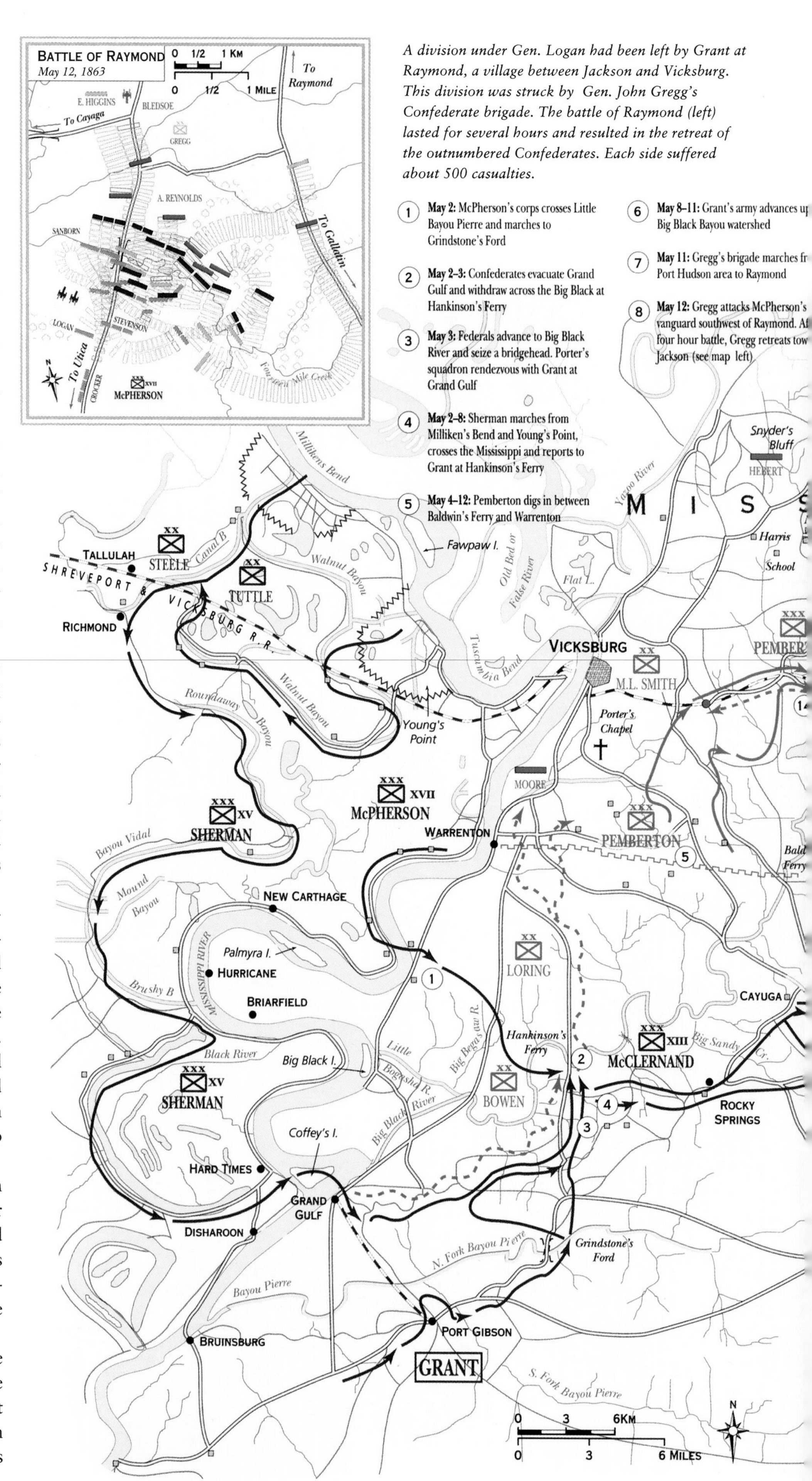
BATTLE OF RAYMOND
May 12, 1863
0 1/2 1 KM
0 1/2 1 MILE
To Raymond
E. HIGGINS
BLEDSOE
To Cayaga
GREGG
A. REYNOLDS
SANBORN
To Gallatin
LOGAN
STEVENSON
To Utica
CROCKER
McPHERSON
XVII
Fourteen Mile Creek
1 May 2: McPherson's corps crosses Little Bayou Pierre and marches to Grindstone's Ford
2 May 2–3: Confederates evacuate Grand Gulf and withdraw across the Big Black at Hankinson's Ferry
3 May 3: Federals advance to Big Black River and seize a bridgehead. Porter's squadron rendezvous with Grant at Grand Gulf
4 May 2–8: Sherman marches from Milliken's Bend and Young's Point, crosses the Mississippi and reports to Grant at Hankinson's Ferry
5 May 4–12: Pemberton digs in between Baldwin's Ferry and Warrenton
6 May 8–11: Grant's army advances up Big Black Bayou watershed
7 May 11: Gregg's brigade marches from Port Hudson area to Raymond
8 May 12: Gregg attacks McPherson's vanguard southwest of Raymond. After four hour battle, Gregg retreats toward Jackson (see map left)
Snyder's Bluff
HEBERT
Yazoo River
M I S
Harris
School
Milliken's Bend
Fawpaw I.
TALLULAH
STEELE
TUTTLE
SHREVEPORT & VICKSBURG R.R.
RICHMOND
Walnut Bayou
Roundaway Bayou
Old Bed or False River
Flat L.
Tuscumbia Bend
VICKSBURG
M.L. SMITH
PEMBERTON
Young's Point
Porter's Chapel
MOORE
XVII
McPHERSON
XV
SHERMAN
WARRENTON
Bayou Vidal
Mound Bayou
NEW CARTHAGE
Palmyra I.
HURRICANE
BRIARFIELD
Mississippi River
Brushy B
LORING
CAYUGA
Hankinson's Ferry
XIII
McCLERNAND
Big Sandy Cr.
Black River
Big Black I.
Little Bogue Sha R.
Big Bayou R.
Big Black River
BOWEN
ROCKY SPRINGS
Coffey's I.
HARD TIMES
GRAND GULF
DISHAROON
N. Fork Bayou Pierre
Grindstone's Ford
Bayou Pierre
PORT GIBSON
BRUINSBURG
GRANT
S. Fork Bayou Pierre
0 3 6KM
0 3 6 MILES
N

*A division under Gen. Logan had been left by Grant at Raymond, a village between Jackson and Vicksburg. This division was struck by Gen. John Gregg's Confederate brigade. The battle of Raymond (left) lasted for several hours and resulted in the retreat of the outnumbered Confederates. Each side suffered about 500 casualties.*

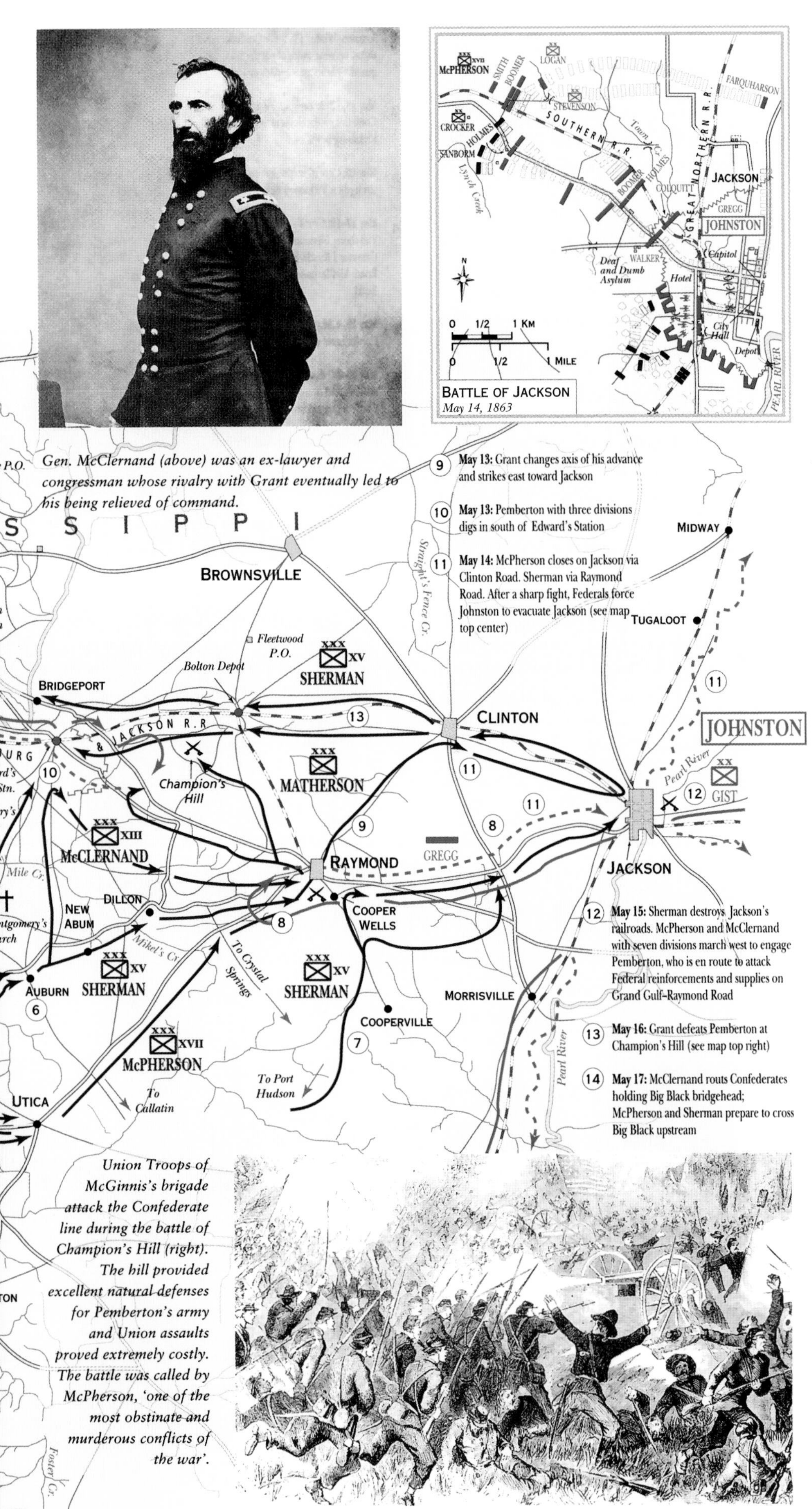
BATTLE OF JACKSON
May 14, 1863
McPHERSON
LOGAN
SMITH
BOOMER
STEVENSON
CROCKER
SANBORN
HOLMES
SOUTHERN R.R.
FARQUHARSON
GREAT NORTHERN R.R.
JACKSON
GREGG
JOHNSTON
COLQUITT
WALKER
Deaf and Dumb Asylum
Hotel
Capitol
City Hall
Depot
PEARL RIVER
Lynch Creek
0 1/2 1 KM
0 1/2 1 MILE
9 May 13: Grant changes axis of his advance and strikes east toward Jackson
10 May 13: Pemberton with three divisions digs in south of Edward's Station
11 May 14: McPherson closes on Jackson via Clinton Road. Sherman via Raymond Road. After a sharp fight, Federals force Johnston to evacuate Jackson (see map top center)
12 May 15: Sherman destroys Jackson's railroads. McPherson and McClernand with seven divisions march west to engage Pemberton, who is en route to attack Federal reinforcements and supplies on Grand Gulf-Raymond Road
13 May 16: Grant defeats Pemberton at Champion's Hill (see map top right)
14 May 17: McClernand routs Confederates holding Big Black bridgehead; McPherson and Sherman prepare to cross Big Black upstream
S I P P I
BROWNSVILLE
MIDWAY
TUGALOOT
Fleetwood P.O.
Bolton Depot
SHERMAN
BRIDGEPORT
JACKSON R.R.
CLINTON
JOHNSTON
GIST
Champion's Hill
MATHERSON
McCLERNAND
RAYMOND
GREGG
JACKSON
DILLON
NEW ABURN
COOPER WELLS
AUBURN
SHERMAN
MORRISVILLE
COOPERVILLE
McPHERSON
UTICA
To Crystal Springs
To Port Hudson
To Callatin
Pearl River
Straight's Fence Cr.

Gen. McClernand (above) was an ex-lawyer and congressman whose rivalry with Grant eventually led to his being relieved of command.

Union Troops of McGinnis's brigade attack the Confederate line during the battle of Champion's Hill (right). The hill provided excellent natural defenses for Pemberton's army and Union assaults proved extremely costly. The battle was called by McPherson, 'one of the most obstinate and murderous conflicts of the war'.

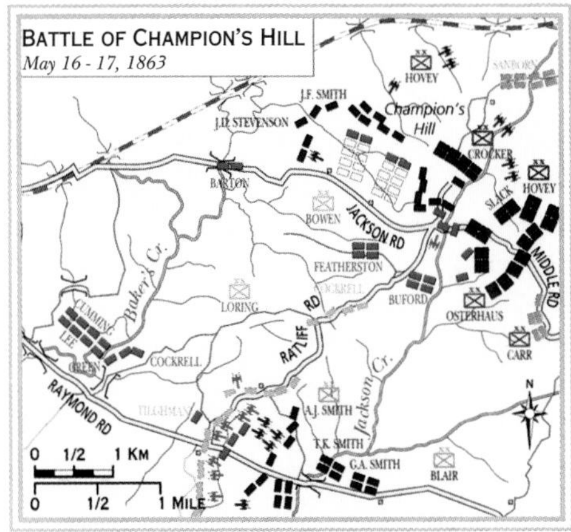
BATTLE OF CHAMPION'S HILL
May 16 - 17, 1863
HOVEY
J.E. SMITH
J.D. STEVENSON
Champion's Hill
CROCKER
BARTON
BOWEN
JACKSON RD
FEATHERSTON
LORING
BUFORD
OSTERHAUS
MIDDLE RD
CARR
COCKRELL
RATLIFF RD
RAYMOND RD
A.J. SMITH
G.A. SMITH
BLAIR
Jackson Cr.
Bakers Cr.
0 1/2 1 KM
0 1/2 1 MILE

*Realising the futility of attempting to halt Grant with only 12,000 men, Johnston began to evacuate to the north. Two brigades were left to delay the Federals at Jackson (above left) but these were easily defeated by McPherson's and Sherman's corps.*

*To meet the threat posed by Grant's advance from Jackson, Pemberton deployed one division to defend his left, and two divisions to oppose McClernand. But the Confederates were blocked by Federal forces at Champion's Hill (above). In the subsequent fighting the hill changed hands three times. The Confederates failed to rally and began to withdraw toward Vicksburg.*

passed before it was concentrated.

Johnston ordered Pemberton east to attack the Federals near Clinton, but one of the Confederate couriers was a Federal agent who delivered the order to Grant. Grant ordered Sherman to destroy Jackson's railroads, while McPherson and McClernand marched to intercept Pemberton.

On May 16, with Grant's columns advancing, Johnston ordered Pemberton to concentrate north of the Southern Railroad. He was too late: McPherson was approaching Champion Hill. Pemberton deployed Major General Carter L. Stevenson's division to meet this threat to his left, while General Bowen's and Major General William W. Loring's divisions faced east to oppose McClernand. At 10.30 a.m. Grant launched an attack on Stevenson. In the bitter fighting, Champion Hill, changed hands three times, and Bowen made one of the great charges of the war. By 5 p.m. the Confederates retreated across Bakers Creek.

Grant resumed the advance on May 17. Sherman and McPherson were to cross the Big Black upstream from the Confederate bridgehead – a line of breastworks held by Bowen. But before they could cross, one of McClernand's brigades charged the breastworks. A Confederate brigade panicked, and those to its left and right fled west across the river. Pemberton retired into Vicksburg with more of a mob than an army.

# Siege and Capture of Vicksburg 18 MAY – 4 JULY 1863

The Army of the Tennessee crossed the Big Black River on May 17, and closed on Vicksburg. On May 19, Grant re-established contact with Admiral Porter's fleet on the Yazoo River. Unaware that Pemberton had two fresh divisions in Vicksburg, Grant thought the victories at Champion Hill's and Big Black River had crushed the Confederates' morale. But Pemberton's units held in force earthworks guarding Graveyard, Jackson and Baldwin's Ferry roads, along which the Federals were approaching.

On May 19, Sherman advanced against the Graveyard road defenses, but rugged terrain disrupted the lines. The Confederates riddled the Federals with shot, pinning them down. On Sherman's left McPherson and McClernand seized ground near the Vicksburg perimeter. After dark, Sherman withdrew his men.

Now painfully aware that Pemberton's army had not been badly damaged, Grant regrouped and prepared for an all-out offensive. On May 22, cannon hammered the Confederate works, while Porter's ironclads bombarded the river forts south of the city. At 10 a.m. the artillery fell silent, and Grant's three corps charged. Sherman's and McPherson's rushes were blunted, but McClernand stormed Railroad Redoubt. Grant renewed the assaults. Pemberton's fresh reserves counterattacked, driving the Federals back. Grant suspended the attack. The Union had lost 3,199 men; the Confederates less than 500.

On May 25, Grant began siege operations to cut off supplies and reinforcements to the city. Porter controlled the Mississippi, and the Louisiana shore was occupied. Trenches were pushed toward the Confederate defenses and breaching batteries established. Federal artillery hurled thousands of shells into the city. To escape the horrors of the bombardment, citizens dug caves into the hillsides. On June 25 and July 1, mines were exploded under the Third Lousiana Redan, but a Federal attack following the first mine failed.

Grant called for reinforcements, and by the third week of June he had more than 77,000 troops. President Davis provided General Johnston with reinforcements, but Johnston was over–cautious and Confederate attacks on Federal enclaves west of the river were repulsed.

By the end of June rations in Vicksburg were running low; the troops began to eat mules and horses. The heat sapped the men's vigor, and morale sagged as it became clear Johnston was not coming to their relief.

By July 2 Pemberton had two options: to cut his way out or surrender. Realising his men were in no condition to attack, Pemberton met Grant on July 3 to discuss terms. Grant demanded unconditional surrender; Pemberton refused. Grant modified his terms, and on July 4 the Confederate army, 29,500 strong, surrendered. Grant took possession of Vicksburg.

*Union officers relax on the platform of "Whistling Dick" within the defenses of Vicksburg after the city's surrender (below). The rifled eighteen pounder had been one of the Confederates' most feared artillery pieces and had gained its name from the distinctive sound made by its shells as they passed overhead.*

1. **Autumn–Winter 1862–63:** Confederates throw up nine miles of earthworks guarding land approaches to Vicksburg
2. **May 17, 1863:** Pemberton and two Confederate divisions retreat into Vicksburg perimeter
3. **May 18:** Confederates occupy and strengthen Vicksburg defenses
4. **May 17–19:** Grant's army approaches Vicksburg: Sherman via Benton and Graveyard Roads, McPherson via Jackson Road, McClernand via Baldwin's Ferry Road
5. **May 19, 1.30 pm–dark:** Grant assails Confederate works but is repulsed
6. **May 20–22, dawn:** Grant emplaces artillery and occupies ground closer to Confederate perimeter
7. **May 22, 6.10 am:** Federal artillery and Porter's ironclads bombard Vicksburg
8. **May 22, 10 am–dark:** Grant assails Confederate defenses from 26th Louisiana Redoubt to Square Fort. Porter attacks South Fort. Federals repulsed with heavy losses
9. **May 25:** As Confederate defenses prove too strong to storm, Grant calls for siege operations and reinforcements
10. **May 27, am:** *USS Cincinnati* is sunk while attempting to gauge strength of Confederate upper water batteries
11. **May 25–July 3:** Federals forge an iron ring sealing the defenses within the perimeter

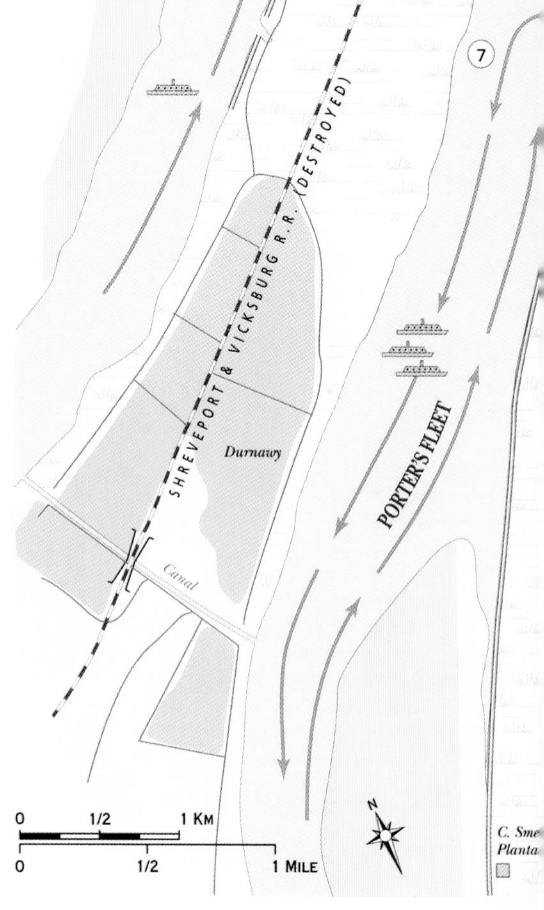
10
Cincinnati Sinks
MISSISSIPPI RIVER
De Soto Peninsular
DE SOTO (Burned)
7
SHREVEPORT & VICKSBURG R.R. (DESTROYED)
Durnawy
Canal
PORTER'S FLEET
0 1/2 1 KM
0 1/2 1 MILE
N

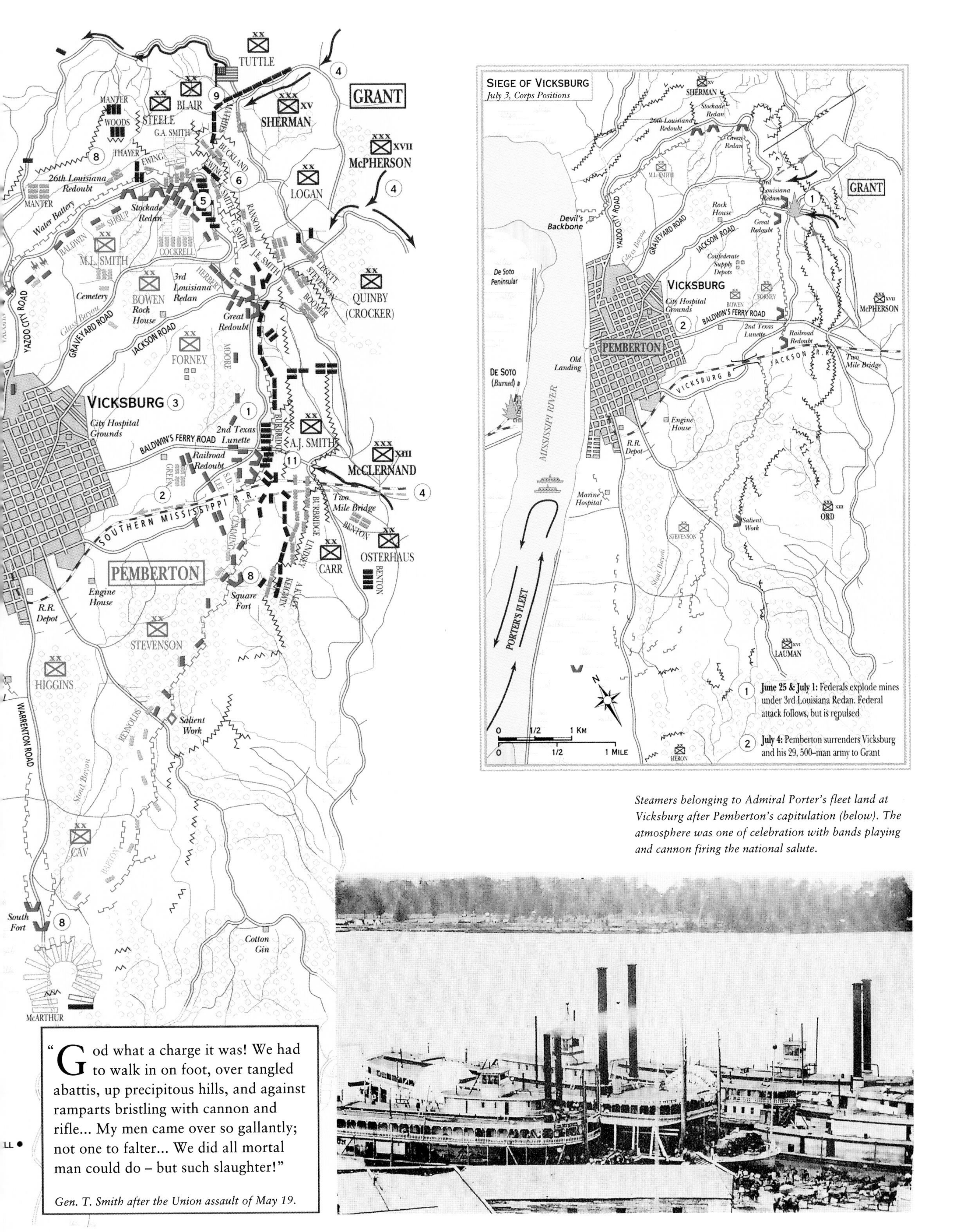
TUTTLE
GRANT
STEELE
BLAIR
SHERMAN
MANTER
WOODS
G.A. SMITH
THAYER
EWING
McPHERSON
26th Louisiana Redoubt
LOGAN
Stockade Redan
Water Battery
SHOUP
COCKRELL
BALDWIN
M.L. SMITH
RANSOM
J.E. SMITH
LEGETT
STEVENSON
BOOMER
QUINBY (CROCKER)
3rd Louisiana Redan
BOWEN
Cemetery
Rock House
Great Redoubt
Glass Bayou
YAZOO CITY ROAD
GRAVEYARD ROAD
JACKSON ROAD
FORNEY
MOORE
VICKSBURG
City Hospital Grounds
2nd Texas Lunette
BALDWIN'S FERRY ROAD
A.J. SMITH
McCLERNAND
Railroad Redoubt
BURBRIDGE
SOUTHERN MISSISSIPPI R.R.
Two Mile Bridge
BENTON
OSTERHAUS
CARR
LINDSEY
PEMBERTON
Engine House
Square Fort
R.R. Depot
STEVENSON
HIGGINS
WARRENTON ROAD
Salient Work
REYNOLDS
Stout Bayou
CAV
South Fort
Cotton Gin
McARTHUR
SIEGE OF VICKSBURG
July 3, Corps Positions
SHERMAN
Stockade Redan
26th Louisiana Redoubt
Green Redan
M.L. SMITH
GRANT
3rd Louisiana Redan
Devil's Backbone
Rock House
Great Redoubt
De Soto Peninsular
Confederate Supply Depots
VICKSBURG
City Hospital Grounds
BOWEN
FORNEY
McPHERSON
BALDWIN'S FERRY ROAD
2nd Texas Lunette
Railroad Redoubt
PEMBERTON
Old Landing
De Soto (Burned)
VICKSBURG & JACKSON R.R.
Two Mile Bridge
MISSISSIPPI RIVER
Engine House
R.R. Depot
Marine Hospital
ORD
STEVENSON
Salient Work
PORTER'S FLEET
LAUMAN
HERON
0 1/2 1 KM
0 1/2 1 MILE
1 June 25 & July 1: Federals explode mines under 3rd Louisiana Redan. Federal attack follows, but is repulsed
2 July 4: Pemberton surrenders Vicksburg and his 29, 500-man army to Grant

*Steamers belonging to Admiral Porter's fleet land at Vicksburg after Pemberton's capitulation (below). The atmosphere was one of celebration with bands playing and cannon firing the national salute.*

> "God what a charge it was! We had to walk in on foot, over tangled abattis, up precipitous hills, and against ramparts bristling with cannon and rifle... My men came over so gallantly; not one to falter... We did all mortal man could do – but such slaughter!"
>
> *Gen. T. Smith after the Union assault of May 19.*

# Port Hudson Campaign MAY 8 – JULY 9 1863

By the spring of 1863, Confederate General Franklin Gardner had transformed Port Hudson into a bastion more formidable than Vicksburg. Hoping to achieve a bloodless victory, Admiral Farragut and General Banks, from March 14 until May 7, attempted to force Gardner to abandon Port Hudson by severing his supply line with the Trans-Mississippi. When this strategy failed, Banks began his encirclement of Port Hudson. Three Federal divisions moved down the Red River to strike from the north, while two others advanced from Baton Rouge and New Orleans to attack from the east and south. By May 22, Banks's 30,000 Federal troops and Farragut's fleet had isolated Gardner's 7,500 men.

Terrain, vague orders, and uncooperative senior officers prevented the five Federal divisons from attacking simultaneously on May 27. Federal troops first struck the fortifications opposite the Confederate left. Reinforcements from the center enabled defenders to repulse the assaults along Little Sandy Creek. The fighting in this sector ended three hours before Sherman's and Augur's divisions assailed the Confederate center. This delay enabled Gardner to abandon his center to reinforce his left without

*Union losses at Port Hudson were increased by the poor judgement of Nathaniel Prentiss Banks (above). Earlier defeats in the Shenandoah Valley and at Cedar Mountain had apparently done little to abate the government's confidence in a man who was essentially a politician rather than a soldier.*

"It gives me pleasure to report that they answered every expectation. In many respects their conduct was heroic. No troops could be more determined or more daring."

*Banks describing the actions of his African American troops.*

Dense Swamp and Cane Brakes

SHE

1. **May 27, 8 am:** Two divisions under Brig. Gen. Godfrey Weitzel reach Confederate breastworks
2. **9 am:** Reinforcements from the center enable Confederates to repulse Weitzel's attack
3. **10 am:** In response to Weitzel's failure to break the Confederate left, Brig. Gen. William Dwight, of Sherman's division, orders two regiments to attack the right
4. **10 am:** Elements of Grover's division join the attack in support of Weitzel
5. **10:30 am:** Four regiments of Grover's division attack Fort Desperate

*A Union assault on the defenses of Port Hudson (below). Besides those lost in the futile attacks, the besiegers were further weakened by disease, dysentery and sunstroke – rife in the Port Hudson swamps.*

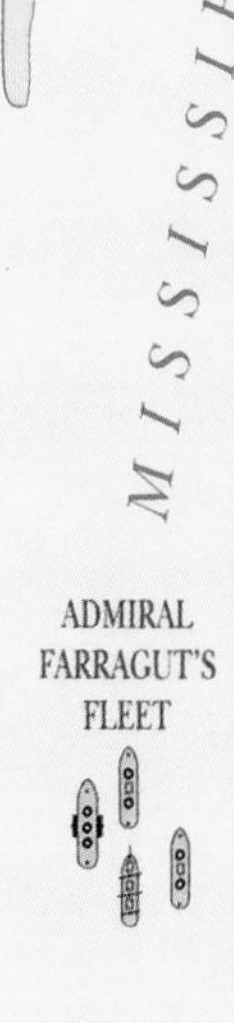
MISSISSIPPI R
ADMIRAL FARRAGUT'S FLEET

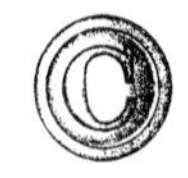

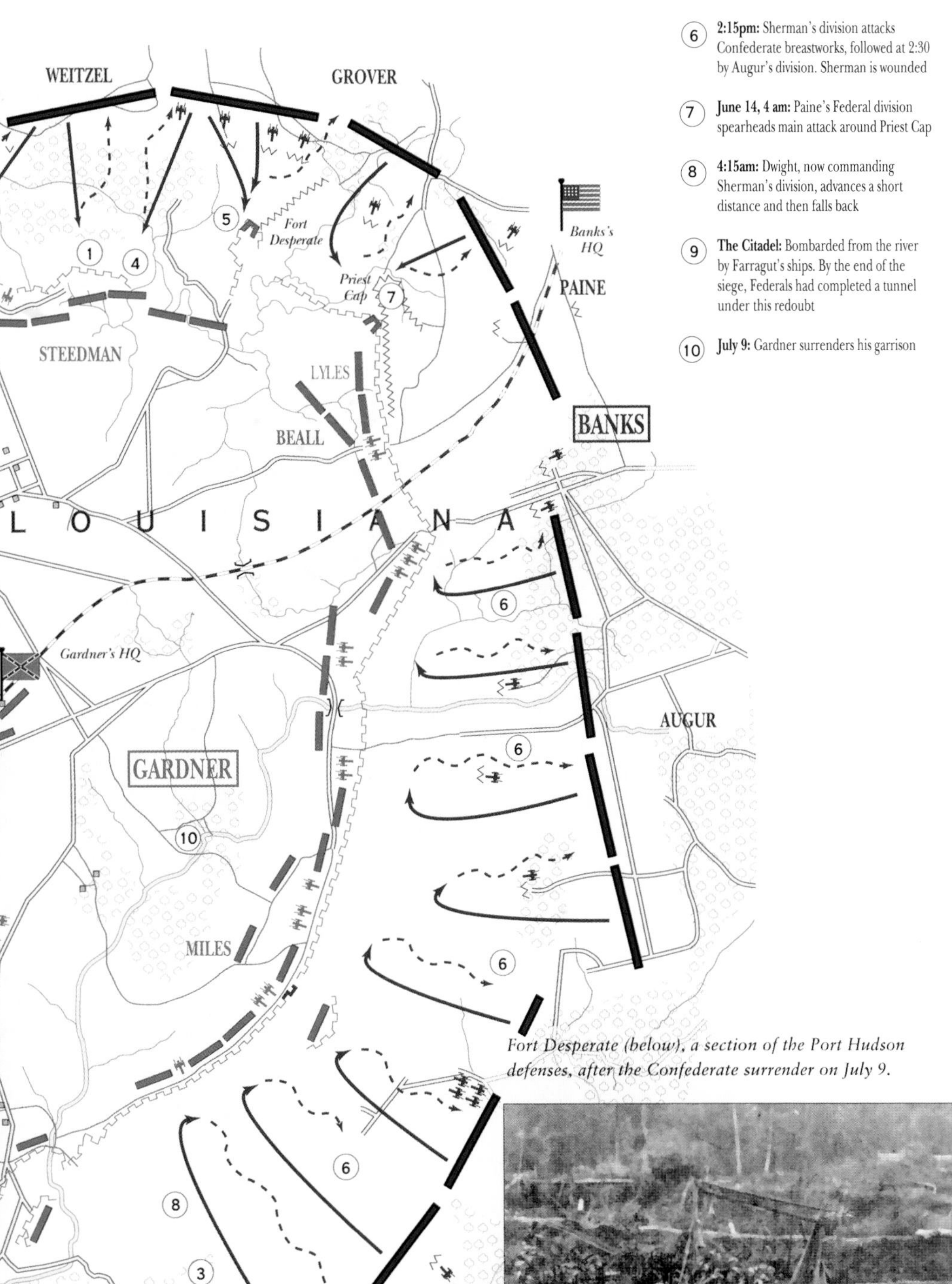

WEITZEL
GROVER
Fort Desperate
Priest Cap
Banks's HQ
PAINE
STEEDMAN
LYLES
BEALL
BANKS
L O U I S I A N A
Gardner's HQ
AUGUR
GARDNER
MILES
SHERMAN (DWIGHT)
N
0 1/4 1/2 KM
0 1/4 1/2 MILE


6 **2:15pm:** Sherman's division attacks Confederate breastworks, followed at 2:30 by Augur's division. Sherman is wounded

7 **June 14, 4 am:** Paine's Federal division spearheads main attack around Priest Cap

8 **4:15am:** Dwight, now commanding Sherman's division, advances a short distance and then falls back

9 **The Citadel:** Bombarded from the river by Farragut's ships. By the end of the siege, Federals had completed a tunnel under this redoubt

10 **July 9:** Gardner surrenders his garrison

"I have to state that my duty requires me to defend this position, and therefore I decline to surrender."

*Gardner in response to Banks's request that the garrison surrender.*

waiting for troops from the right to fill the gap. When the two Federal divisions finally advanced across Farmer Slaughter's field, the gap no longer existed. Not one Federal soldier reached the Confederate breastworks. Federal casualties included 600 African-Americans. These black soldiers had demonstrated that their willingness and ability to fight equaled that of white soldiers.

Unwilling to withdraw, Banks commenced siege operations. On June 13, after bombarding the Confederates for an hour, he demanded their surrender; Gardner refused. The bombardment resumed and Federal troops deployed for an attack the following morning. Paine's division would spearhead the main assault against Gardner's center, supported by diversionary attacks by Dwight's and Weitzel's divisions. Paine's and Weitzel's divisions advanced at 4 a.m., but Dwight's provided little assistance. The attack ended in a bloody repulse for Banks, who sustained 1,805 casualties to less than 200 for Gardner. The siege continued until Gardner learned of Vicksburg's capitulation, which eliminated the strategic importance of Port Hudson to the Confederacy. Gardner surrendered the garrison on July 9.

*Fort Desperate (below), a section of the Port Hudson defenses, after the Confederate surrender on July 9.*

# Chancellorsville: Phase 1 APRIL 26 – MAY 1 1863

Morale in the Federal Army of the Potomac rose with the appointment of Joseph Hooker to command. Hooker reorganized the army and formed a cavalry corps. He wanted to strike at Lee's army while a sizable portion was detached under Longstreet in the Suffolk area. The Federal commander left a substantial force at Fredericksburg to tie Lee to the hills where Burnside had been defeated. Another Union force disappeared westward, crossed the Rapidan and Rappahannock rivers, and converged on Fredericksburg from the west. The Federal cavalry would open the campaign with a raid on Lee's line of communications with the Confederate capital at Richmond. Convinced that Lee would have to retreat, Hooker trusted that his troops could defeat the Confederates as they tried to escape his trap.

On April 29, Hooker's cavalry and three army corps crossed Kelly's Ford. His columns split, with the cavalry pushing to the west while the army corps secured Germanna and Ely's fords. The next day these columns reunited at Chancellorsville. Lee reacted to the news of the Federals in the Wilderness by sending General Richard H. Anderson's division to investigate. Finding the Northerners massing in the woods around Chancellorsville; Anderson com-

1. **April 26:** Federal I and VI Corps cross the river and demonstrate against Lee's Fredericksburg defenses
2. **April 27:** Federal V, XI and XII Corps concentrate, preparing to move upriver
3. **April 29:** Federals cross Kelly's Ford against slight opposition before splitting their columns
4. **April 29:** Meade's V Corps secures Ely's Ford during the evening
5. **April 29:** Howard's XI Corps and Slocum's XII Corps cross Germanna Ford during the evening
6. **April 30:** Sykes's division of V Corps uncovers U.S. Ford for the two Federal corps to cross
7. **April 30:** Hooker reunites the right wing of the Army of the Potomac
8. **April 30:** J.E.B. Stuart's Confederate cavalry clings to the Federal column, sending Lee information as to its strength
9. **April 30-May 1:** Lee divides his army, leaving Early at Fredericksburg while the remainder move toward Chancellorsville

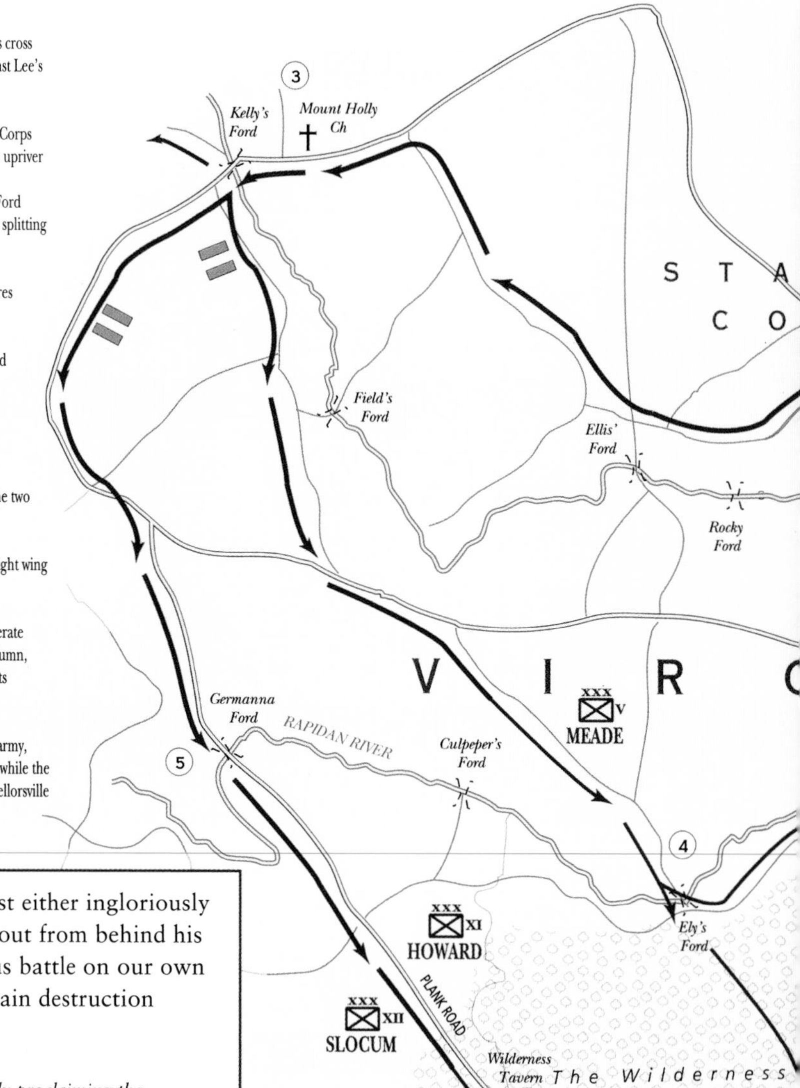
MORRISVILLE
Kelly's Ford
Mount Holly Ch
3
S T A
C O
Field's Ford
Ellis' Ford
Rocky Ford
V I R
XXX V MEADE
Germanna Ford
RAPIDAN RIVER
Culpeper's Ford
5
4
Ely's Ford
XXX XI HOWARD
XXX XII SLOCUM
PLANK ROAD
Wilderness Tavern
The Wilderness
Wilderness Ch
OLD PIKE
PLANK ROAD
BROCK ROAD
S P O T
Todd's Tavern
8
Shady Grove Ch

> "Our enemy must either ingloriously fly, or come out from behind his defenses and give us battle on our own ground, where certain destruction awaits him."
>
> *Gen. Hooker, confidently proclaiming the advantages of his plan on April 30, immediately prior to the Chancellorsville fiasco.*

*Stoneman's Union raiders cross the Rappahannock at Kelly's Ford (below). Stoneman's efforts were acclaimed by the Washington press at a time when news of Union success was much needed, but though the raids accomplished their immediate goals, the division of his forces would cost Hooker dear.*

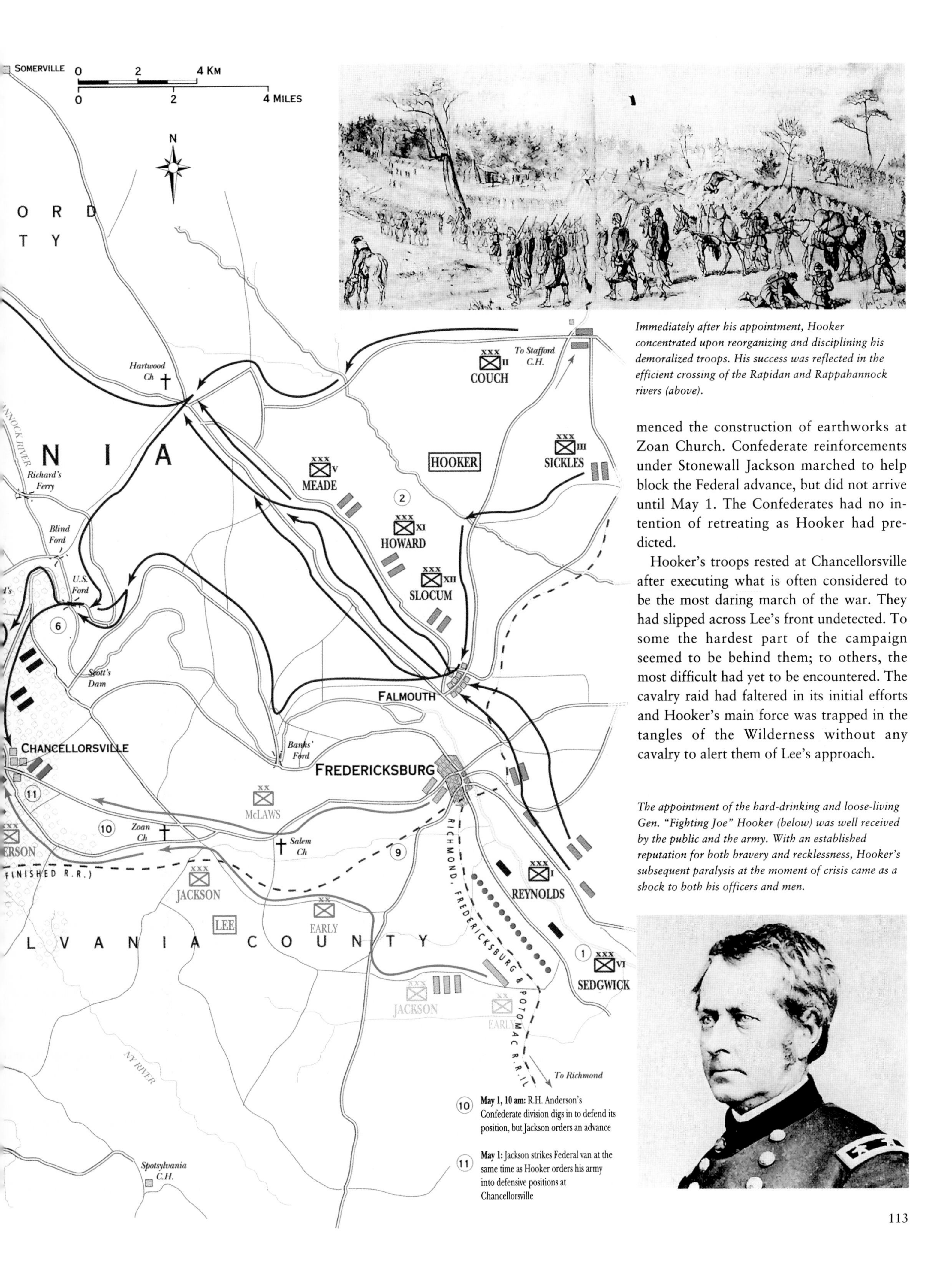

SOMERVILLE
0 2 4 KM
0 2 4 MILES
N
Hartwood Ch
To Stafford C.H.
COUCH
SICKLES
HOOKER
MEADE
HOWARD
SLOCUM
Richard's Ferry
Blind Ford
U.S. Ford
Scott's Dam
FALMOUTH
CHANCELLORSVILLE
Banks' Ford
FREDERICKSBURG
McLAWS
Zoan Ch
Salem Ch
JACKSON
LEE
EARLY
REYNOLDS
SEDGWICK
RICHMOND, FREDERICKSBURG & POTOMAC R.R.
To Richmond
Spotsylvania C.H.
May 1, 10 am: R.H. Anderson's Confederate division digs in to defend its position, but Jackson orders an advance
May 1: Jackson strikes Federal van at the same time as Hooker orders his army into defensive positions at Chancellorsville


*Immediately after his appointment, Hooker concentrated upon reorganizing and disciplining his demoralized troops. His success was reflected in the efficient crossing of the Rapidan and Rappahannock rivers (above).*

menced the construction of earthworks at Zoan Church. Confederate reinforcements under Stonewall Jackson marched to help block the Federal advance, but did not arrive until May 1. The Confederates had no intention of retreating as Hooker had predicted.

Hooker's troops rested at Chancellorsville after executing what is often considered to be the most daring march of the war. They had slipped across Lee's front undetected. To some the hardest part of the campaign seemed to be behind them; to others, the most difficult had yet to be encountered. The cavalry raid had faltered in its initial efforts and Hooker's main force was trapped in the tangles of the Wilderness without any cavalry to alert them of Lee's approach.

*The appointment of the hard-drinking and loose-living Gen. "Fighting Joe" Hooker (below) was well received by the public and the army. With an established reputation for both bravery and recklessness, Hooker's subsequent paralysis at the moment of crisis came as a shock to both his officers and men.*

# Chancellorsville: Phase 2 MAY 1 – 6 1863

As the Federal army converged on Chancellorsville, General Hooker expected Lee to retreat from his forces, which totaled nearly 115,000. Although heavily outnumbered – with just under 60,000 troops – Lee had no intention of retreating. The Confederate commander divided his army: one part remained to guard Fredericksburg, while the other raced west to meet Hooker's advance. When the van of Hooker's column clashed with the Confederates on May 1, Hooker pulled his troops back to Chancellorsville, a lone tavern at a crossroads in a dense wood known locally as The Wilderness. Here Hooker took up a defensive line, hoping Lee's need to carry out an uncoordinated attack through the dense undergrowth would leave the Confederate forces disorganized and vulnerable.

To retain the initiative, Lee risked dividing his forces still further, retaining two divisions to focus Hooker's attention, while Stonewall Jackson marched the bulk of the Confederate army west across the front of the Federal line to a position opposite its exposed right flank. Jackson executed this daring and dangerous maneuver throughout the morning and afternoon of May 2. Striking two hours before dusk, Jackson's

*Confederate dead of Jubal Early's command lie in the rifle pit at the base of Marye's Heights (below). After a long and bloody engagement, Gen. Sedgwick and his three Federal divisions took the Heights with a bayonet charge, thereby diverting Confederate attention from the beleaguered Hooker.*

"Using my musket for a crutch, I began to pull away the burning brushwood, and got some of [the wounded] out. One of the wounded Johnnies... began to help... We were trying to rescue a young fellow in grey. The fire was all around him. The last I saw of that fellow was his face... His eyes were big and blue, and his hair like raw silk surrounded by a wreath of fire. I heard him scream, "O, Mother! O, God!" It left me trembling all over, like a leaf. After it was over my hands were blistered and burned so I could not open and shut them; but me and them rebs tried to shake hands."

*Anonymous Union soldier*

1. **May 1, pm:** Hooker's Federal army assumes a strong defensive position around Chancellorsville
2. **May 1, late pm:** Lee receives news that Hooker's left is weak, and plans an attack with Jackson for May 2
3. **May 2, 7 am-5 pm:** Jackson marches 27,000 troops around Hooker while Lee keeps pressure on the Federals with the remaining 13,000
4. **May 2, 12 noon-5 pm:** Sickles's III Corps attempts to attack Jackson's column but tangles with Lee's force instead

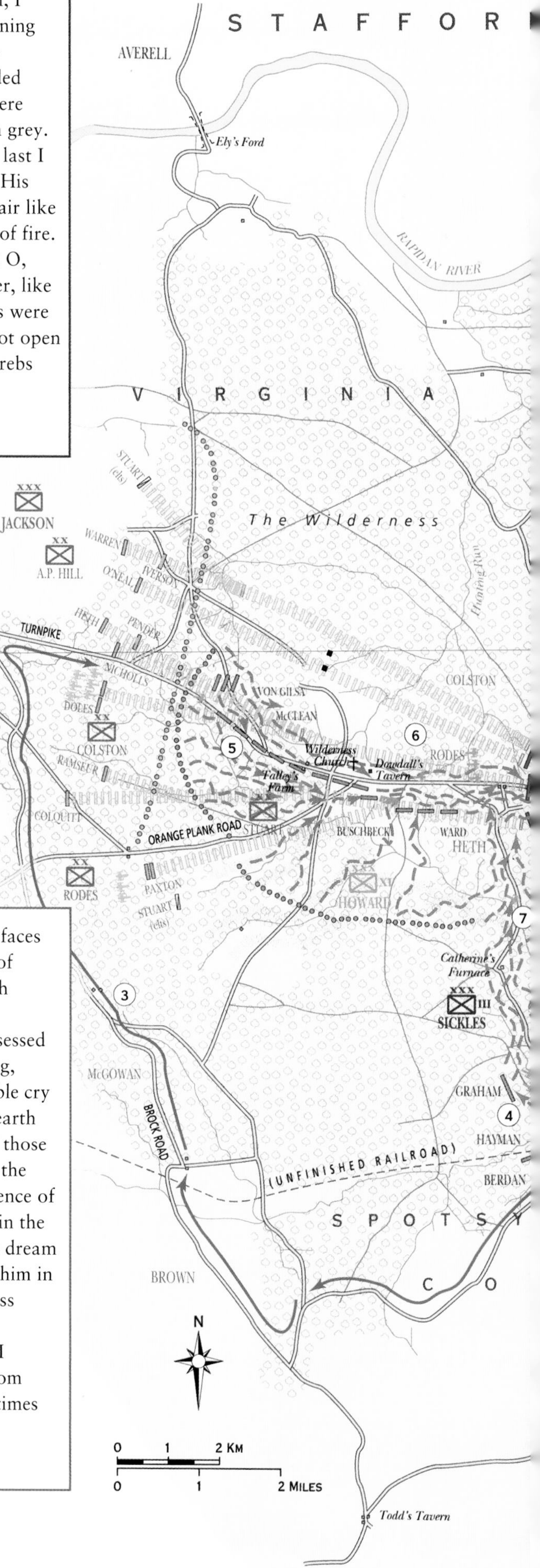
S T A F F O R
AVERELL
Ely's Ford
RAPIDAN RIVER
V I R G I N I A
The Wilderness
STUART (cvls)
JACKSON
A.P. HILL
Wilderness Tavern
WARREN
IVERSON
O'NEAL
TURNPIKE
HETH
PENDER
NICHOLLS
VON GILSA
McCLEAN
DOLES
COLSTON
RAMSEUR
Wilderness Church
Dowdall's Tavern
Talley's Farm
COLQUITT
ORANGE PLANK ROAD
STUART
BUSCHBECK
WARD
HETH
RODES
PAXTON
STUART (cvls)
HOWARD
Catherine's Furnace
SICKLES
McGOWAN
BROCK ROAD
GRAHAM
HAYMAN
BERDAN
(UNFINISHED RAILROAD)
S P O T S Y
C O
BROWN
N
0 1 2 KM
0 1 2 MILES
Todd's Tavern

"The fierce soldiers with their faces blackened with the smoke of battle, the wounded crawling with feeble limbs from the fury of the devouring flames, all seemed possessed with a common impulse. One long, unbroken cheer, in which the feeble cry of those who lay helpless on the earth blended with the strong voices of those who still fought, rose high above the roar of battle and hailed the presence of the victorious chief [Lee]. He sat in the full realization of all that soldiers dream of – triumph... As I looked upon him in the complete fruition of the success which his genius, courage, and confidence in his army had won, I thought that it must have been from such a scene that men in ancient times rose to the dignity of gods."

*An adjutant riding with Lee.*

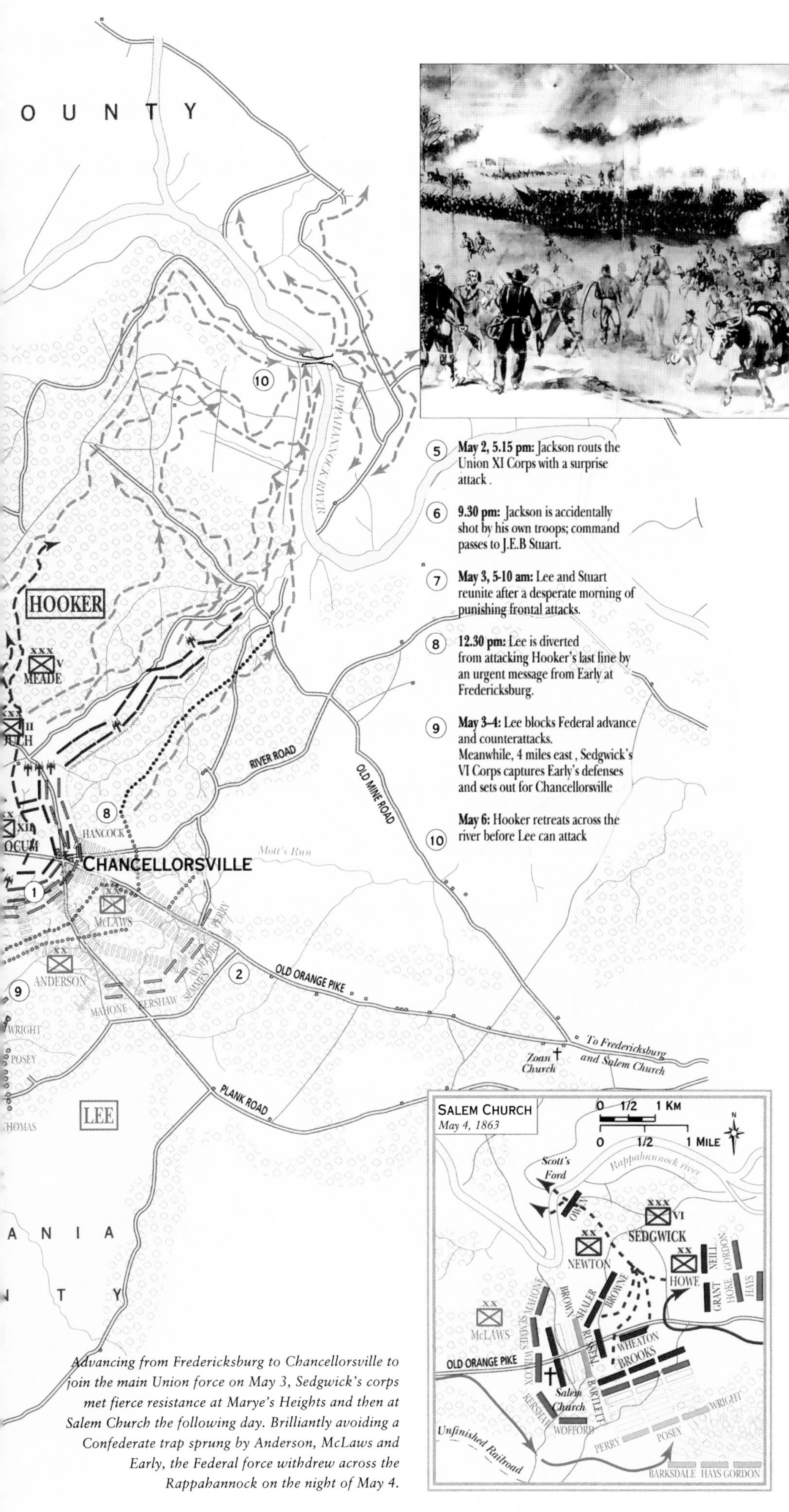

5 May 2, 5.15 pm: Jackson routs the Union XI Corps with a surprise attack.
6 9.30 pm: Jackson is accidentally shot by his own troops; command passes to J.E.B Stuart.
7 May 3, 5-10 am: Lee and Stuart reunite after a desperate morning of punishing frontal attacks.
8 12.30 pm: Lee is diverted from attacking Hooker's last line by an urgent message from Early at Fredericksburg.
9 May 3-4: Lee blocks Federal advance and counterattacks. Meanwhile, 4 miles east, Sedgwick's VI Corps captures Early's defenses and sets out for Chancellorsville
10 May 6: Hooker retreats across the river before Lee can attack
HOOKER
MEADE
HANCOCK
CHANCELLORSVILLE
McLAWS
ANDERSON
MAHONE
KERSHAW
WOFFORD
SEMMES
PERRY
WRIGHT
POSEY
LEE
RAPPAHANNOCK RIVER
RIVER ROAD
OLD MINE ROAD
OLD ORANGE PIKE
PLANK ROAD
Mott's Run
Zoan Church
To Fredericksburg and Salem Church
SALEM CHURCH
May 4, 1863
0 1/2 1 KM
0 1/2 1 MILE
Scott's Ford
Rappahannock river
SEDGWICK
NEWTON
HOWE
NEILL
GRANT
HOKE
GORDON
HAYS
MAHONE
SEMMES
WILCOX
BROWN
SHALER
BROWNE
RUSSELL
WHEATON
BROOKS
BARTLETT
McLAWS
OLD ORANGE PIKE
Salem Church
KERSHAW
WOFFORD
PERRY
POSEY
WRIGHT
Unfinished Railroad
BARKSDALE
HAYS
GORDON


*Advancing from Fredericksburg to Chancellorsville to join the main Union force on May 3, Sedgwick's corps met fierce resistance at Marye's Heights and then at Salem Church the following day. Brilliantly avoiding a Confederate trap sprung by Anderson, McLaws and Early, the Federal force withdrew across the Rappahannock on the night of May 4.*

*Caught completely by surprise when Jackson launched his attack on the Union Flank, Howard's XI Corps fled. Their retreat was covered by Couch's II Corps (above).*

men routed the astonished Federals in their camps. In the gathering darkness, amid the brambles of the Wilderness, the Confederate line became confused and halted at 9 pm to regroup. Riding in front of the lines to reconnoiter, Stonewall Jackson was accidentally shot and seriously wounded by his own men. Later that night, his left arm was amputated just below the shoulder.

On May 3, Jackson's successor, General J.E.B. Stuart, initiated the bloodiest day of the battle when attempting to reunite his troops with Lee's. Despite an obstinate defense by the Federals, Hooker ordered them to withdraw north of the Chancellor House. The Confederates were converging on Chancellorsville to finish Hooker when a message came from Jubal Early that Federal troops had broken through at Fredericks burg. At Salem Church, Lee threw a cordon around these Federals, forcing them to retreat across the Rappahannock. Disappointed, Lee returned to Chancellorsville, only to find that Hooker had also retreated across the river.

Chancellorsville is considered Lee's great est victory, although the Confederate com mander's daring and skill met little resistence from the inept generalship of Joseph Hooker. Using cunning, and dividing their forces repeatedly, the massively outnum bered Confederates drove the Federal army from the battlefield. The cost had been frightful. The Confederates suffered 14,000 casualities, while inflicting 17,000. Perhaps the most damaging loss to the Confederacy was the death of Lee's "right arm," Stonewall Jackson, who died of pneumonia on May 10 while recuperating from his wounds.

# Gettysburg Campaign: The Invasion of Pennsylvania

## JUNE 1863

Following his victory at Chancellorsville in May, 1863, General Lee received approval from his government to invade the north. Lee hoped an invasion would fuel the northern peace movement and, at least, disrupt the Union war effort. After the death of Stonewall Jackson, Lee's Army of Northern Virginia, 75,000-strong, had been reorganized into three army corps under Longstreet, Ewell, and A.P. Hill, with a cavalry division under J.E.B. Stuart. On June 3, advance troops of the Confederate army left their camps near Fredericksburg and marched west toward the Shenandoah Valley.

The 95,000-strong Federal Army of the Potomac, under General Hooker, was initially uncertain of Lee's intentions. On June 9, Hooker ordered cavalry general Alfred Pleasonton to conduct a reconnaissance with 11,000 men across the Rappahannock River toward Brandy Station. Pleasonton ran into Stuart's cavalry, and the largest cavalry battle of the war ensued. The result was a standoff, but the Federals were now alerted to the Confederate army's movements.

By June 13, elements of Ewell's corps appeared before Winchester. On the same day, Hooker with-drew the Army of the Potomac from the Rappahannock and ordered it north. On June 14–15, Ewell attacked the 9,000-strong Federal garrison at Winchester and

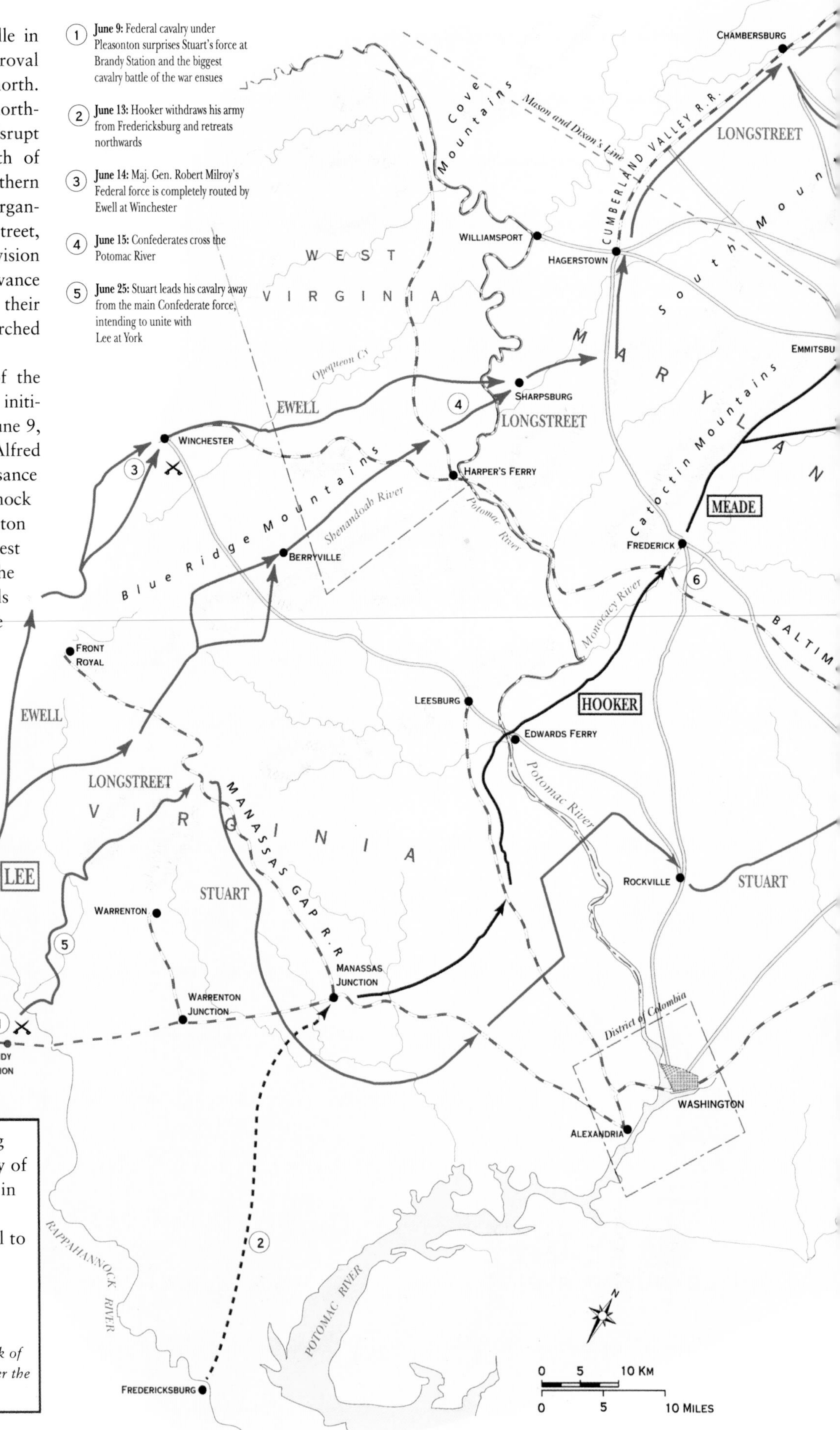

1 June 9: Federal cavalry under Pleasonton surprises Stuart's force at Brandy Station and the biggest cavalry battle of the war ensues
2 June 13: Hooker withdraws his army from Fredericksburg and retreats northwards
3 June 14: Maj. Gen. Robert Milroy's Federal force is completely routed by Ewell at Winchester
4 June 15: Confederates cross the Potomac River
5 June 25: Stuart leads his cavalry away from the Confederate force, intending to unite with Lee at York
CHAMBERSBURG
Mason and Dixon's Line
CUMBERLAND VALLEY R.R.
LONGSTREET
Cove Mountains
WILLIAMSPORT
HAGERSTOWN
South Mountains
WEST VIRGINIA
MARYLAND
EMMITSBURG
Opequon Cr.
SHARPSBURG
EWELL
LONGSTREET
WINCHESTER
Blue Ridge Mountains
HARPER'S FERRY
Shenandoah River
Potomac River
Catoctin Mountains
MEADE
FREDERICK
BERRYVILLE
Monocacy River
BALTIMORE
FRONT ROYAL
LEESBURG
HOOKER
EDWARDS FERRY
EWELL
LONGSTREET
VIRGINIA
MANASSAS GAP R.R.
Potomac River
LEE
STUART
ROCKVILLE
STUART
WARRENTON
MANASSAS JUNCTION
WARRENTON JUNCTION
BRANDY STATION
District of Columbia
WASHINGTON
ALEXANDRIA
RAPPAHANNOCK RIVER
POTOMAC RIVER
FREDERICKSBURG
0 5 10 KM
0 5 10 MILES


6 **June 28:** Hooker is relieved of command and Meade is nominated as his successor

7 **June 30:** Cavalry skirmish at Hanover

8 **June 30:** Meade orders Reynolds to occupy Gettysburg

9 **July 2:** Stuart's cavalry finally rejoins Lee on the second day of the battle at Gettysburg

> "Under the leadership of 'fighting Joe Hooker' the glorious Army of the Potomac is becoming more slow in its movements, more unwieldy, less confident of itself, more of a football to the enemy, and less an honor to the country than any army we have yet raised."
>
> *The* Chicago Tribune *reflecting the general lack of confidence in the abilities of Gen. Hooker, after the Chancellorsville fiasco.*

EWELL
CARLISLE
Conodoguinet Creek
LEE
STUART
CASHTOWN
ETTYSBURG
Conewago Creek
YORK
LITTLESTOWN
HANOVER
HANOVER
JUNCTION
TANNEYTOWN
UNION MILLS
WESTMINSTER

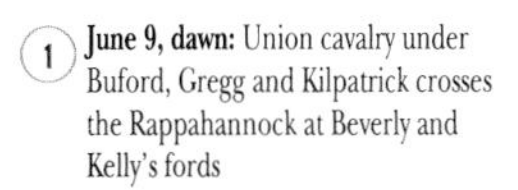

1. **June 9, dawn:** Union cavalry under Buford, Gregg and Kilpatrick crosses the Rappahannock at Beverly and Kelly's fords
2. Union advance from Beverly Ford is contested by Rooney Lee and William E. Jones
3. Fleetwood Hill changes hands four times, the Confederates eventually holding it against continued Union assaults
4. **Dark:** Pleasonton withdraws across the Rappahannock after spotting advancing Confederate infantry

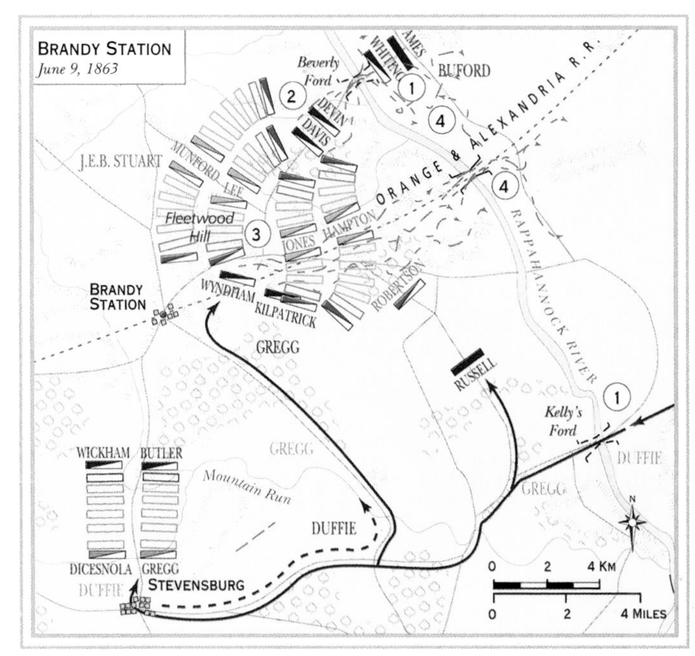
BRANDY STATION
June 9, 1863
Beverly Ford
AMES
WHITING
BUFORD
DEVIN
DAVIS
J.E.B. STUART
MUNFORD
LEE
Fleetwood Hill
JONES
HAMPTON
ORANGE & ALEXANDRIA R.R.
BRANDY STATION
WYNDHAM
KILPATRICK
ROBERTSON
GREGG
RUSSELL
RAPPAHANNOCK RIVER
Kelly's Ford
DUFFIE
WICKHAM
BUTLER
Mountain Run
DUFFIE
DICESNOLA
GREGG
STEVENSBURG
0 2 4 KM
0 2 4 MILES

defeated it, inflicting heavy losses and capturing much valuable war material.

After Winchester, Lee's army moved unchecked into the Cumberland Valley of Pennsylvania. On June 25, Lee agreed to Stuart's plan to take three brigades of cavalry across the Potomac east of the Blue Ridge Mountains, and cut across the rear of the Federal army. Stuart's march encountered frequent delays and detours and an in creasingly aggressive Federal cavalry, and was unable to rejoin Lee until July 2.

By June 28, Longstreet and Hill's corps were at Chambersburg. Divisions of Ewell's corps had crossed the mountains to York and Carlisle, and were preparing to move against Harrisburg. However, Lee learned on this day that the Federal army was at Frederick, and that Hooker had been replaced by General Meade. Lee decided to bring his entire army east of the mountains and offer battle. At the same time, Meade moved his army north. By June 30, both armies were converging upon Gettysburg and the battle, which would be the turning point of the war, was set to commence.

> "Hereabouts we shall probably meet the enemy and fight a great battle, and if God gives us the victory, the war will be over and we shall achieve the recognition of our independence."
>
> *General Lee to Maj. Gen. Isaac Trimble on the approach to Gettysburg, June 27.*

*Though the Confederate cavalry was successful in holding its position at Brandy Station, J.E.B Stuart (below) was much criticized for the unpreparedness of his command. Furthermore, it was believed by some that the parades and war games, suggested by his love of pomp and pageantry, had tired his men and considerably reduced their fighting capacity.*

*Federal cavalry under the command of John Buford attacks the Confederate defenders of Fleetwood Hill in the last charge of the Battle of Brandy Station (below).*

# Gettysburg JULY 1 1863

After the discovery on June 30 that Gettysburg was occupied by Brigadier General John Buford's division of Federal cavalry, the Confederates on July 1 sent the divisions of Major General Henry Heth and Major General William Pender of Hill's Corps, down the Chambersburg Road to drive Buford away and occupy Gettysburg.

The battle began at 5.30 a.m., when shots were exchanged over Marsh Creek. In the face of Buford's resistance, Heth pushed on cautiously until he reached a point about two miles west of Gettysburg. Here he deployed two brigades in line, and pressed ahead; it was nearly 10 a.m. Federal General John F. Reynolds, commanding I Corps, arrived on the field at this point, and determined to engage Heth. He ordered I Corps and Major General Oliver O. Howard's XI Corps to march to Gettysburg.

Soon after 10.30 a.m., I Corps arrived and engaged Heth along McPherson's Ridge. By 11.30 a.m., Heth had been defeated and forced to withdraw to Herr Ridge. Early in the action, Reynolds was killed, and field command devolved upon Howard. A lull now settled over the field as both sides brought up reinforcements. The Federal I Corps deployed to defend the western approaches to Gettysburg, while XI Corps formed up north of the town. Buford's cavalry covered the flanks. Howard left one division in reserve on Cemetery Hill. His strategy was simple: delay the Confederates long enough to enable the rest of the Federal army to concentrate.

Lee arrived on the field after noon. He had initially hoped to avoid a general engagement since the strength of the enemy was unknown, and the terrain in the Gettysburg

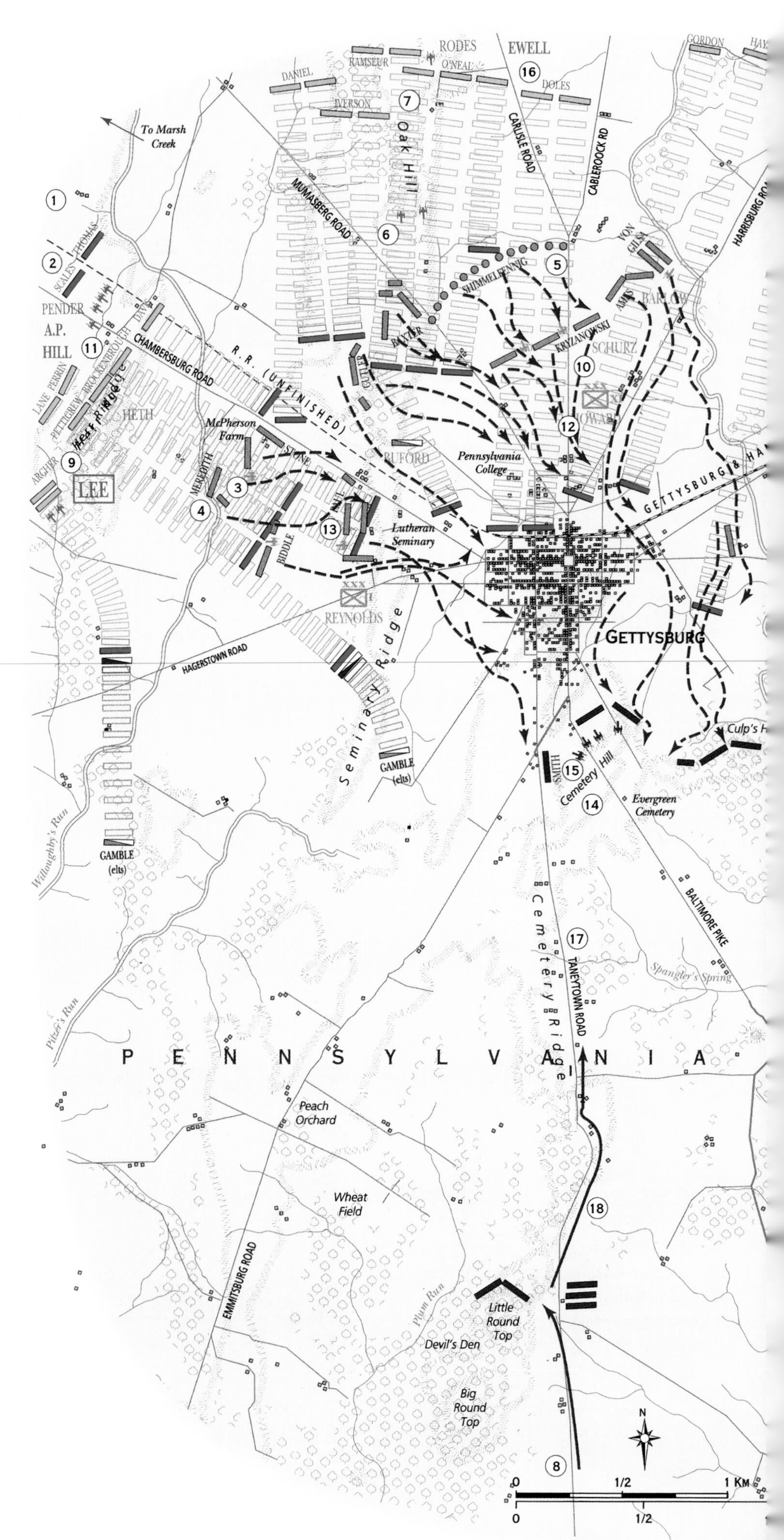
RODES
EWELL
RAMSEUR
O'NEAL
DANIEL
DOLES
IVERSON
GORDON
To Marsh Creek
Oak Hill
MUMASBERG ROAD
CARLISLE ROAD
CABLEROOCK RD
HARRISBURG RD
THOMAS
SCALES
PENDER
A.P. HILL
LANE
PERRIN
PETTIGREW
BROCKENBROUGH
DAVIS
CHAMBERSBURG ROAD
R. R. (UNFINISHED)
SCHIMMELFENNIG
KRZYZANOWSKI
SCHURZ
VON GILSA
AMES
BARLOW
HOWARD
HETH
Herr Ridge
ARCHER
LEE
McPherson Farm
MEREDITH
STONE
BIDDLE
BUFORD
Pennsylvania College
Lutheran Seminary
REYNOLDS
GETTYSBURG & HANOVER
GETTYSBURG
HAGERSTOWN ROAD
Seminary Ridge
GAMBLE (elts)
Culp's H
Cemetery Hill
Evergreen Cemetery
Willoughby's Run
BALTIMORE PIKE
Spangler's Spring
Cemetery Ridge
TANEYTOWN ROAD
Pitzer's Run
P E N N S Y L V A N I A
Peach Orchard
Wheat Field
EMMITSBURG ROAD
Plum Run
Little Round Top
Devil's Den
Big Round Top
N
0
1/2
1 KM

*Maj. Gen. John F. Reynolds (below) – considered to be one of the Union's most competent commanders – was killed when a musket ball struck him in the neck whilst he attempted to succor Buford's cavalry division.*

"In the absence of reports from him, [Stuart] I am in ignorance of what we have in front of us here. It may be the whole Federal Army, or it may be only a detachment. If it is the whole Federal force, we must fight a battle here."

*Lee talking to Gen R.H. Anderson on the approach to Gettysburg.*

ETOWN ROAD

YORK PIKE

Benner's Hill

1. **July 1, 5.30 am:** Opening shots fired
2. **8 am:** Archer and Davis of Heth's division begins advance on Gettysburg
3. **10 am:** Reynolds killed by Confederate marksman: succeeded by Doubleday
4. **Mid am:** Meredith's Iron Brigade turns back Archer's troops; Archer taken prisoner
5. **12 noon:** XI Corps under Schurz arrives
6. **12 noon:** Confederate artillery fires on Union lines from Oak Hill
7. **2 pm:** Rodes advances on the Union right.
8. **Dec 13, 2 pm:** Meade dispatches Hancock from Taneytown to replace Reynolds
9. **2.30 pm:** Lee arrives on Herr Ridge to survey the battlefield
10. **2.30 pm:** Schurz's division crumbles under Early's attack
11. **2.30 pm:** Lee sends in Heth and Pender; Heth wounded
12. **3.30 pm:** Under Early's onslaught, Schurz's line flees south into Gettysburg
13. **4 pm:** Pender's troops force Union retreat to Gettysburg and Cemetery Hill
14. **4 pm:** Hancock arrives on Cemetery Hill
15. **4.30 pm:** Union troops retreat to Cemetery Hill and begin entrenching
16. **4.30 pm:** Lee gives Ewell discretionary orders to attack Cemetery Hill; Ewell declines
17. **6 pm:** Sickles's corps arrives
18. **7 pm:** Hancock leaves for Taneytown to summon

"For seven or eight minutes ensued probably the most desperate fight ever waged between artillery and infantry at close range without a particle of cover on either side, bullets hissing, humming and whistling everywhere; cannon roaring; all crash on crash and peal on peal, smoke, dust, splinters, blood, wreck and carnage indescribable."

*A Union gunner of Meredith's Iron Brigade.*

*The tiny farmhouse (above) to the rear of Cemetery Ridge which was used by Gen. Meade as his Union headquarters throughout most of the conflict at Gettysburg.*

area unfamiliar. But, soon after noon, Rodes's division of Ewell's Corps arrived on Oak Hill and attacked the right of I Corps. At 2 p.m. Heth's division joined the attack on I Corps. At 3 p.m., the battle spread north of the town when Jubal Early's division of Ewell's Corps attacked down the Harrisburg Road and crushed the flank of XI Corps. At about the same time, west of Gettysburg, Pender's division relieved Heth and assaulted I Corps' position along Seminary Ridge. By 4 p.m., both Federal corps were in retreat through Gettysburg to Cemetery Hill. Federal losses numbered slightly over 9,000, including some 3,000 captured, compared with Confederate losses of about 6,500.

The day's action had resulted in a Confederate victory, but Federal forces held onto the high ground south of Gettysburg, where their position was soon strengthened by reinforcements.

*The appointment of George Meade (above) as Hooker's replacement on June 27 was welcomed by the Union army. Meade himself was reluctant to accept command and Gettysburg was the first test of his leadership.*

*Union casualties (below) on the first day of the battle were severe; the 24th Michigan infantry lost 399 of its 496 men. In total, I Corps had lost nearly 5,700 men. Confederate losses were nearly as numerous.*

# Gettysburg JULY 2 1863

The success of his army in the fighting on July 1 encouraged Lee to renew the battle on July 2. An early morning reconnaissance of the Federal left revealed that their line did not extend as far south as Little Round Top. Lee directed Longstreet to take two divisions of I Corps and march south until they reached the flank of the Federal forces. They would attack from this point, supported by a division of A.P. Hill's corps – a total force of nearly 20,000 men. While Longstreet carried out the main offensive, Ewell was ordered to conduct a demonstration against the Federal right. However, he was given discretion to mount a full-scale attack should the opportunity present itself.

The Federal army was well prepared for Lee's offensive. Six of its seven corps had arrived on the battlefield, and VI Corps was making a thirty-six-mile forced march to reach it. Meade had deployed his army in a fish-hook-shaped formation, with the right on Culp's Hill and Cemetery Hill, the center

> "It was a hard fight. The Confederates appeared to have the devil in them. My men did not flinch and then, when their assailants descended into the ravine and crossed the creek, they were received with a deadly volley, every shot of which was effective... On both sides, each one aimed at his man, and men fell dead and wounded with frightful rapidity."
>
> *Col. Regis de Trobriand, present during Sickles's defense of the Peach Orchard.*

*The flamboyant Gen. Sickles (below) anchored the left of his advance line at Devil's Den, a rocky terrain exposed to attack on three sides. The successful Confederate advance almost turned the Union left flank.*

Oak Hill
CARLISLE ROAD
CABLEROCK ROAD
HARRISBURG ROAD
MUMASBERG ROAD
R.R. (UNFINISHED)
CHAMBERSBERG ROAD
Herr Ridge
HETH
McPherson Farm
LEE
Pennsylvania College
O'NEAL
EWELL
G'BURG & H'OVER R.R.
GORDON
DANIELS
Lutheran Seminary
RODES
RAMSEUR
IVERSON
DOLES
HAYS
EARLY
AVERY
GETTYSBURG
HAGERSTOWN ROAD
PENDER
A.P. HILL
PERRIN
LANE
SCALES
MAHONE
THOMAS
Seminary Ridge
ANDERSON
POSEY
WRIGHT
SCHURZ
XI
HOWARD
Culp's Hill
WADSWORTH
AMES
Cemetery Hill
Evergreen Cemetery
GEARY
XII
RUGER
HAYS
I
DOUBLEDAY
BALTIMORE PIKE
MEADE
Cemetery Ridge
HARROW
WEBB
TANEYTOWN ROAD
Willoughby's Run
LANG
WILCOX
Pitzer's Run
HUMPHREYS
BROOKE
ZOOK
CROSS
KELLY
CALDWELL
BARNES
BREWSTER
Peach Orchard
III
SICKLES
McLAWS
WOFFORD
BARKSDALE
AYRES
GRAHAM
Wheat Field
BIRNEY
DE TROBRIAND
V
SYKES
SEMMES
KERSHAW
LONGSTREET
EMMITSBURG ROAD
WARD
Plum Run
Little Round Top
VINCENT
Devil's Den
Big Round Top
ANDERSON
ROBERTSON
BENNING
LAW
N
0
1/2
1 KM

On the evening of July 2, Jubal Early's Confederates attacked the Union defensive line on Cemetery Hill, overlooking the town of Gettysburg (above). The rebels experienced severe losses in their abortive assault.

Confederate division commander, John B. Hood, whose troops were involved in the assault upon Little Round Top (above). Hood was incapacitated early in the fighting of July 2 by a shot which shattered his left arm.

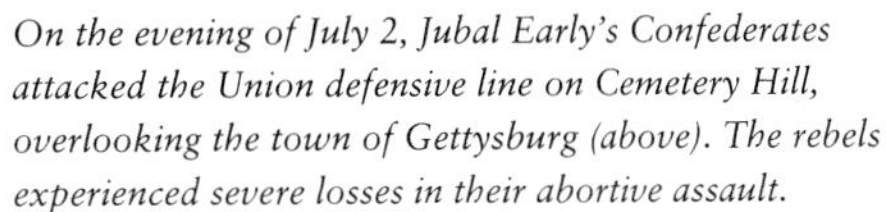

WILLIAMS

JOHNSON

STEUART

WALKER

enner's Hill

① **July 2, c 4 pm to dusk:** Devil's Den/Little Round Top, Confederates attack lightly-defended Union left; Confederates take Devil's Den but are held off at the base of Little Round Top

② **After Devil's Den to dusk:** Wheat Field/Peach Orchard, McLaws enters.
c. 5.30 pm In this hotly contested battle, the Wheat Field changes hands four times. Sickles's Union III Corps are finally routed to the base of Little Round Top, where continued Confederate assault is held off

③ **c. 6.30 pm to after dark:** Culp's Hill/Cemetery Hill, Hay's and Avery's Confederate divisions enter; c. 8 pm Ewell's Confederate divisions gain some ground but fail to take Union positions on Culp's and Cemetery Hill

SEDGEWICK

along Cemetery Ridge, and the left on Little Round Top. The left of the Federal line was held by Major General Daniel Sickles's III Corps. Sickles was dissatisfied with his assigned position and in the early afternoon, without orders, he advanced his line nearly half a mile west in order to take advantage of the high open ground around a nearby peach orchard.

Soon after Sickles took up this new position, Longstreet attacked. Third Corps was hard pressed and Meade sent V Corps and part of II Corps to reinforce Sickles in the Peach Orchard. But, after furious fighting, Longstreet's forces broke through, causing Sickles's entire line to collapse. The Confederates pursued to the base of Little Round Top, but Federal reinforcements, including elements of VI Corps, checked their advance. Farther north, elements of a division of the Confederate III Corps advanced to the slopes of Cemetery Ridge before they too were forced to retire.

On the Federal right, Ewell did not attack until evening, after Longstreet's onslaught had subsided. The effort to storm Cemetery Hill was ultimately unsuccessful. Ewell's attacks were also repulsed at Culp's Hill, although a foothold was gained near the base of the hill.

The second day's fighting had cost each army some 9,000 casualties. Lee's forces had again gained ground, but had failed to dislodge the Federal army from its strong position.

Little Round Top, a hill of vital strategic importance to any army that held it (below). The troops positioned in this boulder-strewn landscape were involved in some of the bitterest fighting in the entire war, and the Confederate assault was costly to both sides.

"Within half an hour I could convert Little Round Top into a Gibraltar that I could hold against ten times the number of men that I had."

*Colonel William C. Oates, of the Confederate 15th Alabama Regiment.*

# Gettysburg JULY 3 1863

Lee's confidence was unshaken by the events of July 2. That night, he ordered Longstreet, who had been reinforced by Major General George Pickett's division, to renew his assault on the Federal left. Simultaneously, Ewell, who had also been reinforced, was to storm Culp's Hill. Stuart's cavalry, which had rejoined the army late that day, was ordered to march well east of Gettysburg, and attempt to penetrate to the Federal rear where they might disrupt communications and distract Meade.

Meanwhile, Meade had determined to hold his position and await Lee's attack. However, at Culp's Hill he authorized XII Corps to drive Ewell's forces out of the captured Federal trenches at daylight. The Federal effort opened with a concentrated artillery bombardment which precipitated a tremendous musketry battle.

With Ewell already engaged, Lee rode to Longstreet's headquarters to observe his preparations for the attack on the Federal left. Longstreet misunderstood his orders and was planning instead a movement to turn the Federal left. With the hope of a coordinated attack now lost, Lee was forced to modify his plans. He determined to shift his

*Gen. Pickett (below), chosen by Lee to lead the charge against Cemetery Ridge, accepted his responsibility enthusiastically. Lee was later to tell him, "the men and officers of your command have written the name of Virginia as high today as it has ever been written."*

"I can still hear them cheering as I gave the order, 'Forward!' The thrill of their joyous voices as they called out, 'We'll follow you, Marse George, we'll follow you!' On, how faithfully they followed me on, on to their death, and I led them on, on, on; Oh God!"

*Gen. George Pickett in a letter to his fiancée on the evening of July 3.*

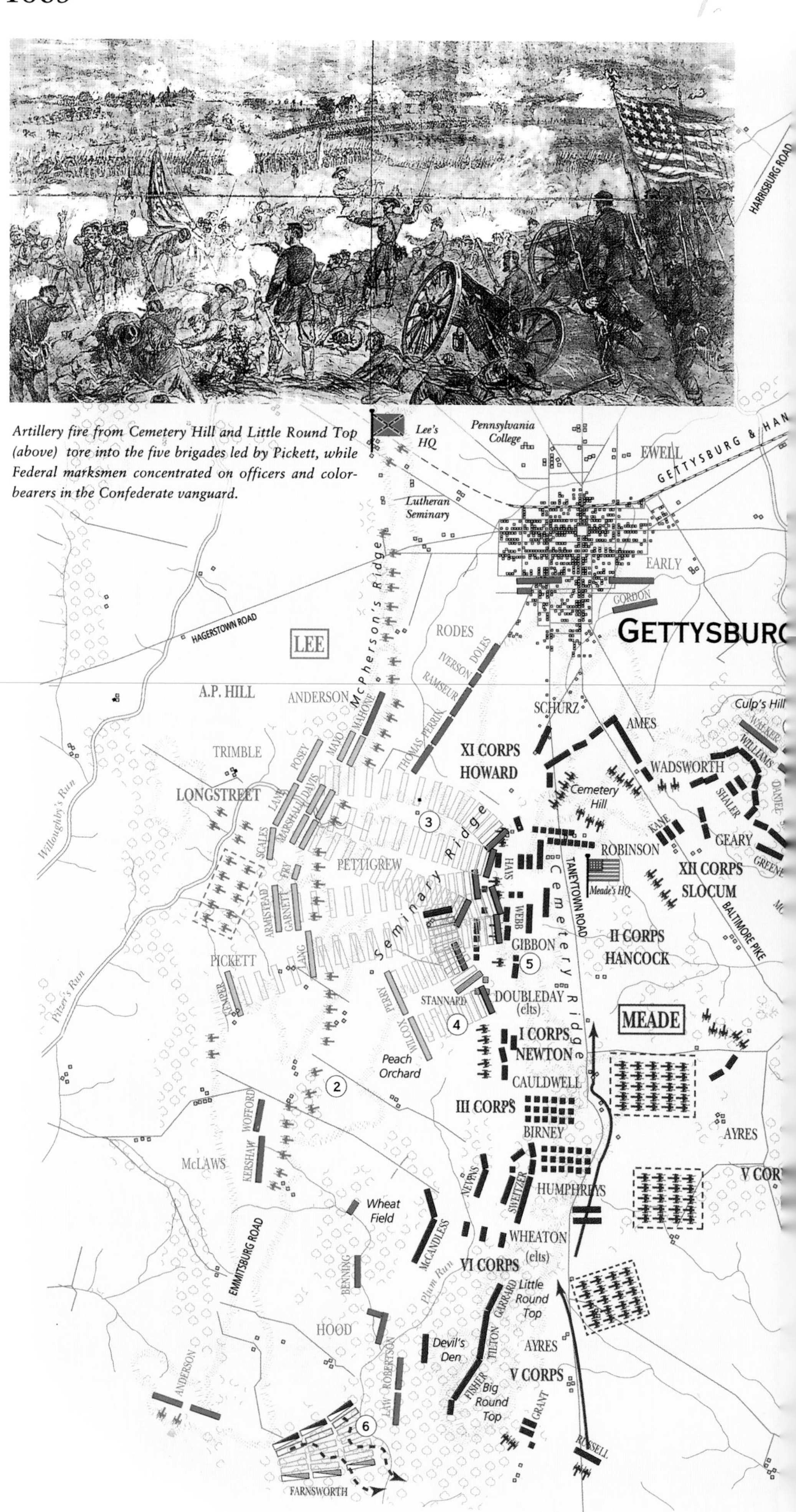
Lee's HQ
Pennsylvania College
EWELL
GETTYSBURG & HAN
HARRISBURG ROAD
Lutheran Seminary
EARLY
GORDON
GETTYSBURG
HAGERSTOWN ROAD
LEE
RODES
IVERSON
DOLES
RAMSEUR
A.P. HILL
ANDERSON
McPherson's Ridge
MAHONE
THOMAS
PERRIN
SCHURZ
AMES
Culp's Hill
WALKER
WILLIAMS
TRIMBLE
XI CORPS
HOWARD
WADSWORTH
LONGSTREET
POSEY
MAYO
DAVIS
LANE
MARSHALL
Cemetery Hill
SHALER
DANIEL
KANE
GEARY
Willoughby's Run
SCALES
FRY
PETTIGREW
Seminary Ridge
ROBINSON
HAYS
TANEYTOWN ROAD
Cemetery Ridge
XII CORPS
SLOCUM
GREENE
ARMISTEAD
GARNETT
Meade's HQ
WEBB
BALTIMORE PIKE
LANG
GIBBON
II CORPS
HANCOCK
PICKETT
KEMPER
PERRY
STANNARD
DOUBLEDAY (elts)
MEADE
Pitzer's Run
WILCOX
I CORPS
NEWTON
Peach Orchard
CAULDWELL
III CORPS
WOFFORD
BIRNEY
AYRES
McLAWS
KERSHAW
NEVINS
SWEITZER
HUMPHREYS
V CORPS
Wheat Field
McCANDLESS
WHEATON (elts)
EMMITSBURG ROAD
VI CORPS
Plum Run
BENNING
Little Round Top
GARRARD
HOOD
TILTON
Devil's Den
AYRES
V CORPS
ROBERTSON
ANDERSON
LAW
FISHER
Big Round Top
GRANT
RUSSELL
FARNSWORTH

*Artillery fire from Cemetery Hill and Little Round Top (above) tore into the five brigades led by Pickett, while Federal marksmen concentrated on officers and color-bearers in the Confederate vanguard.*

YORK PIKE
Benner's Hill
PENNSYLVANIA
Rock Creek
1 July 3, 5.30am-10am: Johnson's division of Ewell's corps launches repeated but unsuccessful attacks on Culp's Hill
2 1pm: Artillery cannonade begins with signal gun at the Peach Orchard
3 3pm: Pickett's, Pettigrew's and Trimble's Confederate infantry advance for attack
4 3.30pm: Stannard's Federal brigade attacks flank of Pickett's division
5 3.45pm: Limit of Confederate infantry attacks at the Angle and clump of trees
6 5 .30pm: Farnsworth's cavalry charge against Confederate right meets with disaster
N
0
1/2
1 KM
0
1/2
1 MILE

main attack to the Federal center on Cemetery Ridge. Longstreet was placed in command of the effort. The plan was first to subject the Federal position to bombardment by nearly 140 cannon, then to send Pickett, Pettigrew and half of Trimble's divisions (formerly Heth's and Pender's) – nearly 12,000 men – forward to smash the Federal center.

While Longstreet made his preparations during the morning, Ewell's forces were defeated in their counterattacks on Culp's Hill, and withdrew around 11 a.m.

At 1pm, Longstreet opened the great bombardment of the Federal line. The Federal army replied with approximately 80 cannon and a giant duel ensued which lasted for nearly two hours. After the bombardment subsided, the infantry went forward. Federal artillery, followed by musketry, cut their formations to pieces and inflicted devastating losses. A small Confederate force effected one small penetration of the Federal line, but was overwhelmed. The attack ended in disaster, with nearly 5,600 Confederate casualties. Meanwhile, three miles east of Gettysburg, Stuart's cavalry was engaged by Federal cavalry under Brigadier General David Gregg. The cavalry clash was indecisive, but Stuart was neutralized and posed no threat to the Federal rear.

The battle was effectively over. Federal losses numbered approximately 23,000, while estimates of Confederate losses range between 20,000 and 28,000.

*The defenders of Cemetery Ridge included a brigade commanded by Brig. Gen. Alexander Webb (above). Webb's leadership in rallying his men during the Confederate assault won him the Medal of Honor.*

> "It seems likely we shall be led in a plodding, ordinary sort of way, neither giving nor receiving any serious blows, a great pity."
>
> *One of Meade's lieutenants expressing the disappointment which many felt at his failure to immediately follow up the Gettysburg victory.*

*Confederate prisoners of war await transferral to Northern jails (below). When he began his retreat to Virginia on the night of July 4, Lee's combat strength had been reduced by nearly one third.*

# Two Raids JULY–AUGUST 1863

To support the operations of Confederate General Bragg in the Middle Tennessee, Brigadier General John H. Morgan led 2,500 cavalrymen from Tennessee into Kentucky in early July, 1863. Morgan eluded Federal forces, crossed the Ohio River and entered Indiana on July 8. The raiders then moved east, passing north of Cincinnati, Ohio, destroying railroads and private property, and causing panic throughout the midwest.

Federal pursuit, directed by General Burnside, was clumsy but ultimately effective, as fatigue slowed Morgan's march. Late on July 18, the Confederates reached Buffington Island, intending to recross the Ohio, but were attacked and scattered the following morning by a Federal force. Badly outnumbered, almost a third of Morgan's men were captured; some escaped across the river, but the majority continued east under Morgan. They surrendered near West Point on July 26. Although they had caused widespread damage, Federal operations were not significantly disrupted.

William Clarke Quantrill was a veteran of the guerilla fighting between free state and slave state forces along the Kansas-Missouri border during the 1850s. After the outbreak of the Civil War, he led a band of partisans in raids on Kansas. In the summer of 1863, he targeted the town of Lawrence. His men considered their raid to be a retaliation for the deaths in August of five women held by Federal authorities in Kansas City who had died when their prison cell collapsed.

On August 19, Quantrill headed west with over 300 men. He crossed into Kansas, arriving on the outskirts of Lawrence near dawn on August 2l. Encountering no organized resistance, the raiders burned over 100 homes, looted banks and stores, and killed some 150 male civilians.

Alerted belatedly to the raid, Federal forces in Kansas gave chase, skirmishing with Quantrill's men on the 2lst as they headed for Missouri. A brief fight occurred on August 22, but the raiders escaped.

Principally as a result of the Lawrence Massacre, Federal Thomas Ewing ordered the forced evacuation of four Missouri counties bordering Kansas. This "Order No.11" made Ewing as infamous in Missouri as Quantrill was in Kansas.

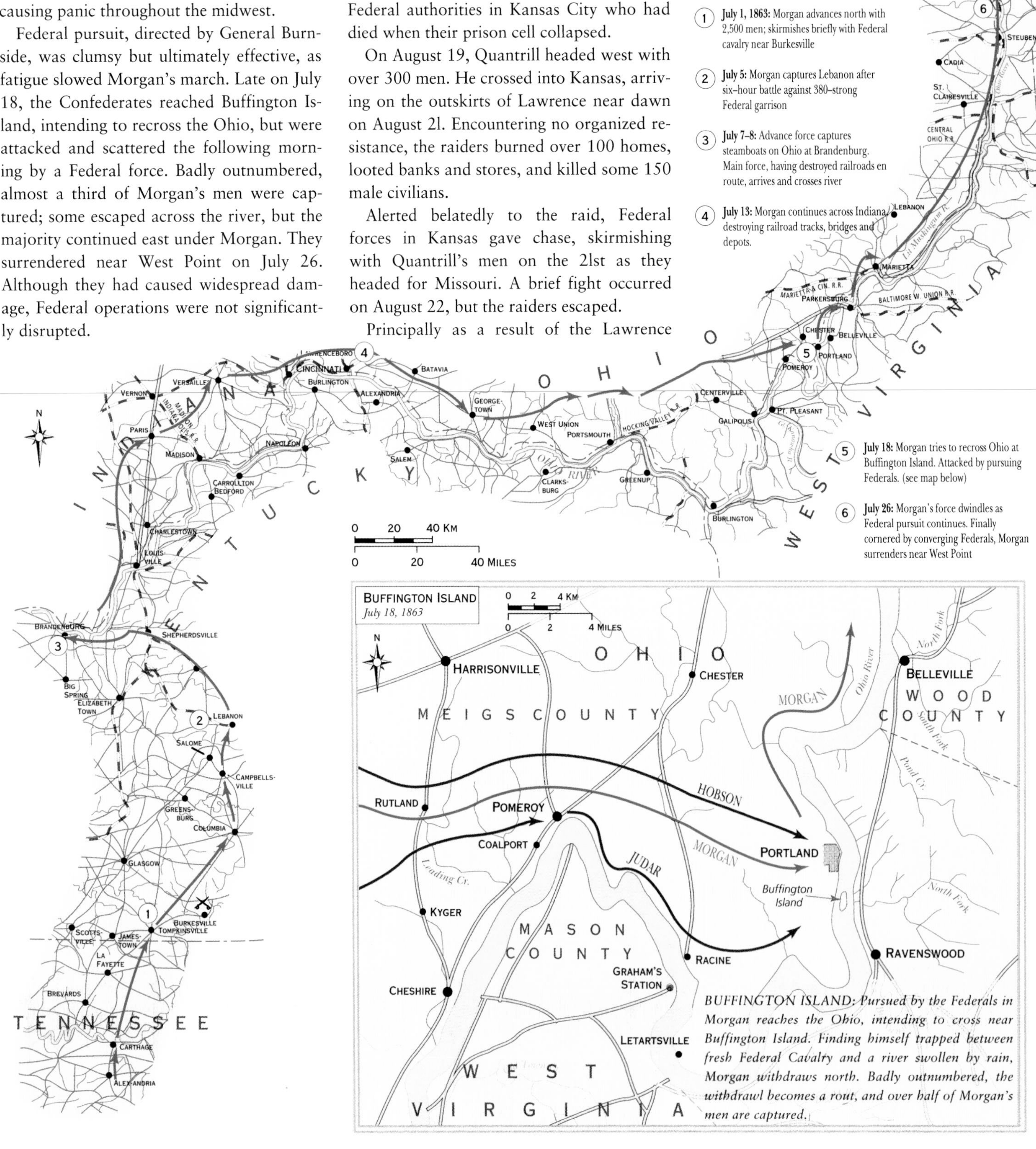
MORGAN'S RAID
July 1–26, 1863
1 July 1, 1863: Morgan advances north with 2,500 men; skirmishes briefly with Federal cavalry near Burkesville
2 July 5: Morgan captures Lebanon after six-hour battle against 380-strong Federal garrison
3 July 7–8: Advance force captures steamboats on Ohio at Brandenburg. Main force, having destroyed railroads en route, arrives and crosses river
4 July 13: Morgan continues across Indiana, destroying railroad tracks, bridges and depots.
5 July 18: Morgan tries to recross Ohio at Buffington Island. Attacked by pursuing Federals. (see map below)
6 July 26: Morgan's force dwindles as Federal pursuit continues. Finally cornered by converging Federals, Morgan surrenders near West Point
I N D I A N A
K E N T U C K Y
O H I O
W E S T V I R G I N I A
T E N N E S S E E
N
0 20 40 KM
0 20 40 MILES
STEUBENVILLE
CADIA
ST. CLAIRESVILLE
CENTRAL OHIO R.R.
LEBANON
MARIETTA
MARIETTA & CIN. R.R.
PARKERSBURG
BALTIMORE W. UNION R.R.
CHESTER
BELLEVILLE
PORTLAND
POMEROY
CENTERVILLE
GALIPOLIS
PT. PLEASANT
HOCKING VALLEY R.R.
GREENUP
BURLINGTON
WEST UNION
PORTSMOUTH
OHIO RIVER
CLARKSBURG
GEORGETOWN
BATAVIA
ALEXANDRIA
SALEM
CINCINNATI
BURLINGTON
VERSAILLE
VERNON
PARIS
MADISON
NAPOLEON
CARROLLTON BEDFORD
CHARLESTOWN
LOUISVILLE
BRANDENBURG
SHEPHERDSVILLE
BIG SPRING
ELIZABETHTOWN
LEBANON
SALOME
CAMPBELLSVILLE
GREENSBURG
COLUMBIA
GLASGOW
BURKESVILLE
TOMPKINSVILLE
SCOTTSVILLE
JAMESTOWN
LA FAYETTE
BREVARDS
CARTHAGE
ALEXANDRIA
BUFFINGTON ISLAND
July 18, 1863
0 2 4 KM
0 2 4 MILES
N
HARRISONVILLE
O H I O
MEIGS COUNTY
CHESTER
BELLEVILLE
WOOD COUNTY
MORGAN
Ohio River
North Fork
South Fork
Pond Cr.
HOBSON
RUTLAND
POMEROY
COALPORT
MORGAN
PORTLAND
JUDAH
Buffington Island
Leading Cr.
KYGER
MASON COUNTY
RACINE
RAVENSWOOD
GRAHAM'S STATION
CHESHIRE
LETARTSVILLE
WEST VIRGINIA

*BUFFINGTON ISLAND: Pursued by the Federals in Morgan reaches the Ohio, intending to cross near Buffington Island. Finding himself trapped between fresh Federal Cavalry and a river swollen by rain, Morgan withdraws north. Badly outnumbered, the withdrawl becomes a rout, and over half of Morgan's men are captured.*

*In an attempt to curb the activities of Confederate guerillas, Brig. Gen. Thomas Ewing (right) ordered that any women aiding and abetting the raiders should be detained in Kansas City. The building in which the women were held collapsed, killing five and injuring many others, and spurring-on Quantrill and his men to their act of 'revenge'.*

*Although he obtained a captain's commission early in the war, William Clarke Quantrill (far right) was denied promotion because of his reputation for ferocity. The son of an Ohio schoolteacher, Quantrill appears to have sided with the Confederacy because such an allegiance offered greater opportunities for the exercise of his own particular brand of offensive warfare.*

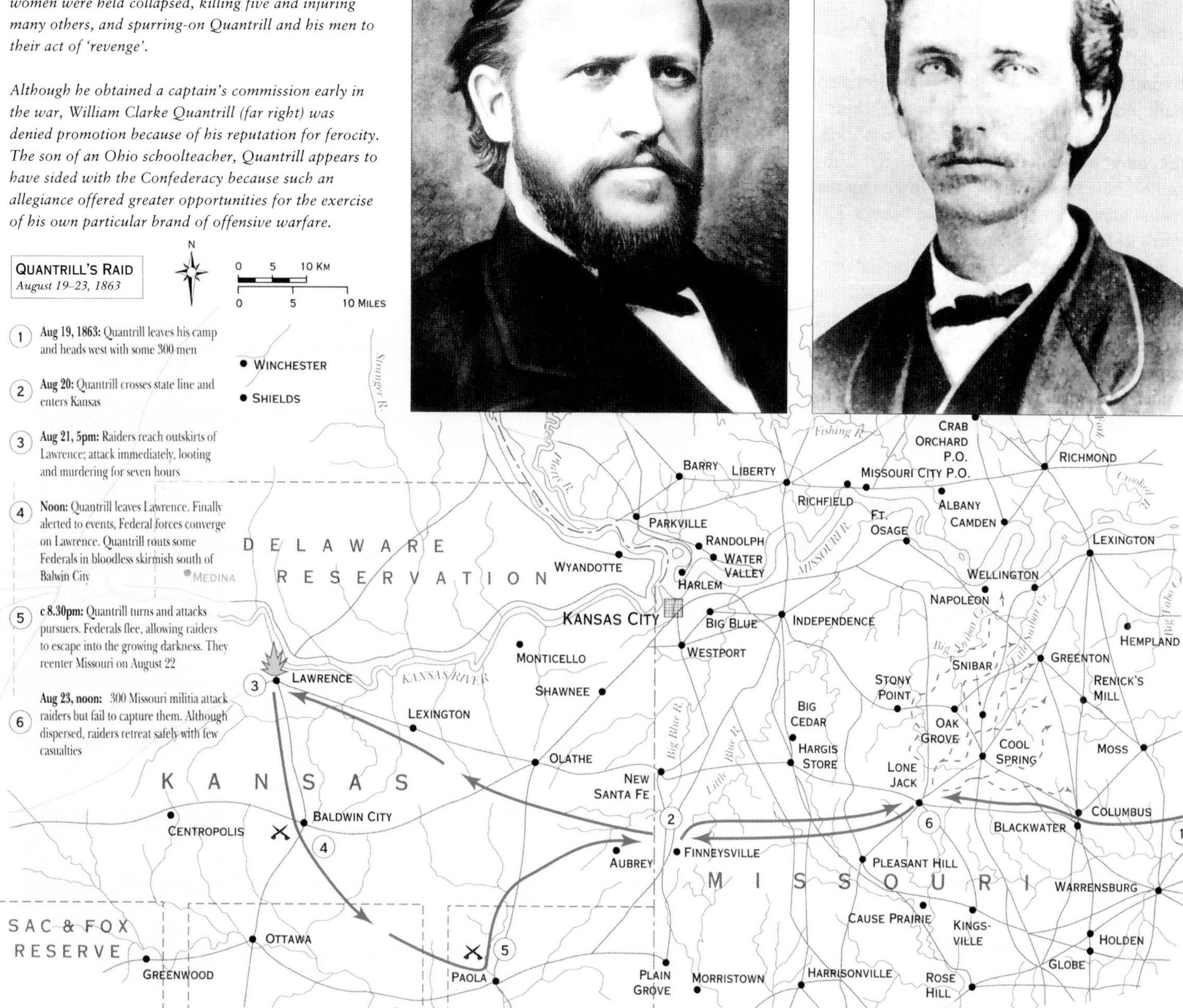

QUANTRILL'S RAID
August 19–23, 1863
N
0 5 10 KM
0 5 10 MILES
1 Aug 19, 1863: Quantrill leaves his camp and heads west with some 300 men
2 Aug 20: Quantrill crosses state line and enters Kansas
3 Aug 21, 5pm: Raiders reach outskirts of Lawrence; attack immediately, looting and murdering for seven hours
4 Noon: Quantrill leaves Lawrence. Finally alerted to events, Federal forces converge on Lawrence. Quantrill routs some Federals in bloodless skirmish south of Balwin City
5 c 8.30pm: Quantrill turns and attacks pursuers. Federals flee, allowing raiders to escape into the growing darkness. They reenter Missouri on August 22
6 Aug 23, noon: 300 Missouri militia attack raiders but fail to capture them. Although dispersed, raiders retreat safely with few casualties
DELAWARE RESERVATION
KANSAS
MISSOURI
SAC & FOX RESERVE
WINCHESTER
SHIELDS
MEDINA
LAWRENCE
KANSAS RIVER
LEXINGTON
CENTROPOLIS
BALDWIN CITY
OTTAWA
GREENWOOD
PAOLA
WYANDOTTE
MONTICELLO
SHAWNEE
OLATHE
NEW SANTA FE
AUBREY
FINNEYSVILLE
PLAIN GROVE
MORRISTOWN
BARRY
LIBERTY
PARKVILLE
RANDOLPH
WATER VALLEY
HARLEM
KANSAS CITY
BIG BLUE
WESTPORT
INDEPENDENCE
RICHFIELD
FT. OSAGE
MISSOURI CITY P.O.
CRAB ORCHARD P.O.
ALBANY
CAMDEN
RICHMOND
LEXINGTON
WELLINGTON
NAPOLEON
HEMPLAND
GREENTON
SNIBAR
RENICK'S MILL
STONY POINT
BIG CEDAR
HARGIS STORE
OAK GROVE
COOL SPRING
MOSS
LONE JACK
COLUMBUS
BLACKWATER
PLEASANT HILL
WARRENSBURG
CAUSE PRAIRIE
KINGSVILLE
HOLDEN
GLOBE
HARRISONVILLE
ROSE HILL


*Quantrill's men, who included the future outlaws Frank and Jesse James, were ordered to kill every male capable of bearing arms, and to burn every house; they were not, however, as this Northern engraving implies, authorized to harm Lawrence's female residents (right).*

> "I would give my farm in White County, Tennessee, and all the salt in Kentucky, if I had it, to stand once more safe and sound on the banks of Calf-Killer Creek."
>
> *A farmer travelling in company with Morgan's raiders.*

# The Campaign Against Charleston

## APRIL 7 – SEPTEMBER 24 1863

On January 6, Rear Admiral Samuel F. du Pont was ordered to capture Charleston, South Carolina.

By early April du Pont was off Charleston. Shortly before the attack, he moved his command to *New Ironsides*, an ironclad built like a conventional steamer, with heavy iron plating and guns mounted in broadside, the only non-monitor in the force. Altogether, nine vessels would pass in line up the main channel, and close to within about 700 yards of Fort Sumter.

Aware the Confederates had placed torpedoes in the channel, du Pont ordered the ship *Weehawken* to take the lead. Fitted with a clumsy raft-like device on her bow, she was expected to sweep the "infernal machines" aside. On April 7, du Pont's line began to move, but was delayed some minutes by *Weehawken's* raft fouling her anchor.

When *Weehawken* resumed her course, she struck one of the torpedoes. An explosion shook her, but did no damage. However, while trying to avoid the torpedoes, *New Ironsides* ran aground. As she struggled to free herself, she twice collided with neighboring monitors.

Hampered by their awkwardness and slow rate of fire, the monitors were hit repeatedly by Confederate shore batteries, the shot dislodging plates and damaging turrets. In the face of such effective fire, du Pont withdrew. Having failed to capture Charleston, he was replaced by Rear Admiral John Dahlgren.

Dahlgren launched a combined attack on Morris Island and Battery Wagner on July 10. The army supplied 5,000 troops, led by Brigadier General Quincy Gillmore. During the day, Dahlgren's monitors hurled shot and shells at Wagner, but with few discernible results; after many seaborne attacks on their positions, the Confederates had learned how to protect themselves. When the ships lifted their fire, Gillmore attacked, but was repulsed with heavy losses.

Undaunted, Dahlgren tried again. On July 18, Battery Wagner was bombarded for over eleven hours. The Confederates retreated to their bomb proofs, but when Federal troops attacked, they were again repulsed.

On September 4, Dahlgren attacked for

*Rear Admiral John Dahlgren on board the* U.S.S. Pawnee *in Charleston Harbor (left). Dahlgren had already proved his ability as an efficient chief of naval ordnance but he had not been tested as a fleet commander. Renowned as an innovative artillery expert, Dahlgren's tactics were to prove essential to the capture of Battery Wagner.*

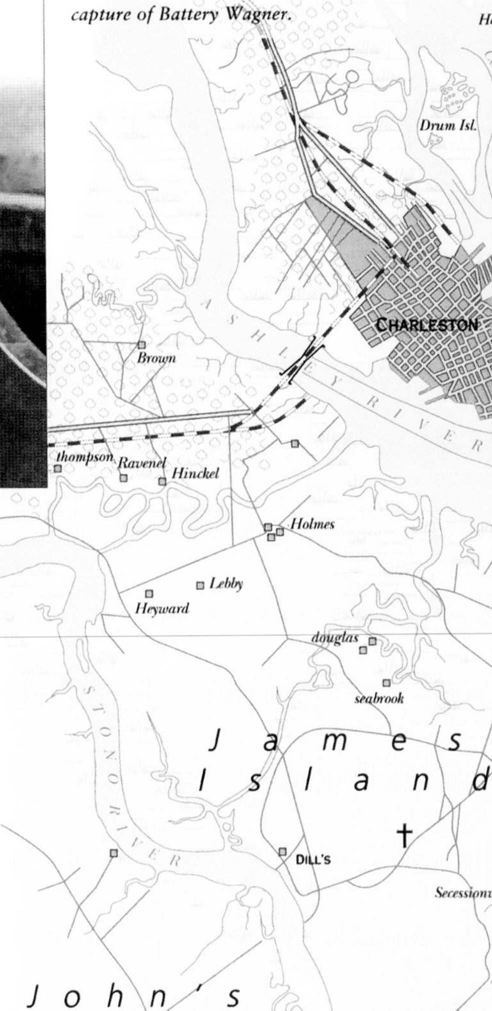

Hobca
COOPER
Drum Isl.
CHARLESTON
ASHLEY RIVER
Brown
thompson
Ravenel
Hinckel
Holmes
Lebby
Heyward
douglas
seabrook
STONO RIVER
James Island
DILL'S
Secessionville
John's Island
Tom Grimball
Legaré
FOLLY
STONO INLET
Kiawah Island

> "It would appear, sir, that despairing of reducing these works, you now resort to the novel measure of turning your guns against the old men, the women and children... of a sleeping city."
>
> *Gen. P.G.T. Beauregard to Gillmore after the bombardment of Charleston.*

*The deck of the* U.S.S. Catskill *which had been numbered among du Pont's attack force (left). During du Pont's assault five of the nine ironclads were partially or wholly disabled, exploding the myth of their invulnerability.*

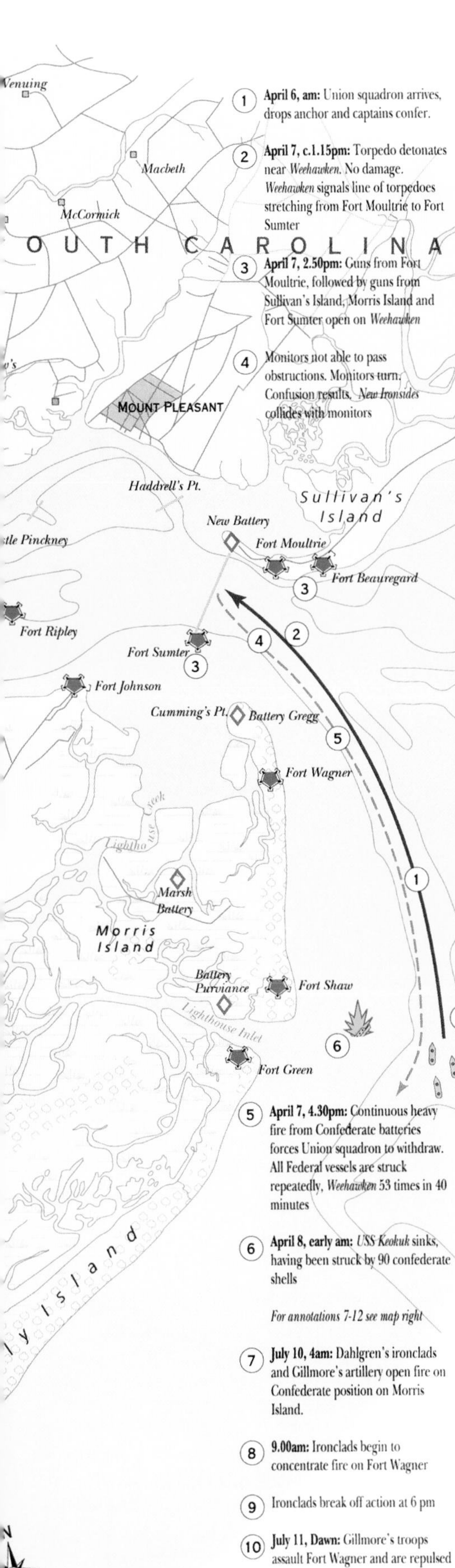

Macbeth
McCormick
OUTH CAROLINA
MOUNT PLEASANT
Haddrell's Pt.
Sullivan's Island
New Battery
Fort Moultrie
Fort Beauregard
Fort Ripley
Fort Sumter
Fort Johnson
Cumming's Pt.
Battery Gregg
Fort Wagner
Marsh Battery
Morris Island
Battery Purviance
Fort Shaw
Fort Green
Lighthouse Inlet
1 April 6, am: Union squadron arrives, drops anchor and captains confer.
2 April 7, c.1.15pm: Torpedo detonates near Weehawken. No damage. Weehawken signals line of torpedoes stretching from Fort Moultrie to Fort Sumter
3 April 7, 2.50pm: Guns from Fort Moultrie, followed by guns from Sullivan's Island, Morris Island and Fort Sumter open on Weehawken
4 Monitors not able to pass obstructions. Monitors turn. Confusion results. New Ironsides collides with monitors
5 April 7, 4.30pm: Continuous heavy fire from Confederate batteries forces Union squadron to withdraw. All Federal vessels are struck repeatedly, Weehawken 53 times in 40 minutes
6 April 8, early am: USS Keokuk sinks, having been struck by 90 confederate shells
For annotations 7-12 see map right
7 July 10, 4am: Dahlgren's ironclads and Gillmore's artillery open fire on Confederate position on Morris Island.
8 9.00am: Ironclads begin to concentrate fire on Fort Wagner
9 Ironclads break off action at 6 pm
10 July 11, Dawn: Gillmore's troops assault Fort Wagner and are repulsed
11 July 18: Ironclads again bombard Fort Wagner. Land attack repulsed
12 September 6: Having endured 60 days of near constant bombardment, Confederate forces evacuate Morris Island under cover of darkness
N
2 KM
1
2 MILES

*Marsh battery (nicknamed "The Swamp Angel"), the Union battery designed and constructed by Col. E. W. Sewell of the New York Engineers, which fired 35 missiles into Charleston. On the thirty sixth round the eight-inch 200-pound Parrott rifle gun exploded (above).*

> "As an experiment with heavy guns, to test their endurance under the severest trial to which they could possibly be subjected in actual service, the results were ... highly interesting and novel."
>
> *Gen. Q. Gillmore commenting on the operation of the "Swamp Angel."*

the third time, bombarding the Confederates for nearly two full days. Realizing their position was hopeless, the Confederates retired under cover of darkness, leaving the Federals to occupy the fort. Despite this Fed -eral victory, Charleston continued to hold out.

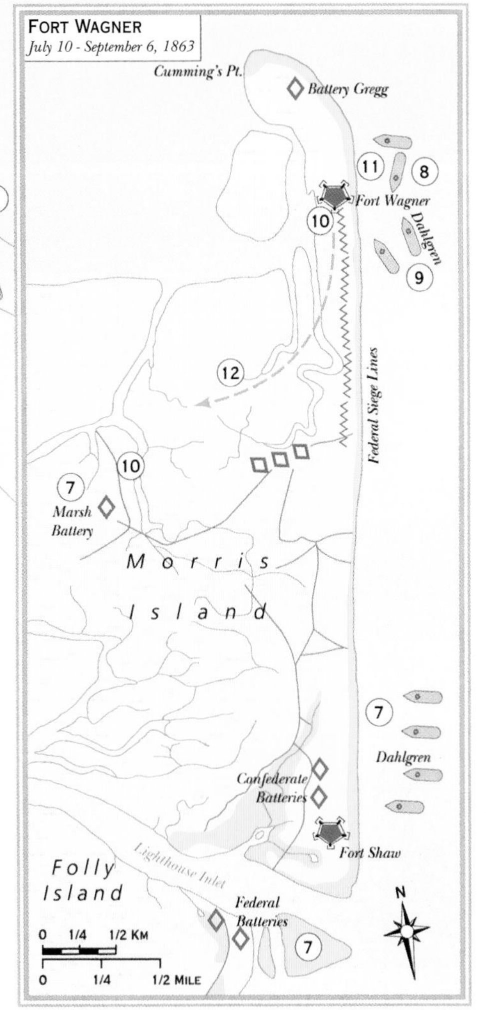

FORT WAGNER
July 10 - September 6, 1863
Cumming's Pt.
Battery Gregg
Fort Wagner
Dahlgren
Federal Siege Lines
Marsh Battery
Morris Island
Dahlgren
Confederate Batteries
Fort Shaw
Folly Island
Lighthouse Inlet
Federal Batteries
0 1/4 1/2 KM
0 1/4 1/2 MILE
N


*Brig. Gen. Quincy Gillmore, commander of the Union land forces, who ordered the construction of the "Swamp Angel" in an attempt to compel the surrender of Fort Wagner (below).*

# Tullahoma Campaign & Capture of Chattanooga

## JUNE 24 – SEPTEMBER 9 1863

Following its defeat at Stones River (called Murfreesboro by the Confederacy) in January 1863, General Braxton Bragg's Confederate Army of Tennessee withdrew to a defensive line which had been established northwest of Tullahoma, Tennessee. Weak and easily turned, Bragg's position nevertheless permitted him to retain control of a rich agricultural area for the Confederacy. For six months, General Rosecrans's Army of the Cumberland remained in its camps around Murfreesboro, reorganizing and refitting before resuming its advance toward Chattanooga on June 23.

Feinting toward Bragg's left at Shelbyville, Rosecrans used the bulk of his infantry to outflank the Confederate right beyond Wartrace. Undeterred by massive rainstorms, within two days the Army of the Cumberland had broken through Confederate forward defenses at Hoover's Gap. Learning that his right was turned, Bragg fell back on his base at Tullahoma. Again Rosecrans feinted directly toward Tullahoma while simultaneously turning Bragg's right. In response, Bragg again withdrew, this time all the way to Chattanooga, while Rosecrans halted to consolidate his gains. Rosecrans's masterful campaign had secured virtually all of Tennessee's most productive region at a cost of only 560 Federal casualties. Confederate losses are unknown, but are estimated at more than 1,600.

On August 16, the Army of the Cumberland resumed its advance. Feinting north of Chattanooga to distract Bragg's attention, Rosecrans sent his army across the Tennessee River at four places. Once across, the Federals began a broad-front advance through rugged mountainous terrain toward the southeast. While XXI Corps threatened Chattanooga directly, forty miles to the south XX Corps and the Cavalry Corps drove toward Bragg's railroad lifeline to Atlanta. In the center, XIV Corps attempted to maintain a tenuous connection between Rosecrans's wings.

By September 8 much of the Federal army was crossing Lookout Mountain, outflanking Bragg, and forcing him to evacuate the city. While the Confederates withdrew southward, XXI Corps entered Chattanooga on September 9. Believing that Bragg's army was in hasty retreat toward Rome, Georgia, Rosecrans brushed aside suggestions to pause and consolidate his scattered units. Instead, he ordered a general pursuit of the Confederates by all three corps. Bragg, meanwhile, had retreated no farther than La Fayette, Georgia, where he concentrated his units for a possible counterstroke.

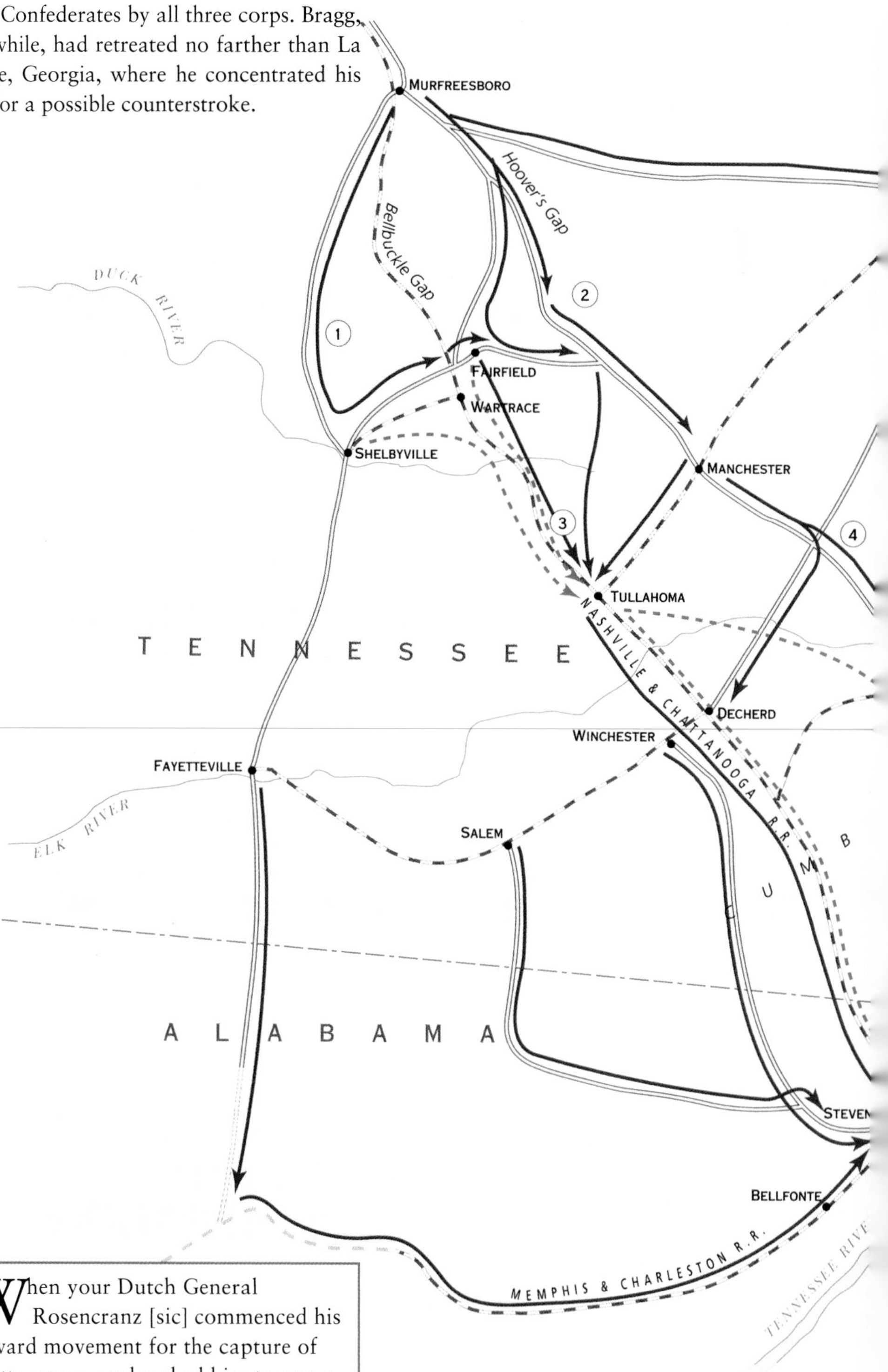

MURFREESBORO
Hoover's Gap
Bellbuckle Gap
DUCK RIVER
1
2
3
4
FAIRFIELD
WARTRACE
SHELBYVILLE
MANCHESTER
TULLAHOMA
T E N N E S S E E
NASHVILLE & CHATTANOOGA R.R.
DECHERD
WINCHESTER
FAYETTEVILLE
ELK RIVER
SALEM
A L A B A M A
STEVEN
BELLFONTE
MEMPHIS & CHARLESTON R.R.
TENNESSEE RIVER


"When your Dutch General Rosencranz [sic] commenced his forward movement for the capture of Chattanooga, we laughed him to scorn. We believed that the black brow of Lookout Mountain would frown him out of existence... and that the northern people and the goverment at Washington would perceive how hopeless were their efforts when they came to attack the real South."

*A Confederate officer in communication with a Federal correspondent.*

"While few persons exhibited more estimable qualities, I have never seen a public man possessing talent with less administrative power, less clearness and steadiness in difficulty, and greater practical incapacity than General Rosecrans."

*Charles A. Dana, in a report to the War Department.*

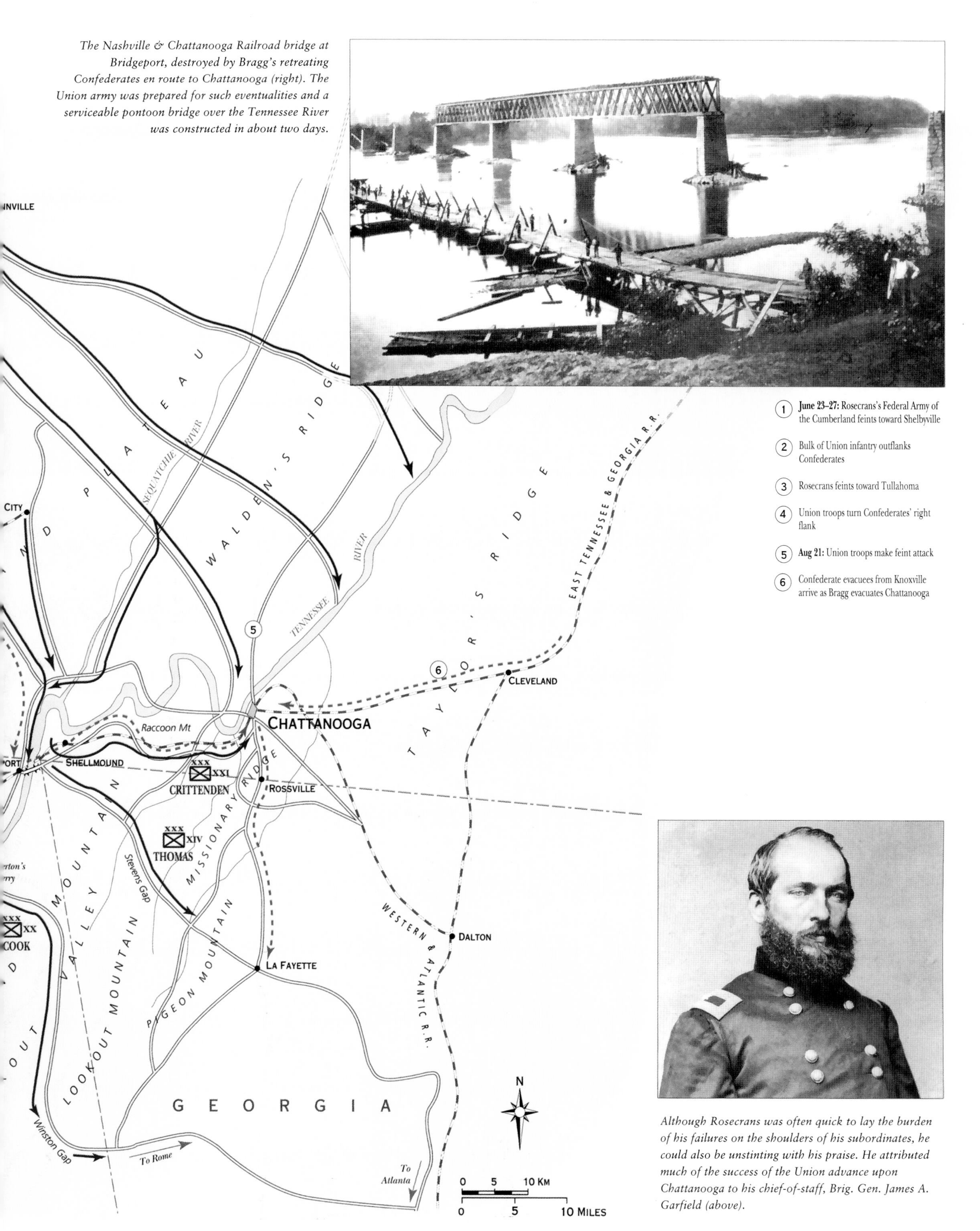
1 June 23–27: Rosecrans's Federal Army of the Cumberland feints toward Shelbyville
2 Bulk of Union infantry outflanks Confederates
3 Rosecrans feints toward Tullahoma
4 Union troops turn Confederates' right flank
5 Aug 21: Union troops make feint attack
6 Confederate evacuees from Knoxville arrive as Bragg evacuates Chattanooga
CHATTANOOGA
CLEVELAND
ROSSVILLE
DALTON
LA FAYETTE
SHELLMOUND
Raccoon Mt
CRITTENDEN
XXI
THOMAS
XIV
COOK
XX
WALDEN'S RIDGE
TAYLOR'S RIDGE
MISSIONARY RIDGE
LOOKOUT MOUNTAIN
PIGEON MOUNTAIN
VALLEY
SEQUATCHIE RIVER
TENNESSEE RIVER
EAST TENNESSEE & GEORGIA R.R.
WESTERN & ATLANTIC R.R.
Stevens Gap
Winston Gap
To Rome
To Atlanta
GEORGIA
N
0 5 10 KM
0 5 10 MILES

*The Nashville & Chattanooga Railroad bridge at Bridgeport, destroyed by Bragg's retreating Confederates en route to Chattanooga (right). The Union army was prepared for such eventualities and a serviceable pontoon bridge over the Tennessee River was constructed in about two days.*

*Although Rosecrans was often quick to lay the burden of his failures on the shoulders of his subordinates, he could also be unstinting with his praise. He attributed much of the success of the Union advance upon Chattanooga to his chief-of-staff, Brig. Gen. James A. Garfield (above).*

# Chickamauga Campaign SEPTEMBER 10 – 18 1863

General Rosecrans's decision to pursue the Confederate Army of Tennessee into the mountains of northern Georgia was based upon the overt testimony of deserters, and his own intuitive belief that Bragg was beaten. Even though his infantry corps were scattered widely over the rough terrain, Rosecrans believed that the risk to his army was minimal. Therefore, he ordered each corps to pursue the enemy without reference to other Federal units. Accordingly, XXI Corps headed east from Chattanooga toward Ringgold, Georgia; XIV Corps advanced across McLemore's Cove toward La Fayette; XX Corps and the Cavalry Corps moved toward Rome.

Unknown to Rosecrans, the testimony of the Confederate deserters – upon which he had relied – was intentionally deceptive. Bragg's army was neither retreating nor de moralized. Concentrated around La Fayette, the Army of Tennessee daily gained strength as reinforcements arrived from East Tennessee and Mississippi. Upon learning that the Federal corps were widely scattered, Bragg saw an opportunity to defeat them individually. When two XIV Corps divisions incautiously approached the eastern exit of McLemore's Cove at Dug Gap, Bragg ordered several Confederate divisions to converge upon them. However, the Confederates proved unable to trap the Federals, who rapidly withdrew to safety on September 11. Two days later, a similar effort by Bragg to defeat a detached part of XXI Corps also failed near Lee and Gordon's Mill.

The near-disaster in McLemore's Cove caused Rosecrans to recognize that the Army of the Cumberland was in grave danger, and he immediately ordered it to concentrate just south of Chattanooga. Unfortunately, the XX Corps was forty miles distant at Alpine, Georgia, and did not join XIV Corps in McLemore's Cove until September 17. Only then did the two corps march north to unite with XXI Corps on Chickamauga Creek. While the Federals marched, Bragg waited briefly for the arrival of reinforcements from Virginia. Then, on September 18, he ordered an offensive movement to turn the Federal left flank north of Lee and Gordon's Mill. By the end of the day Confederate units had seized crossings of Chickamauga Creek at Reed's Bridge, Alexander's Bridge, and Thedford's Ford, driving away Federal pickets in the process. Bragg was now poised to interpose his army between Rosecrans and Chattanooga.

*Confederate artillery opens fire on the Union troops holding Reed's Bridge over the Chickamauga on September 18 (below). During the engagement the main body of the Confederate army crossed the river at different points.*

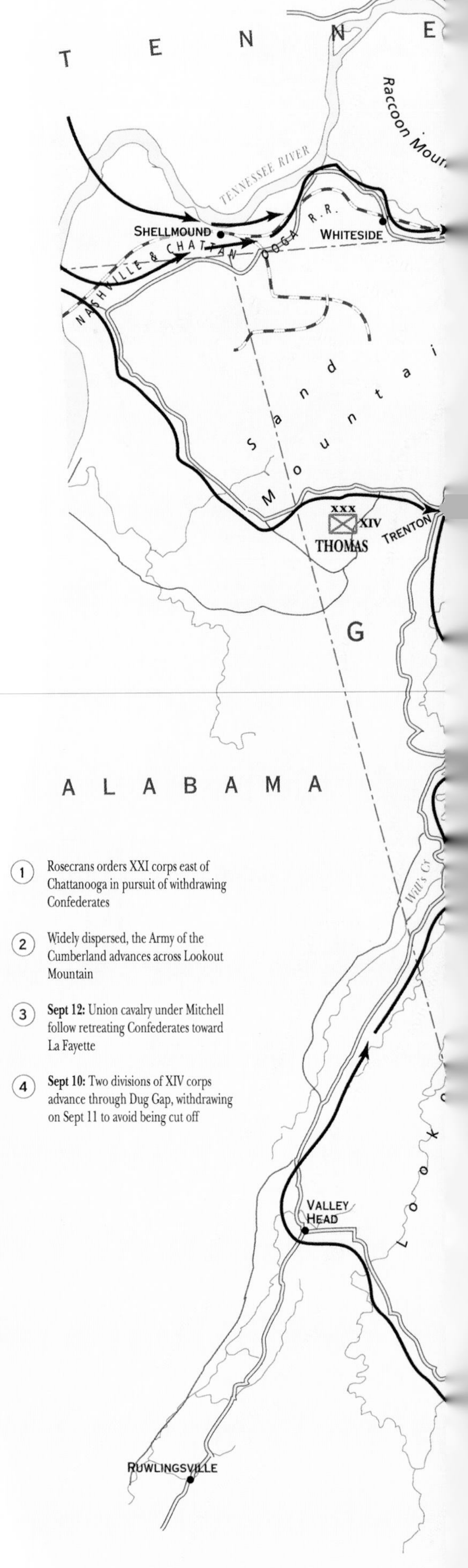
T E N N E
Raccoon Moun
TENNESSEE RIVER
SHELLMOUND
WHITESIDE
NASHVILLE & CHATTANOOGA R.R.
Sand Mountain
XXX
XIV
THOMAS
TRENTON
G
ALABAMA
Will's Cr
Look
VALLEY HEAD
RAWLINGSVILLE
1 Rosecrans orders XXI corps east of Chattanooga in pursuit of withdrawing Confederates
2 Widely dispersed, the Army of the Cumberland advances across Lookout Mountain
3 Sept 12: Union cavalry under Mitchell follow retreating Confederates toward La Fayette
4 Sept 10: Two divisions of XIV corps advance through Dug Gap, withdrawing on Sept 11 to avoid being cut off

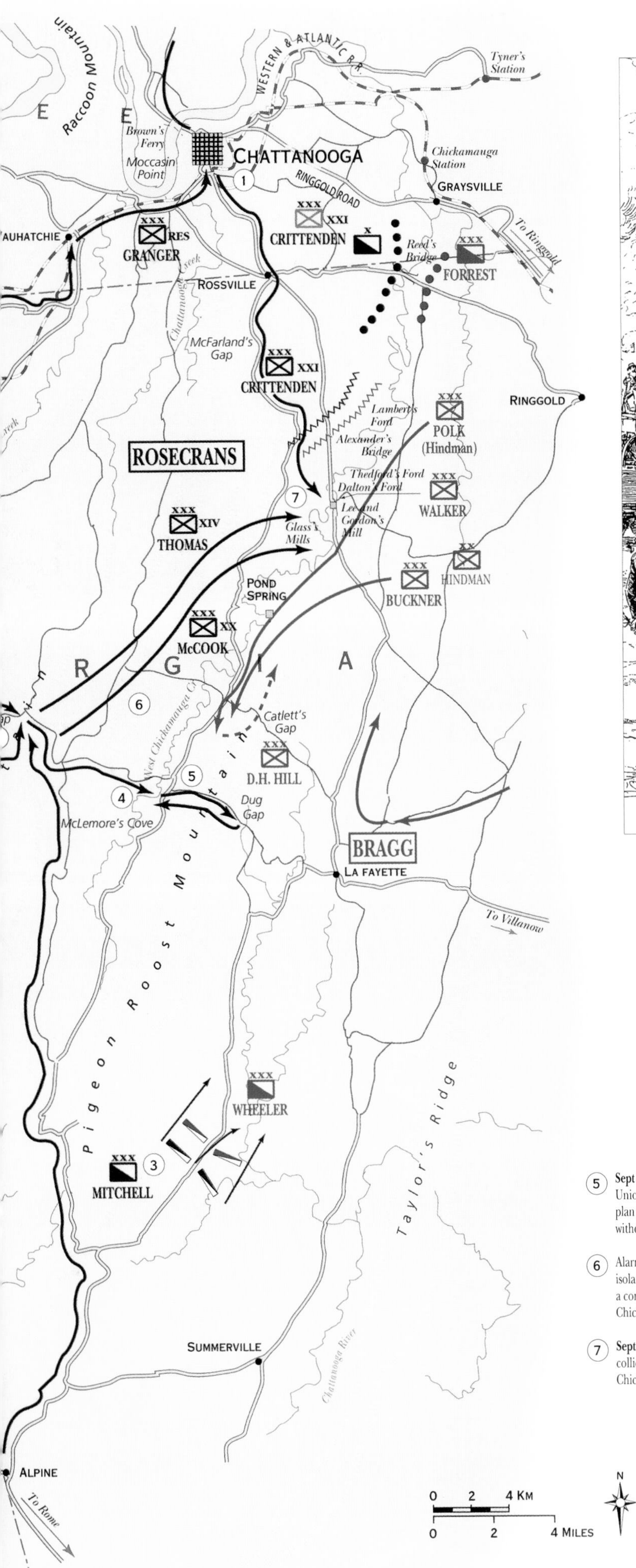

Raccoon Mountain
WESTERN & ATLANTIC R.R.
Tyner's Station
Brown's Ferry
Moccasin Point
CHATTANOOGA
Chickamauga Station
RINGGOLD ROAD
GRAYSVILLE
AUHATCHIE
GRANGER
RES
CRITTENDEN
XXI
Reed's Bridge
FORREST
To Ringgold
ROSSVILLE
McFarland's Gap
Chattanooga Creek
CRITTENDEN
RINGGOLD
Lambert's Ford
Alexander's Bridge
POLK (Hindman)
ROSECRANS
Thedford's Ford
Dalton's Ford
WALKER
Lee and Gordon's Mill
Glass's Mills
THOMAS
XIV
HINDMAN
POND SPRING
BUCKNER
McCOOK
XX
Catlett's Gap
West Chickamauga Cr.
D.H. HILL
McLemore's Cove
Dug Gap
BRAGG
LA FAYETTE
To Villanow
Pigeon Roost Mountain
Taylor's Ridge
WHEELER
MITCHELL
SUMMERVILLE
Chattanooga River
ALPINE
To Rome
0 2 4 KM
0 2 4 MILES
N

W&A RR


*Confederate reinforcements arrive by rail to support Braxton Bragg's advance (above). Although the regional rail network was in poor condition it proved an essential supply route, vital for Bragg's intended maneuvers.*

"God bless the gallant fellows, not one man intoxicated, not one rude word did I hear. It was a strange sight. What seemed miles of platform cars, and soldiers rolled in their blankets lying in rows... In their grey blankets packed in regular order, they looked like swathed mummies."

*A Southern matron upon viewing the reinforcements advancing by rail to support Bragg.*

5. **Sept 10:** Bragg orders attack on isolated Union units in McLemore's Cove; the plan fails and Union forces are able to withdraw

6. Alarmed by Confederate attempt to isolate part of his forces, Rosecrans orders a concentration in the valley of West Chickamauga Creek

7. **Sept 19, early am:** Opposing patrols collide, precipitating the battle of Chickamauga

"I had my horse to leap from the train, mounted with one arm in a sling, and about 3pm, joined our forces ... in line of battle... in a short time we arrived at Reed's Bridge, across the Chickamauga, and discerned the Federals drawn up in battle array beyond the bridge, which they had partially destroyed. I ordered some... artillery to open fire... our opponents quickly retreated. We repaired the bridge, and continued to advance."

*Brig. Gen. John B. Hood.*

# Chickamauga SEPTEMBER 19 – 20 1863

General Bragg's plan on September 19 called for his army to complete its crossing of Chickamauga Creek and commence a drive south toward Lee and Gordon's Mill, where he believed the left flank of the Federal Army of the Cumberland was located. The Confederate army could then drive the Federals southward into the wilds of McLemore's Cove, away from both Chattanooga and their supply line to Nashville. When the plan was formulated, the Federal left indeed lay at Lee and Gordon's Mill, but during the early hours of September 19 General George Thomas's XIV Corps had passed behind XXI Corps and extended the Federal line northward beyond the Confederate position.

After sunrise on September 19, Thomas sent a division eastward in search of a Confederate brigade rumored to be isolated west of Chickamauga Creek. The division encountered Confederate cavalry covering the rear of Bragg's army, and the resulting action soon drew both armies to the scene. Throughout the day, both commanders threw more and more units into the fight, but neither side gained a decisive advantage in the struggle which raged in the tangled thickets and small fields just west of the creek. After dark, the two armies drew apart slightly, but each prepared to renew the contest at first light.

That night, the Army of Tennessee received additional reinforcements from Virginia. Dividing his forces into two wings

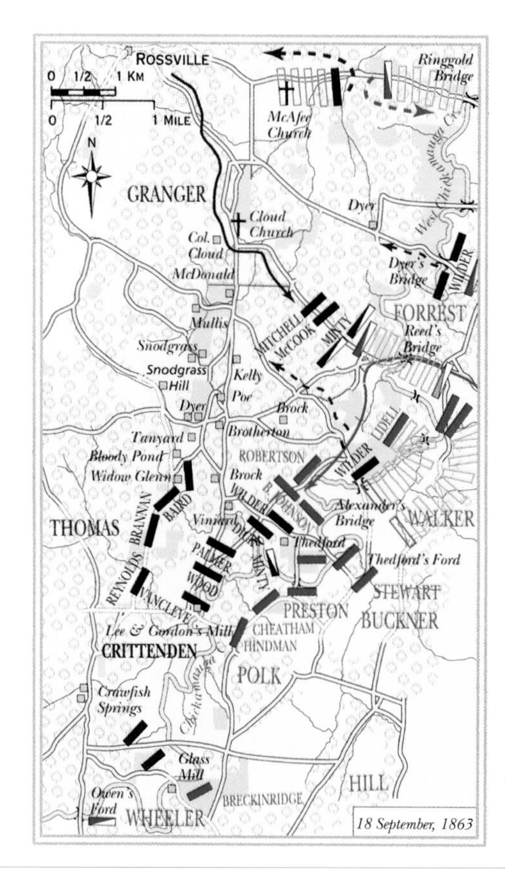
ROSSVILLE
Ringgold Bridge
McAfee Church
GRANGER
Cloud Church
Dyer
Col. Cloud
McDonald
Dyer's Bridge
WILDER
FORREST
Reed's Bridge
MITCHELL
McCOOK
MINTY
Mullis
Snodgrass
Snodgrass Hill
Kelly
Poe
Brock
Dyer
Tanyard
Brotherton
LIDELL
Bloody Pond
Widow Glenn
ROBERTSON
Brock
THOMAS
BAIRD
BRANNAN
Alexander's Bridge
WALKER
Thedford
Thedford's Ford
REYNOLDS
PALMER
WOOD
VANCLEVE
STEWART
PRESTON
BUCKNER
Lee & Gordon's Mill
CHEATHAM
HINDMAN
CRITTENDEN
POLK
Crawfish Springs
Chickamauga
Glass Mill
Owen's Ford
WHEELER
BRECKINRIDGE
HILL
18 September, 1863

To interpose his forces between Rosecrans and Chattanooga, Bragg marched the bulk of his army north across Chickamauga Creek (above). Heavy skirmishing broke out with Federal Cavalry, beginning at Reed's Bridge. Rosecrans sent Thomas's Corps northeast to prevent the Confederates outflanking his Federal forces toward Chattanooga. A major battle threatened.

*"The Rock of Chickamauga", Maj. Gen. George Thomas (below). A total rout of Rosecrans's army was only prevented by Thomas's insistence upon a reinforcement of the Union left and the spirit of resistance with which he inspired his men.*

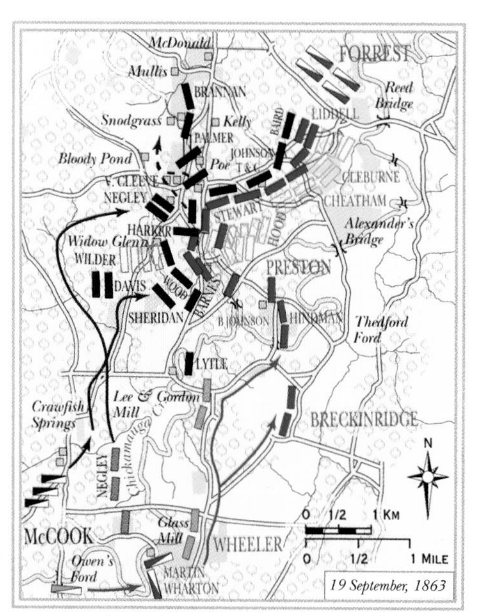
McDonald
FORREST
Mullis
Reed Bridge
BRANNAN
Snodgrass
Kelly
LIDDELL
PALMER
Bloody Pond
Poe
V. CLEVE
NEGLEY
CLEBURNE
CHEATHAM
HARKER
STEWART
Alexander's Bridge
Widow Glenn
WILDER
PRESTON
DAVIS
SHERIDAN
B. JOHNSON
HINDMAN
Thedford Ford
LYTLE
Crawfish Springs
Lee & Gordon Mill
BRECKINRIDGE
NEGLEY
Chickamauga
McCOOK
Glass Mill
WHEELER
Owen's Ford
MARTIN
WHARTON
19 September, 1863

*With neither side aware of the precise position of its opponent, Thomas sent a division forward to reconnoiter the Chickamauga (above). The advancing Federals drove back Forrest's Cavalry, but were in turn hard hit by two of Walker's Confederate divisions. The battle swung back and forth, losses were high, but the results by nightfall were negligible.*

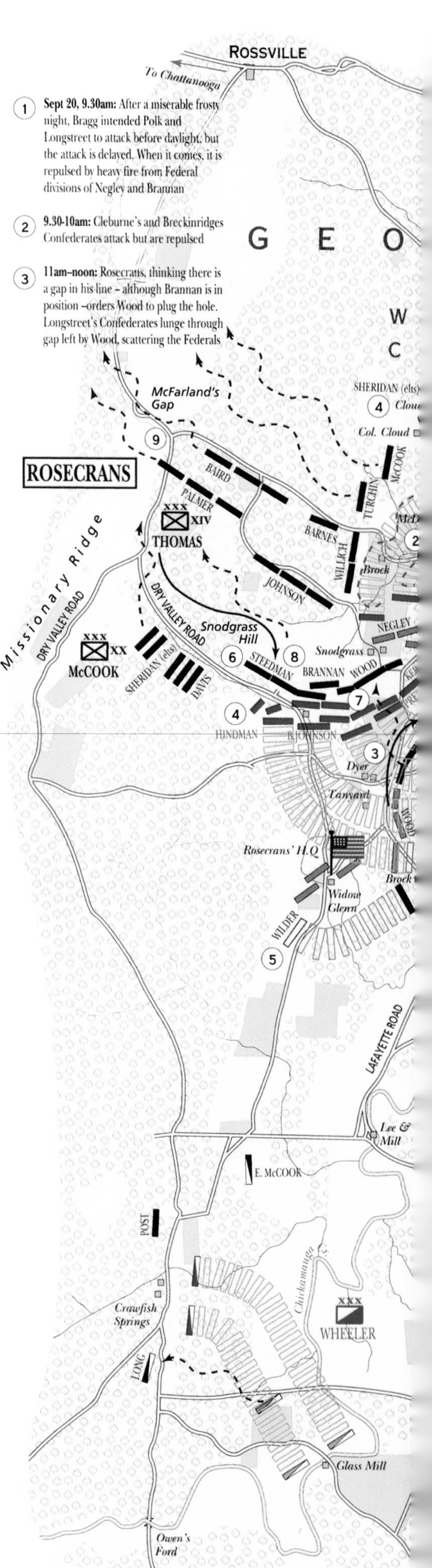
ROSSVILLE
To Chattanooga
1 Sept 20, 9.30am: After a miserable frosty night, Bragg intended Polk and Longstreet to attack before daylight, but the attack is delayed. When it comes, it is repulsed by heavy fire from Federal divisions of Negley and Brannan
2 9.30-10am: Cleburne's and Breckinridges Confederates attack but are repulsed
3 11am–noon: Rosecrans, thinking there is a gap in his line – although Brannan is in position –orders Wood to plug the hole. Longstreet's Confederates lunge through gap left by Wood, scattering the Federals
G E O
McFarland's Gap
ROSECRANS
BAIRD
PALMER
THOMAS
XIV
Missionary Ridge
DRY VALLEY ROAD
McCOOK
SHERIDAN (elts)
DAVIS
Snodgrass Hill
STEEDMAN
BRANNAN
WOOD
JOHNSON
BARNES
TURCHIN
HINDMAN
B. JOHNSON
Dyer
Tanyard
Rosecrans' H.Q
Widow Glenn
WILDER
LAFAYETTE ROAD
Lee & Mill
E. McCOOK
POST
Crawfish Springs
Chickamauga
WHEELER
LONG
Glass Mill
Owen's Ford

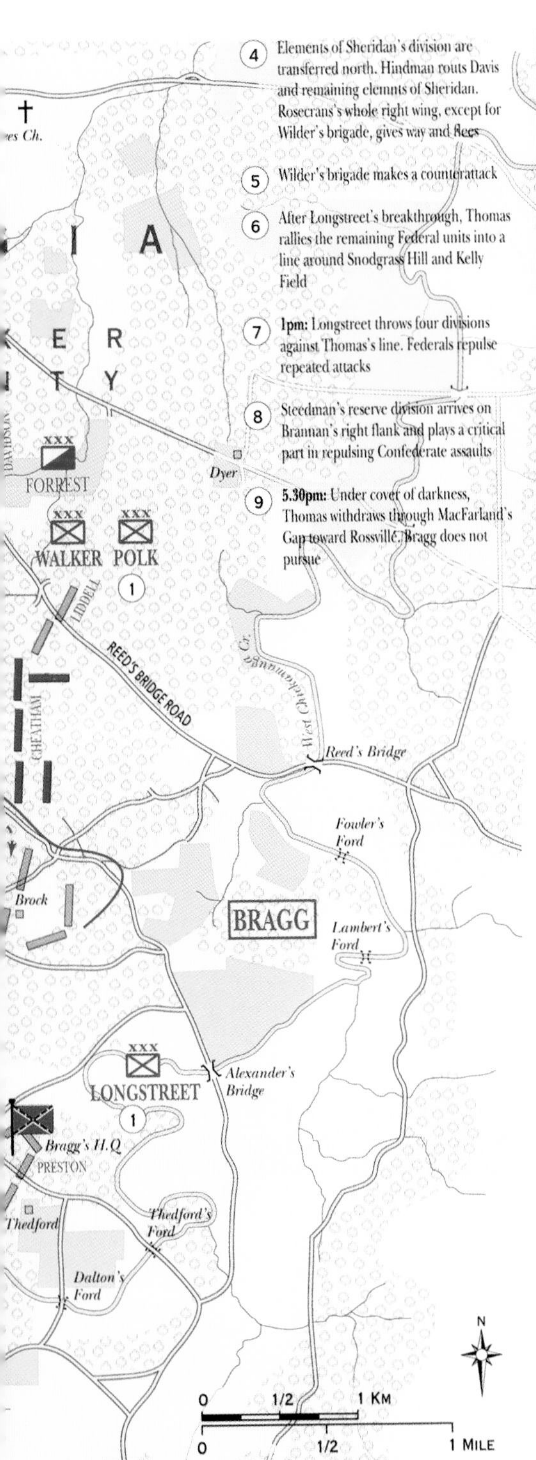
4 Elements of Sheridan's division are transferred north. Hindman routs Davis and remaining elemnts of Sheridan. Rosecrans's whole right wing, except for Wilder's brigade, gives way and flees
5 Wilder's brigade makes a counterattack
6 After Longstreet's breakthrough, Thomas rallies the remaining Federal units into a line around Snodgrass Hill and Kelly Field
7 1pm: Longstreet throws four divisions against Thomas's line. Federals repulse repeated attacks
8 Steedman's reserve division arrives on Brannan's right flank and plays a critical part in repulsing Confederate assaults
9 5.30pm: Under covet of darkness, Thomas withdraws through MacFarland's Gap toward Rossvillé. Bragg does not pursue
FORREST
WALKER
POLK
LONGSTREET
BRAGG
Bragg's H.Q.
PRESTON
REED'S BRIDGE ROAD
Reed's Bridge
Fowler's Ford
Lambert's Ford
Alexander's Bridge
Thedford's Ford
Dalton's Ford
Dyer
Brock
Thedford
LIDDELL
CHEATHAM
0 1/2 1 KM
0 1/2 1 MILE

the left under General Longstreet and the right under General Polk – Bragg ordered a coordinated attack with the right wing to commence at dawn. Delayed by poor staff work for four hours, the attack stalled after achieving momentary success. The movement of Federal reserves northward, however, opened a gap in Rosecrans's line that was exploited by the Confederate left wing. The Federal right collapsed, forcing Rosecrans and other senior commanders from the field. Undismayed, Thomas rallied the remaining Federals around Snodgrass Hill and Kelly Field, and held his ground until nightfall. Battered, but not decisively beaten, the Army of the Cumberland withdrew into Chattanooga. Bragg had inflicted 16,170 casualties on the Federal army and had held the field, but in turn he had paid a heavy price – 18,454 Confederate casualties – and despite all the determination and sacrifice of the Army of the Tennessee, Chatanooga still remained firmly in Federal hands.

*While encouraging his men to hold their positons, Gen. John B. Hood was struck in the leg by a Union minie ball. Already crippled by a wound from Gettysburg, this further injury necessitated the amputation of a leg.*

*Twelve-year-old drummer-boy Johnny Clem who was promoted to the rank of sergeant for felling a rebel colonel who was pursuing him on horseback (above).*

"We see a flash, as of sheet lightning; we hear the report; we see our comrades falling, some never to rise again; some mortally wounded, weltering in blood, others crippled and writhing in agony... The battle is on in earnest.

*Union Private A. van Lisle at Chickamauga.*

*The Battle of Chickamauga (right). One Union officer reported that the dead lay in such masses that the ground beneath them was scarcely visible and that the Chickamauga ran red with blood.*

# Chattanooga SEPTEMBER 21 – NOVEMBER 25 1863

After Chickamauga, the Confederate Army of Tennessee followed the Federals to Chattanooga. Unable to hold Lookout Mountain and Missionary Ridge – which overlooked the city – General Rosecrans established his defenses on the valley floor. When the Confederates occupied the two mountain ridges, Rosecrans's only access to his supply base at Stevenson, Alabama, was a long, tortuous wagon road over Walden's Ridge. Within a month, the Army of the Cumberland was in danger of being starved out of Chattanooga. In haste, the Lincoln administration transferred troops from Mississippi and Virginia to reinforce the Army of the Cumberland. Given overall command in the Chattanooga area, General U.S. Grant replaced Rosecrans with General Thomas, and ordered General Sherman's and General Hooker's units to assist Thomas in breaking Bragg's siege of Chattanooga.

Long before Chickamauga, the Confederate Army of Tennessee had been noted for factional squabbling within its command structure. By October, virtually the only point of agreement among the army's senior commanders was their dislike for General Bragg. Even after Bragg forced the departure of Generals Polk and D.H. Hill, dissension continued to be fueled by General Longstreet. Matters came to a head when Longstreet failed to prevent the Federals from opening a direct supply line to Chattanooga at the battle of Wauhatchie on October 28. Rather than break openly with Longstreet, Bragg permitted him to take his corps north to besiege the Federal garrison at Knoxville, Tennessee.

Reduced in strength, Bragg's army lost Lookout Mountain to Hooker on November 24, and concentrated on Missionary Ridge for a final stand. The following day, Grant sent Sherman and Hooker to envelop the flanks of the new Confederate position. With their attacks stalled, Grant asked Thomas to mount a demonstration in the center of the Federal line. Anxious to eradicate the stain of Chickamauga, the Army of the Cumberland turned the demonstration into a smashing attack, which gained the crest of Missionary Ridge and shattered Bragg's army. Federal losses were only 5,815, while Confederate casualties were not much larger at 6,667. Chattanooga, however, was irretrievably lost to the Confederacy. After briefly pursuing Bragg's army into northern Georgia, the Federals returned to Chattanooga; both sides ceased operations for the winter.

*Unpopular with both his officers and men, Gen. Braxton Bragg (above) was argumentative, irritable and extremely critical of his subordinates. His friendship with Jefferson Davis and the reputation which he gained during the Mexican War, ensured his commission as a brigadier general at the outbreak of hostilities. Despite his less attractive characteristics, Grant would later call him " remarkably intelligent and well-informed".*

*Owing his appointment to his political background rather than to proven military ability, the initial Northern confidence in the abilities of William Starke Rosecrans (above) rapidly began to decline. Though conscientious and well-intentioned, Rosecrans lacked the resolution and decisiveness essential in a general officer. His replacement by the capable George Thomas came as a relief to his subordinates.*

1. **Sept 21:** Unable to hold lookout Mountain and Missionary Ridge, Rosecrans withdraws his forces to Chattanooga
2. **Sept 21:** Bragg's Confederates occupy Lookout Mountain and Missionary Ridge
3. **Sept 25-Oct 24:** To aid Rosecrans, Federal XI and XII Corps of Army of the Potomac are transferred 1,233 miles by rail from Culpeper to Chattanooga
4. **Oct 23:** Grant, given overall command in the west, arrives in Chattanooga
5. **Oct 26-27, night:** Gen. Smith and 3,500 men sail down Tennessee River and march across Moccasin Point to Brown's Ferry, chase-off Confederate pickets and erect pontoon bridge
6. **Oct 28-29:** At Wauhatchie, Longstreet's troops attack Federal XI Corps, but fail to drive them from the newly-opened Union supply route from Bridgeport, nick-n amed the "Cracker-line"
7. **Nov 23-24:** Grant sends Sherman and Hooker to envelop the flanks of Confederate position

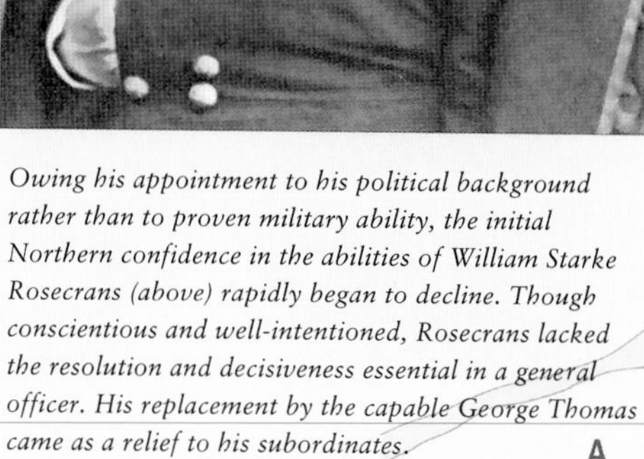
A L
*To Bridgeport Union Army Railhead & Stevenson*

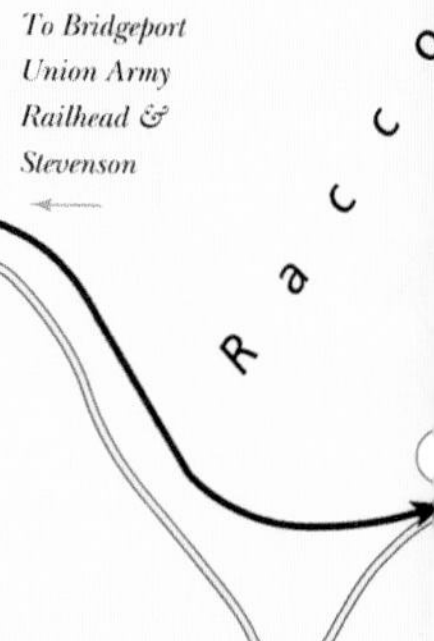

*Having positioned Gen. Hooker at the foot of Lookout Mountain, Gen. Grant was able to succor the besieged Chattanooga with food and supplies sent by rail and steamboat from Bridgeport. Although the needs of Thomas's beleaguered army could not be met in full, the establishment of this tortuous supply route did prevent actual starvation and in itself constituted a masterpiece of military transportation (below).*

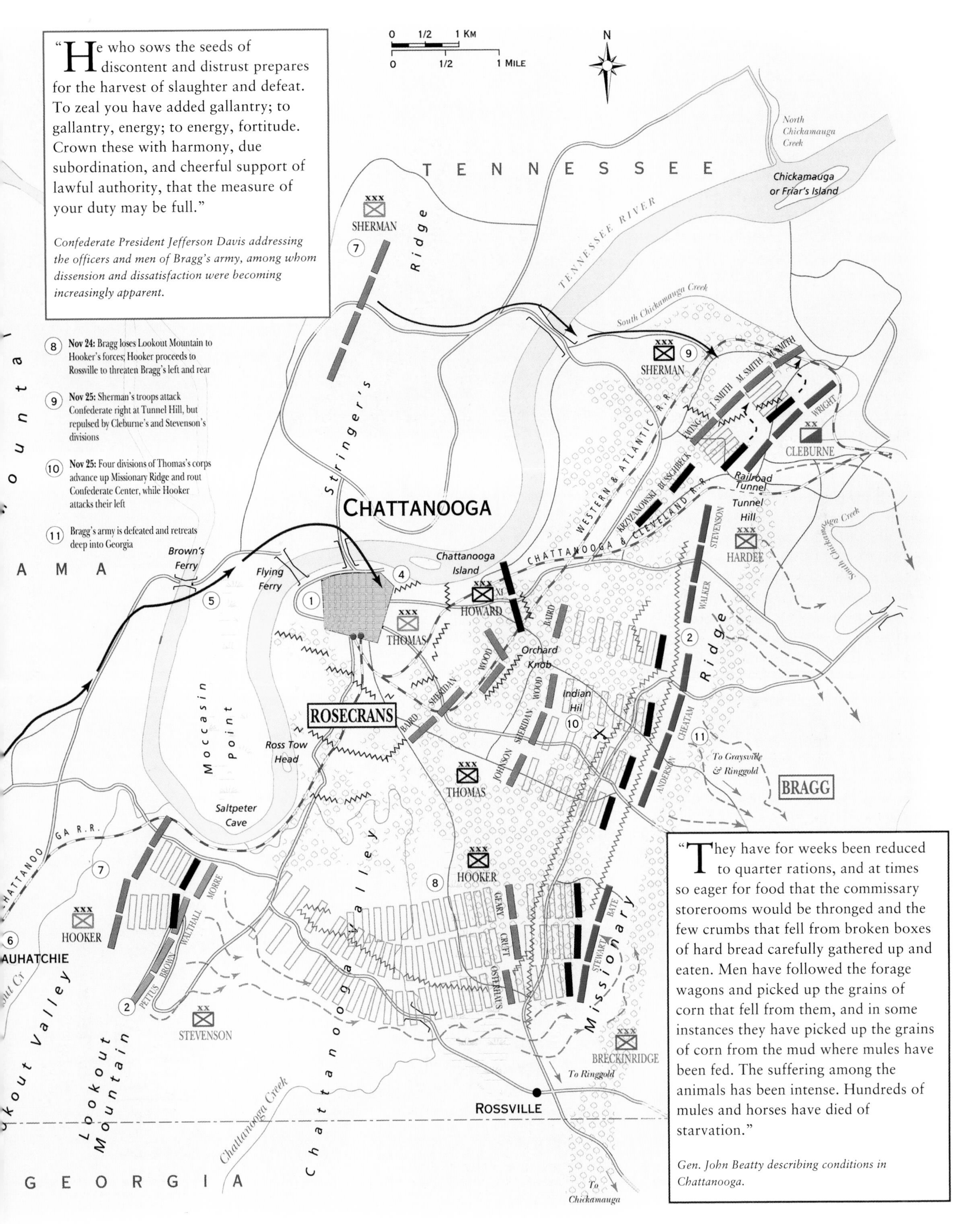
0 1/2 1 KM
0 1/2 1 MILE
N
Nov 24: Bragg loses Lookout Mountain to Hooker's forces; Hooker proceeds to Rossville to threaten Bragg's left and rear
Nov 25: Sherman's troops attack Confederate right at Tunnel Hill, but repulsed by Cleburne's and Stevenson's divisions
Nov 25: Four divisions of Thomas's corps advance up Missionary Ridge and rout Confederate Center, while Hooker attacks their left
Bragg's army is defeated and retreats deep into Georgia
TENNESSEE
TENNESSEE RIVER
North Chickamauga Creek
Chickamauga or Friar's Island
South Chickamauga Creek
SHERMAN
Stringer's Ridge
CHATTANOOGA
Brown's Ferry
Flying Ferry
Chattanooga Island
HOWARD
THOMAS
ROSECRANS
Moccasin Point
Ross Tow Head
Saltpeter Cave
CLEBURNE
Railroad Tunnel
Tunnel Hill
HARDEE
WESTERN & ATLANTIC R.R.
CHATTANOOGA & CLEVELAND R.R.
Orchard Knob
Indian Hil
Missionary Ridge
To Graysville & Ringgold
BRAGG
HOOKER
STEVENSON
BRECKINRIDGE
To Ringgold
ROSSVILLE
Chattanooga Valley
Chattanooga Creek
Lookout Valley
Lookout Mountain
AUHATCHIE
GEORGIA
To Chickamauga
AMA

"He who sows the seeds of discontent and distrust prepares for the harvest of slaughter and defeat. To zeal you have added gallantry; to gallantry, energy; to energy, fortitude. Crown these with harmony, due subordination, and cheerful support of lawful authority, that the measure of your duty may be full."

*Confederate President Jefferson Davis addressing the officers and men of Bragg's army, among whom dissension and dissatisfaction were becoming increasingly apparent.*

"They have for weeks been reduced to quarter rations, and at times so eager for food that the commissary storerooms would be thronged and the few crumbs that fell from broken boxes of hard bread carefully gathered up and eaten. Men have followed the forage wagons and picked up the grains of corn that fell from them, and in some instances they have picked up the grains of corn from the mud where mules have been fed. The suffering among the animals has been intense. Hundreds of mules and horses have died of starvation."

*Gen. John Beatty describing conditions in Chattanooga.*

# Knoxville Campaigns 15 AUGUST – 4 DECEMBER 1863

Because of its large loyalist population, the mountainous area of East Tennessee had long been an objective for the Federal armies. In mid-August, 1863, Major General Ambrose Burnside began an advance from Kentucky toward Knoxville, East Tennessee's most important city. His opponent was Confederate Major General Simon Buckner, who garrisoned East Tennessee with a few scattered units. When the simultaneous advance by the Federal Army of the Cumberland toward Chattanooga threatened General Bragg's army, Buckner's small force was consolidated and sent to its aid. The Confederate withdrawal permitted Burnside to enter Knoxville on September 2.

By taking Knoxville, Burnside severed the direct rail route between Bragg's army at Chattanooga, and Virginia. Thereafter, the Federals used Knoxville as a base from which to expand their control into all parts of East Tennessee. Most significantly, they forced the surrender of the 2,500-man Confederate garrison at Cumberland Gap on September 9. By the end of the month, Burnside's command had grown to two army corps, the IX and XXIII, and had advanced as far as Jonesboro and Loudon, Tennessee.

Following the fiasco at Wauhatchie in late October, on November 4 Bragg permitted James Longstreet to leave the siege lines at Chattanooga and take two divisions northward to deal with Burnside. In the face of Longstreet's advance, Burnside fell back gradually until he entered the defenses of Knoxville on November 18. Unable to mount a formal siege because of inadequate numbers, Longstreet resolved to carry the city by direct assault. Waiting more than a week for reinforcements from Bragg, he finally attacked a salient known as Fort Sanders on November 29. The result was a complete and bloody repulse, with the Confederate loss of 813 men far exceeding the Federal total of 100.

By this time Bragg had been defeated at Missionary Ridge, and large Federal reinforcements under General Sherman were en route to relieve Burnside. Before they could arrive, Longstreet on December 4 withdrew eastward into the mountains and went into winter quarters. There was no significant Federal pursuit, and Longstreet's continued presence in the area caused the Federals to garrison East Tennessee heavily until the following spring.

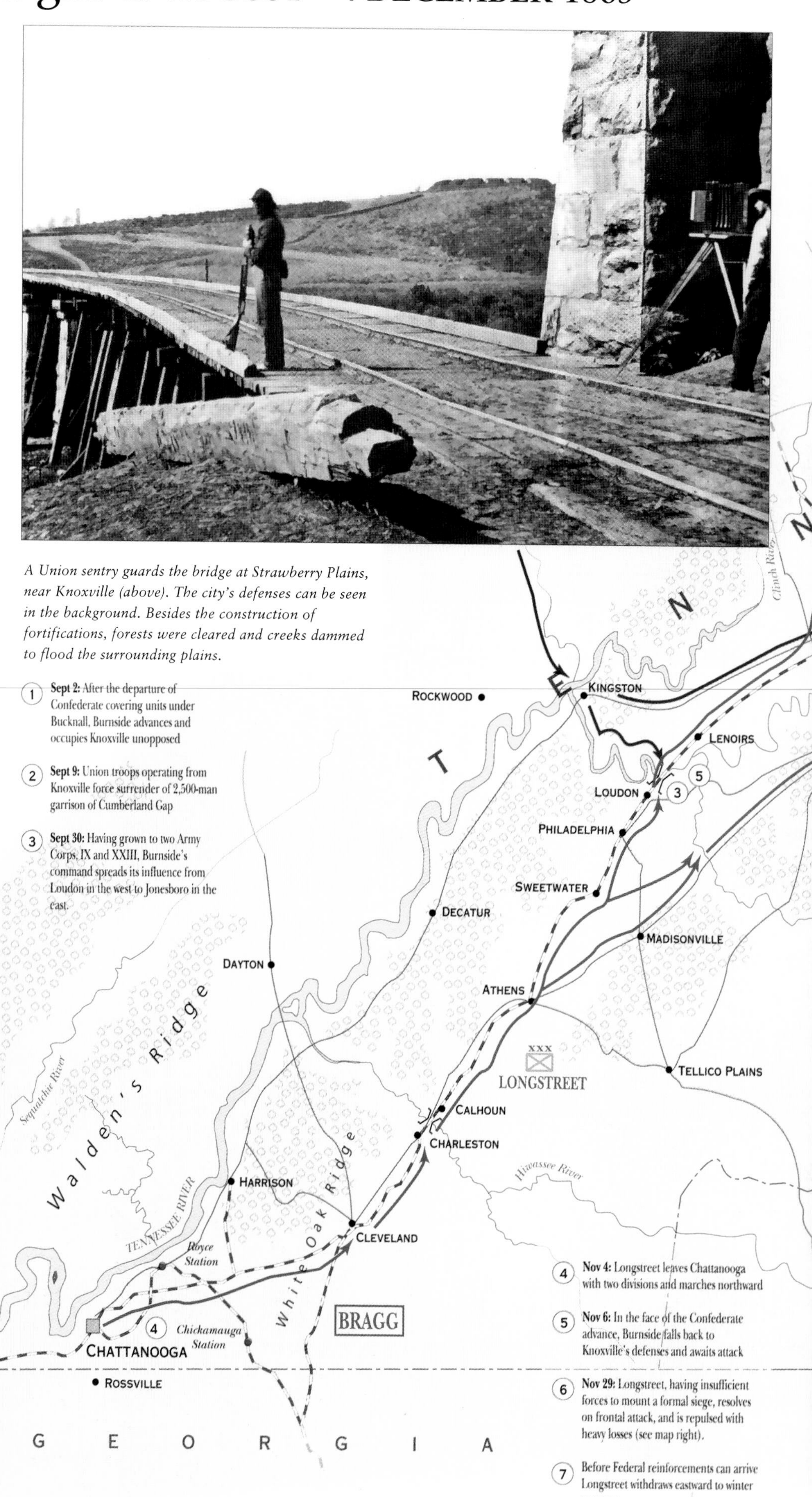
1 Sept 2: After the departure of Confederate covering units under Bucknall, Burnside advances and occupies Knoxville unopposed
2 Sept 9: Union troops operating from Knoxville force surrender of 2,500-man garrison of Cumberland Gap
3 Sept 30: Having grown to two Army Corps, IX and XXIII, Burnside's command spreads its influence from Loudon in the west to Jonesboro in the east.
ROCKWOOD
KINGSTON
LENOIRS
LOUDON
PHILADELPHIA
SWEETWATER
DECATUR
MADISONVILLE
DAYTON
ATHENS
Walden's Ridge
XXX
LONGSTREET
TELLICO PLAINS
CALHOUN
CHARLESTON
Hiwassee River
HARRISON
White Oak Ridge
TENNESSEE RIVER
Royce Station
CLEVELAND
BRAGG
Chickamauga Station
CHATTANOOGA
ROSSVILLE
T E N
G E O R G I A
4 Nov 4: Longstreet leaves Chattanooga with two divisions and marches northward
5 Nov 6: In the face of the Confederate advance, Burnside falls back to Knoxville's defenses and awaits attack
6 Nov 29: Longstreet, having insufficient forces to mount a formal siege, resolves on frontal attack, and is repulsed with heavy losses (see map right).
7 Before Federal reinforcements can arrive Longstreet withdraws eastward to winter quarters

*A Union sentry guards the bridge at Strawberry Plains, near Knoxville (above). The city's defenses can be seen in the background. Besides the construction of fortifications, forests were cleared and creeks dammed to flood the surrounding plains.*

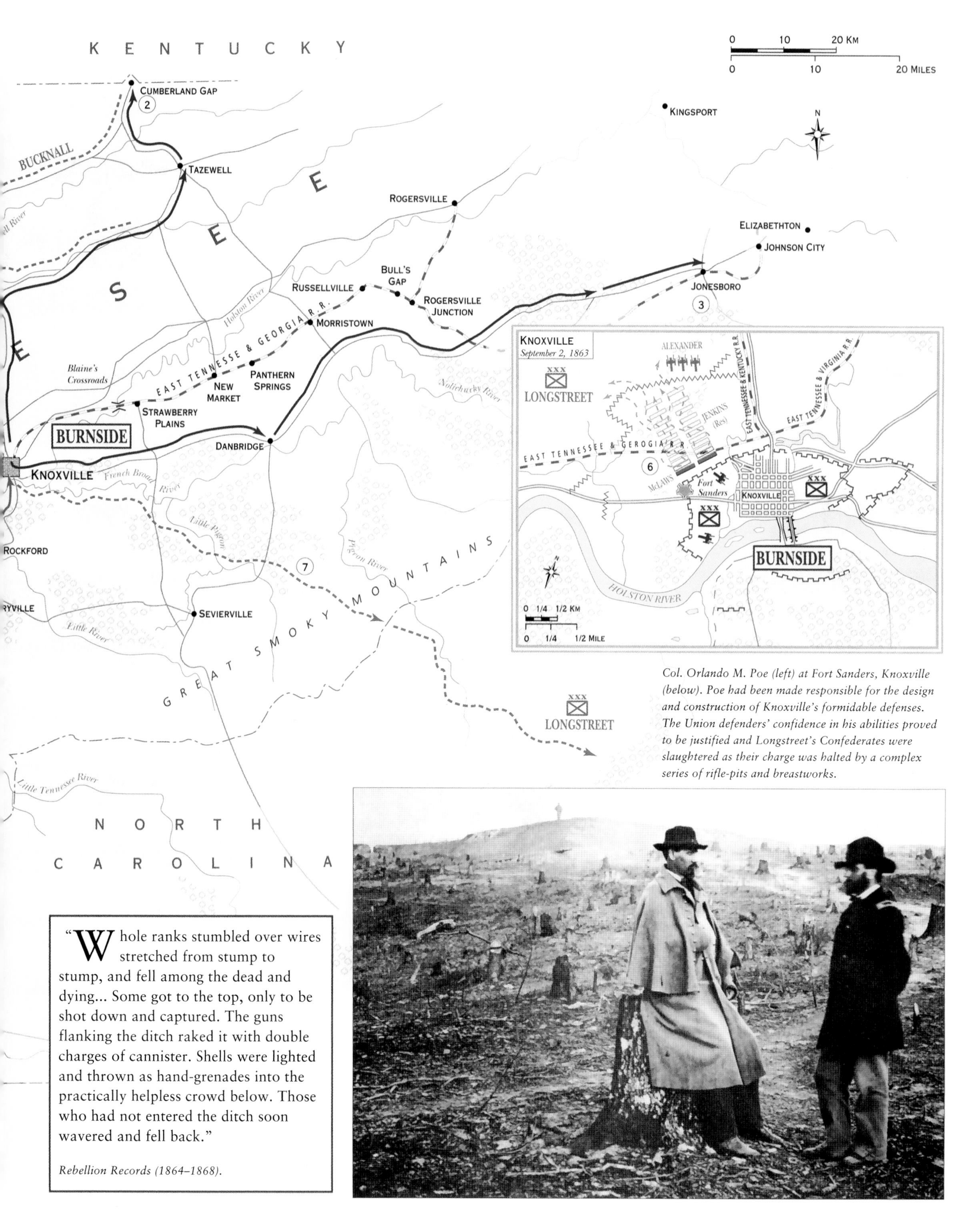
KENTUCKY
0 10 20 KM
0 10 20 MILES
CUMBERLAND GAP
2
KINGSPORT
N
BUCKNALL
TAZEWELL
ROGERSVILLE
ELIZABETHTON
JOHNSON CITY
BULL'S GAP
RUSSELLVILLE
ROGERSVILLE JUNCTION
JONESBORO
3
MORRISTOWN
Holston River
EAST TENNESSEE & GEORGIA R.R.
Blaine's Crossroads
NEW MARKET
PANTHERN SPRINGS
STRAWBERRY PLAINS
BURNSIDE
DANBRIDGE
KNOXVILLE
French Broad River
Nolichucky River
Little Pigeon
Pigeon River
ROCKFORD
7
SEVIERVILLE
Little River
GREAT SMOKY MOUNTAINS
LONGSTREET
Little Tennessee River
NORTH CAROLINA
KNOXVILLE
September 2, 1863
LONGSTREET
ALEXANDER
JENKINS (Res)
EAST TENNESSEE & KENTUCKY R.R.
EAST TENNESSEE & VIRGINIA R.R.
EAST TENNESSEE & GEOROGIA R.R.
6
McLAWS
Fort Sanders
KNOXVILLE
BURNSIDE
HOLSTON RIVER
0 1/4 1/2 KM
0 1/4 1/2 MILE

*Col. Orlando M. Poe (left) at Fort Sanders, Knoxville (below). Poe had been made responsible for the design and construction of Knoxville's formidable defenses. The Union defenders' confidence in his abilities proved to be justified and Longstreet's Confederates were slaughtered as their charge was halted by a complex series of rifle-pits and breastworks.*

"Whole ranks stumbled over wires stretched from stump to stump, and fell among the dead and dying... Some got to the top, only to be shot down and captured. The guns flanking the ditch raked it with double charges of cannister. Shells were lighted and thrown as hand-grenades into the practically helpless crowd below. Those who had not entered the ditch soon wavered and fell back."

*Rebellion Records (1864–1868).*

# Operations in Virginia OCTOBER 9 – NOVEMBER 26 1863

In the aftermath of Gettysburg, the opposing armies maneuvered back and forth across Virginia looking for a decisive victory. Lee held a defensive line behind the Rapidan River, sending Longstreet's Confederates west to join General Bragg's army. Following the Federal disaster at Chickamauga, Meade detached XI and XII Corps to the West also. Facing a weakened adversary, Lee sensed an opportunity to strike.

On October 10, Confederates spilled across the Rapidan in an attempt to turn Meade's right. The Federals abandoned their line and fell back. Both armies raced northward toward Centreville. On October 14, Lee's advance ran into a gap in the Federal march at Bristoe Station. General A. P. Hill attacked Major General George Sykes's V Corps, only to be savaged by Major General Gouverneur K. Warren's II Corps, which suddenly appeared along a railroad cut on his right. A.P. Hill had 1,360 Confederate losses, while Warren lost only 350 men. Finding Meade strongly posted, and Confederate supply lines overextended, Lee withdrew on October 18, tearing up the Orange & Alexandria Railroad as he went.

Meade followed Lee, and captured two Confederate strongholds on the Rappahannock. Attacking both Rappahannock Station and Kelly's Ford on November 7, the Federals captured 1,800 prisoners, and compelled Lee to fall back behind the Rapidan.

On November 26, Meade's 60,000 troops attempted to flank Lee's right. Delayed by

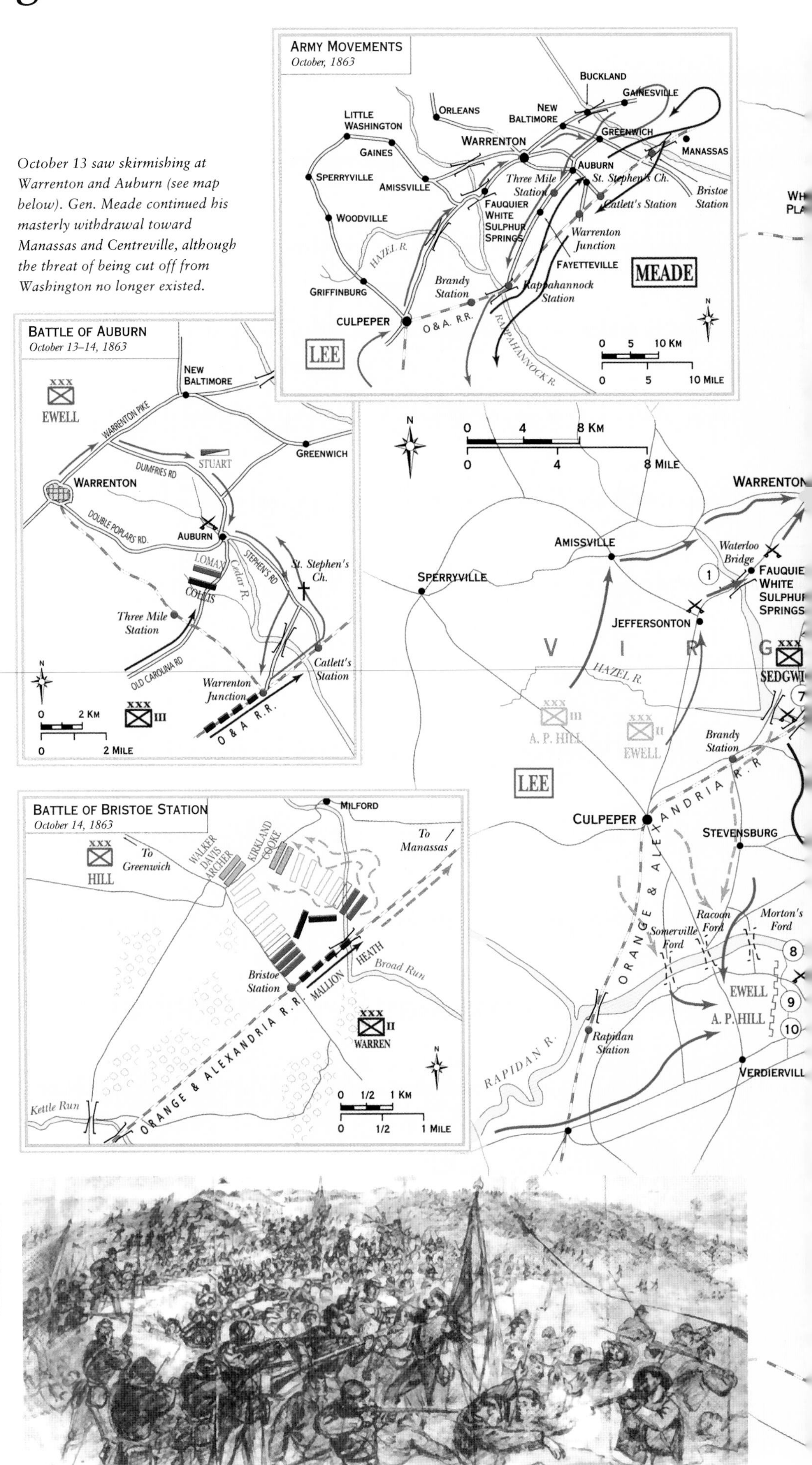
ARMY MOVEMENTS
October, 1863
BUCKLAND
GAINESVILLE
ORLEANS
NEW BALTIMORE
LITTLE WASHINGTON
GREENWICH
WARRENTON
MANASSAS
GAINES
AUBURN
SPERRYVILLE
AMISSVILLE
Three Mile Station
St. Stephen's Ch.
Bristoe Station
Catlett's Station
FAUQUIER WHITE SULPHUR SPRINGS
WOODVILLE
Warrenton Junction
HAZEL R.
FAYETTEVILLE
MEADE
Brandy Station
Rappahannock Station
GRIFFINBURG
CULPEPER
O & A. R.R.
RAPPAHANNOCK R.
LEE
BATTLE OF AUBURN
October 13–14, 1863
NEW BALTIMORE
EWELL
WARRENTON PIKE
DUMFRIES RD
STUART
GREENWICH
WARRENTON
DOUBLE POPLARS RD.
AUBURN
LOMAX
COLLIS
Cedar R.
STEPHEN'S RD
St. Stephen's Ch.
Three Mile Station
OLD CAROLINA RD
Warrenton Junction
Catlett's Station
III
O & A R.R.
BATTLE OF BRISTOE STATION
October 14, 1863
MILFORD
HILL
To Greenwich
WALKER
DAVIS
ARCHER
KIRKLAND
COOKE
To Manassas
HEATH
Broad Run
Bristoe Station
MALLION
ORANGE & ALEXANDRIA R.R.
II
WARREN
Kettle Run
WARRENTON
AMISSVILLE
Waterloo Bridge
FAUQUIER WHITE SULPHUR SPRINGS
SPERRYVILLE
JEFFERSONTON
V I R
HAZEL R.
SEDGWICK
A. P. HILL
EWELL
Brandy Station
LEE
CULPEPER
ORANGE & ALEXANDRIA R.R.
STEVENSBURG
Racoon Ford
Morton's Ford
Somerville Ford
EWELL
A. P. HILL
Rapidan Station
RAPIDAN R.
VERDIERVILLE

*October 13 saw skirmishing at Warrenton and Auburn (see map below). Gen. Meade continued his masterly withdrawal toward Manassas and Centreville, although the threat of being cut off from Washington no longer existed.*

*As it retreated toward Manassas, the Army of the Potomac was attacked at Bristoe Station by the leading corps of Lee's Army of Northern Virginia, under A. P. Hill. The attacks lacked significant strength to succeed and the Confederates failed to strike the center of the Union column (see map right).*

*The capture of the Confederate fortifications at Rappahanock Station (right). Lee had intended to hold the Union advance at Rappahanock while attacking in superior force at Kelly's Ford, but a surprise Federal attack resulted in the capture of most of the Confederates north of the river.*

> "I am convinced that I made the attack too hastily, and at the same time that a delay of half an hour, and there would have been no enemy to attack. In that event I believe I should equally have blamed myself for not attacking at once ."
>
> *Gen. A. P. Hill, after the battle of Bristoe Station.*

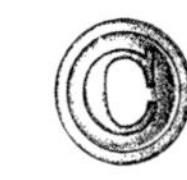

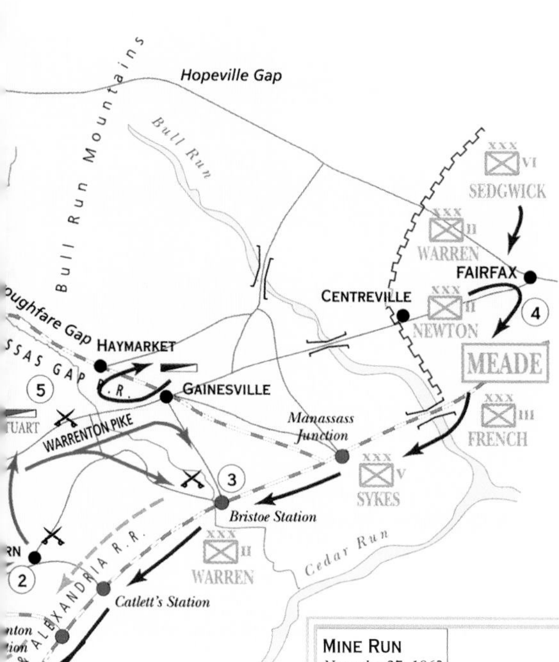

Hopeville Gap
Bull Run Mountains
Bull Run
SEDGWICK
WARREN
FAIRFAX
CENTREVILLE
NEWTON
MEADE
HAYMARKET
GAINESVILLE
Manassas Junction
FRENCH
WARRENTON PIKE
SYKES
Bristoe Station
WARREN
Cedar Run
Catlett's Station
N I A
MEADE
FRENCH
RAPPAHANNOCK R.


*Past experience had shown that the weakest part of a defensive line at Rappahannock was Kelly's Ford, where the south bank was lower than the north. On November 7, the Federals constructed a pontoon bridge which enabled them to send substantial reinforcements to the south bank, capturing two regiments assigned to picket duty by Rodes. (see map right).*

poor weather, difficult river crossings, troops losing their way, and a surprise battle at Payne's Farm, Meade could not advance his army before Lee concentrated his forces behind powerful works along Mine Run. Faced with Lee's extensive fortifications, Meade canceled his offensive and withdrew across the Rapidan on December 1.

Neither army could gain a decisive advantage in the stalemated Virginia theater. Federal and Confederate soldiers retired into their winter encampments, where they rested while their generals prepared for the next campaign.

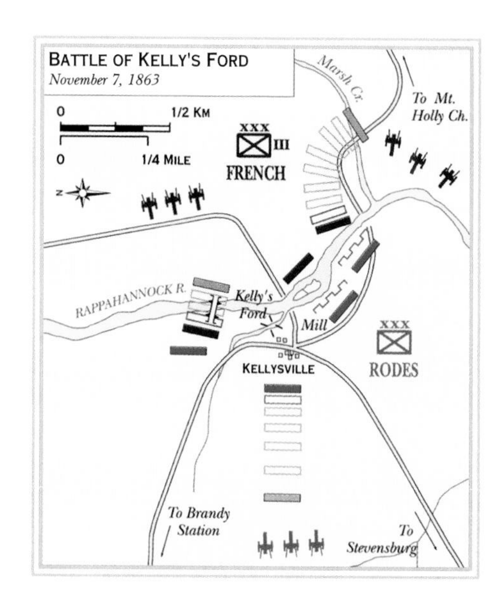

BATTLE OF KELLY'S FORD
November 7, 1863
0 1/2 KM
0 1/4 MILE
FRENCH
Marsh Cr.
To Mt. Holly Ch.
RAPPAHANNOCK R.
Kelly's Ford
Mill
KELLYSVILLE
RODES
To Brandy Station
To Stevensburg


"His object is to prevent my advance, and in the meantime send more troops to Bragg. This was [sic] a deep game, and I am free to admit that in the playing of it he has got the advantage of me."

*Gen. Meade speaking of Lee before Bristoe Station.*

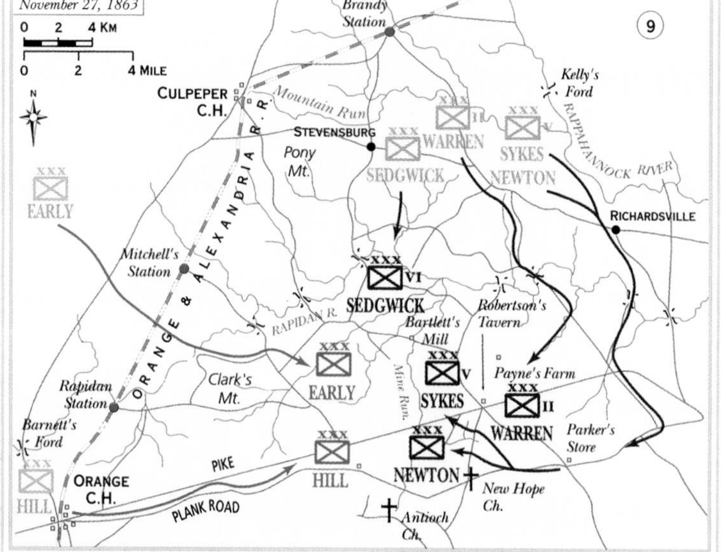

MINE RUN
November 27, 1863
0 2 4 KM
0 2 4 MILE
Brandy Station
Kelly's Ford
CULPEPER C.H.
Mountain Run
STEVENSBURG
Pony Mt.
WARREN
SYKES
SEDGWICK
NEWTON
RAPPAHANNOCK RIVER
EARLY
RICHARDSVILLE
Mitchell's Station
ORANGE & ALEXANDRIA R.R.
SEDGWICK
RAPIDAN R.
Bartlett's Mill
Robertson's Tavern
Payne's Farm
Clark's Mt.
EARLY
Mine Run
SYKES
WARREN
Rapidan Station
Barnett's Ford
Parker's Store
PIKE
HILL
NEWTON
New Hope Ch.
ORANGE C.H.
PLANK ROAD
Antioch Ch.
HILL


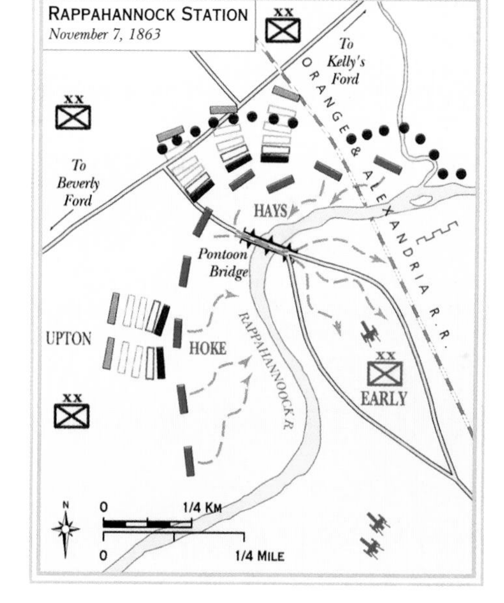

RAPPAHANNOCK STATION
November 7, 1863
To Kelly's Ford
ORANGE & ALEXANDRIA R.R.
To Beverly Ford
HAYS
Pontoon Bridge
UPTON
HOKE
RAPPAHANNOCK R.
EARLY
0 1/4 KM
0 1/4 MILE


1. **Oct 12:** Ewell's Confederates overwhelm Federal cavalry at Jeffersonton and Fauquier White Sulphur Springs, and flank the Union's Rappahannock defenses
2. **Oct 13-14:** Stuart's Confederate cavalry is cut off at Auburn while probing Meade's withdrawal and forced to hack their way to freedom (see map center left)
3. **Oct 14:** A. P. Hill's Confederates recklessly pitch into the Federals at Bristoe Station; repulsed with severe losses (see map left)
4. **Oct 15-18:** Meade selects a strong defensive line; Lee refuses to attack and returns to the Rappahannock
5. **Oct 20:** Stuart ambushes the Federal cavalry at Buckland Mills, calling the rout "the Buckland Races"
6. **Nov 7:** Meade sends French's III Corps to demonstrate against Kelly's Ford to hold Lee's attention (see map top right)
7. **Nov 7:** Meade overruns the Confederate defenders at Rappahannock Station capturing 1,800 Confederates (see map right)
8. **Nov 27:** Advancing on Lee's right flank, French's III Corps blunders into an unexpected battle with Johnson's division of Ewell's Corps at Payne's Farm
9. **Nov 27:** Meade's offensive against Lee's right stalls when Confederates at Robertson's Tavern fall back behind Mine Run and dig in (see map above)
10. **Nov 30:** Meade cancels a major attack in the face of strengthened Confederate fortifications, and withdraws behind the Rapidan River

*News of the Federal attack at Kelly's Ford became relatively insignificant when Early reported that Union troops had successfully stormed and overrun the north bank entrenchments at Rappahannock Station, killing or capturing two Confederate brigades as well as negating his plans for a counteroffensive and necessitating a Confederate withdrawal (see map right).*

*Union artillery in action at Kelly's Ford (right). The Union batteries succeeded in breaking the Confederate line and the pursuing Federals waded across the river in their anxiety to crush the retreating enemy.*

# 1864: Total War

Defeatism infected many Southern people during the winter of 1863–1864. "I have never actually despaired of the cause," wrote a Confederate War Department official in November, but "steadfastness is yielding to a sense of hopelessness." Inflation galloped out of control. A Richmond diarist recorded this incident in October 1863: A merchant told a poor woman that the price of a barrel of flour was $70. "My God!" exclaimed she, "how can I pay such prices? I have seven children; what shall I do?" "I don't know, madam," said he, coolly, "unless you eat your children."

The Davis administration, like the Lincoln administration a year earlier, had to face congressional elections during a time of public discontent, for the Confederate Constitution mandated such elections in odd-numbered years. Political parties had ceased to exist in the Confederacy after Democrats and former Whigs had tacitly declared a truce in 1861 to form a united front for the war effort. Many congressmen had been elected without opposition in 1861. By 1863, however, significant hositility to Davis had emerged. Though it was not channeled through an organized party, it took on partisan trappings. An inchoate anti-Davis faction appeared in Congress and in the election campaign of 1863.

Ominously, some anti-administration candidates ran on a quasi-peace platform (analogous to that of the Copperheads in the North) calling for an armistice and peace negotiations to end the killing. Left unresolved were the terms of such negotiations – reunion or independence. But any peace overture from a position of weakness was tantamount to conceding defeat. The peace movement was particularly strong in North Carolina, where for a time it appeared that the next governor would be elected on a peace platform (in the end, the peace candidate was defeated). Still, anti-administration candidates made significant gains in the 1863 Confederate elections, though they fell about 15 seats short of a majority in the House and two seats short in the Senate.

Shortages, inflation, political discontent, military defeat, high casualties, and the loss of thousands of slaves (some of them now in arms against the Confederacy) bent but did not break the Southern spirit. A religious revival in Confederate armies helped to sustain morale. As the spring of 1864 approached, a renewed determination infused both home front and battle front. No longer would Confederate armies be powerful enough to invade the North or try to win the war with a knockout blow, as they had tried to do in 1862 and 1863. But they were still strong enough to fight a war of attrition, as the patriots had done in the war of 1775–1783 against Britain. If they could hold out long enough, and inflict sufficient casualties on Union armies, they might weaken the Northern will to continue fighting. In particular, if they could hold out until the Union presidential election in November, Northern voters might reject Lincoln and elect a Peace Democrat. "If we can break up the enemy's arrangements early and throw him back," wrote General James Longstreet, "he will not be able to recover his position or his morale until the Presidential election is over, and then we shall have a new President to deal with."

*General Ulysses Simpson Grant, appointed by President Lincoln as General-in-Chief of the Armies of the United States on March 10, 1864*

Northerners were vulnerable to this strategy of psychological attrition. Military success had created a mood of confidence in the North. The mood was fed by Grant's appointment as General-in-Chief. When Grant decided to remain in Virginia with the Army of the Potomac, leaving Sherman in command of the Union forces in Georgia, Northerners expected these two heavyweights to floor the Confederacy with a one-two punch. Lincoln was alarmed by this euphoria. "The people are too sanguine," he told a reporter. "They expect too much at once." Disappointment might trigger despair.

It almost happened that way. Grant started the military campaign of 1864 with a strategic plan, elegant in its simplicity. While smaller Union armies in peripheral theaters were to carry out auxiliary campaigns, the two principal armies in Virginia and Georgia would attack the main Confederate forces under Lee and Johnston. Convinced that in previous years the Union armies in various theaters had "acted independently and without concert, like a balky team, no two ever pulling together," Grant ordered simultaneous offensives on all fronts to prevent Confederates from shifting reinforcements from one theater to another.

Grant's offensives began the first week of May. The heaviest fighting occurred in Virginia. When the Army of the Potomac crossed the Rapidan River, Lee decided to attack it in the flank while it was still in the Wilderness, a thick scrub forest where Lee had defeated Hooker a year earlier in the battle of Chancellorsville. Lee's action precipitated two days (May 5–6) of the most confused and frenzied fighting the war had yet seen. Having apparently halted Grant's offensive, Confederates claimed a victory.

But Grant did not admit defeat. Instead, he moved to his left toward Spotsylvania Courthouse, a key crossroads hamlet ten miles closer to Richmond. Lee pulled back to cover the road junction and entrenched. Repeated Union assaults during the next twelve days achieved only minor Federal gains at a high cost in casualties. The Confederates fought from an elaborate network of trenches and log breastworks they had constructed virtually overnight. Civil War soldiers had learned the advantages of entrenchment, which gave the defense an enormous advantage and made frontal assaults almost suicidal. By the time the war was over, Virginia – and northern Georgia – would look as if giant moles had burrowed their way across the countryside.

Having achieved no better than stalemate around Spotsylvania, Grant again moved south around Lee's right flank in an effort to force the outnumbered Confederates into an open fight. But Lee anticipated Grant's moves, confronting him from behind formidable defenses at the North Anna River, Totopotomoy Creek, and near the crossroads inn of Cold Harbor only ten miles northeast of Richmond. Presuming the Confederates to be exhausted and demoralized by their repeated retreats, Grant decided to attack at Cold Harbor on June 3 – a costly mistake, as it turned out.

On June 12, for the fifth time Grant shifted to the left in an attempt to circumvent Lee's flank. This time he moved all the way across the James River to strike at Petersburg, an industrial city 20 miles south of Richmond where several rail lines came together. If Petersburg fell, the Confederates could not hold Richmond. But once more Lee's troops raced southward on the inside track and blocked Grant's advance. Four days of Union assaults (June 15–18) against the Petersburg trenches produced no breakthrough.

During six weeks of marching and fighting, Union losses had been so high – some 65,000 killed, wounded, and captured compared with 37,000 Confederate casualties – that the Army of the Potomac had lost its fighting power. This was a new kind of war, unprecedented in its relentless intensity and duration. In previous years these two armies had fought each other in several big battles, each followed by the retreat of one or the other behind the nearest river. Both armies would then rest and recuperate for a month or more before resuming the fight (the sole exception was the two and a half week interval between Second Manassas and Antietam). In 1864, the armies were never out of contact with each other. Some form of fighting, plus a great deal of marching and digging, took place virtually every day and a great many nights as well. Mental and physical exhaustion began to take a toll; officers and men suffered from what in later wars would be called shell shock or combat fatigue. One Union officer noted that in three weeks men "had grown thin and haggard. The experience of those twenty days seemed to have added twenty years to their age." Captain Oliver Wendell Holmes, Jr., observed that "many a man has gone crazy since this campaign began from the terrible pressure on mind and body." Grant reluctantly settled down for a siege along the Petersburg–Richmond front that would last nine gruelling months, punctuated by frequent battles and skirmishes that would force Lee to lengthen his supply lines until they grew so thin that they would break.

Meanwhile, other Union operations in Virginia had achieved little success. General Benjamin Butler bungled an attack up the James River against Richmond and was stopped by a scraped-together army under General P.G.T. Beauregard. A Union thrust up the Shenandoah Valley was blocked at Lynchburg by Jubal Early, commanding Stonewall Jackson's old corps which Lee had detached from the Cold Harbor trenches. Early then led a raid on July 11–12 down the Valley, across the Potomac, and all the way to the outskirts of Washington, before being driven back to Virginia. Union cavalry raids, commanded by General Philip Sheridan, inflicted considerable damage on Confederate resources in Virginia – including the mortal wounding of Jeb Stuart in the battle of Yellow Tavern on May 11 – but they did not strike a crippling blow. In the North, frustration rose over the failure to win the quick, decisive victory the public had expected when the campaign began.

In Georgia, Sherman seemed to accomplish more at less cost than Grant did in Virginia. But there too, Union efforts bogged down in apparent stalemate by August. The strategy of both Sherman and Johnston in Georgia contrasted with that of Grant and Lee in Virginia. Grant repeatedly forced Lee back by flanking moves to the Union left, but only after bloody battles. Sherman forced Johnston south toward Atlanta by constantly flanking him to the Union right, generally without major battles. By the end of June, Sherman had advanced 80 miles at the cost of 17,000 casualties to Johnston's 14,000 – one-third the combined losses of Grant and Lee in Virginia.

The Davis administration grew alarmed by Johnston's apparent willingness to yield territory without much of a

*General William Tecumseh Sherman directing the bombardment of Atlanta, from a painting by Thure de Thulstrup*

fight. Sherman again flanked the Confederate defenses (after a failed attack) at Kennesaw Mountain in the first week of July. Crossing the Chattahoochee River, he drove Johnston back to Peachtree Creek less than five miles from Atlanta, a major rail and manufacturing center. Fearing that Johnston would abandon the city, Davis on July 17 replaced him with John Bell Hood.

A fighting general from Lee's army who had come south with Longstreet for the battle of Chickamauga (where he lost a leg), Hood immediately prepared to counterattack against the Yankees closing in on Atlanta. He did so three times, bringing on the battles of Peachtree Creek (July 20), Atlanta (July 22), and Ezra Church (July 28). Each time the Confederates reeled back in defeat, suffering a total of 15,000 casualties to Sherman's 6,000. At last, Hood retreated into the formidable earthworks ringing Atlanta, and launched no more attacks. But his army did manage to keep Sherman from taking the two railroads leading into Atlanta from the south. Like Grant at Petersburg, Sherman seemed to settle down for a siege.

By August, the Confederate strategy of attrition – to hold out until the Union presidential election and to exhaust the Northern will to continue fighting – seemed to be working. "Who shall revive the withered hopes that bloomed at the opening of Grant's campaign?" asked Democratic newspapers. "STOP THE WAR! ... All are tired of this damnable tragedy. If nothing else would impress upon people the absolute necessity of stopping this war, its utter failure to accomplish any results would be sufficient."

Public pressure in both the North and South compelled Lincoln and Davis to explore the possibility of peace negotiations – or at least to appear to do so. Each president knew that his conditions for peace were unacceptable to the other: union and emancipation (Lincoln); Confederate independence with slavery (Davis). But in a flurry of peace moves during the summer, Democratic propagandists managed to convince many Northern voters that eventual reunion through negotiations would be possible if Lincoln would drop his insistence on emancipation. "Tens of thousands of white men must yet bite the dust to allay the negro mania of the President," ran a typical Democratic editorial. By August, even staunch Republicans, like party chairman Henry Raymond and his associate, Thurlow Weed, were convinced that "the desire for peace" and the impression that Lincoln "is fighting not for the Union but for the abolition of slavery" made his reelection "an impossibility." Lincoln thought so too. "I am going to be beaten," he told a friend in August, "and unless some great change takes place, badly beaten."

Lincoln nevertheless resisted pressures to drop emancipation as a condition of peace so the onus could be shifted to Jefferson Davis's insistence on independence. Some 130,000 black soldiers and sailors were fighting for the Union, noted Lincoln in August 1864. They would not do so if they thought the North intended to betray them. If they stake their lives for us they must be prompted by the strongest motive...the promise of freedom. And the promise being made, must be kept. ... There have been men who proposed to me to return to slavery the[se] black warriors. ... I should be damned in time and eternity for so doing. The world shall know that I will keep my faith to friends and enemies, come what will.

At the end of August the Democrats nominated none other than George B. McClellan for president on a platform which declared that "after four years of failure to restore the Union by the experiment of war... [we] demand that immediate efforts be made for a cessation of hositilities." Southerners were jubilant. Democratic victory on that platform, said the Charleston *Mercury*, "must lead to peace and our independence" if "for the next two months we hold our own and prevent military success by our foes."

But three days after McClellan's nomination came news of Sherman's capture of Atlanta. It produced the biggest reversal of momentum in the war. "VICTORY!" blazoned Republican headlines. "IS THE WAR A FAILURE? OLD ABE'S REPLY TO THE [DEMOCRATIC] CONVENTION." Confederates were stunned. The "disaster at Atlanta," lamented the Richmond *Examiner*, came "in the very nick of time" to "save the party of Lincoln from irretrievable ruin. It will obscure the prospect of peace, late so bright. It will also diffuse gloom over the South."

If Atlanta was not enough to transform the prospects for Lincoln's reelection, events in Virginia's Shenandoah Valley were. During the month from September 19 to October 19, Philip Sheridan's Army of the Shenandoah won three spectacular victories over Jubal Early's small Confederate army there. Northern opinion catapulted from the depths of despair in August to the heights of renewed confidence in November. Lincoln won reelection by a margin of 212 to 21 in the electoral college. Soldiers played a notable role in the balloting. Every Northern state except three, whose legislatures were controlled by Democrats, had passed laws allowing absentee voting by soldiers at the front. Despite the lingering affection of some soldiers for their old commander McClellan, 78% of the army vote went to Lincoln – compared with 54% of the civilian vote. The men who were doing the fighting sent a clear message that they meant to finish the job.

Many Southerners got the message. But not Jefferson Davis. The Confederacy remained "as erect and defiant as ever," he told his Congress in November 1864. "Nothing has

changed in the purpose of its Government, in the indomitable valor of its troops, or in the unquenchable spirit of its people." It was this last-ditch defiance that William T. Sherman set out to break in his famous march from Atlanta to the sea.

Sherman had long pondered the nature of the war. He had concluded that "we are not only fighting hostile armies, but a hostile people." Defeat of Confederate armies was not enough to win the war; the railroads, factories, and farms that supported those armies must be destroyed. The will of the civilian population that sustained the war must be crushed. Sherman was ahead of his time in his understanding of psychological warfare. "We cannot change the hearts of those people of the South," he said, "but we can make war so terrible and make them so sick of war that generations would pass away before they would again appeal to it."

Sherman urged Grant to let him cut loose from his base in Atlanta and march through the heart of Georgia, living off the land and destroying all the resources not needed by his army. Grant and Lincoln were reluctant to authorize such a risky move, especially with Hood's army of 40,000 men still intact and hovering in northern Alabama, ready to move into Tennessee if Sherman marched off in the opposite direction. But Sherman promised to send George Thomas – who would be more than a match for Hood – to take command of a force of 60,000 men in Tennessee. With another 60,000, Sherman could "move through Georgia, smashing things to the sea. ... I can make the march, and make Georgia howl!"

Lincoln and Grant finally consented. On November 16, Sherman's avengers marched out of Atlanta after burning a third of the city, including some non-military property. Sherman and his men cared little; in their view the rebels had sown the wind and deserved to reap the whirlwind. Southward they marched, 280 miles to Savannah, wrecking everything in their path that could, by any stretch of the imagination, be considered of military value. As they marched South, Hood marched north in a quixotic invasion of Tennessee that culminated in the crippling of his army at Franklin on November 30, and its destruction by Thomas's Union forces at Nashville on December 15–16.

The fourth Christmas season of the war was bleaker in the South than the first two had been in the North. "The deep waters are closing over us," wrote the Southern diarist Mary Boykin Chesnut on December 19. "We are going to be wiped off the earth."

*The battle of Kennesaw Mountain, Georgia, fought on June 27, 1864, from a lithograph by Kurz and Allison*

# Operations in Mississippi and Florida

## FEBRUARY – MARCH 1864

With the armies in winter quarters, Sherman returned to Vicksburg on January 30, 1864. Four days later he crossed the Big Black River with two corps. His goal was Meridian, now the state's most important city still in Confederate hands. Brigadier General William Sooy Smith's 7,000 cavalry at Memphis were to join Sherman at Meridian. On February 5 the Federals occupied Jackson for the third time in nine months, and on February 7, Sherman entered Brandon.

Confederate general Leonidas Polk called up infantry and artillery from Mobile, Alabama, and ordered them to dig in on the hills southwest of Morton. But on learning that his force was outnumbered, Polk retreated to Meridian. Sherman entered Morton on the 8th and Meridian on the 14th. Polk had evacuated the town earlier in the day, falling back on Demopolis.

Sherman ordered the destruction of the town's railroads and warehouses, then waited for Smith to join him from Memphis. By the 20th, hearing nothing from Smith, Sherman returned to Vicksburg where he learned that Smith had left Memphis late and did not cross the Tallahatchie until February 16.

On the 20th, Smith encountered one of Major General Forrest's cavalry brigades, and drove it through West Point. But Smith had now lost his nerve: believing Sherman was on his way back to Vicksburg, and that Forrest had

1. **Feb 3, 1864:** Two Federal corps cross Big Black River
2. **Feb 4–6:** Despite clashes between Confederate and Federal cavalry, Federal columns converge on Jackson
3. **Feb 7–8:** Sherman's army crosses Pearl River and advances toward Meridian. To challenge Sherman, Polk digs in near Morton
4. **Feb 8:** Confederates evacuate their Morton line; begin retreat toward Tombigee River
5. **Feb 9–14:** Sherman passes through Hillsboro and Decatur; occupies Meridian - a strategic Confederate communications and supply center - on 14th
6. **Feb 15–20:** Federals wreak havoc on Meridian's railroads
7. **Feb 20–26:** Unable to wait further for Sooy Smith to join him from Memphis, Sherman evacuates Meridian and starts back to Vicksburg
8. **Feb 22–25:** Sherman sends Col. Winslow's cavalry to find Sooy Smith; fails to locate him and returns to Sherman at Canton
9. Sherman marches from Canton to Vicksburg
10. **Feb 11, 1864:** Sooy Smith's cavalry leaves Memphis for Meridian
11. **Feb 12–20:** Smith crosses Tallahatchie, rides down the Pontotoc Ridge. Meets one of Gen. Forrest's brigades near West Point and drives it through town.
12. **Feb 21–22:** Smith looses his nerve; forced back by Forrest to Okolona and then to Pontotoc
13. **Feb 23–27:** Smith withdraws to Memphis
14. **Feb 1–28:** Federal amphibious expedition under Col. J.H. Coates ascends Yazoo river; disembarks at Yazoo City
15. **Mar 5:** Coates's force is attacked by two Confederate brigades under Brig. Gen. Lawrence S. Ross
16. **Mar 6:** Coates's force evacuates Yazoo City and returns to Vicksburg

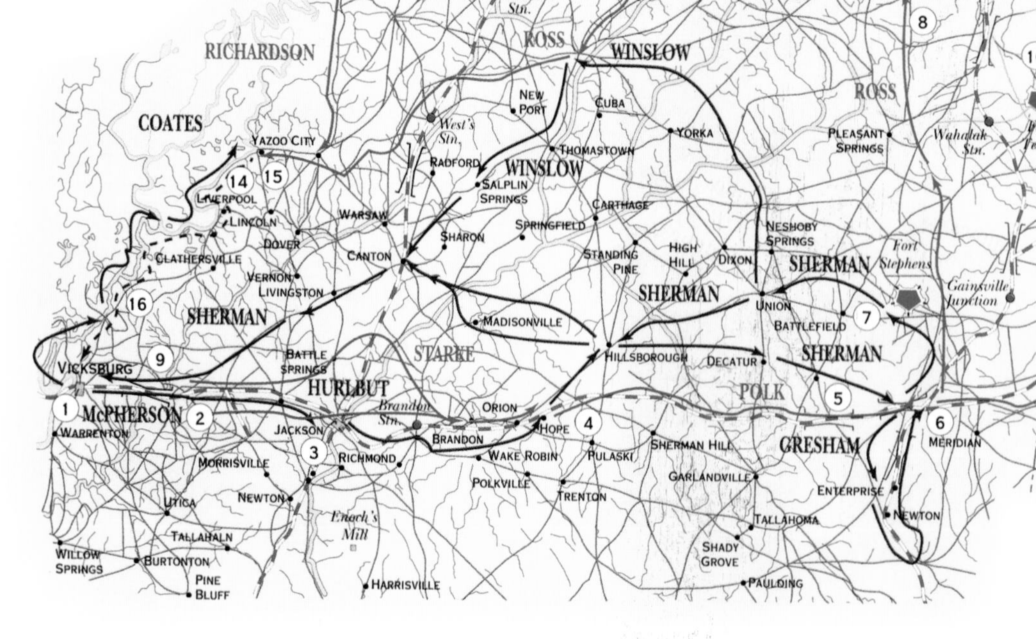
RICHARDSON
ROSS
WINSLOW
COATES
YAZOO CITY
NEW PORT
CUBA
YORKA
PLEASANT SPRINGS
Wahalak Stn.
West's Stn.
RADFORD
THOMASTOWN
WINSLOW
SALPLIN SPRINGS
LIVERPOOL
LINCOLN
WARSAW
CARTHAGE
SPRINGFIELD
SHARON
DOVER
CLATHERSVILLE
CANTON
STANDING PINE
HIGH HILL
DIXON
NESHOBY SPRINGS
Fort Stephens
SHERMAN
VERNON
LIVINGSTON
SHERMAN
UNION
Gainsville Junction
MADISONVILLE
BATTLEFIELD
BATTLE SPRINGS
STARKE
HILLSBOROUGH
DECATUR
SHERMAN
VICKSBURG
HURLBUT
POLK
Brandon Stn.
ORION
McPHERSON
WARRENTON
JACKSON
BRANDON
HOPE
SHERMAN HILL
GRESHAM
MERIDIAN
RICHMOND
WAKE ROBIN
PULASKI
MORRISVILLE
POLKVILLE
TRENTON
GARLANDVILLE
ENTERPRISE
NEWTON
UTICA
Enoch's Mill
TALLAHALN
TALLAHOMA
WILLOW SPRINGS
BURTONTON
PINE BLUFF
SHADY GROVE
HARRISVILLE
PAULDING

0 10 20 KM
0 10 20 MILES
N

been reinforced, Smith retired to Okolona on the 21st. Forrest followed, assailed the Union rearguard, and in the ensuing battle, defeated Smith who hurried back to Memphis.

The Lincoln administration now made plans to occupy Jacksonville and the west bank of the St. Johns River: Lincoln saw the expedition as a step toward the restoration of Florida to the Union, while the military wanted to recruit soldiers into their black units.

On February 5, Brigadier General Truman Seymour's Federals sailed from Port Royal Sound and occupied Jacksonville without opposition. On the 8th, Seymour's cavalry, led by Colonel Guy Henry, struck inland and after twice engaging the Confederates, reached Baldwin. Following a bitter skirmish, Henry crossed St. Mary's River at Barber's Ferry, but was opposed near Lake City by some 600 Confederates. Although he outnumbered the enemy, Henry retreated to Barber's Ferry. While awaiting reinforcements, the Confederates advanced to Olustee on the 13th, and threw up earthworks.

On February 20, ignoring orders not to advance beyond Barber's Ferry, Seymour - with 5,500 men - set off to capture Lake City. The Confederates, now reinforced to 5,200, intercepted and attacked the Federals near Olustee. The battle continued until dusk, when the Federals retreated to Baldwin. Federal losses were 1,861; Confederates 934.

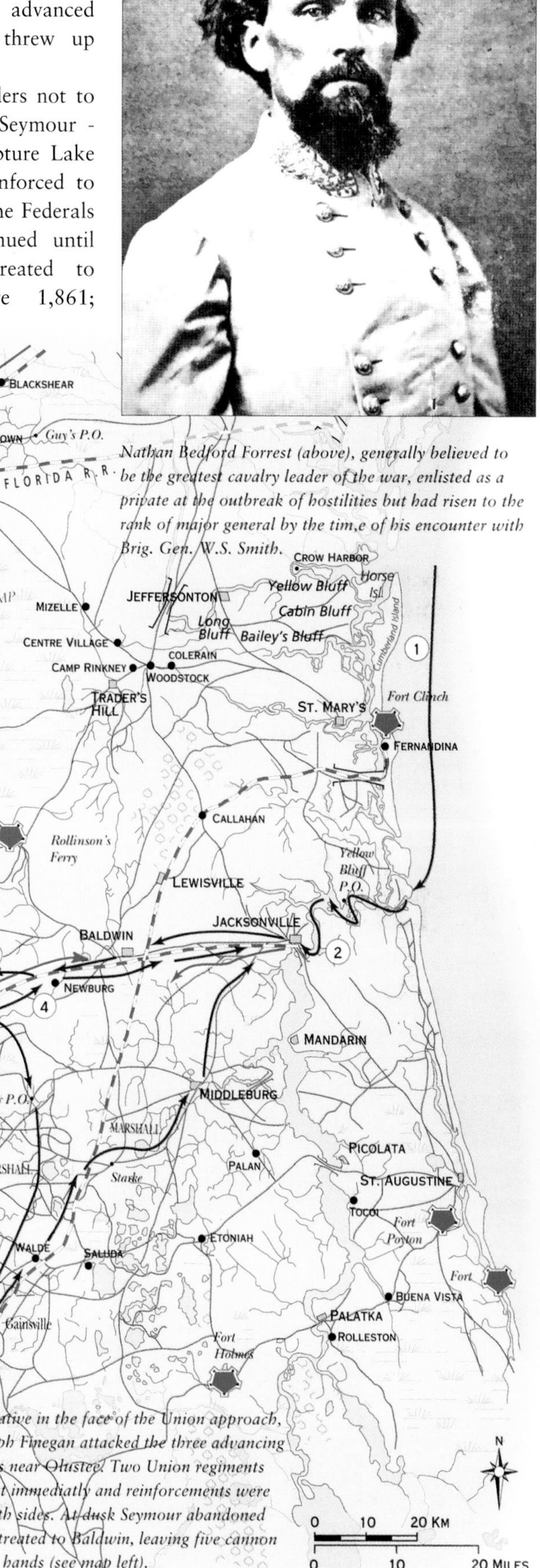

*Nathan Bedford Forrest (above), generally believed to be the greatest cavalry leader of the war, enlisted as a private at the outbreak of hostilities but had risen to the rank of major general by the time of his encounter with Brig. Gen. W.S. Smith.*

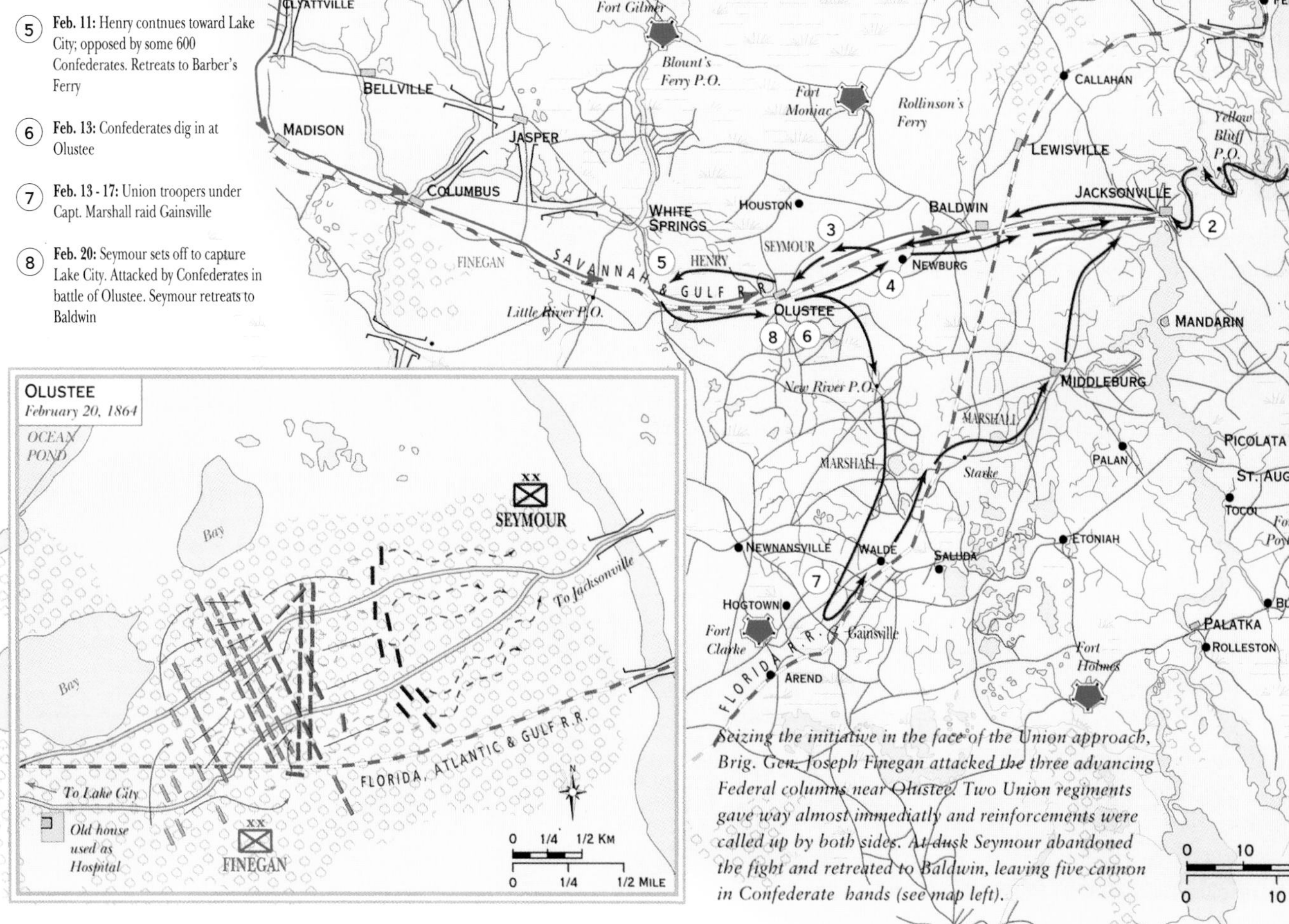

1 Feb 5: Gen. Seymour's 6,000 Federals sail from Port Royal, S. Carolina
2 Feb 7: Seymour enters St. John's River, lands and occupies Jacksonville
3 Feb. 8: Federal cavalry, led by Col. Henry, advances inland, skirmishes with Confederates; occupies Camp Finegan
4 Feb. 10: Henry clashes with Confederates at Barber's Ferry; crosses South Fork of St. Mary's River
5 Feb. 11: Henry contnues toward Lake City; opposed by some 600 Confederates. Retreats to Barber's Ferry
6 Feb. 13: Confederates dig in at Olustee
7 Feb. 13 - 17: Union troopers under Capt. Marshall raid Gainsville
8 Feb. 20: Seymour sets off to capture Lake City. Attacked by Confederates in battle of Olustee. Seymour retreats to Baldwin
PLEASANT VALLEY
BLACKSHEAR
RED BLUFF
YANKEETOWN
Guy's P.O.
B'WICK & FLORIDA R.R.
McDONALD
HOMERVILLE
GUEST
MILLTOWN
MAGNOLIA
OKEFENOKEE SWAMP
GRIFFIS
TROMPYVILLE
COW CREEK
VALDOSTA
CLYATTVILLE
CROW HARBOR
Yellow Bluff
Horse Isl.
JEFFERSONTON
Cabin Bluff
MIZELLE
Long Bluff
Bailey's Bluff
Cumberland Island
CENTRE VILLAGE
COLERAIN
CAMP RINKNEY
WOODSTOCK
TRADER'S HILL
ST. MARY'S
Fort Clinch
FERNANDINA
Fort Gilmer
Blount's Ferry P.O.
Fort Moniac
Rollinson's Ferry
CALLAHAN
BELLVILLE
MADISON
JASPER
LEWISVILLE
Yellow Bluff P.O.
COLUMBUS
JACKSONVILLE
HOUSTON
BALDWIN
WHITE SPRINGS
SEYMOUR
FINEGAN
SAVANNAH & GULF R.R.
HENRY
NEWBURG
OLUSTEE
Little River P.O.
MANDARIN
New River P.O.
MIDDLEBURG
MARSHALL
PICOLATA
PALAN
Starke
ST. AUGUSTINE
TOCOI
Fort Peyton
NEWNANSVILLE
WALDE
SALUDA
ETONIAH
Fort
HOGTOWN
BUENA VISTA
Fort Clarke
Gainsville
PALATKA
ROLLESTON
FLORIDA R.R.
Fort Holmes
AREND
OLUSTEE
February 20, 1864
OCEAN POND
Bay
SEYMOUR
To Jacksonville
FLORIDA, ATLANTIC & GULF R.R.
To Lake City
Old house used as Hospital
FINEGAN
0 1/4 1/2 KM
0 1/4 1/2 MILE
0 10 20 KM
0 10 20 MILES


*Seizing the initiative in the face of the Union approach, Brig. Gen. Joseph Finegan attacked the three advancing Federal columns near Olustee. Two Union regiments gave way almost immediatly and reinforcements were called up by both sides. At dusk Seymour abandoned the fight and retreated to Baldwin, leaving five cannon in Confederate hands (see map left).*

# The Red River Campaign MARCH 10 – MAY 22 1864

By 1864, the French invasion of Mexico, the shortage of cotton, and the desire to expand political reconstruction in Louis iana, compelled Federal officials to send an expedition up the Red River. On March 10, Brigadier General Andrew J. Smith's 10,000 troops left Vicksburg. Landing at Simsport next day, Smith's men joined with Rear Admiral David D. Porter's fleet to capture Fort de Russy. After occupying Alexandria on the 15th, the Federals scoured the countryside for cotton. Ten days later, General Nathaniel Banks's 30,000-man army arrived. This force would be joined near Shreveport by Major General Frederick Steele and 10,000 men from Little Rock.

Although Confederate Lieutenant General E. Kirby Smith ruled the semi-autonomous Trans-Mississippi Department from Shreveport, Major General Richard Taylor had immediate command of troops in Louisiana. His field army of 7,000 could only withdraw before Banks, who marched along the river protected by Porter's guns. Taylor's opportunity to offer resistance came when Banks marched west from Grand Ecore.

On the morning of April 8, Taylor deployed 8,800 troops east of Mansfield. He attacked when Banks's vanguard of 7,000 infantry and cavalry appeared at 4 p.m. Within minutes, the dazed Federals began falling back. The Confederates drove them for two miles before 5,000 Federal reinforcements checked their pursuit. After darkness ended the fighting, Banks withdrew to Pleasant Hill and went on the defensive. Bolstered by reinforcements,

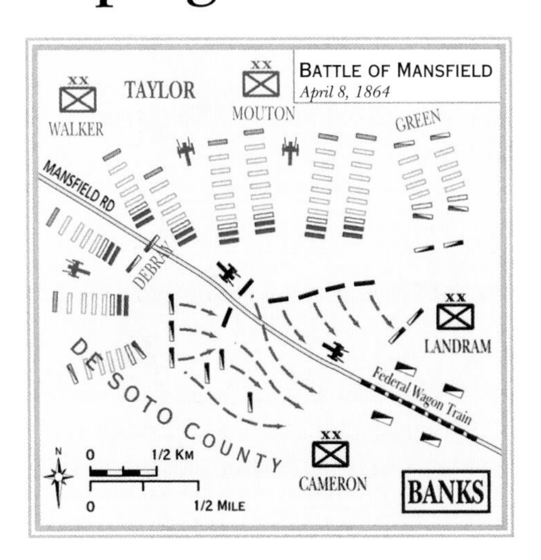
BATTLE OF MANSFIELD
April 8, 1864
TAYLOR
WALKER
MOUTON
GREEN
MANSFIELD RD
DEBRAY
LANDRAM
DE SOTO COUNTY
Federal Wagon Train
CAMERON
BANKS
0 1/2 KM
0 1/2 MILE
N

*Halting to consolidate his forces, which were strung out along the Mansfield Road, Banks was assailed by Taylor's Confederates. Though somewhat disjointed, the attack succeeded in routing the Federals whose precipitous flight was hampered by an injudiciously placed waggon train. Besides heavy casualties the Federals lost 20 cannon and 200 waggons (above).*

Taylor attacked the following afternoon, but failed to dislodge Banks. Kirby Smith responded by foolishly ordering most of Taylor's infantry to Arkansas, which eliminated any possiblity of Taylor cutting off Banks and capturing or destroying Porter's fleet.

Unusually low water in the Red River influenced Banks to continue his withdrawal. He reached Alexandria on April 26, where shallow water trapped Porter. After constructing a dam to save the fleet, and setting the town alight, Banks proceeded down river on May 13. The campaign ended a week later, when the Federals escaped beyond the Atchafalaya River.

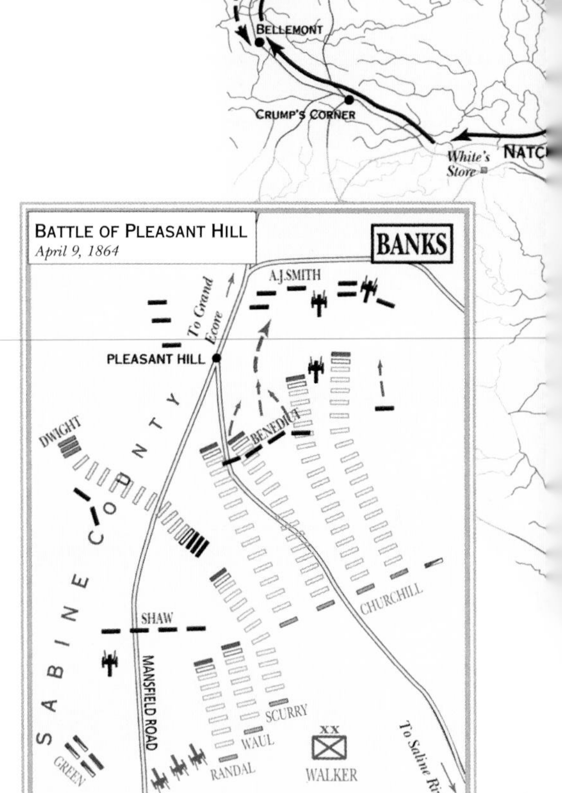
To Shreveport & Little Rock
To Vicksburg
KEATCHIE
RED BLUFF
GRAND CANE
Lake Cannisnia
SPRINGFIELD
AUGUSTA
SMITHPORT
MANSFIELD
Bayou Pierre Lake
Carroll's Mill
Wilson's Farm
Blair's Landing
PLEASANT HILL
BELLEMONT
CRUMP'S CORNER
White's Store
RED RIVER
BATTLE OF PLEASANT HILL
April 9, 1864
BANKS
A.J.SMITH
To Grand Ecore
PLEASANT HILL
DWIGHT
BENEDICT
SABINE COUNTY
SHAW
CHURCHILL
MANSFIELD ROAD
SCURRY
WAUL
RANDAL
GREEN
BEE
WALKER
TAYLOR
To Sabine River
0 1/4 1/2 KM
0 1/4 1/2 MILE

*After the Union defeat the previous day, Banks anticipated further Confederate assaults and drew up his forces in a defensive line at Pleasant Hill. Light skirmishing was succeeded by a full scale Confederate attack in the late afternoon. Initially pushed back, the Federals rallied and counterattacked eventually forcing the Confederates from the field. Union casualties were 1,369, Confederate 1,626.*

*The U.S.S. Osage (left), part of the fleet which Admiral Porter boasted could operate "wherever the sand was damp". In fact, the fleet came perilously close to being stranded when the Red River's water levels began to fall with unexpected rapidity.*

(1) **March 13:** Smith's forces land at Simsport

(2) **March 14:** Federals capture Fort de Russy

(3) **March 16:** Alexandria occupied

(4) **March 21:** Federals surprise and capture Vincent's Confederate cavalry and a 4-gun battery at Henderson's Hill

(5) **April 4:** Confederate Brig. Gen. St. John R. Liddell's men skirmish with Porter's fleet at Campti

(6) **April 7:** Union Brig. Gen. Albert L. Lee's cavalry skirmishes with Brig. Gen. Thomas Green's cavalry at Wilson's Farm

(7) **April 7:** Lee's Union cavalry skirmishes with Green's Confedereate troopers at Carroll's Mill

(8) **April 8, pm:** Battle of Mansfield

(9) **April 9, pm:** Battle of Pleasant Hill

(10) **April 12, pm:** Supported by one battery, Green's Confederate cavalry attacks Porter's fleet at Blair's Landing in an unsuccessful attempt to block Porter's retreat

(11) **April 24:** Confederate Brig. Gen. Hamilton P. Bee is unsuccessful in his attempt to block Banks's retreat at Monnet's Ferry on the Cane River

(12) **May 16:** Banks continues to retreat by pushing through Taylor's main force near Mansura

(13) **May 18:** In a final effort to prevent Banks from escaping, Taylor attacks at Yellow Bayou. Outnumbered more than four to one, Taylor's vain attack costs his army 600 casualties

(14) **May 20:** The Federals complete their crossing of the Atchafalaya River. Lacking the capability to cross, Taylor could pursue no further

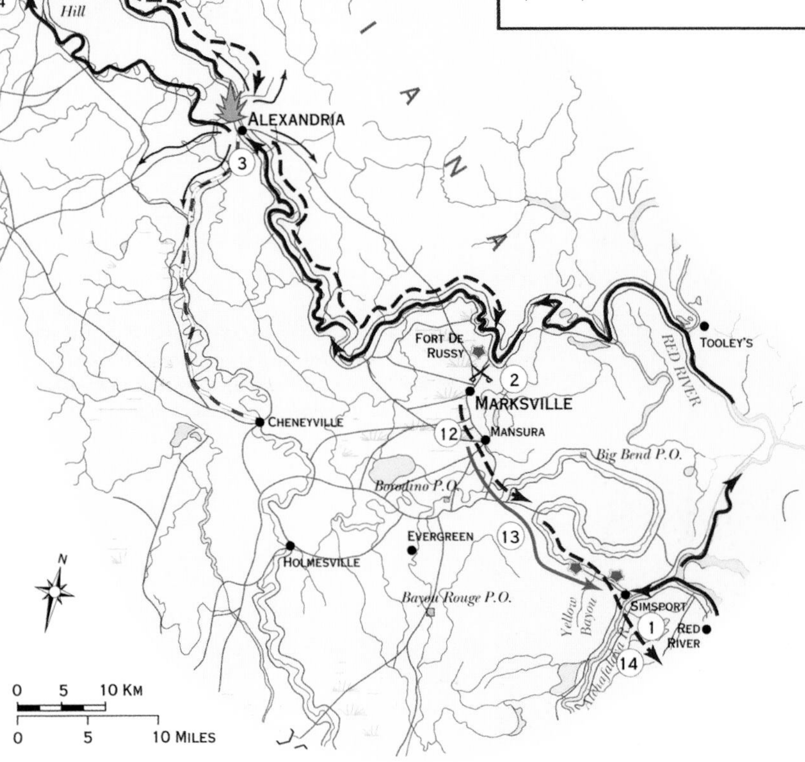
Black Lake
Campti
Saline Lake
Grand Ecore
Lanark
St. Maurice
Wheeling
Cloutierville
Rigolet du Bon Dieu
Louisiana
Big Creek
Henderson's Hill
Alexandria
Fort De Russy
Marksville
Mansura
Red River
Tooley's
Cheneyville
Big Bend P.O.
Evergreen
Holmesville
Bayou Rouge P.O.
Yellow Bayou
Simsport
Red River
N
0 5 10 Km
0 5 10 Miles

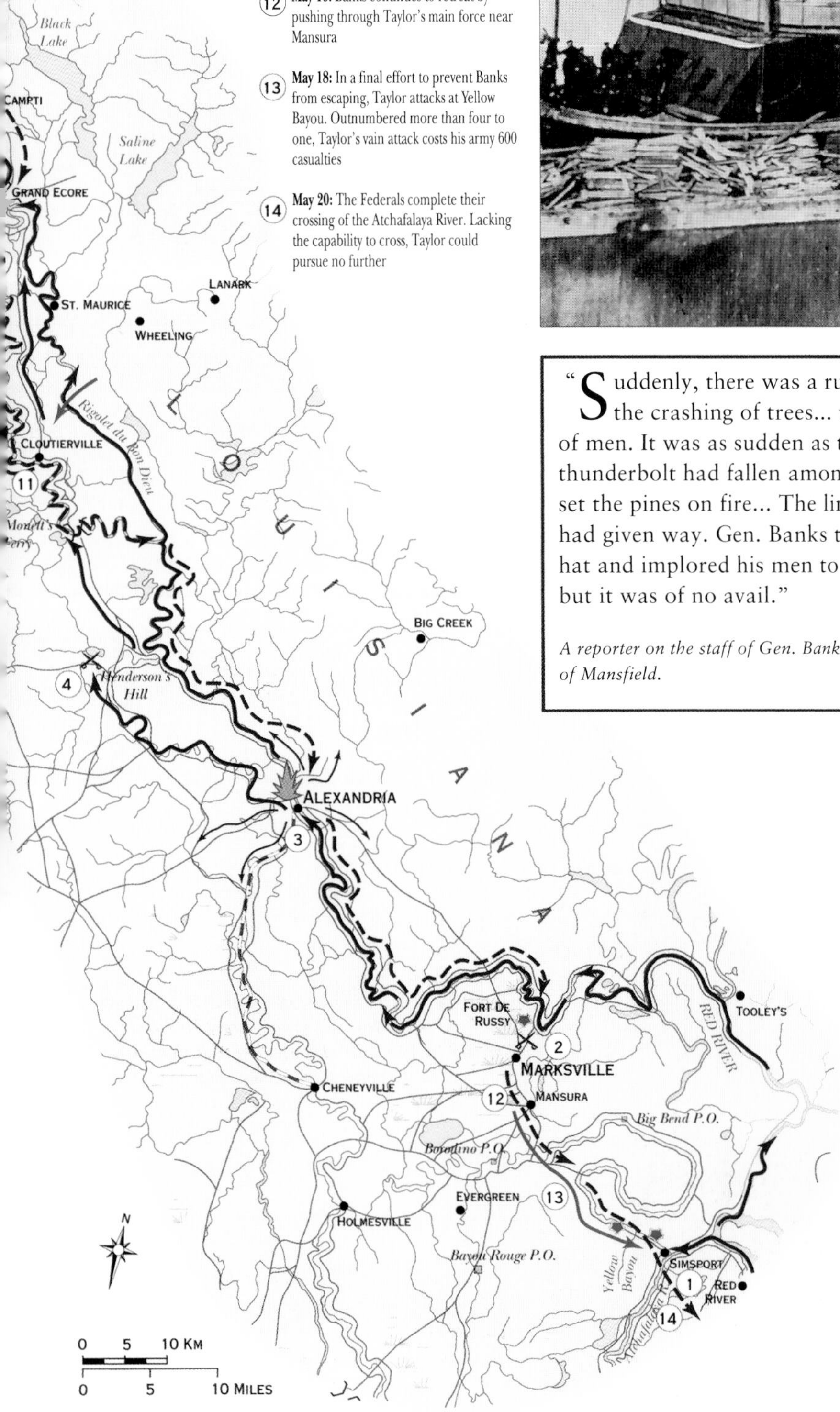

> "Suddenly, there was a rush, a shout, the crashing of trees... the scamper of men. It was as sudden as though a thunderbolt had fallen among us, and set the pines on fire... The line of battle had given way. Gen. Banks took off his hat and implored his men to remain... but it was of no avail."
>
> *A reporter on the staff of Gen. Banks at the battle of Mansfield.*

*Damming material is towed into place by the U.S.S.* Signal *(above). Designed by the XX Corps engineer, Lt. Col. Joseph Bailey, the dam raised the waters sufficiently for Porter's squadron to run the rapids at Alexandria, thereby escaping the Confederate pursuit.*

*Lt. Gen. Edmund Kirby Smith, commander of all Confederate forces in the Trans-Mississippi (below). Upon Banks's withdrawal, Kirby Smith's army remained practically unmolested for the remainder of the war. The Confederate department eventually became known as "Kirby Smithdom," after its virtually autonomous ruler.*

# Steele's Arkansas Campaign MARCH 1 – MAY 3 1864

To distract Confederate forces away from General Nathaniel Banks's Red River cam paign in Louisiana, Major General Frederick Steele led 8,500 men south from Little Rock on March 23. Forces led by General Sterling Price, who commanded the Confederate District of Arkansas, unsuccessfully opposed his crossing of the Little Missouri River at Elkins' Ferry on April 3. Casualties were insignificant. Reinforced by 5,000 men from Fort Smith under Brigadier General John Milton Thayer, Steele next outmaneuvered Price and occupied Camden on April 15.

As there were no railroads linking his base at Little Rock with the rugged region through which he traveled, Steele had to sustain his force by long wagon trains and by foraging – both vulnerable to enemy attack. On April 18, just west of Camden at Poison Spring, Confederates encountered a foraging party of over 1,000 Federals, many of whom were African Americans – former slaves enlisted in the Federal army. The Federals were decisively defeated, sustaining 300 casualties. Due to intense racial animosity, the Confederates killed a large number of African American soldiers as they attempted to surrender. A similar incident occurred on April 25, near Mark's Mills, when Confederates captured a Federal supply train returning from Camden to Pine Bluff. Over 1,000 Federals were taken prisoner. But, as at Poison Spring, many of the African American soldiers were simply massacred.

Learning that Banks's expedition had failed, Steele withdrew from Camden on April 27, to return to Little Rock. On April 30, at Jenkin's Ferry on the Saline River, his rearguard of 4,000 men repelled a determined attack by 8,000 Confederates. These men, led by Trans-Mississippi Department commander General Edmund Kirby Smith, sustained over 1,000 casualties, while Steele's losses were considerably fewer. Steele reached Little Rock on May, ending the campaign.

Although intended as a diversion only, Steele's campaign was perceived as a major defeat: in part because he lost over 600 wagons in various Confederate attacks on his supply line; partly because his total casualties – nearly 2,500 – were twice that of the Confederates. In reality, logistical difficulties were the principal reason the Federals failed to gain complete control over southern Arkansas.

*Confederate assault upon a Federal supply train (above). In the attack at Mark's Mill the Confederates encountered stronger opposition than they had anticipated and suffered 300 casualties as a result, but they succeeded in capturing a total of 1,300 Federals and 300 wagons.*

*By April 30, both Price's and Steele's men were exhausted – the former as a result of their forced march, the latter from hauling their wagon train through rivers of mud. By 7.30 a.m. Marmaduke's troopers were skirmishing with Federal infantry two miles in front of Jenkin's Ferrry. Price committed his forces as rapidly as possible but the Federals were ensconced behind log breastworks and thick fog and mud hampered the assault. The Federals withdrew across the river once the wagon train had passed over. Confederate casualties 1,000, Union 700.*

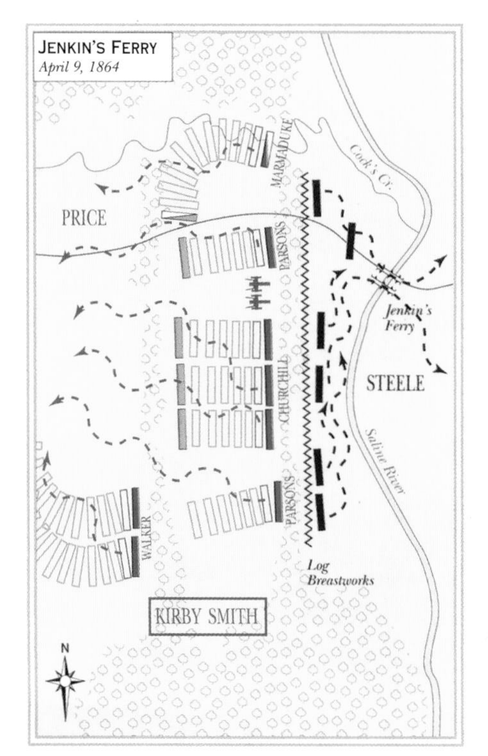
JENKIN'S FERRY
April 9, 1864
PRICE
MARMADUKE
PARSONS
CHURCHILL
PARSONS
WALKER
Cock's Cr.
Jenkin's Ferry
STEELE
Saline River
Log Breastworks
KIRBY SMITH
N

"I determined to charge them first, last, and all the time."

*Brig. Gen. James Fagan, Confederate cavalry commander at Mark's Mills.*

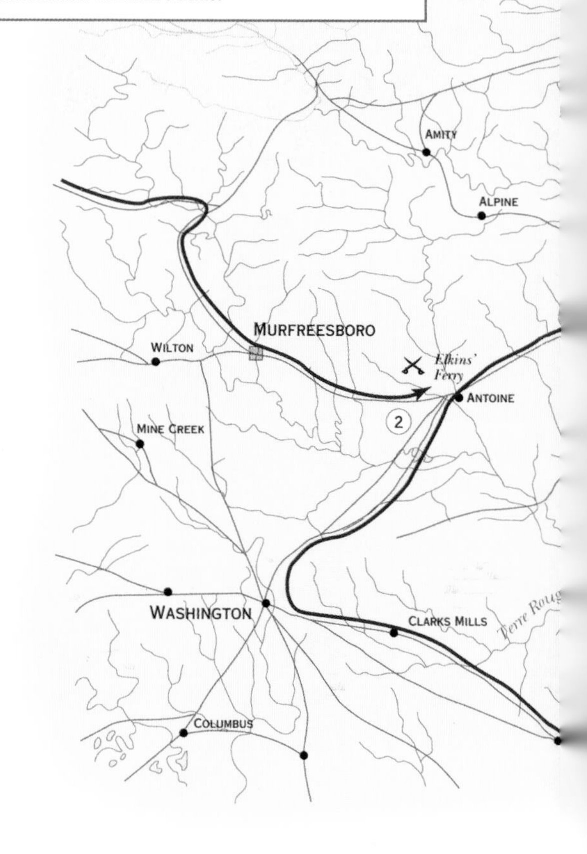
Harold P.O.
AMITY
ALPINE
MURFREESBORO
WILTON
Elkins' Ferry
ANTOINE
2
MINE CREEK
WASHINGTON
CLARKS MILLS
COLUMBUS

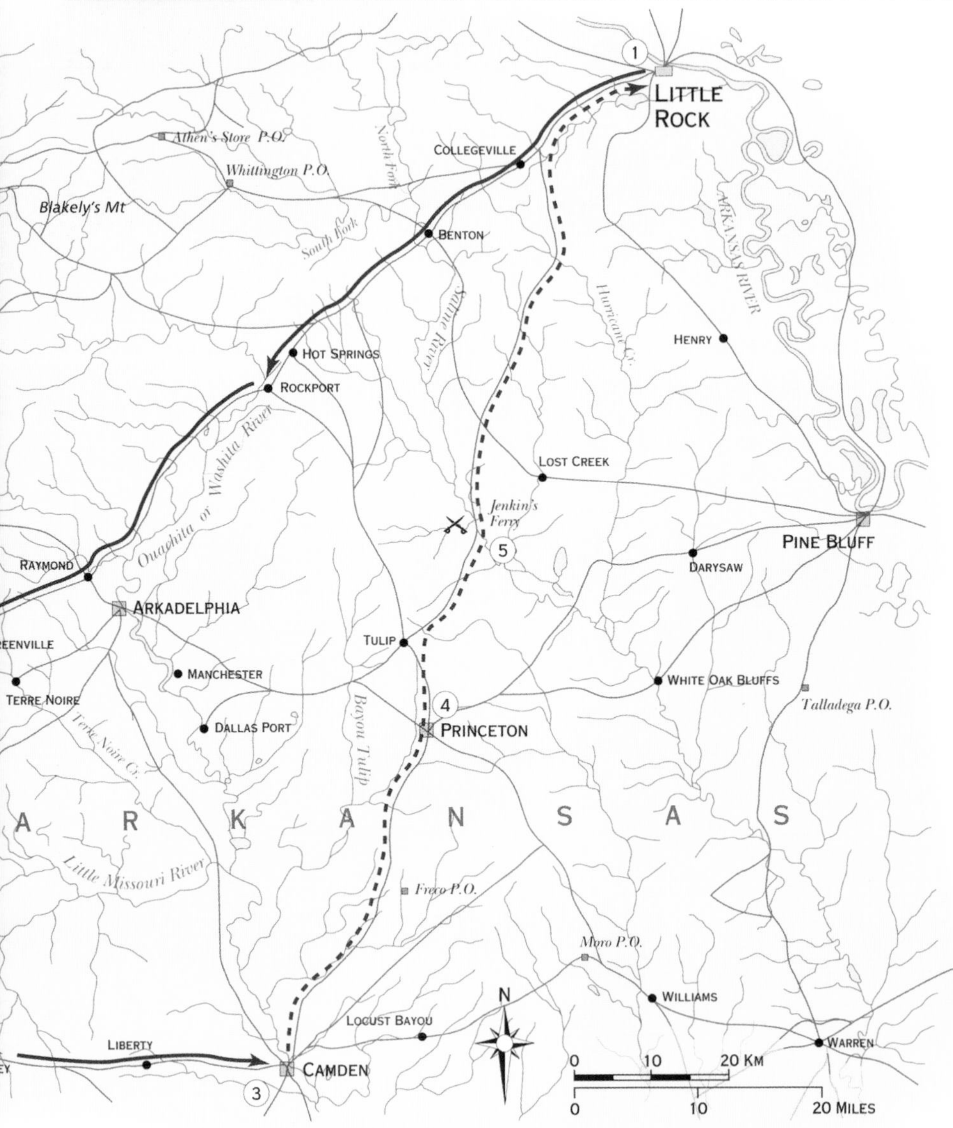
1
LITTLE ROCK
Allen's Store P.O.
Whittington P.O.
COLLEGEVILLE
Blakely's Mt
BENTON
South Fork
North Fork
ARKANSAS RIVER
Saline River
Hurricane Cr.
HENRY
HOT SPRINGS
ROCKPORT
Ouachita or Washita River
LOST CREEK
Jenkin's Ferry
5
PINE BLUFF
DARYSAW
RAYMOND
ARKADELPHIA
TULIP
MANCHESTER
WHITE OAK BLUFFS
TERRE NOIRE
Talladega P.O.
4
DALLAS PORT
PRINCETON
Bayou Tulip
Terre Noire Cr.
A R K A N S A S
Little Missouri River
Freeo P.O.
Moro P.O.
WILLIAMS
LOCUST BAYOU
N
WARREN
LIBERTY
CAMDEN
3
0 10 20 KM
0 10 20 MILES

*Company E of the 4th U.S. Colored Infantry. Many of the Union soldiers killed at Poison Spring were wounded men murdered where they lay. One regimental commander reported that of his 182 casualties, 117 were dead.*

1. **March 23:** Steele leaves Little Rock, marching into southern Arkansas to divert the attention of local Confederate forces away from Banks's Red River campaign in Louisiana
2. **April 3:** Steele drives off Confederate forces under Brig. Gen. J. S. Marmaduke which had disputed his crossing of the Little Missouri River at Elkins' Ferry
3. **April 15:** After feinting towards Washington (the Confederate capital of Arkansas since the occupation of Little Rock by Steele the previous September), Steele occupies Camden without a fight
4. **April 25:** Confederate cavalry under Brig. Gen. J. F. Fagan captures a Federal supply train at Mark's Mills, massacring a number of African American soldiers
5. **April 30:** Following Banks's defeat in Louisiana, Steele returns to Little Rock. En-route he defeats a Confederate attack on his rear guard at Jenkin's Ferry

"While we lay here the long-looked-for and much-talked-of reinforcement of 'Thayer's command' arrived... A nondescript style of reinforcement it was too, numbering almost every kind of soldier... and accompanied by multitudinous vehicles, of all descriptions, which had been picked up along the roads."

*A member of Steele's command upon the arrival of Brig. Gen. J. M. Thayer's Frontier Division.*

# The Wilderness MAY 5–7 1864

On May 4, 1864, the Federal Army of the Potomac, 120,000-strong, plunged across the Rapidan River and disappeared into the forbidding Wilderness. Federal generals Grant and Meade wanted to press southward using the dense undergrowth to screen their movements.

General Lee, however, had other plans: his 61,000 Confederates advanced into the Wilderness with the express intention of engaging the Federal column while it was in the confines of the woods. Lee also hoped that conditions in the Wilderness would negate the Federals' two-to-one numerical advantage.

On May 5, now aware of Lee's ap proach, Grant and Meade halted, and turned to meet the challenge. The Federal V Corps struck first, but were repulsed by Ewell's Corps at Saunders Field on the Orange Turnpike. The battle spread two and a half miles south of the turnpike; there, Federal troops thwarted A.P. Hill's thrust along the Orange Plank Road. Intense fighting raged into the night, but without any decisive result.

Attempting to break the stalemate, the Federals launched a massive attack at dawn on May 6. Federal troops broke Lee's lines along the Orange Plank Road, and assaulted the Confederate rear. All seemed lost for Lee until Longstreet's timely arrival paralyzed the Federal drive and re established Lee's lines. Lee and Longstreet now seized the initiative with a daring flank attack that surprised the Federals – as it had at Chancellorsville the previous year – but it soon unraveled when Longstreet was accidentally wounded by his own men. The Confederates shifted their focus back to the north with another bold flanking attack led by Brigadier General John B. Gordon. But Gordon's initially successful attack gave way under swarms of Federal reinforcements, and the second day ended in deadlock.

After two days of fighting, the Federals had lost 18,000 men; the Confederates about 12,000. Raging fires, ignited by musket flashes, consumed many of the dead and wounded trapped in the thick undergrowth, adding a special horror to the battle.

Grant and Meade refused to retreat. Shortly after dark on May 7, the Federals abandoned their lines and, side-stepping Lee, continued their progress south.

*The Army of the Potomac crossing the Rapidan on the pontoon bridges of Germanna Ford, May 4 (left). Meade commanded the army, but Grant accompanied it on it's advance into the Wilderness.*

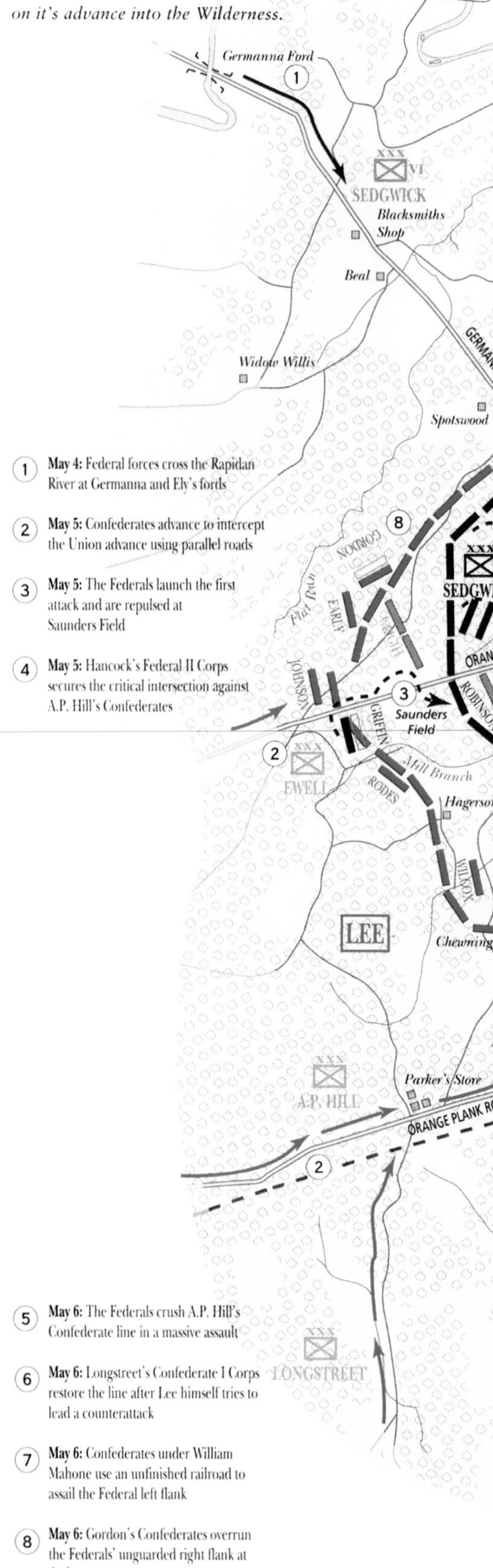

Germanna Ford
SEDGWICK
Blacksmiths Shop
Beal
GERMANNA PL
Widow Willis
Spotswood
GORDON
EARLY
JOHNSON
Flat Run
SEDGWICK
ORANGE &
Saunders Field
ROBINSON
GRIFFIN
EWELL
RODES
Mill Branch
Hagerson
WILCOX
LEE
Chewning
Parker's Store
A.P. HILL
ORANGE PLANK ROAD
LONGSTREET
1 May 4: Federal forces cross the Rapidan River at Germanna and Ely's fords
2 May 5: Confederates advance to intercept the Union advance using parallel roads
3 May 5: The Federals launch the first attack and are repulsed at Saunders Field
4 May 5: Hancock's Federal II Corps secures the critical intersection against A.P. Hill's Confederates
5 May 6: The Federals crush A.P. Hill's Confederate line in a massive assault
6 May 6: Longstreet's Confederate I Corps restore the line after Lee himself tries to lead a counterattack
7 May 6: Confederates under William Mahone use an unfinished railroad to assail the Federal left flank
8 May 6: Gordon's Confederates overrun the Federals' unguarded right flank at dusk

*During their march through the Wilderness the men of Grant's and Lee's armies were reminded of Chancellorsville by the bones of the men who had died during the earlier campaign (left).*

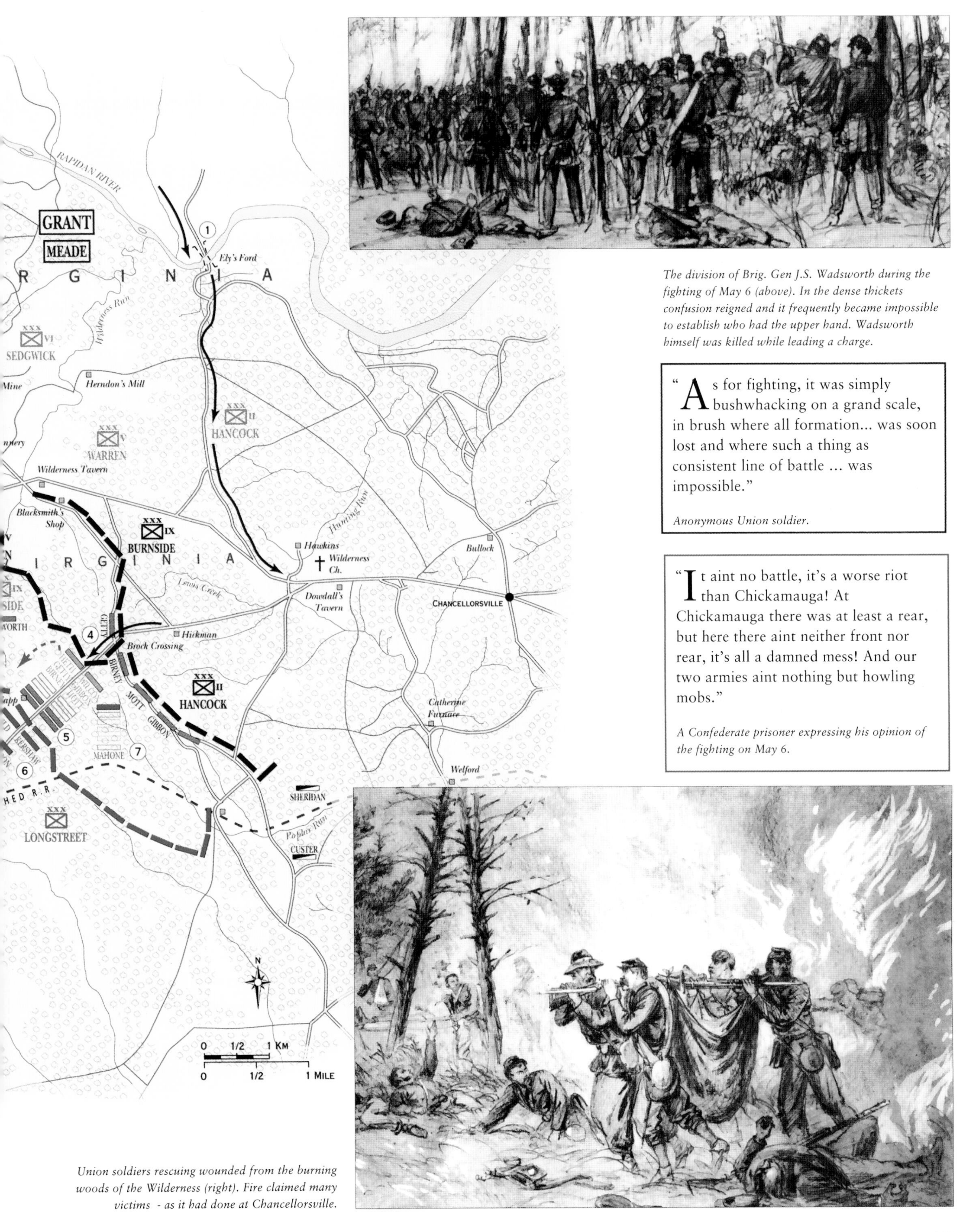
RAPIDAN RIVER
GRANT
MEADE
Ely's Ford
R G I N I A
Wilderness Run
SEDGWICK
Herndon's Mill
HANCOCK
WARREN
Wilderness Tavern
Blacksmith's Shop
BURNSIDE
Hunting Run
Hawkins
Wilderness Ch.
Bullock
Lewis Creek
Dowdall's Tavern
CHANCELLORSVILLE
Hickman
Brock Crossing
GETTY
BIRNEY
MOTT
GIBBON
HANCOCK
Catherine Furnace
MAHONE
Welford
SHERIDAN
Poplar Run
CUSTER
LONGSTREET
N
0 1/2 1 KM
0 1/2 1 MILE

*The division of Brig. Gen J.S. Wadsworth during the fighting of May 6 (above). In the dense thickets confusion reigned and it frequently became impossible to establish who had the upper hand. Wadsworth himself was killed while leading a charge.*

"As for fighting, it was simply bushwhacking on a grand scale, in brush where all formation... was soon lost and where such a thing as consistent line of battle ... was impossible."

*Anonymous Union soldier.*

"It aint no battle, it's a worse riot than Chickamauga! At Chickamauga there was at least a rear, but here there aint neither front nor rear, it's all a damned mess! And our two armies aint nothing but howling mobs."

*A Confederate prisoner expressing his opinion of the fighting on May 6.*

*Union soldiers rescuing wounded from the burning woods of the Wilderness (right). Fire claimed many victims - as it had done at Chancellorsville.*

# Spotsylvania: Phase 1 MAY 8–12 1864

Determined to press on with their campaign against Lee, Grant and Meade ordered the Federal army south to Spotsylvania Court House. Leaving their Wilderness trenches, the Federals marched through the night of May 7. Unknown to them, Lee's Confederates were also heading for Spotsylvania on a parallel road. Confederate cavalry delayed the Federal advance, giving Lee time to reach the key road intersection at the Courthouse. There, the Confederates raced into line along Brock Road at Spindle Farm, and with only seconds to spare, fired into the oncoming Federals. Throughout May 8, both sides brought up reinforcements, and dug in.

Grant and Meade now looked for a way to overcome the Confederate entrenchments. On May 9, they tried flanking them by crossing the Po River, but the Confederates stopped them at Block House Bridge, counterattacking on May 10. Abandoning that plan, the Federals launched a series of piecemeal attacks against Laurel Hill and Confederate lines along "the Mule Shoe" salient. At dusk, an enterprising colonel, Emory Upton, unleashed a surprise assault with 12 regiments, penetrating the trenches with ease. But, lacking support, the attack failed. Grant and Meade determined to try Upton's tactics again, this time on a grander scale.

The Federals launched a concentrated attack on the apex of the Mule Shoe-shaped line. Crashing over the fortifications at 5a.m. on May 12, the Federal soldiers captured 3,000 prisoners, including two generals. Lee lashed back with a series of brutal counterattacks that escalated into a massive struggle for a bend in the earth-works, known as "the Bloody Angle." Both sides fed fresh troops into the melée. The fierce hand-to-hand fighting continued uninterrupted for 22 hours. By 3a.m. on May 13, the Confederates had constructed a new and stronger line. Lee ordered the survivors of his army to abandon the Bloody Angle and fall back behind the new line.

The Federals had wrested the strategically important position away from the Confederates, but were unable to turn the victory to their advantage. The battle continued.

*Emory Upton (left), whose daring attack on the Confederate lines on May 10 resulted in his promotion from colonel to brigadier general. Upton's men were hand-picked and carried the enemy works with a bayonet charge, without firing a shot.*

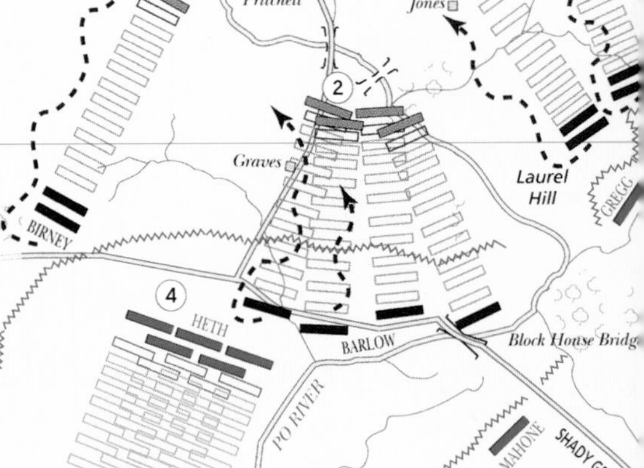

1 May 8: Confederates block the road south to Spotsylvania seconds before the Federals arrive
2 May 9: Federals attempt to flank Lee's left by crossing the Po River
3 May 10: Gen. John Sedgwick killed by sniper's bullet whilst reassuring his troops that Confederate snipers "could not hit an elephant at this distance"
4 May 10: Confederates stop the Federal advance and drive them back across the Po River
5 May 10: Federals initiate a series of ineffective assaults against the Confederate works at Laurel Hill
6 May 10: Upton's attack pierces the Confederate lines but is left unsupported and fails
WARREN
CRAWFORD
BIRNEY
Pritchett
Jones
Graves
Laurel Hill
HETH
BARLOW
Block House Bridge
PO RIVER
MAHONE
BROCK RD
GLADY RUN
7 May 12: Federals overrun the Confederate salient in a massive dawn assault (see map right)
8 May 12: Confederates counterattack, leading to intense hand-to-hand action at the "Bloody Angle" (see map far right)
9 May 12: Wright's VI Corps reinforces the battle at the "Bloody Angle"
10 May 12: Burnside's IX Corps tries to support the Federal attack but is flanked and falls back


*Grant's veterans cheer their general en route to Spotsylvania Court House (left). The expression of confidence was not welcomed by Grant who stated, "this is most unfortunate. The sound will reach the ears of the enemy, and I fear it may reveal our movement".*

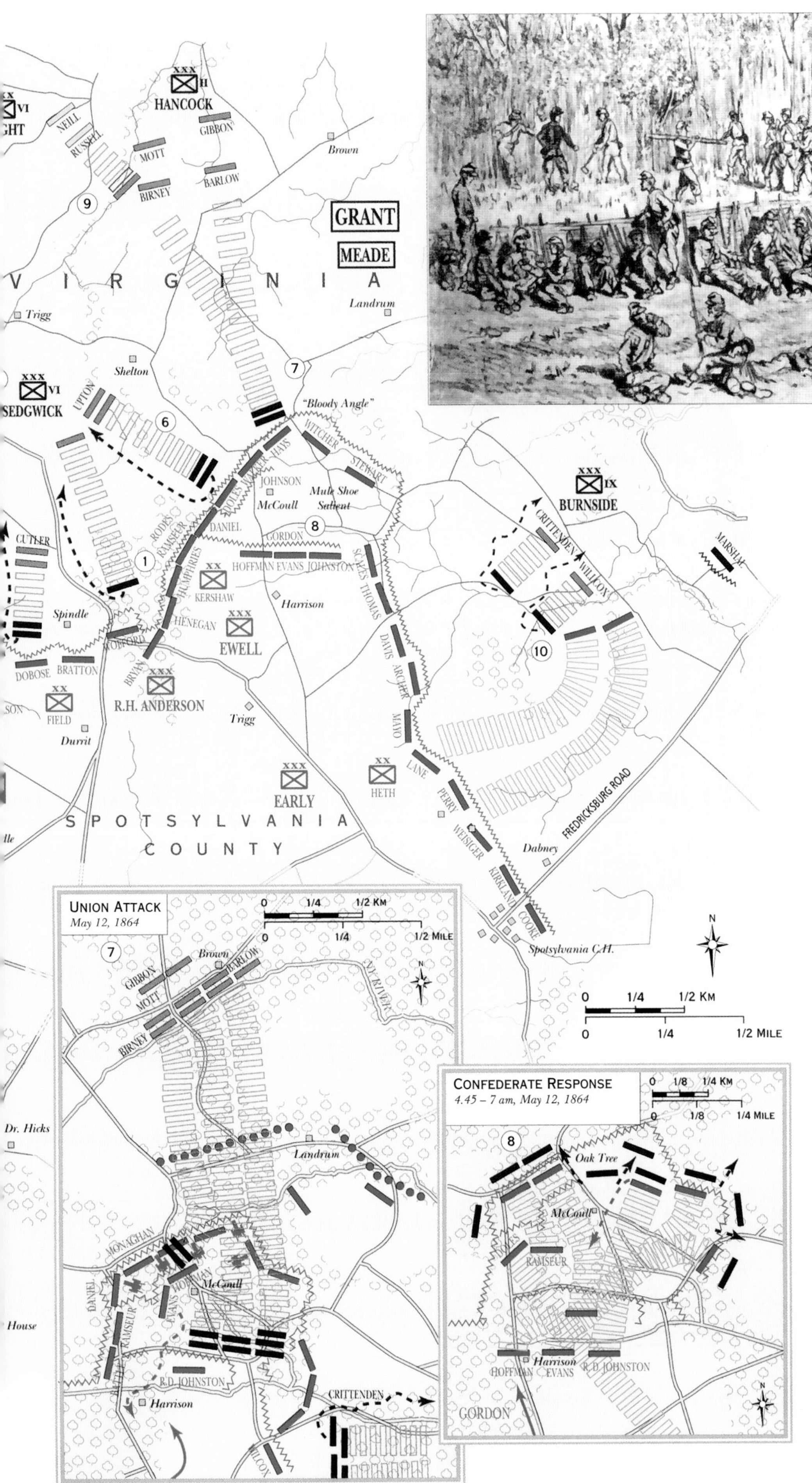
HANCOCK
NEILL
RUSSELL
MOTT
GIBBON
BIRNEY
BARLOW
Brown
GRANT
MEADE
V I R G I N I A
Trigg
Landrum
Shelton
SEDGWICK
UPTON
"Bloody Angle"
WITCHER
HAYS
JONES WALKER
JOHNSON
McCoull
Mule Shoe Salient
STEWART
GORDON
DANIEL
RODES
RAMSEUR
CUTLER
HUMPHRIES
HOFFMAN EVANS JOHNSTON
SCALES THOMAS
KERSHAW
Harrison
Spindle
HENEGAN
EWELL
WOFFORD
DOBOSE
BRATTON
BRYAN
R.H. ANDERSON
FIELD
Durrit
Trigg
DAVIS
ARCHER
MAYO
LANE
PERRY
WEISIGER
KIRKLAND
COOKE
HETH
EARLY
BURNSIDE
CRITTENDEN
WILLCOX
MARSHALL
FREDRICKSBURG ROAD
Dabney
Spotsylvania C.H.
S P O T S Y L V A N I A
C O U N T Y
0 1/4 1/2 KM
0 1/4 1/2 MILE
UNION ATTACK
May 12, 1864
0 1/4 1/2 KM
0 1/4 1/2 MILE
Brown
GIBBON
BARLOW
MOTT
BIRNEY
NY RIVER
Landrum
MONAGHAN
DANIEL
RAMSEUR
HOFFMAN
EVANS
McCoull
R.D. JOHNSTON
Harrison
CRITTENDEN
WILLCOX
Dr. Hicks
House
CONFEDERATE RESPONSE
4.45 – 7 am, May 12, 1864
0 1/8 1/4 KM
0 1/8 1/4 MILE
Oak Tree
McCoull
RAMSEUR
Harrison
HOFFMAN
EVANS
R.D. JOHNSTON
GORDON

*In the Wilderness, Union troops clear trees and undergrowth in an attempt to increase the effectiveness of their guns against any attempted Confederate assault (above).*

"The confusion was indescribable; it was a struggle at close quarters, a hand-to-hand conflict, resembling a mob... The air was filled with... shouts, cheers, commands, oaths, the sharp report of rifles, the hissing shot, dull, heavy thuds of clubbed muskets, the swish of swords and sabres, groans and prayers."

*A Maine infantryman describing combat conditions in the Wilderness.*

*The struggle for the salient, May 12 (below). Following the example set by Emory Upton, Winfield S. Hancock charged the salient and succeeded in cutting Lee's army in two. At the moment of crisis Lee himself arrived on the battlefield and actually attempted to lead the Confederate counterattack himself. Though prevented by his staff, the general's example inspired his men and John B. Gordon successfully repulsed the Federals.*

# Spotsylvania: Phase 2 MAY 13–19 1864

As Lee improved his new defenses along the base of the Mule Shoe salient, Grant and Meade shifted the bulk of the Federal army east to attack Spotsylvania from another angle. Federal V and VI Corps abandoned their lines less than a day after they had captured the salient, and marched to their left. On May 15, II Corps joined them, extending the Federal left eastward. The Confederates responded by concentrating their strength opposite the Federal build-up.

Suspecting that the Confederates had stripped their old defenses to meet the new Federal move, Grant and Meade decided to double back to the salient, and launch another attack. Starting from the Bloody Angle, Federal II and VI Corps assailed the Confederates on May 18, but Confederate artillery decimated the attackers and drove them back. Grant and Meade quickly abandoned the assault and marched the troops back to the left. Convinced they could never gain the advantage, the Federals once again prepared to circumvent Lee's army.

Believing the Federals were abandoning the area, Lee ordered Ewell's II Corps to probe the Federal right for information. Pushing northeast on May 19, Ewell struck the Federals at Harris Farm. Finding a fresh division of heavy artillery, the Confederate drive ground to a halt. After a vicious and confused battle, the Confederates withdrew. The bloody encounter added a further 900 Confederate and 1,500 Federal casualties to the already burgeoning total.

The Federal army left Spotsylvania the following day, maneuvering further to the south and east. On May 2l, Lee also left, racing to beat the Federals to the North Anna River. Neither side had achieved their respective goals during the fighting around Spotsylvania Court House. Grant and Meade had been unable either to defeat Lee or destroy his army, while Lee still had not found a way of preventing the Federal drive southward. In the course of two weeks the Union had lost 18,000 men; the Confederates between 11,000 and 12,000.

> "The men in the ranks did not look as they did... Their uniforms were now torn, ragged, and stained with mud; the men had grown thin and haggered. The experience of those twenty days seemed to have added twenty years to their age."
>
> *An anonymous Union soldier at Spotsylvania.*

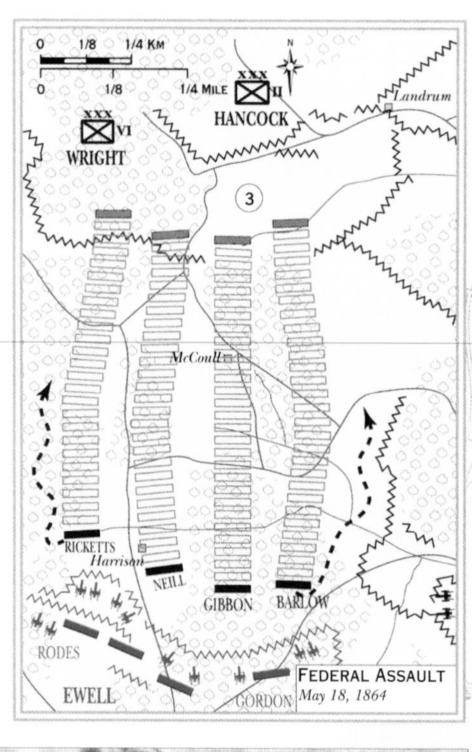
0 1/8 1/4 KM
0 1/8 1/4 MILE
N
HANCOCK
II
VI
WRIGHT
Landrum
3
McCoull
RICKETTS
Harris
NEILL
GIBBON
BARLOW
RODES
EWELL
GORDON
FEDERAL ASSAULT
May 18, 1864

*After days of comparative quiet around Spotsylvania Court House, Grant ordered the corps of Hancock and Wright to make a dawn assault on the Confederate left, mainly against Ewell's men, fighting from behind strong, newly built lines. After several unsuccessful charges, General Meade abandoned the attack (right).*

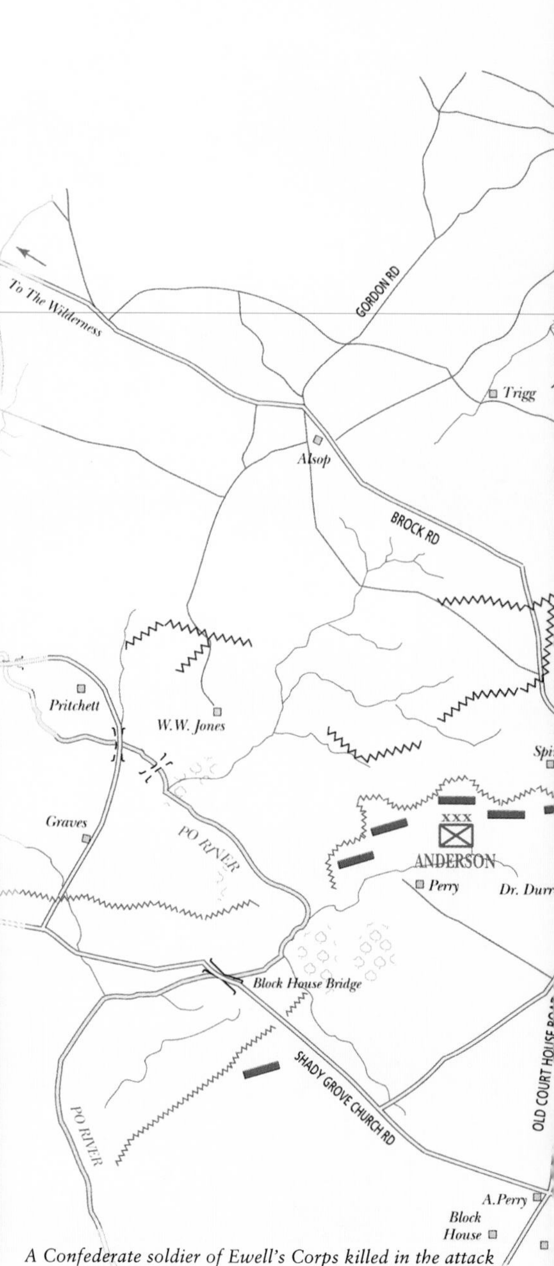
0 1/4 1/2 KM
0 1/4 1/2 MILE
N
To The Wilderness
GORDON RD
Trigg
Alsop
BROCK RD
Pritchett
W.W. Jones
Graves
PO RIVER
ANDERSON
Perry
Block House Bridge
SHADY GROVE CHURCH RD
PO RIVER
OLD COURT HOUSE ROAD
A.Perry
Block House

*A Confederate soldier of Ewell's Corps killed in the attack of May 19 (left). So devastating had been the Union fire that many of the Confederate dead lay in orderly rows, the alignment of their ranks perfectly preserved.*

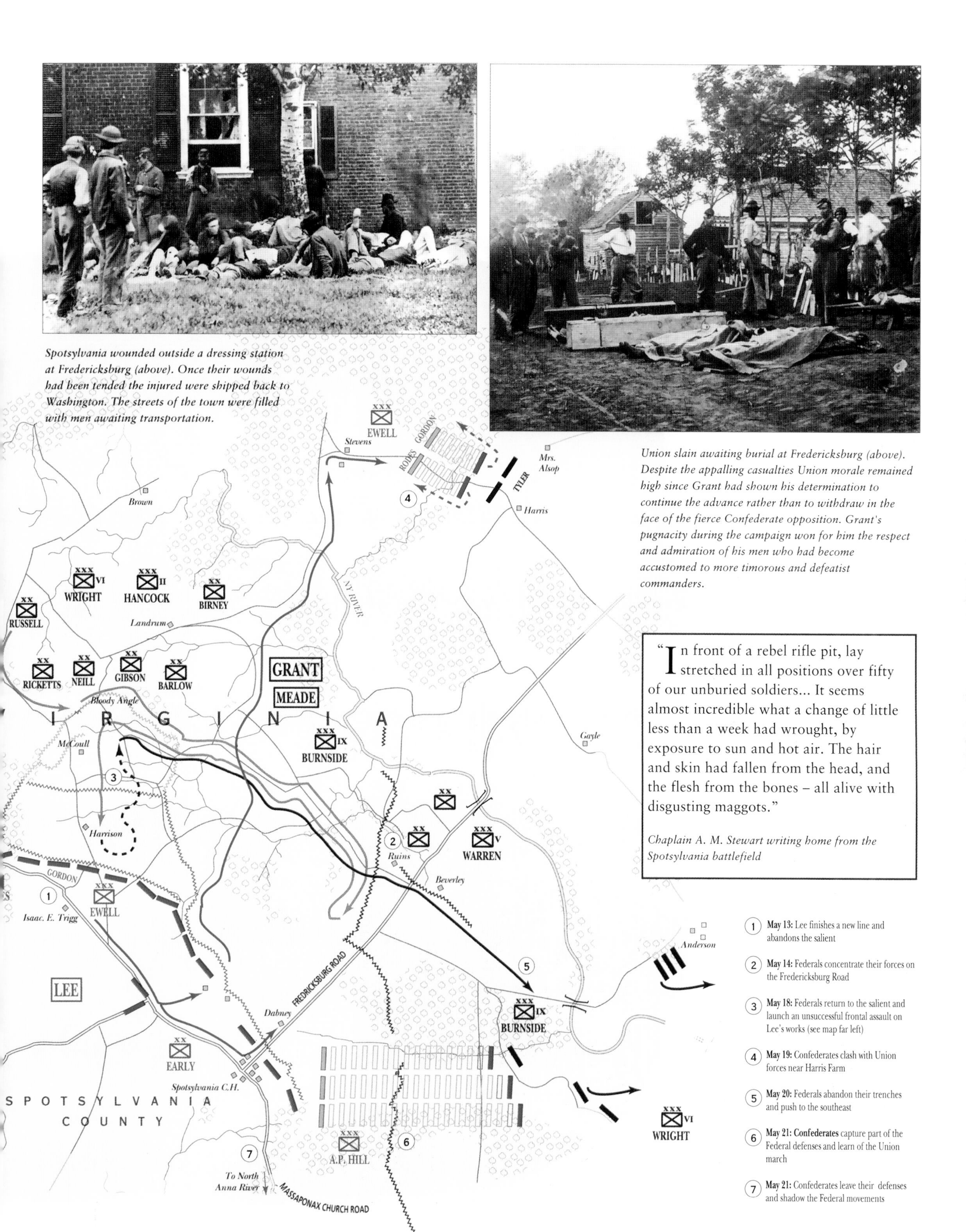
EWELL
Stevens
RODES
GORDON
Mrs. Alsop
TYLER
Harris
Brown
NY RIVER
WRIGHT
VI
HANCOCK
II
BIRNEY
RUSSELL
Landrum
RICKETTS
NEILL
GIBSON
BARLOW
GRANT
MEADE
Bloody Angle
V I R G I N I A
McCoull
BURNSIDE
IX
Gayle
Harrison
WARREN
V
Ruins
Beverley
GORDON
EWELL
Isaac. E. Trigg
FREDRICKSBURG ROAD
Anderson
LEE
Dabney
BURNSIDE
IX
EARLY
Spotsylvania C.H.
SPOTSYLVANIA
COUNTY
WRIGHT
VI
A.P. HILL
To North Anna River
MASSAPONAX CHURCH ROAD
1 May 13: Lee finishes a new line and abandons the salient
2 May 14: Federals concentrate their forces on the Fredericksburg Road
3 May 18: Federals return to the salient and launch an unsuccessful frontal assault on Lee's works (see map far left)
4 May 19: Confederates clash with Union forces near Harris Farm
5 May 20: Federals abandon their trenches and push to the southeast
6 May 21: Confederates capture part of the Federal defenses and learn of the Union march
7 May 21: Confederates leave their defenses and shadow the Federal movements

*Spotsylvania wounded outside a dressing station at Fredericksburg (above). Once their wounds had been tended the injured were shipped back to Washington. The streets of the town were filled with men awaiting transportation.*

*Union slain awaiting burial at Fredericksburg (above). Despite the appalling casualties Union morale remained high since Grant had shown his determination to continue the advance rather than to withdraw in the face of the fierce Confederate opposition. Grant's pugnacity during the campaign won for him the respect and admiration of his men who had become accustomed to more timorous and defeatist commanders.*

"In front of a rebel rifle pit, lay stretched in all positions over fifty of our unburied soldiers... It seems almost incredible what a change of little less than a week had wrought, by exposure to sun and hot air. The hair and skin had fallen from the head, and the flesh from the bones – all alive with disgusting maggots."

*Chaplain A. M. Stewart writing home from the Spotsylvania battlefield*

# North Anna and Cold Harbor MAY 21 – JUNE 3 1864

Anticipating another Federal move to the southeast, General Lee's 55,000 troops left Spotsylvania and headed for the North Anna River, arriving on May 22. Grant and Meade attacked Lee on May 23, and seized Telegraph Bridge over the North Anna. Other Federal troops crossed the river near Jericho Mill, and repulsed a savage assault from A. P. Hill. Despite being ill and bed-ridden, Lee planned a unique, inverted V-shaped line that separated the Federal crossing points. On May 24, the Federals sustained heavy losses at Ox Ford, and failed to unite their lines. Unable to breach Lee's defenses, the Federal army abandoned the North Anna on May 27, and again ventured to the southeast.

On May 28, the Federals slipped across the Pamunkey River at Hanovertown. Their cavalry screened the move by engaging the Confederates in an intense battle at Haw's Shop (or Enon Church). Lee guessed the Federals would drive west against the Richmond railroads, and on May 29 he assumed a defensive line along Totopotomoy Creek. The armies skirmished for three days until, on June 1, General Sheridan's cavalry seized the important crossroads at Cold Harbor, only ten miles from Richmond. Both armies at once converged on this strategic point. The Confederates spent all the following day constructing a strong line of fortifications.

Grant and Meade attacked on June 3. In a series of frontal assaults, the Federals were slaughtered, sustaining approximately 7,000 casualties compared to Confederate losses of 1,500. Grant always regretted ordering the assault at Cold Harbor.

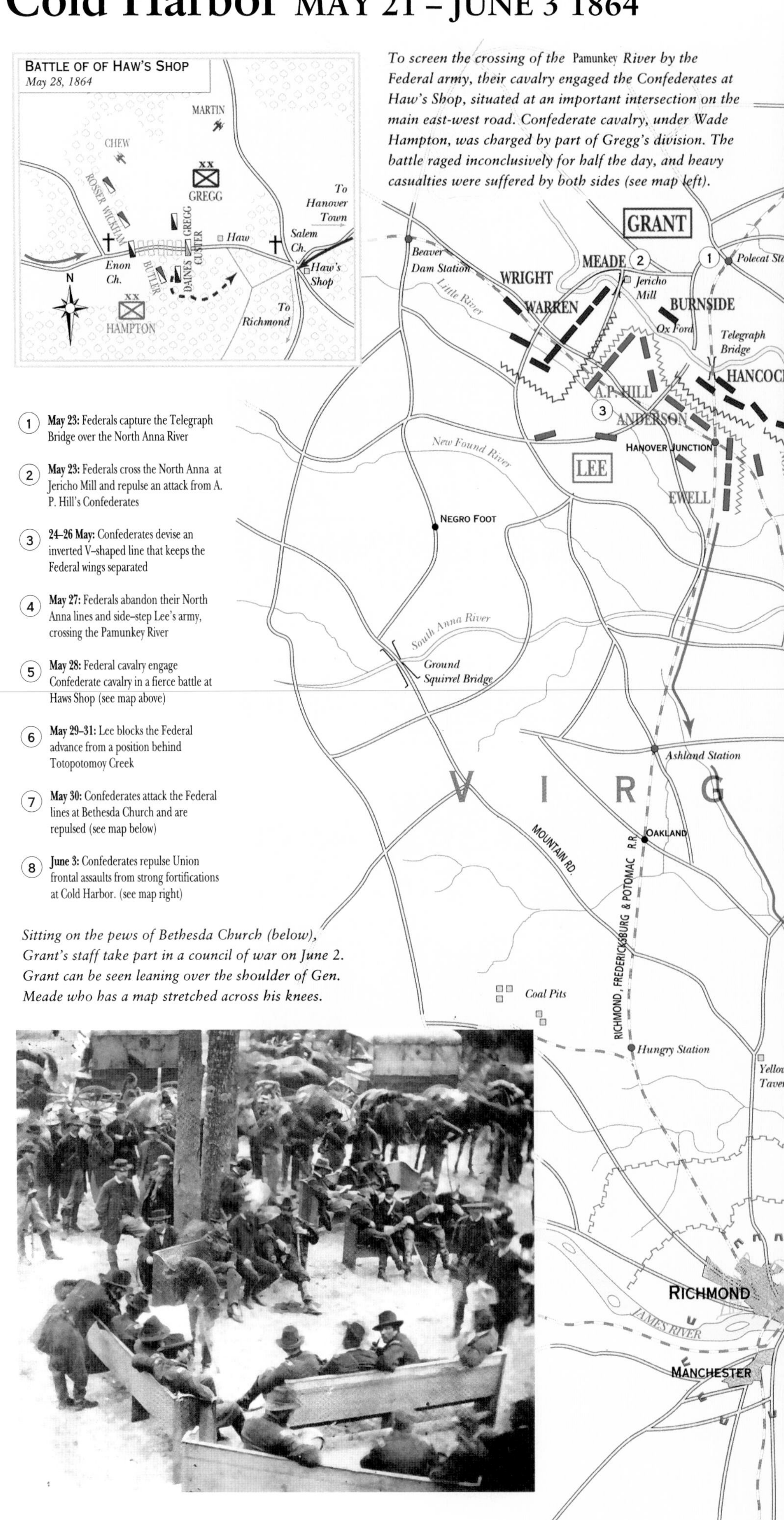

BATTLE OF OF HAW'S SHOP
May 28, 1864
MARTIN
CHEW
GREGG
ROSSER
WICKHAM
GREGG
CUSTER
DAINES
BUTLER
HAMPTON
Enon Ch.
Haw
Salem Ch.
Haw's Shop
To Hanover Town
To Richmond
N
GRANT
MEADE
WRIGHT
WARREN
BURNSIDE
HANCOCK
A.P. HILL
ANDERSON
LEE
EWELL
Beaver Dam Station
Little River
Jericho Mill
Ox Ford
Telegraph Bridge
Polecat Sta
HANOVER JUNCTION
New Found River
NEGRO FOOT
South Anna River
Ground Squirrel Bridge
Ashland Station
V I R G
MOUNTAIN RD.
OAKLAND
RICHMOND, FREDERICKSBURG & POTOMAC R.R.
Coal Pits
Hungry Station
Yellow Tavern
RICHMOND
JAMES RIVER
MANCHESTER


*To screen the crossing of the Pamunkey River by the Federal army, their cavalry engaged the Confederates at Haw's Shop, situated at an important intersection on the main east-west road. Confederate cavalry, under Wade Hampton, was charged by part of Gregg's division. The battle raged inconclusively for half the day, and heavy casualties were suffered by both sides (see map left).*

1. **May 23:** Federals capture the Telegraph Bridge over the North Anna River
2. **May 23:** Federals cross the North Anna at Jericho Mill and repulse an attack from A. P. Hill's Confederates
3. **24–26 May:** Confederates devise an inverted V-shaped line that keeps the Federal wings separated
4. **May 27:** Federals abandon their North Anna lines and side-step Lee's army, crossing the Pamunkey River
5. **May 28:** Federal cavalry engage Confederate cavalry in a fierce battle at Haws Shop (see map above)
6. **May 29–31:** Lee blocks the Federal advance from a position behind Totopotomoy Creek
7. **May 30:** Confederates attack the Federal lines at Bethesda Church and are repulsed (see map below)
8. **June 3:** Confederates repulse Union frontal assaults from strong fortifications at Cold Harbor. (see map right)

*Sitting on the pews of Bethesda Church (below), Grant's staff take part in a council of war on June 2. Grant can be seen leaning over the shoulder of Gen. Meade who has a map stretched across his knees.*

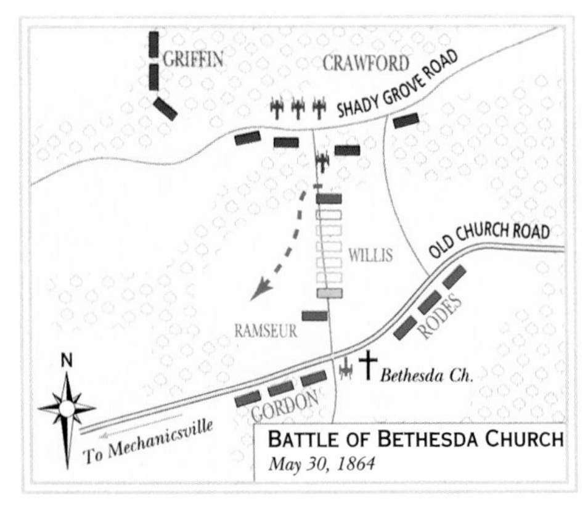

GRIFFIN
CRAWFORD
SHADY GROVE ROAD
WILLIS
OLD CHURCH ROAD
RAMSEUR
RODES
Bethesda Ch.
GORDON
To Mechanicsville
N
BATTLE OF BETHESDA CHURCH
May 30, 1864


*Early sent Rodes's division forward towards Bethesda Church on the Old Church Road. There the Confederates collided with part of Crawford's division and began to drive it north toward Shady Grove Road. Union artillery, however, pounded the Confederates and Willis's brigade was badly mauled and Willis mortally wounded (see map above).*

"Our men have... been foolishly and wantonly sacrificed ... We were recklessly ordered to assault the Enemy's intrenchments [sic], knowing neither their strength nor position. Our loss was heavy, and to no purpose."

*Brig. Gen. Upton after the Union assault, June 3.*

*Throwing up breastworks in the vicinity of Cold Harbor (above). One Confederate reported that his brigade "worked all night with only bayonets, cups, two or three picks, and as many shovels."*

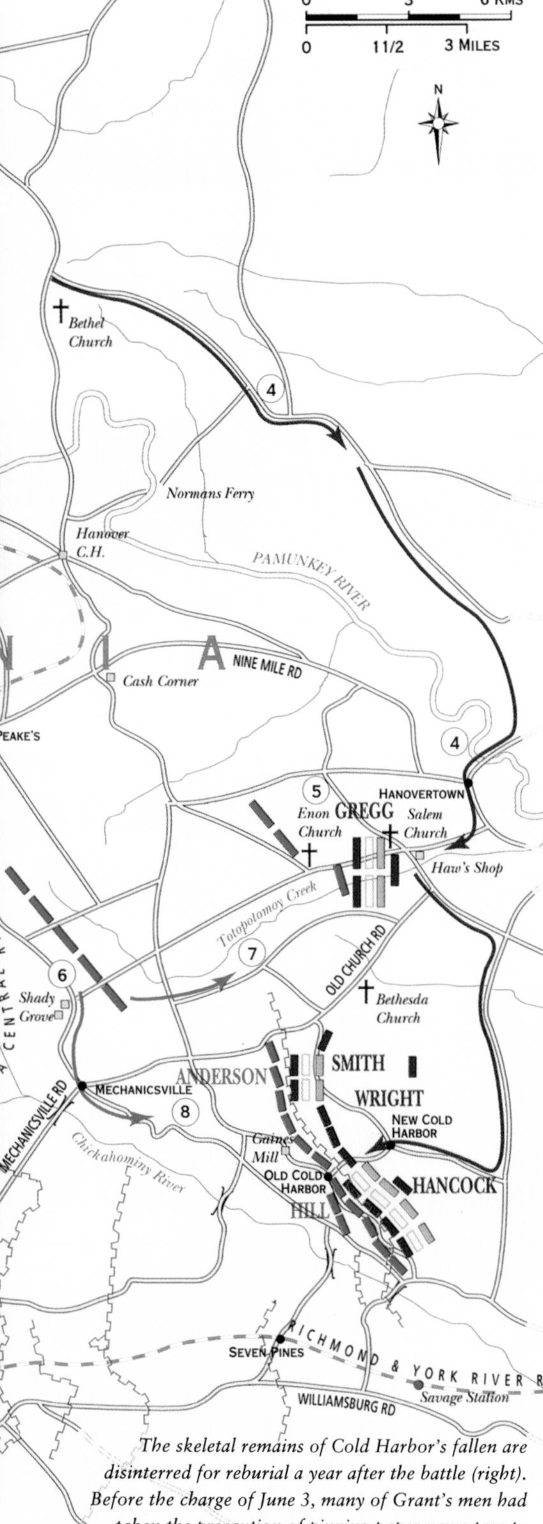
0 3 6 KMS
0 11/2 3 MILES
N
Bethel Church
Normans Ferry
Hanover C.H.
PAMUNKEY RIVER
NINE MILE RD
Cash Corner
PEAKE'S
HANOVERTOWN
Enon Church
GREGG
Salem Church
Haw's Shop
Totopotomoy Creek
OLD CHURCH RD
Bethesda Church
Shady Grove
CENTRAL R.R.
MECHANICSVILLE
MECHANICSVILLE RD
ANDERSON
SMITH
WRIGHT
NEW COLD HARBOR
Gaines Mill
OLD COLD HARBOR
HILL
HANCOCK
Chickahominy River
RICHMOND & YORK RIVER R.R.
SEVEN PINES
WILLIAMSBURG RD
Savage Station

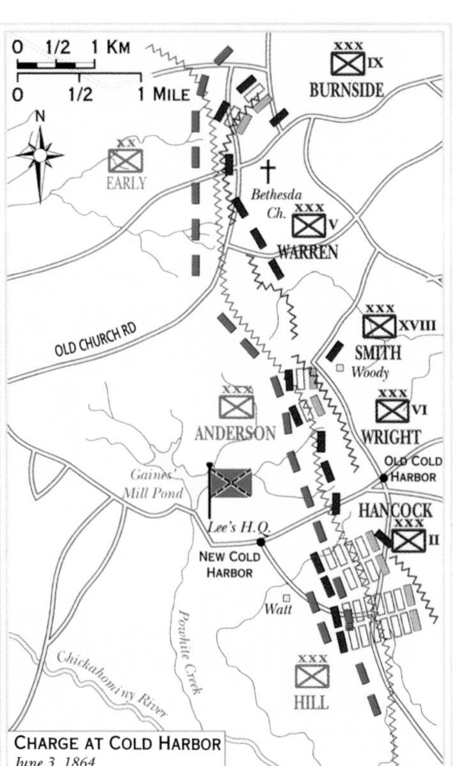
0 1/2 1 KM
0 1/2 1 MILE
N
IX
BURNSIDE
EARLY
Bethesda Ch.
V
WARREN
XVIII
SMITH
Woody
OLD CHURCH RD
ANDERSON
VI
WRIGHT
OLD COLD HARBOR
Gaines' Mill Pond
Lee's H.Q.
HANCOCK
II
NEW COLD HARBOR
Watt
Powhite Creek
Chickahominy River
HILL
CHARGE AT COLD HARBOR
June 3, 1864

The Federals dug in. For ten days, the two antagonists remained in their trenches. But the Federal commanders had no intention of fighting a stagnant war of siege and attrition, and now proposed one final, side-stepping movement around Lee's Confederates. The Federal army would direct its next move against Richmond's railroad center, at Petersburg. Lee, his army lacking the strength to take the initiative, was compelled to wait for the next Federal attack to come.

*After the seizure by Sheridan's cavalry of the vital crossroads at Cold Harbor, Grant decided to launch a massive frontal assault on the Confederates, who had constructed a strong line of fortifications running from north to south. At 4.30am, Federal II, VI and XVIII corps – a total of 50,000 men – advanced on the lines of Hill & Anderson. But the Federals came under thunderous fire from the Confederates, and were decimated.*

*The skeletal remains of Cold Harbor's fallen are disinterred for reburial a year after the battle (right). Before the charge of June 3, many of Grant's men had taken the precaution of pinning paper name tags to their uniforms so that they might be identified if slain.*

# Cold Harbor to Petersburg JUNE 4–15 1864

Following its repulse at Cold Harbor on June 3, the Army of the Potomac entrenched opposite its opponent. While both Grant and Lee pondered their options, the men in the ranks endured the heat and dust of a Virginia summer as well as the daily wastage of life at the hands of merciless sharpshooters. Meanwhile, in the Shenandoah Valley, a small Federal army under Brigadier General David Hunter advanced to Staunton. Believing it imperative to protect the Valley and its resources from Federal depredations, Lee detached Jubal Early's II Corps and sent it to the Valley on June 12.

Early's departure was not immediately apparent to Grant, who was simultaneously considering a change in the line of operations of the Army of the Potomac. Finding no opening on his immediate front, Grant eventually determined to move his army across the James River and seize Petersburg, a critical railroad hub connecting Richmond with the lower South. Such a movement required the Army of the Potomac to break contact with the Confederates, march around Lee's flank, cross the Chickahominy and James rivers, and assault Petersburg before the Army of Northern Virginia could react in strength.

Commencing on the evening of June 12, II and VI Corps occupied a shortened trench line while V Corps established a new position covering the Federal routes to the James. At the same time, XVIII Corps marched east to White House, and embarked on ships for transfer to Bermuda Hundred, a peninsula formed by the James and Appomattox rivers less than 20 miles from Richmond. As soon as V Corps was in

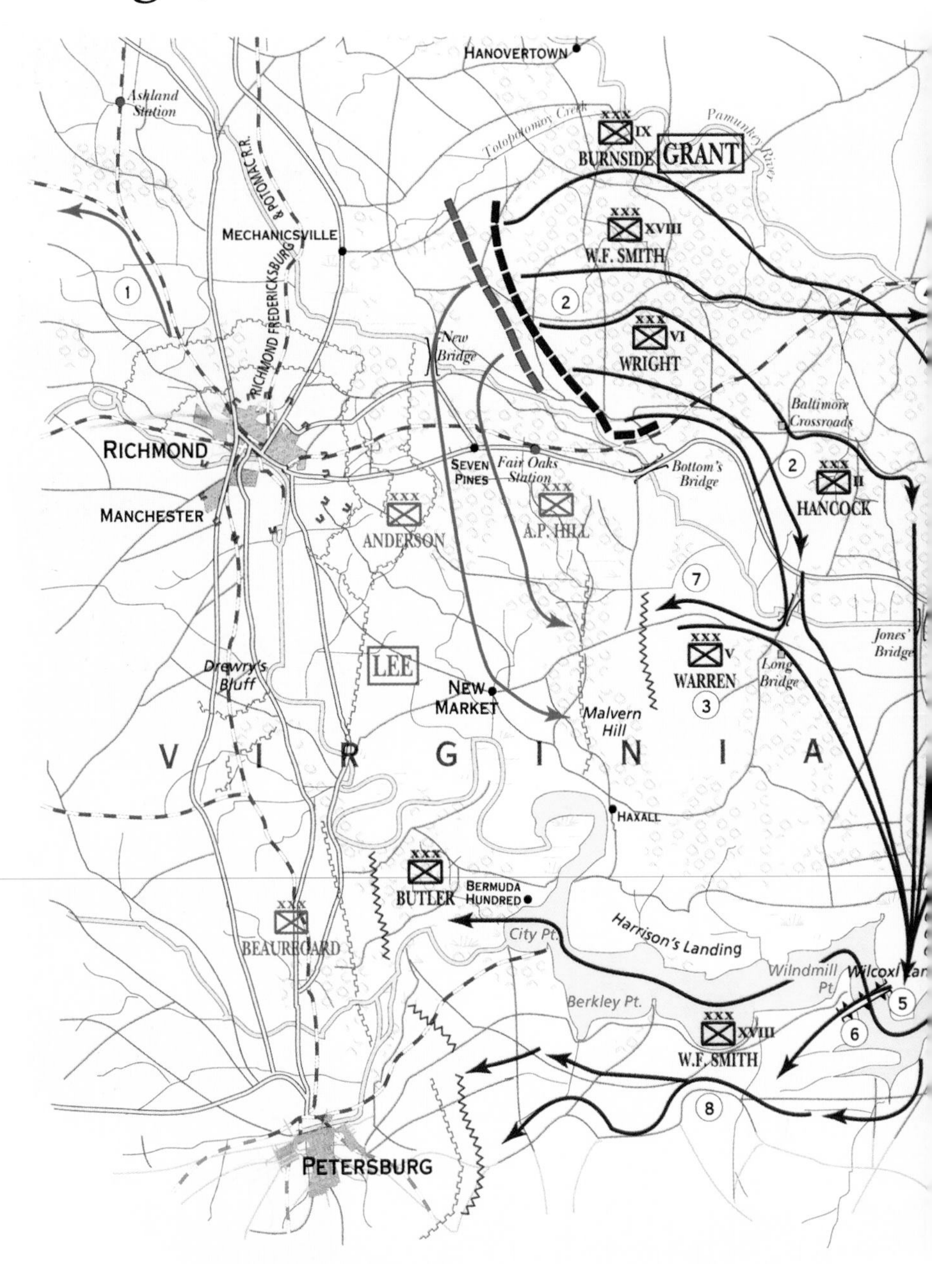
HANOVERTOWN
Ashland Station
RICHMOND FREDERICKSBURG & POTOMAC R.R.
IX BURNSIDE
GRANT
XVIII W.F. SMITH
MECHANICSVILLE
1
2
New Bridge
VI WRIGHT
Baltimore Crossroads
RICHMOND
SEVEN PINES
Fair Oaks Station
Bottom's Bridge
II HANCOCK
MANCHESTER
ANDERSON
A.P. HILL
7
Jones' Bridge
V WARREN
Long Bridge
LEE
Drewry's Bluff
NEW MARKET
Malvern Hill
3
V I R G I N I A
HAXALL
BUTLER
BERMUDA HUNDRED
BEAUREGARD
City Pt.
Harrison's Landing
Windmill Pt.
Berkley Pt.
5
6
XVIII W.F. SMITH
8
PETERSBURG

66
TIN
STOVE WARE-HOUSE
TIN WARE
WHOLESALE & RETAIL

"Grant must do with Lee what he did with Pemberton, he must outgeneral him and force him to fight him on his own ground. This all of us uninformed think he could accomplish by crossing the James and taking Richmond in the rear, and accordingly we are most eager that that should be done."

*Capt. Charles F. Adams, Jr., attached to the staff of Gen. Meade.*

*The Courthouse in Petersburg (left), photographed before the Union advance upon the city. At the time the James was crossed, Petersburg was practically defenseless and open to a Union assault.*

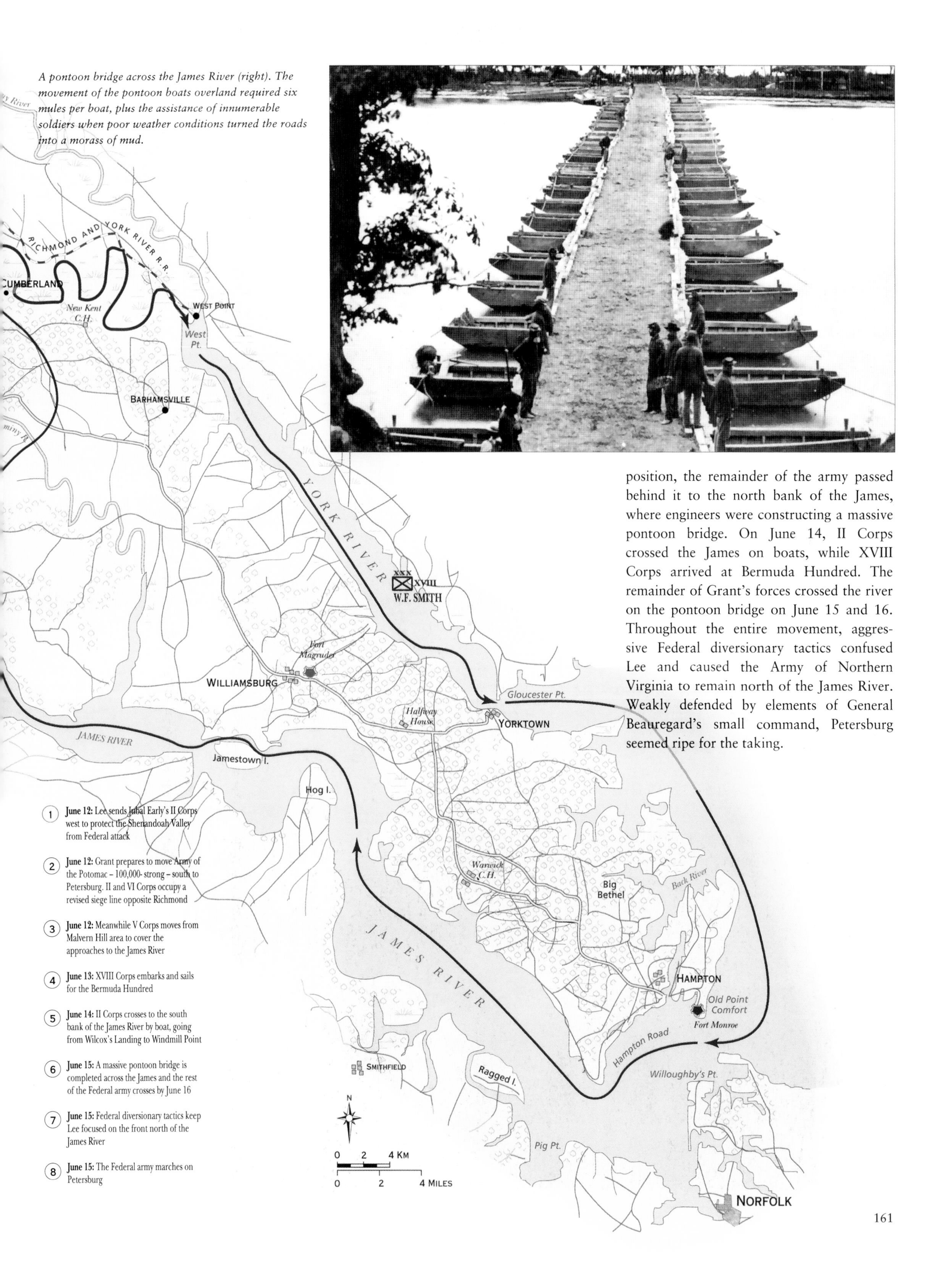
RICHMOND AND YORK RIVER R.R.
CUMBERLAND
New Kent C.H.
WEST POINT
West Pt.
BARHAMSVILLE
YORK RIVER
XXX
XVIII
W.F. SMITH
Fort Magruder
WILLIAMSBURG
Gloucester Pt.
Halfway House
YORKTOWN
JAMES RIVER
Jamestown I.
Hog I.
Warwick C.H.
Big Bethel
Back River
JAMES RIVER
HAMPTON
Old Point Comfort
Fort Monroe
Hampton Road
Willoughby's Pt.
SMITHFIELD
Ragged I.
N
Pig Pt.
0 2 4 KM
0 2 4 MILES
NORFOLK
1 June 12: Lee sends Jubal Early's II Corps west to protect the Shenandoah Valley from Federal attack
2 June 12: Grant prepares to move Army of the Potomac – 100,000- strong – south to Petersburg. II and VI Corps occupy a revised siege line opposite Richmond
3 June 12: Meanwhile V Corps moves from Malvern Hill area to cover the approaches to the James River
4 June 13: XVIII Corps embarks and sails for the Bermuda Hundred
5 June 14: II Corps crosses to the south bank of the James River by boat, going from Wilcox's Landing to Windmill Point
6 June 15: A massive pontoon bridge is completed across the James and the rest of the Federal army crosses by June 16
7 June 15: Federal diversionary tactics keep Lee focused on the front north of the James River
8 June 15: The Federal army marches on Petersburg

*A pontoon bridge across the James River (right). The movement of the pontoon boats overland required six mules per boat, plus the assistance of innumerable soldiers when poor weather conditions turned the roads into a morass of mud.*

position, the remainder of the army passed behind it to the north bank of the James, where engineers were constructing a massive pontoon bridge. On June 14, II Corps crossed the James on boats, while XVIII Corps arrived at Bermuda Hundred. The remainder of Grant's forces crossed the river on the pontoon bridge on June 15 and 16. Throughout the entire movement, aggressive Federal diversionary tactics confused Lee and caused the Army of Northern Virginia to remain north of the James River. Weakly defended by elements of General Beauregard's small command, Petersburg seemed ripe for the taking.

# Assaults at Petersburg JUNE 16–18 1864

The Federal plan for the capture of Petersburg called for General William Smith's XVIII Corps to advance from its Bermuda Hundred enclave, and assault the city's defenses from the east. Smith was to be supported by II Corps of the Army of the Potomac. Although only 2,200 of General Beauregard's 5,400 troops were in the Petersburg lines, Smith delayed his attack until late on June 15. The 16,000 Federals quickly broke through the Confederate fortifications, causing the defenders to withdraw behind Harrison's Creek. Another push would probably have secured Petersburg, but Smith simply replaced his tired troops with II Corps units, and consolidated his gains.

While the Federals rested, Beauregard abandoned the Bermuda Hundred lines, and rushed his remaining units to Petersburg. On the following day, these units lost additional ground to assaults by II, IX, and XVIII Corps, but again managed to keep the Federals out of Petersburg.

With the arrival of V Corps, the Federal attacks resumed on June 17, but vigorous Confederate counterattacks regained some of the lost ground before the day was over. That night, Beauregard withdrew slightly to a hastily constructed defensive line. Meanwhile, Lee had begun to shift part of the Army of Northern Virginia southward to Beauregard's aid. When the Federals resumed their advance on the morning of June 18, their attacks were disorganized and uncoordinated. By evening, as Lee's army poured into the Petersburg defenses, the Federal assault stalled. That night, Grant ordered an end to the frontal attacks on Petersburg, and briefly resumed trench warfare.

> "Our officers are, during this temporary quiet, freely indulging in those refined tastes which army life is so well calculated to develop, by engaging in such innocent amusements... as horse racing, gambling, and their usual accompaniments, commissary whisky, midnight revels and broken noses."
>
> *A Michigan volunteer during the siege of Petersburg.*

> "I shall hold as long as practicable, but without reinforcements I may have to evacuate the city very shortly."
>
> *Gen. Pierre G.T. Beauregard during the Union assaults of June, 1864.*

**FEDERAL ASSAULTS**
*June 16, 1864*

1. **June 15:** Smith's Federal XVIII Corps advances from its Bermuda Hundred base
2. **June 15:** XVIII Corps attacks Petersburg defenses; takes Battery 5
3. **June 15-16, night:** Beauregard's forces entrench a new line and are reinforced by two divisions from north of the James River
4. **June 16:** II, IX and XVIII Corps attack and seize part of Petersburg defenses but do not achieve a breakthrough
5. **June 16, midnight:** Warren's V Corps arrives

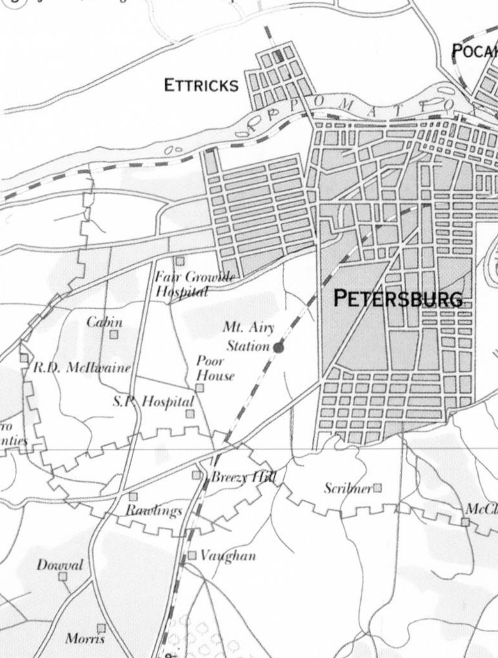
8
LEE
POCAHON
ETTRICKS
PETERSBURG
Fair Grounds Hospital
Cabin
Mt. Airy Station
R.D. McIlwaine
Poor House
S.P. Hospital
Negro Shanties
Breezy Hill
Scribner
McClean
Rawlings
Vaughan
Dowval
Kayton

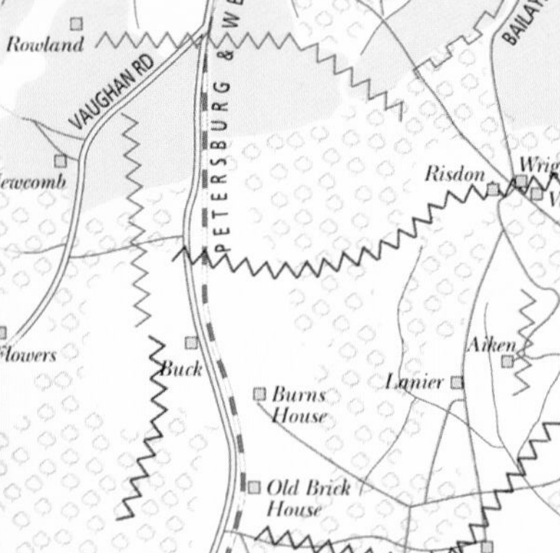
Morris
Johnston
Bailey
SQUIRREL LEVEL RD
PETERSBURG & WELDON R.R.
BAILAYS RD
VAUGHAN RD
Davis
Rowland
Newcomb
Risdon
Wright
Vaughan

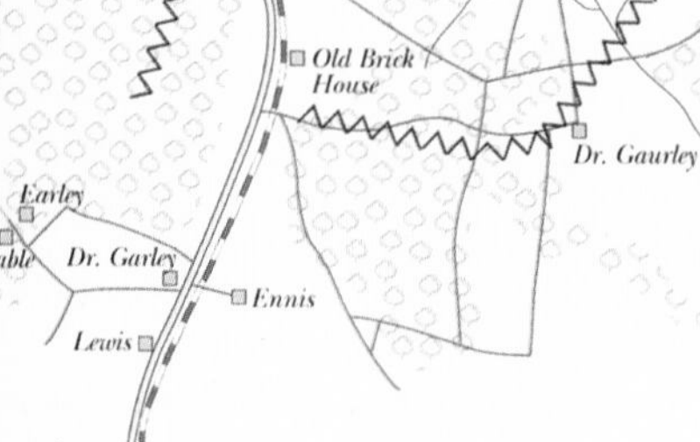
Flowers
Buck
Burns House
Lanier
Aiken
Old Brick House
Dr. Gaurley
Earley
Cable
Dr. Garley
Ennis
Lewis
Perkins
N

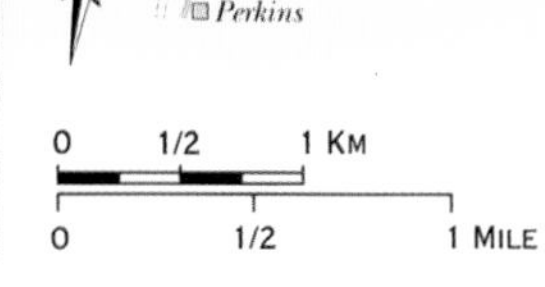
0
1/2
1 KM
0
1/2
1 MILE

*Gen. Orlando Wilcox and staff pass the time with a cockfight before the Petersburg defenses (below). Gambling in its various forms became a popular means by which to pass the time. Enduring long periods of inactivity, dissatisfaction and war weariness were part of the way of life among the soldiery on both sides.*

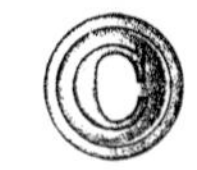

Chaplains of the IX Corps pose for the camera outside Petersburg (right). Both armies witnessed religious revivals during the conflict – most notably that of 1863 in which Lee himself participated. The inmates of the military hospitals were particularly receptive to the ministrations of the chaplains.

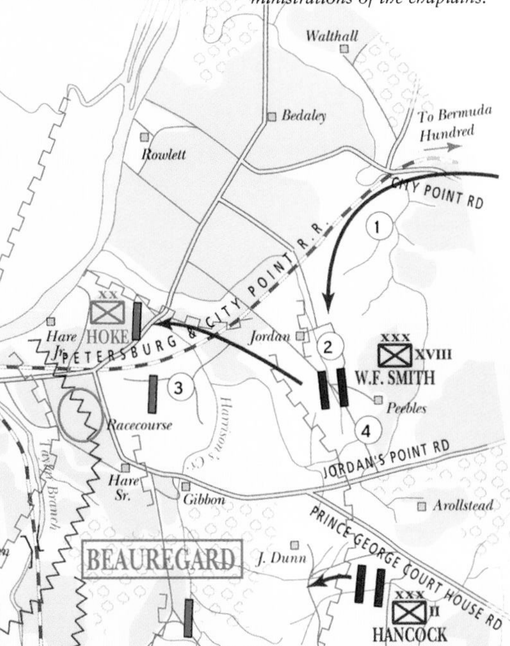

Walthall
Bedaley
To Bermuda Hundred
Rowlett
CITY POINT RD
PETERSBURG & CITY POINT R.R.
HOKE
Hare
Jordan
XVIII
W.F. SMITH
Peebles
Racecourse
JORDAN'S POINT RD
Hare Sr.
Gibbon
Arollstead
PRINCE GEORGE COURT HOUSE RD
BEAUREGARD
J. Dunn
HANCOCK (BIRNEY)
Webb
MEADE
Barn
JOHNSON
IX
BURNSIDE
Negro Qrs.
WISE
Barns
Bowden Sr.
At. Sally's
V
WARREN
NORFOLK & PETERSBURG R.R.
Daniel's Cabin
DEARING
KAUTZ
Bowden Jr.
Wells
JERUSALEM PLANK RD


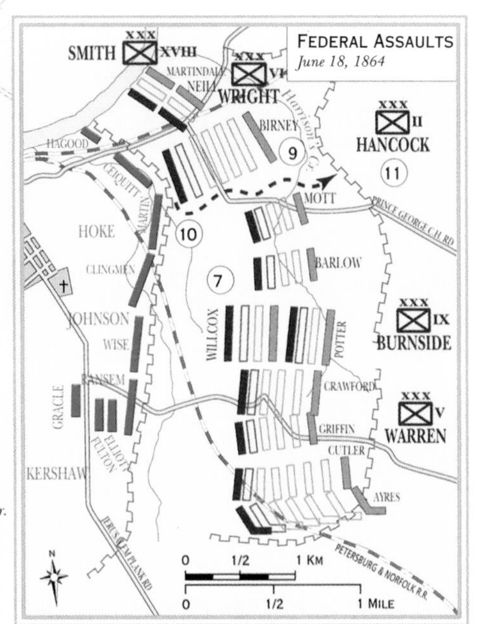

FEDERAL ASSAULTS
June 18, 1864
SMITH
XVIII
WRIGHT
HANCOCK
BIRNEY
MOTT
BARLOW
HOKE
CLINGMEN
JOHNSON
WISE
WILLCOX
POTTER
BURNSIDE
CRAWFORD
GRIFFIN
CUTLER
WARREN
AYRES
GRACIE
RANSEM
ELLIOTT
FULTON
KERSHAW
HAGOOD
0 1/2 1 KM
0 1/2 1 MILE

Officers and men of the 170th New York Infantry rest in front of the Petersburg defenses (above).

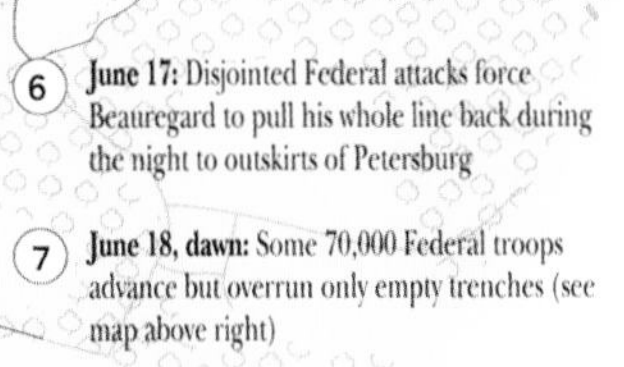

6. **June 17:** Disjointed Federal attacks force Beauregard to pull his whole line back during the night to outskirts of Petersburg
7. **June 18, dawn:** Some 70,000 Federal troops advance but overrun only empty trenches (see map above right)
8. **June 18, 7.30 – 11am:** Lee's divisions begin to arrive, and reinforce Petersburg's defenses
9. **June 18, noon:** Birney attacks Beauregard's new defense line, but is repulsed (see map above right)
10. **June 18, 4pm:** One of Birney's regiments attacks and suffers 632 casualties out of 850, (see map above right) the severest loss suffered in a single engagement by any Union regiment in the course of the war
11. **June 18, late pm:** The Federal assault is discontinued by Grant (see map above right)

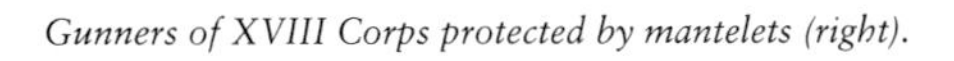

Gunners of XVIII Corps protected by mantelets (right).

# The Drewry's Bluff Campaign MAY 5–16 1864

As part of General Grant's master plan for the defeat of the Confederacy in 1864, General Butler was to lead a small Federal army up the James River toward the southern approaches of Richmond. If all went well, Butler would meet Grant near the Confederate capital within ten days of the opening of the campaign. Styled the Army of the James, Butler's force numbered approximately 40,000 troops. On May 5, convoyed by the navy, Butler sailed up the James and landed at Bermuda Hundred, less than 20 miles from Richmond. Although initially he encountered no significant opposition, Butler spent several days establishing a defensive enclave before attempting major offensive operations.

Because all Confederate units available to reinforce the Virginia theater were under orders to join the Army of Northern Virginia, the south side of the James was defended by only a few troops under General Beauregard. Caught off guard by the Federal landing, which was nearer Richmond than most of his own troops, Beauregard hastily began to concentrate a strike force from the Carolinas by rail. For several days, however, he could do nothing but view Butler's activities with alarm. Fortunately for the Confederates, Butler did little more than lightly damage the Richmond & Petersburg Railroad before he finally began an advance toward Richmond on May 12.

The week Butler spent at Bermuda Hundred permitted him to perfect a fortified base at Bermuda Hundred the City Point, but it also permitted Beauregard to gather an almost equivalent force at Petersburg. Marching northward across Butler's front on May 11, Beauregard joined troops from the Department of Richmond and barred Butler's route to the Confederate capital. By May 15, the Army of the James had gained an outer Confederate defense line, but was unable to advance further. Seizing the initiative, Beauregard on May 16 struck Butler's army a devastating blow in the battle of Drewry's Bluff, and drove it within its Bermuda Hundred entrenchments. This action alone cost 3,004 Federal casualties and 2,966 Confederate. Although Butler and Beauregard retained some freedom of movement, the approach of the main armies soon siphoned troops from the Bermuda Hundred front, which subsided into stalemate for the remainder of the war.

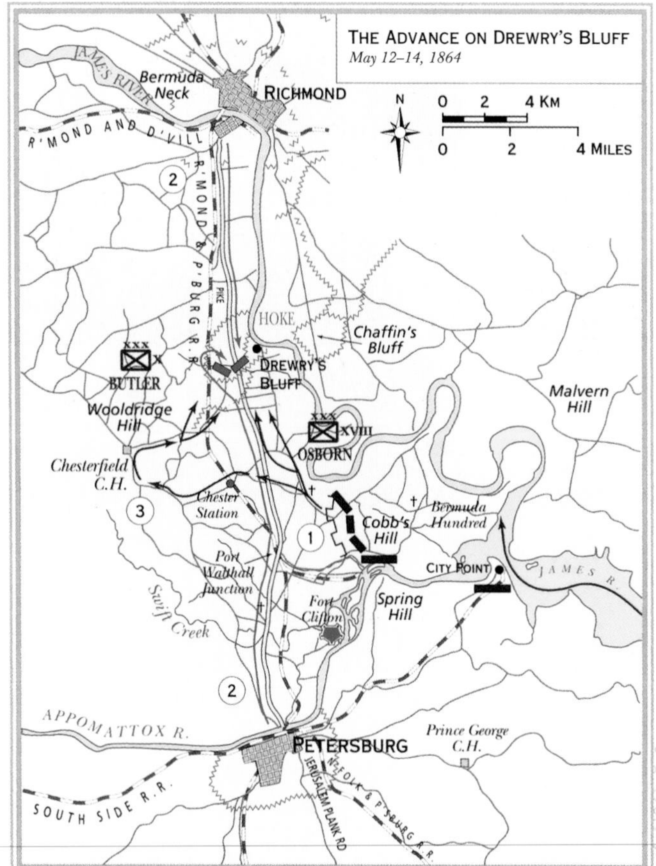
THE ADVANCE ON DREWRY'S BLUFF
May 12–14, 1864
0 2 4 KM
0 2 4 MILES
N
RICHMOND
Bermuda Neck
JAMES RIVER
R'MOND AND D'VILL
R'MOND & P'BURG R.R.
PIKE
HOKE
Chaffin's Bluff
DREWRY'S BLUFF
BUTLER
Wooldridge Hill
XVIII
OSBORN
Malvern Hill
Chesterfield C.H.
Chester Station
Cobb's Hill
Bermuda Hundred
CITY POINT
JAMES R.
Port Walthall Junction
Swift Creek
Fort Clifton
Spring Hill
APPOMATTOX R.
PETERSBURG
Prince George C.H.
SOUTH SIDE R.R.
JERUSALEM PLANK RD
NORFOLK & P'BURG R.R.
1
2
3

*Gen. Butler's Army of the James, positioned on the Peninsula, was to take part in Gen. Grant's master plan to crush the Confederate armies and end the war before November. On May 5, after steaming up the James River, Butler's forces landed midway between Petersburg and Richmond. Instead of moving quickly to cut the railroad between the two cities, and entering Richmond against the opposition, Butler first dug defensive lines then, after a week's delay, began a cautious advance on the Confederate capital. By then, however, Gen. Beauregard's meagre forces had been reinforced, and he was able to meet the Federals at Drewry's Bluff on almost equal terms (see map above).*

*Although he had supported the State's Rights candidate against Lincoln in 1860, Benjamin Butler (below) quickly raised a volunteer regiment for the Union after Fort Sumter fell. An astute politician and lawyer, Butler was a failure in the military and eventually resigned his commission in November 1865.*

1. The Army of the James digs defensive lines (see map left)
2. **May 11:** Having collected reinforcements Beauregard marches north, meeting troops heading south from Richmond
3. **May 12:** Butler advances toward Richmond (see map left)
4. **May 16:** Confederates advance out of early morning mist and clash with Union right
5. Hoke advances against Gilmore's Union divisions. Heavy losses incurred by both sides in confused engagements exacerbated by fog
6. Beauregard orders Colquitt forward into gap between Ransom and Hoke's Confederate divisions
7. After pressing forward most of the day suffering 2500 casualties and not receiving cavalry assistance expected from the south west, Beauregard orders army into night positions
8. Butler withdraws during the night to Bermuda Hundred

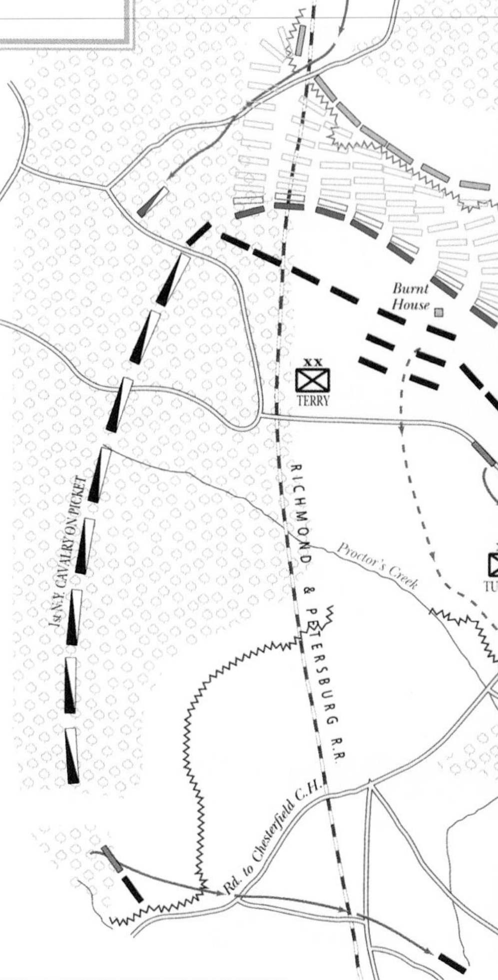
3rd N.C. CAVALRY
Burnt House
TERRY
RICHMOND & PETERSBURG R.R.
Proctor's Creek
1st N.Y. CAVALRY ON PICKET
Rd. to Chesterfield C.H.

BATTLE OF DREWRY'S BLUFF
*May 16, 1864*

*Constructed upon Butler's orders , the "Crows Nest" observation platform looms over an incomplete Union mortar battery (above). The tower served no useful function and pinpointed the location of the batteries at its base, thereby assisting the Confederate gunners.*

*Among other needless activities, Butler ordered that a number of pontoon bridges should be built across the Appomattox River (above). Frustrated by his subordinate's incompetency, Grant complained that the Army of the James might as well be "in a bottle strongly corked".*

"The Hawlett House [Confederate] battery has fired occasionally at our Crow's Nest observation tower, but never seriously damaged it. Our big battery has replied now and then, and when standing behind our guns we could see the shot as they arched to the other side... The enemy tried to destroy our signal tower and we tried to destroy their guns, but nothing much resulted, except a big noise."

*An artilleryman in the Army of the James.*

"We could and should have done more. We could and should have captured Butler's entire army."

*Gen. P.G.T. Beauregard on the failure of the Confederate army to drive Butler's army from the Bermuda Hundred after the battle of Drewry's Bluff.*

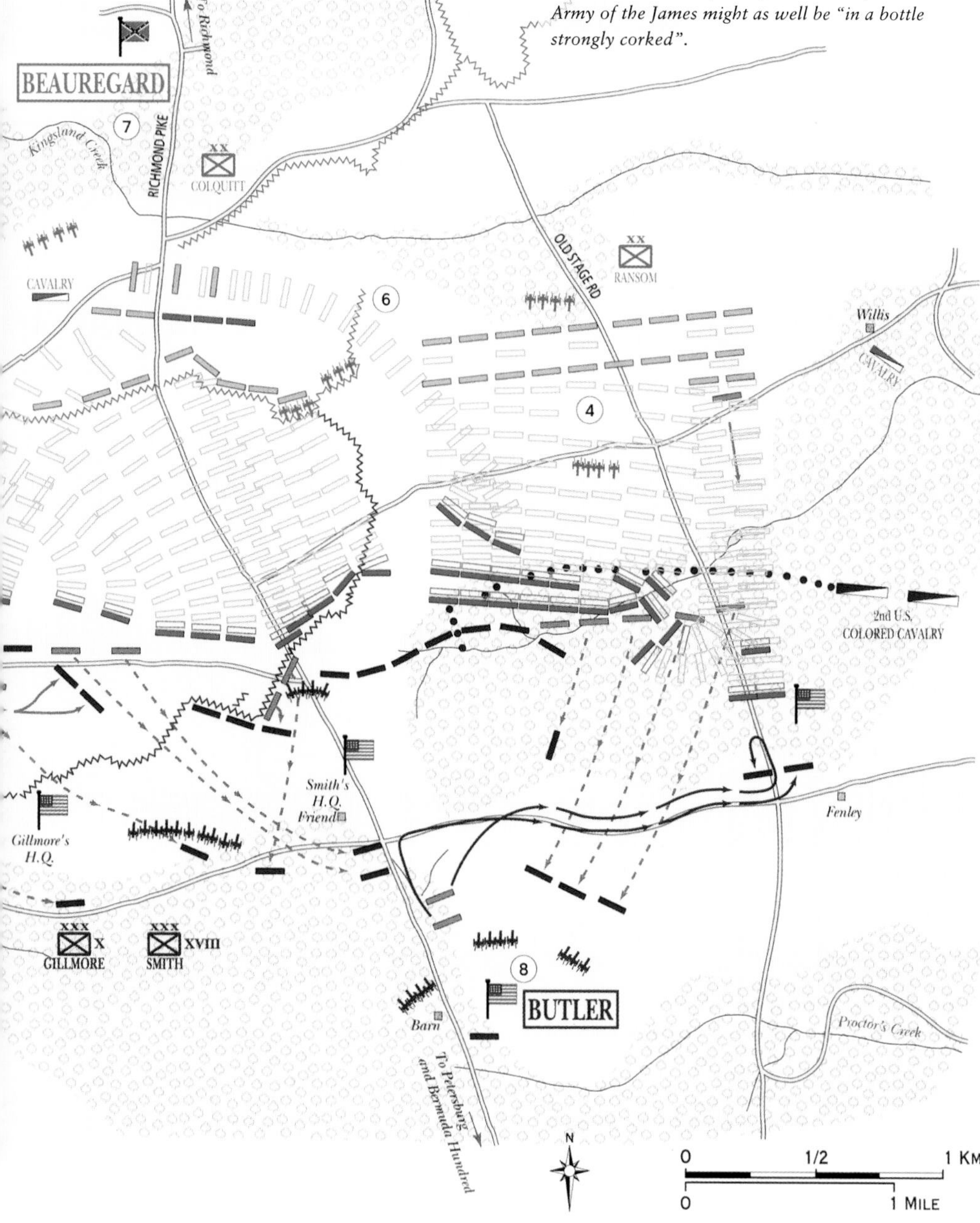

To Richmond
BEAUREGARD
7
Kingsland Creek
RICHMOND PIKE
XX
COLQUITT
CAVALRY
6
OLD STAGE RD
XX
RANSOM
Willis
CAVALRY
4
2nd U.S. COLORED CAVALRY
Smith's H.Q. Friend
Fenley
Gillmore's H.Q.
XXX
X
GILLMORE
XXX
XVIII
SMITH
8
BUTLER
Barn
Proctor's Creek
To Petersburg and Bermuda Hundred
N
0
1/2
1 KM
0
1 MILE


*Butler's headquarters on the James River (below). Though keeping up the pretence of ingenious activity, in failing to advance immediately upon Richmond, Butler had lost his last opportunity to win any laurels during the Civil War.*

# Sheridan's Raids MAY 9 –24 and JUNE 7 –28 1864

As part of his efforts to energize the Army of the Potomac, in the spring of 1864 General Grant appointed General Philip Sheridan to command the army's cavalry. Small in stature and abrasive in personality, Sheridan aggressively sought a larger role for the Federal Cavalry. After repeated clashes between Sheridan and the army commander, General Meade, Grant authorized Sheridan to operate independently to defeat J.E.B. Stuart's Confederate cavalry.

On May 9, Sheridan began the raid with 10,000 troops, while Stuart followed with only 4,500 men. Splitting his command, Sheridan held off Stuart's pursuit with some units while sending others to strike the Virginia Central Railroad at Beaver Dam Station. After damaging the Railroad, Sheridan's reunited force continued southward toward Richmond. Leaving subordinates to follow Sheridan's trail, Stuart led the bulk of his command on a forced march to get between the Federals and the Confederate capital. On May 11, Stuart confronted Sheridan at Yellow Tavern: the outnumbered Confederates were routed, and Stuart himself mortally wounded. Too weak to force his way into Richmond, Sheridan then skirted the city's defenses and reached Federal lines at Haxall's Landing on May 14. Unwilling to assist the Army of the James in its advance, Sheridan soon departed for the Army of the Potomac, reaching it on May 24. Except for the death of Stuart, the raid had accomplished little.

Two weeks later, Grant again sent Sheridan's cavalry to raid the Virginia Central Railroad. This time the raid's primary purpose was to distract Confederate attention from Grant's pending crossing of the James River. Departing with 6,000 horsemen on June 7, Sheridan looped northward around Richmond and four days later struck the railroad far to the west in the vicinity of Trevilian Station. Before the Federals could damage the track significantly, they were attacked by Major General Wade Hampton and 5,000 Confederate cavalrymen. In a confused action around the station that lasted for two days, the Federals inflicted nearly 1,000 casualties upon Hampton's command, but lost an equivalent number of their own. Unable to accomplish more destruction or force Hampton from the field, Sheridan then withdrew, eventually rejoining the Army of the Potomac at Petersburg on June 28.

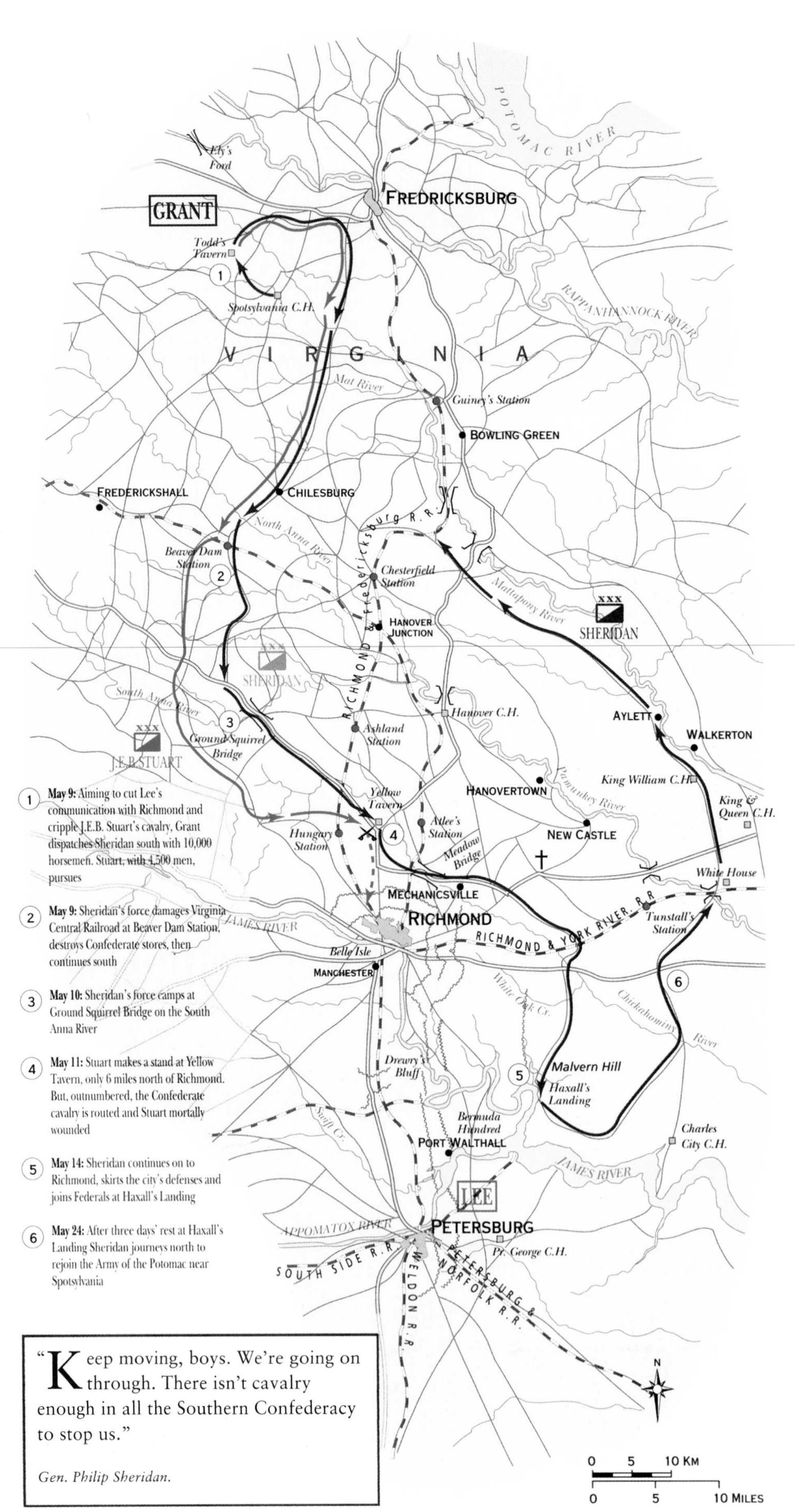
POTOMAC RIVER
Ely's Ford
GRANT
FREDRICKSBURG
Todd's Tavern
Spotsylvania C.H.
RAPPAHANNOCK RIVER
VIRGINIA
Mat River
Guiney's Station
BOWLING GREEN
FREDERICKSHALL
CHILESBURG
North Anna River
Beaver Dam Station
Chesterfield Station
Mattapony River
SHERIDAN
HANOVER JUNCTION
RICHMOND & Fredericksburg R.R.
SHERIDAN
South Anna River
Hanover C.H.
AYLETT
WALKERTON
J.E.B. STUART
Ground Squirrel Bridge
Ashland Station
King William C.H.
HANOVERTOWN
Pamunkey River
King & Queen C.H.
Yellow Tavern
Hungary Station
Atlee's Station
Meadow Bridge
NEW CASTLE
White House
MECHANICSVILLE
RICHMOND
JAMES RIVER
Tunstall's Station
RICHMOND & YORK RIVER R.R.
Belle Isle
MANCHESTER
White Oak Cr.
Chickahominy River
Drewry's Bluff
Malvern Hill
Haxall's Landing
Bermuda Hundred
Swift Cr.
PORT WALTHALL
Charles City C.H.
JAMES RIVER
LEE
APPOMATOX RIVER
PETERSBURG
Pr. George C.H.
SOUTH SIDE R.R.
WELDON R.R.
PETERSBURG & NORFOLK R.R.
1 May 9: Aiming to cut Lee's communication with Richmond and cripple J.E.B. Stuart's cavalry, Grant dispatches Sheridan south with 10,000 horsemen. Stuart, with 4,500 men, pursues
2 May 9: Sheridan's force damages Virginia Central Railroad at Beaver Dam Station, destroys Confederate stores, then continues south
3 May 10: Sheridan's force camps at Ground Squirrel Bridge on the South Anna River
4 May 11: Stuart makes a stand at Yellow Tavern, only 6 miles north of Richmond. But, outnumbered, the Confederate cavalry is routed and Stuart mortally wounded
5 May 14: Sheridan continues on to Richmond, skirts the city's defenses and joins Federals at Haxall's Landing
6 May 24: After three days' rest at Haxall's Landing Sheridan journeys north to rejoin the Army of the Potomac near Spotsylvania
N
0 5 10 KM
0 5 10 MILES

"Keep moving, boys. We're going on through. There isn't cavalry enough in all the Southern Confederacy to stop us."

*Gen. Philip Sheridan.*

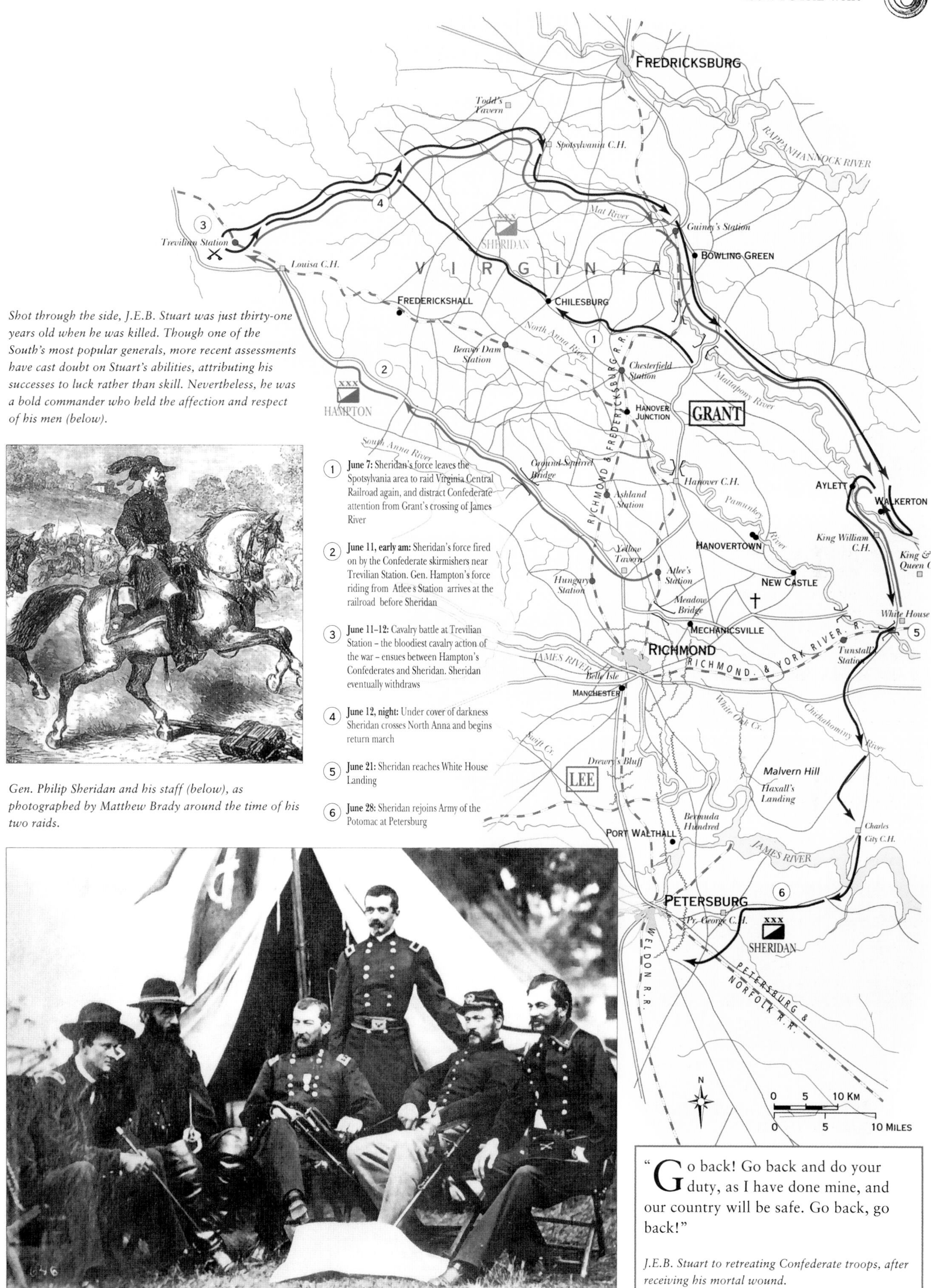
FREDRICKSBURG
Todd's Tavern
Spotsylvania C.H.
RAPPAHANNOCK RIVER
Trevilian Station
Louisa C.H.
SHERIDAN
VIRGINIA
Mat River
Guiney's Station
BOWLING GREEN
FREDERICKSHALL
CHILESBURG
North Anna River
Beaver Dam Station
Chesterfield Station
RICHMOND & FREDERICKSBURG R.R.
HAMPTON
HANOVER JUNCTION
GRANT
Mattapony River
South Anna River
Ground Squirrel Bridge
Ashland Station
Hanover C.H.
Pamunkey River
AYLETT
WALKERTON
King William C.H.
King & Queen C.H.
HANOVERTOWN
Yellow Tavern
Atlee's Station
NEW CASTLE
Hungary Station
Meadow Bridge
MECHANICSVILLE
White House
RICHMOND
JAMES RIVER
Belle Isle
RICHMOND & YORK RIVER R.R.
Tunstall's Station
MANCHESTER
White Oak Cr.
Chickahominy River
Swift Cr.
Drewry's Bluff
LEE
Malvern Hill
Haxall's Landing
Bermuda Hundred
Charles City C.H.
PORT WALTHALL
JAMES RIVER
PETERSBURG
Pr. George C.H.
SHERIDAN
WELDON R.R.
PETERSBURG & NORFOLK R.R.
N
0 5 10 KM
0 5 10 MILES
1 June 7: Sheridan's force leaves the Spotsylvania area to raid Virginia Central Railroad again, and distract Confederate attention from Grant's crossing of James River
2 June 11, early am: Sheridan's force fired on by the Confederate skirmishers near Trevilian Station. Gen. Hampton's force riding from Atlee s Station arrives at the railroad before Sheridan
3 June 11–12: Cavalry battle at Trevilian Station – the bloodiest cavalry action of the war – ensues between Hampton's Confederates and Sheridan. Sheridan eventually withdraws
4 June 12, night: Under cover of darkness Sheridan crosses North Anna and begins return march
5 June 21: Sheridan reaches White House Landing
6 June 28: Sheridan rejoins Army of the Potomac at Petersburg

*Shot through the side, J.E.B. Stuart was just thirty-one years old when he was killed. Though one of the South's most popular generals, more recent assessments have cast doubt on Stuart's abilities, attributing his successes to luck rather than skill. Nevertheless, he was a bold commander who held the affection and respect of his men (below).*

*Gen. Philip Sheridan and his staff (below), as photographed by Matthew Brady around the time of his two raids.*

"Go back! Go back and do your duty, as I have done mine, and our country will be safe. Go back, go back!"

*J.E.B. Stuart to retreating Confederate troops, after receiving his mortal wound.*

# Operations in the Shenandoah Valley MAY – JUNE 1864

The Shenandoah Valley was a thorn in General Grant's side: the Confederates had used the Valley repeatedly as an invasion route into Maryland and Pennsylvania, and the fertile region had also served as the breadbasket of the Confederacy. Grant wanted Federal troops in the Valley to work in conjunction with Meade's Army of the Potomac.

General Franz Sigel led 8,900 Federals up the Valley until Major General John C. Breckinridge's Confederates challenged their advance at New Market on May 15, 1864. Breckinridge attacked with 5,300 troops, including 247 cadets from the Virginia Military Institute. The Confederates battered the Federals, compelling them to retreat and Breckinridge took advantage of the lull to reinforce Lee's army at the North Anna River.

Grant replaced Sigel with Major General David Hunter, who pushed up the Valley with 12,000 soldiers. The Confederates attempted to halt the Federal drive with a small force of 5,600 men under Brigadier General William E. "Grumble" Jones. Hunter savagely attacked the Confederate barricades at the town of Piedmont on June 5. When Jones was killed, the Confederate defense disintegrated. The Confederacy lost 1,600 men; the Union 875. Without meeting any resistance, Hunter pressed on to Lexington, where he set alight the Virginia Military Institute, before starting for Lynchburg.

Breckinridge's command hurried back to Lynchburg, and General Lee sent General Early's Confederate corps after it. Hunter tested the city's defenses on June 17 and 18.

*Maj. Gen. David Hunter (above), who was given the task of subjugating the Shenandoah Valley. Hunter's despoliation of the valley's riches was ferocious, and the general tended to express pride in his achievements rather than regret them as military necessities.*

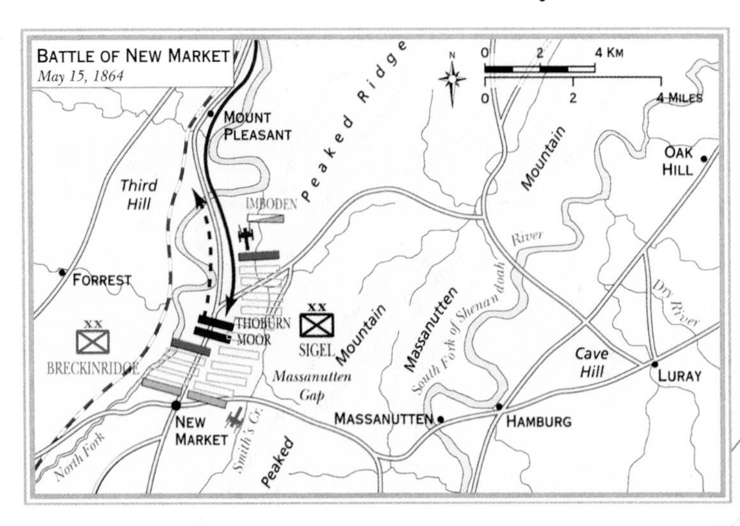
BATTLE OF NEW MARKET
May 15, 1864
0 2 4 KM
0 2 4 MILES
MOUNT PLEASANT
Third Hill
IMBODEN
Peaked Ridge
Mountain
OAK HILL
River
FORREST
Dry River
THOBURN
MOOR
SIGEL
BRECKINRIDGE
Mountain
Massanutten
South Fork of Shenandoah
Massanutten Gap
Cave Hill
LURAY
NEW MARKET
Smith's Cr.
North Fork
Peaked
MASSANUTTEN
HAMBURG

*Threatened by Sigel's advance up the Shenandoah Valley, Breckinridge attacked Sigel north of New Market. By late morning Sigel was falling back, retreating to Strasburg. The Confederates had eased the pressure on their forces in the valley and inflicted a sharp and humiliating defeat on the Federals (above).*

Finding himself outnumbered by Early's 14,000 troops, Hunter decided to retreat. The Federal force fell back to the Kanawha Valley, leaving the Shenandoah Valley clear for Early to advance northward. Early marched down the Valley, intending to menace the Washington defenses and force Grant to transfer Federal forces from the Richmond-Petersburg front. The veteran Confederates of Stonewall Jackson's command, now under Early, hoped their return to the Valley would be attended with the spectacular success that had blessed their famous 1862 campaign.

*In an attempt to stop Gen. Hunter's destructive raids on civilian property in the Shenandoah Valley, Gen. Jones met Hunter's large Federal force at Piedmont, (see map below). In a series of charges and countercharges, Hunter's larger force eventually overran the Confederates. Gen. Jones was killed, and over a thousand of his men captured.*

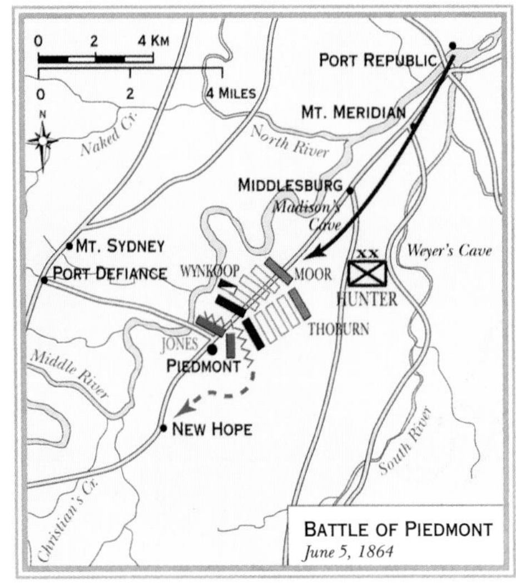
0 2 4 KM
0 2 4 MILES
PORT REPUBLIC
MT. MERIDIAN
Naked Cr.
North River
MIDDLESBURG
Madison's Cave
Weyer's Cave
MT. SYDNEY
PORT DEFIANCE
WYNKOOP
MOOR
HUNTER
THOBURN
JONES
PIEDMONT
Middle River
NEW HOPE
South River
Christian's Cr.
BATTLE OF PIEDMONT
June 5, 1864

> "On the 12th, I... burned the Virginia Military Institute... I found a violent... proclamation from John Hatcher, lately Governor of Virginia, inciting... a guerrilla warfare on my troops, and... I ordered his property to be burned."
>
> *Gen. Hunter's report to Washington in May.*

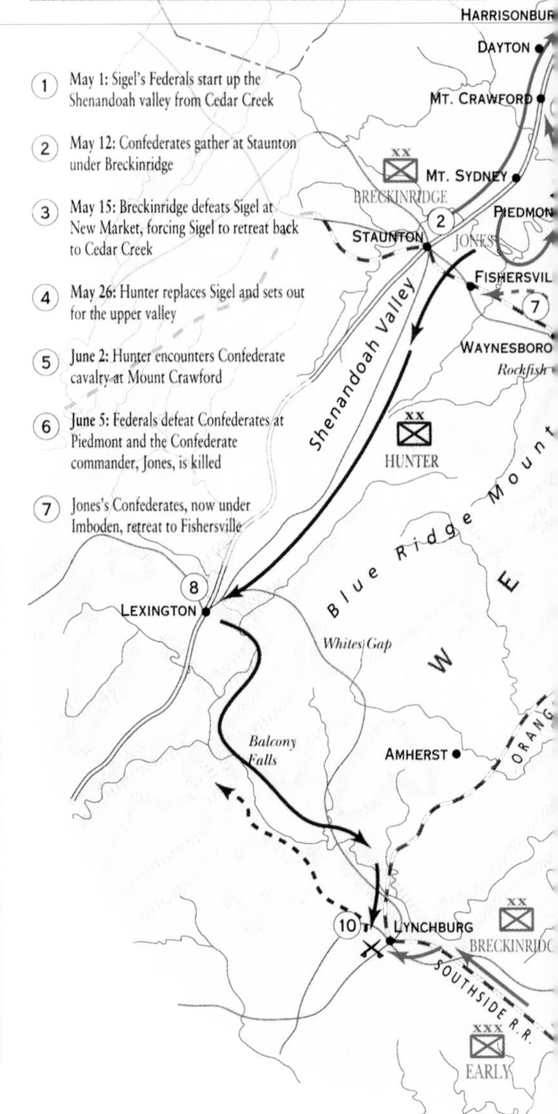
MOOREFIELD
DAYTON
MT. CRAWFORD
1 May 1: Sigel's Federals start up the Shenandoah valley from Cedar Creek
2 May 12: Confederates gather at Staunton under Breckinridge
3 May 15: Breckinridge defeats Sigel at New Market, forcing Sigel to retreat back to Cedar Creek
4 May 26: Hunter replaces Sigel and sets out for the upper valley
5 June 2: Hunter encounters Confederate cavalry at Mount Crawford
6 June 5: Federals defeat Confederates at Piedmont and the Confederate commander, Jones, is killed
7 Jones's Confederates, now under Imboden, retreat to Fishersville
MT. SYDNEY
BRECKINRIDGE
PIEDMONT
STAUNTON
JONES
WAYNESBORO
Rockfish
Shenandoah Valley
HUNTER
Blue Ridge
LEXINGTON
Whites Gap
Balcony Falls
AMHERST
LYNCHBURG
BRECKINRIDGE
SOUTHSIDE R.R.
EARLY

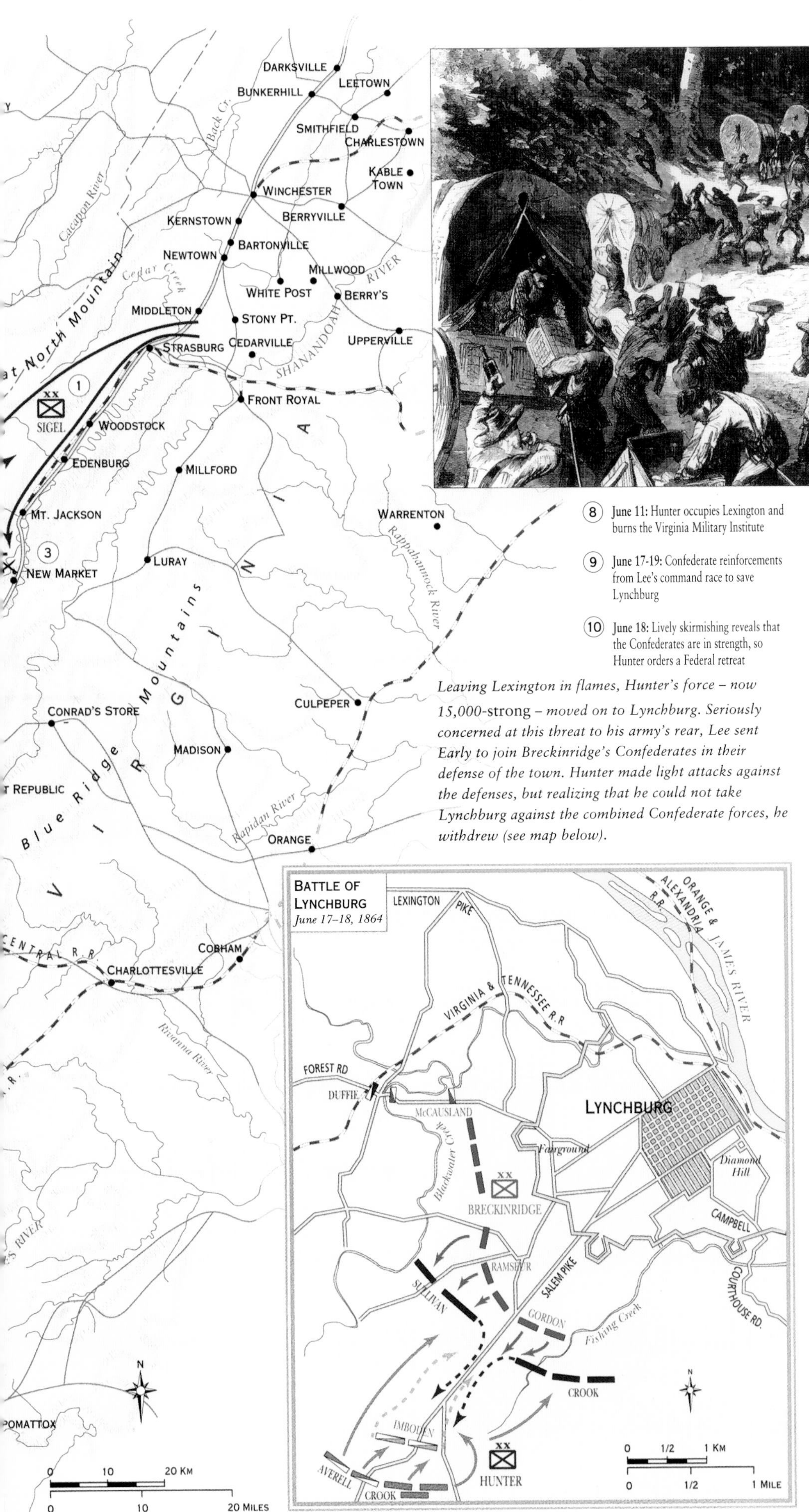
DARKSVILLE
LEETOWN
BUNKERHILL
SMITHFIELD
CHARLESTOWN
KABLE TOWN
WINCHESTER
KERNSTOWN
BERRYVILLE
NEWTOWN
BARTONVILLE
MILLWOOD
WHITE POST
BERRY'S
MIDDLETON
STONY PT.
STRASBURG
CEDARVILLE
UPPERVILLE
FRONT ROYAL
SIGEL
WOODSTOCK
EDENBURG
MILLFORD
MT. JACKSON
NEW MARKET
LURAY
WARRENTON
CULPEPER
CONRAD'S STORE
MADISON
ORANGE
COBHAM
CHARLOTTESVILLE
Back Cr.
Cacapon River
Cedar Creek
SHENANDOAH RIVER
Rappahannock River
Rapidan River
Rivanna River
Blue Ridge Mountains
VIRGINIA
0 10 20 KM
0 10 20 MILES

(8) **June 11:** Hunter occupies Lexington and burns the Virginia Military Institute

(9) **June 17-19:** Confederate reinforcements from Lee's command race to save Lynchburg

(10) **June 18:** Lively skirmishing reveals that the Confederates are in strength, so Hunter orders a Federal retreat

*Leaving Lexington in flames, Hunter's force – now 15,000-strong – moved on to Lynchburg. Seriously concerned at this threat to his army's rear, Lee sent Early to join Breckinridge's Confederates in their defense of the town. Hunter made light attacks against the defenses, but realizing that he could not take Lynchburg against the combined Confederate forces, he withdrew (see map below).*

BATTLE OF LYNCHBURG
June 17–18, 1864
LEXINGTON PIKE
ORANGE & ALEXANDRIA R.R.
JAMES RIVER
VIRGINIA & TENNESSEE R.R.
FOREST RD
DUFFIE
McCAUSLAND
LYNCHBURG
Fairground
Diamond Hill
BRECKINRIDGE
CAMPBELL COURTHOUSE RD.
RAMSEUR
SULLIVAN
SALEM PIKE
GORDON
Fishing Creek
CROOK
IMBODEN
AVERELL
CROOK
HUNTER
0 1/2 1 KM
0 1/2 1 MILE

*Col. John S. Mosby's partisans capture and sack a sutler's wagon train (above).*

*Col. John Singleton Mosby (above), commander of the 43rd Battalion of the Virginia Cavalry. A brilliant guerrilla leader, Mosby continually harassed Hunter by disrupting his supply lines and communications, and by capturing unwary stragglers. So great was Mosby's control of Northern Virginia that the area became known as 'Mosby's Confederacy'.*

"His exploits are not surpassed in daring and enterprise by those of ... any age. Unswerving devotion to duty, self-abnegation, and unflinching courage... are the characteristics of this officer...The Gallant band of Captain Mosby shares his glory, as they did the danger."

*Gen. J.E.B. Stuart reporting to Lee on the actions of Col. J.S. Mosby.*

# Early's Washington Raid JUNE 27 – AUGUST 4 1864

General Jubal Early set out to make a nuisance of himself by harassing the Washington defenses. After General Hunter's Federals retreated from Lynchburg, Early's Army of the Valley raced northward down the Shenandoah Valley. Crossing the Potomac River into Maryland, Early headed for Washington. Federal General Lew Wallace attempted to delay Early's advance near Frederick on July 9, 1864. In a costly, time consuming battle along Monocacy Creek, the Confederates forced Wallace to fall back into the Washington defenses. Early's force arrived outside Washington on July 11, causing a near panic in the capital. Militia units and convalescents manned the fortifications, and Lincoln appealed to General Grant for reinforcements. With the veteran Federal VI and XIX Corps marching into the ring of Washington forts, Early sparred lightly with Fort Stevens, and then fell back into Virginia.

The Federals linked up with Hunter's small command, now under Major General George Crook, and pursued Early to Snickers Gap. Striking Early at Cool Springs on July 18, the Federals watched the Confederates retreat southward, and assumed the worst was over. Crook's force remained to guard the Valley, while the rest of the Federal troops started back to the Richmond-Petersburg front.

Early forced them to return when he attacked Crook on July 23, defeating him at the battle of Second Kernstown. Before the Federal army could return to the Valley, Early kept his Confederate army occupied: quickly following the Kernstown victory, Early advanced to Martinsburg where he destroyed the railroad. At the same time, he dispatched his cavalry on a mission of revenge, sending them into Pennsylvania on a raid to Chambersburg. Arriving on July 30, the Confederates demanded the town pay an indemnity for property destroyed by Hunter's previous campaign in the Valley. When the citizens of Chambersburg failed to pay, the Confederates burned the town.

Grant had had enough of Early's mischief: during August 5, the Federal commander ordered General Philip Sheridan to rid the Valley of Early forever.

*Although he had voted against secession in 1861, Jubal Anderson Early accepted a colonelcy in Virginia's army when his home state did leave the Union. By the time of his Washington raid Early had reached the rank of lieutenant general (above).*

"Quick to decide and almost inflexible in decision, with a boldness to attack that approached rashness and a tenacity in resisting that resembled desperation, [Early] was yet on the field of battle not equal to his own intellect or decision."

*A Confederate officer serving on Early's staff.*

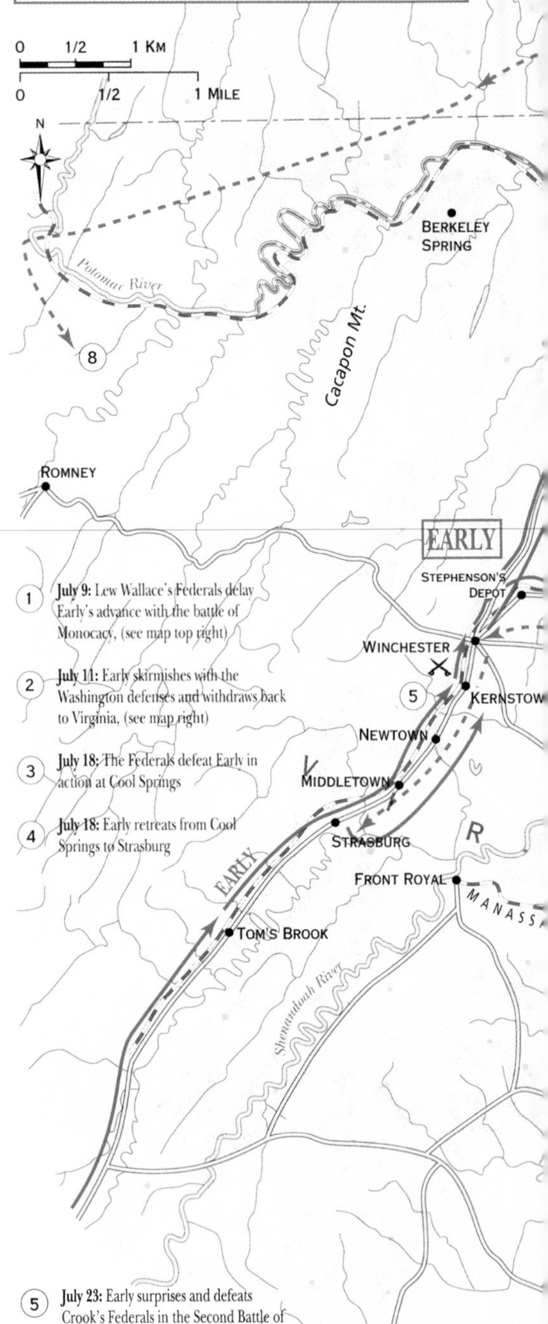
0 1/2 1 KM
0 1/2 1 MILE
N
BERKELEY SPRING
Potomac River
Cacapon Mt.
ROMNEY
EARLY
STEPHENSON'S DEPOT
WINCHESTER
KERNSTOW
NEWTOWN
MIDDLETOWN
STRASBURG
FRONT ROYAL
TOM'S BROOK
Shenandoah River
EARLY
MANASS
1 July 9: Lew Wallace's Federals delay Early's advance with the battle of Monocacy, (see map top right)
2 July 11: Early skirmishes with the Washington defenses and withdraws back to Virginia, (see map right)
3 July 18: The Federals defeat Early in action at Cool Springs
4 July 18: Early retreats from Cool Springs to Strasburg
5 July 23: Early surprises and defeats Crook's Federals in the Second Battle of Kernstown
6 July 28: Early's Confederates destroy the railroad at Martinsburg
7 July 30: Confederate cavalry burns Chambersburg.
8 July 31 – Aug 3: Confederate cavalry withdraw southwestward and escape back to Virginia

*Battery Rodgers on the Potomac River. By 1864, the Washington defenses had been extended to such a degree that their circuit was thirty-three miles. One observer commented that, in comparison, the Richmond defenses were " the merest castle-building of children".*

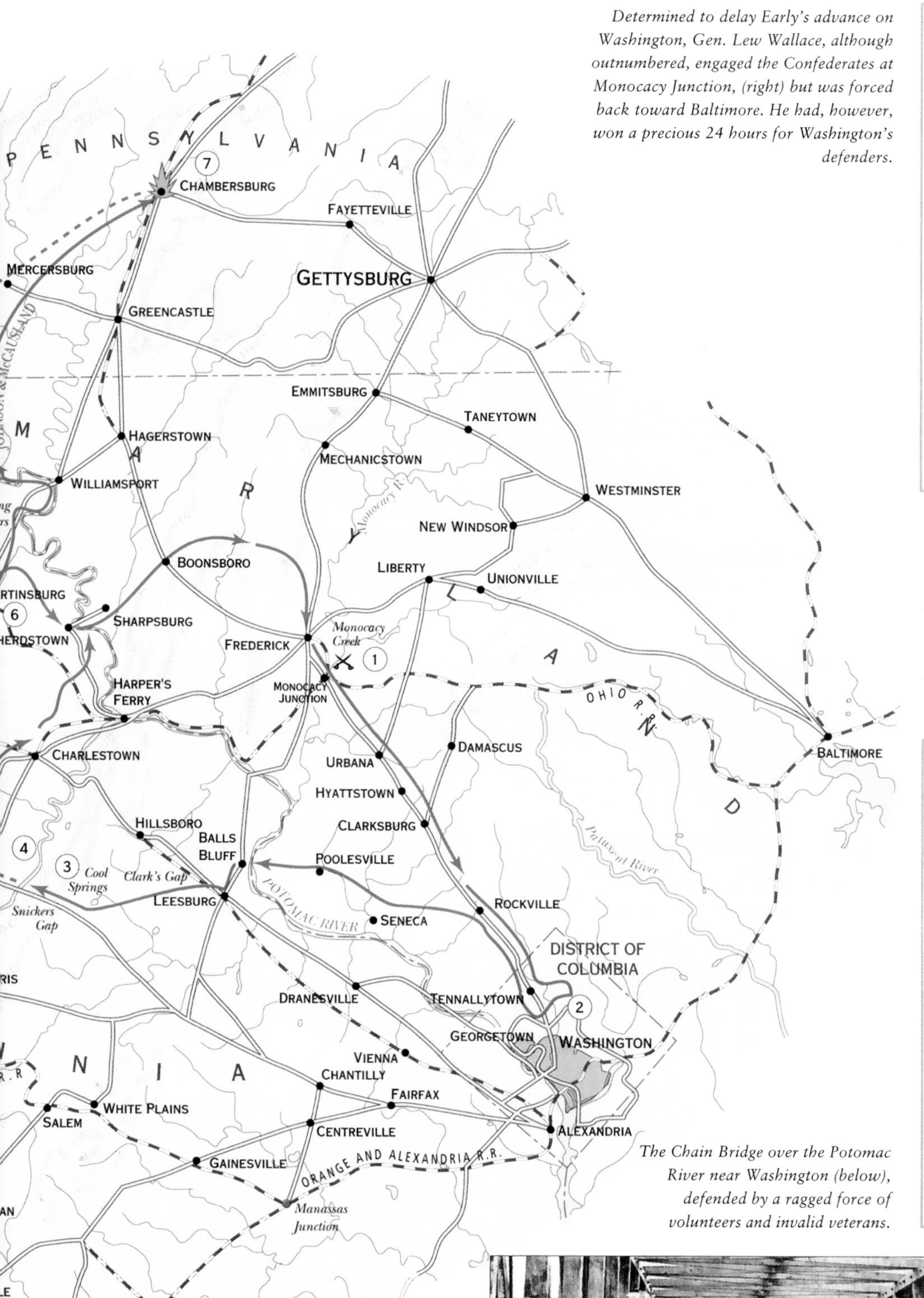
PENNSYLVANIA
CHAMBERSBURG
FAYETTEVILLE
MERCERSBURG
GETTYSBURG
GREENCASTLE
EMMITSBURG
TANEYTOWN
HAGERSTOWN
MECHANICSTOWN
WILLIAMSPORT
WESTMINSTER
NEW WINDSOR
BOONSBORO
LIBERTY
UNIONVILLE
SHARPSBURG
FREDERICK
Monocacy Creek
MONOCACY JUNCTION
HARPER'S FERRY
OHIO R.R.
MARYLAND
CHARLESTOWN
DAMASCUS
BALTIMORE
URBANA
HYATTSTOWN
HILLSBORO
CLARKSBURG
BALLS BLUFF
POOLESVILLE
Cool Springs
Clark's Gap
LEESBURG
Snickers Gap
POTOMAC RIVER
SENECA
ROCKVILLE
DISTRICT OF COLUMBIA
DRANESVILLE
TENNALLYTOWN
GEORGETOWN
WASHINGTON
VIENNA
CHANTILLY
FAIRFAX
WHITE PLAINS
SALEM
CENTREVILLE
ALEXANDRIA
GAINESVILLE
ORANGE AND ALEXANDRIA R.R.
Manassas Junction
VIRGINIA

Determined to delay Early's advance on Washington, Gen. Lew Wallace, although outnumbered, engaged the Confederates at Monocacy Junction, (right) but was forced back toward Baltimore. He had, however, won a precious 24 hours for Washington's defenders.

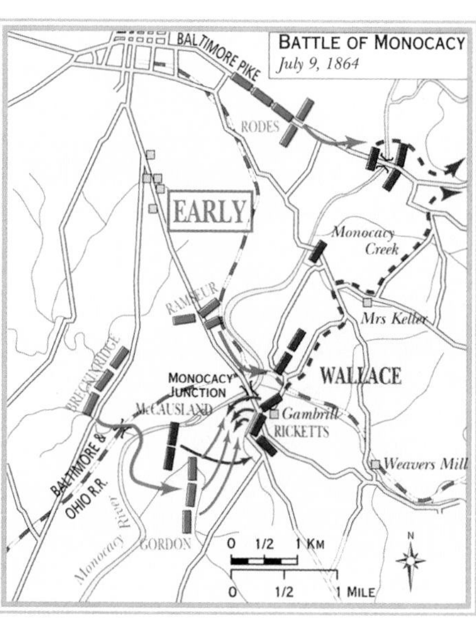
BATTLE OF MONOCACY
July 9, 1864
BALTIMORE PIKE
RODES
EARLY
Monocacy Creek
Mrs Keller
RAMSEUR
BRECKINRIDGE
MONOCACY JUNCTION
McCAUSLAND
WALLACE
Gambrill
RICKETTS
Weavers Mill
BALTIMORE & OHIO R.R.
GORDON
0 1/2 1 KM
0 1/2 1 MILE

After defeating the Federals at Monocacy, Early continued to advance toward Washington, reaching the city's northern environs on July 11. In the ensuing panic, over 20,000 men, including invalids and militia, were rushed into Washington's defenses. Faced with these numbers, Early gave up his plans for assault. After extensive skirmishing, especially at Fort Stevens, the Confederates withdrew (below).

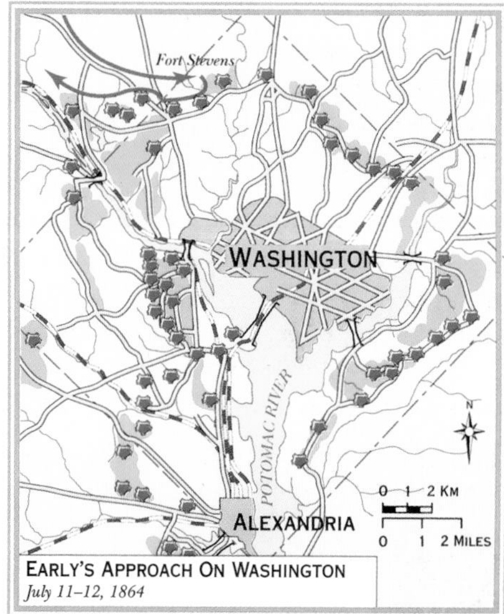
Fort Stevens
WASHINGTON
POTOMAC RIVER
ALEXANDRIA
0 1 2 KM
0 1 2 MILES
EARLY'S APPROACH ON WASHINGTON
July 11–12, 1864

The Chain Bridge over the Potomac River near Washington (below), defended by a ragged force of volunteers and invalid veterans.

"The men raised their battle-cry, sharper, shriller, and more like the composite yelping of wolves than I had ever heard it... The demonstration was more than exciting – it was really fearful."

*Gen. Lew Wallace commenting on the approach of Confederate troops at Monocacy.*

# Sherman's Atlanta Campaign: Phase 1 MAY 7–20 1864

By early May, Sherman had assembled a force of 100,000 men in the Chattanooga area. The force consisted of General Thomas's Army of the Cumberland, McPherson's Army of the Tennessee, and Schofield's Army of the Ohio. Facing it at Dalton, Georgia, were about 55,000 Confederates under General Johnston, organized into two infantry corps under Generals William J. Hardee and John B. Hood, and a cavalry corps led by Major General Joseph Wheeler.

Aware that Johnston's position on Rocky Face Ridge was impregnable to a frontal assault, Sherman outflanked it by sending McPherson via Snake Creek Gap to cut the Western & Atlantic Railroad – Johnston's sole supply line – at Resaca. Although McPherson failed, this move obliged Johnston to retreat to Resaca. There, on May 14 and 15, he repelled Federal attacks on his center and right, but in turn suffered repulses when he twice tried to turn Sherman's left. Learning that a Federal force had crossed the Oostanaula River above Resaca, and was threatening his supply line, Johnston evacuated Resaca on May 15, and retreated toward the Etowah River.

Sherman's army followed in four separate columns. Hoping to take ad-vantage of this dispersal, Johnston – whose strength had been increased by General Polk's Army of Mississippi to about 70,000 – on May 19 concentrated his forces northwest of Cassville, intending to crush the Federal left wing. However, before Johnston could strike, Federal cavalry appeared in the rear of Hood's corps, and Johnston withdrew to a ridge southeast of Cassville. There he planned to make a stand, but when Hood's and Polk's positions were enfiladed by enemy artillery, Johnston retreated south of the Etowah.

During this phase of the campaign, Johnston failed to protect the easily defended Snake Creek Gap and displayed poor judgment in his choice of position at Cassville. Equally, Sherman wasted an opportunity to smash Johnston's army by failing to send a larger force through Snake Creek Gap not only to cut the railroad, but to seize Resaca.

*An audacious cavalry commander, Gen. Wheeler (below) had sixteen horses shot from under him during the war. "Fighting Joe's" primary function throughout the campaign was to observe and inform Johnston of Sherman's flanking maneuvers.*

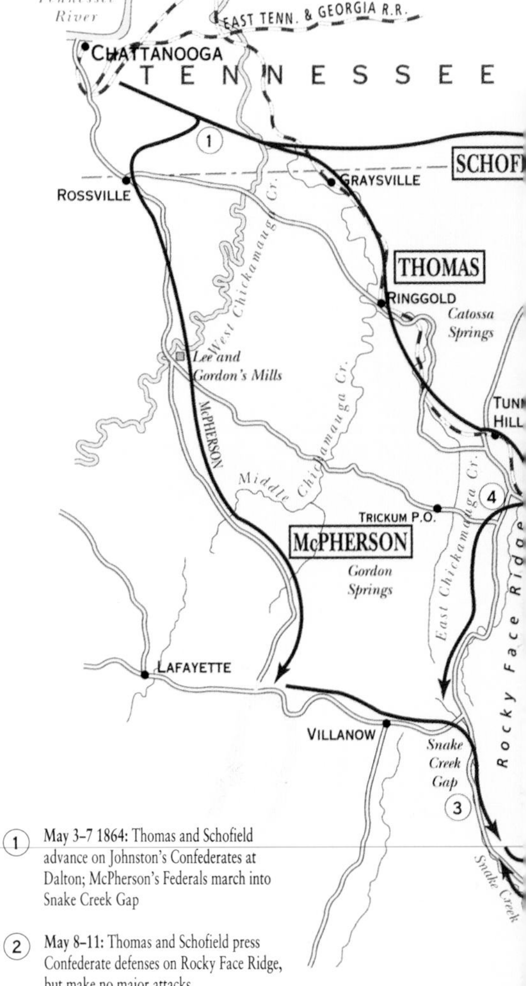

Tennessee River
EAST TENN. & GEORGIA R.R.
CHATTANOOGA
TENNESSEE
SCHOFIELD
ROSSVILLE
GRAYSVILLE
THOMAS
RINGGOLD
Catoosa Springs
Lee and Gordon's Mills
West Chickamauga Cr.
Chickamauga Cr.
Middle Chickamauga Cr.
McPHERSON
TUNNEL HILL
TRICKUM P.O.
McPHERSON
Gordon Springs
East Chickamauga Cr.
Rocky Face Ridge
LAFAYETTE
VILLANOW
Snake Creek Gap
Snake Creek
GEORGIA
OOSTANAULA R.


1. May 3–7 1864: Thomas and Schofield advance on Johnston's Confederates at Dalton; McPherson's Federals march into Snake Creek Gap
2. May 8–11: Thomas and Schofield press Confederate defenses on Rocky Face Ridge, but make no major attacks
3. May 9: McPherson probes Confederate defenses at Resaca, but withdraws to Snake Creek Gap without cutting railroad
4. May 10–13: Sherman leaves IV Corps at Rocky Face Ridge and joins McPherson; Johnston withdraws to Resaca and is reinforced by Polk
5. May 14, late am–early pm: Federal XIV and XXIII Corps make unsuccessful assault on Confederate center at Resaca (see map left)
6. May 14, late pm: Johnston tries to turn Sherman's left but is repelled (see map left)
7. May 14, evening: Union XV Corps storms hill above railroad bridge at Resaca, then repels Confederate counterattacks (see map left)
8. May 14, early pm: Federal XX Corps, supported by IV Corps, makes unsuccessful attack on center of Confederate right wing (see map left)

RESACA May 14, 1864

*Finding Johnston's army occupying a four-mile long defensive line at Resaca, Sherman ordered McPherson to again attempt outflanking his opponent while a number of divisions tested Johnston's strength. The assaults of the first day resulted in bloody repulse, followed by a Confederate counterattack (see map left). On the second day, Hooker was ordered to support Howard in further attacks. Heavy losses were suffered by both sides but the outcome of the battle was decided by McPherson's success in outflanking Johnston, who was forced into further withdrawal (see map below).*

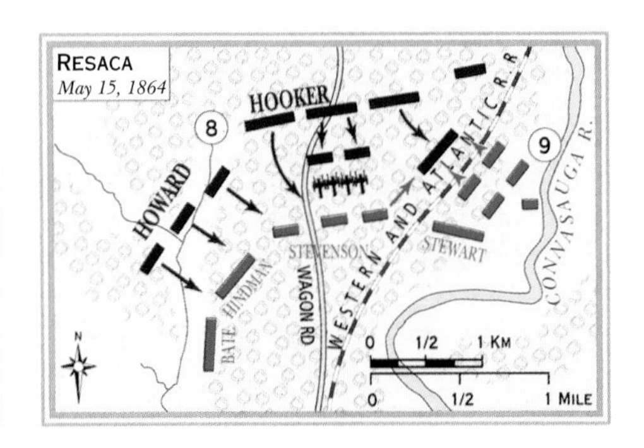

RESACA
May 15, 1864
HOOKER
HOWARD
BATE
HINDMAN
STEVENSON
STEWART
WAGON RD
WESTERN AND ATLANTIC R.R.
CONASAUGA R.
0 1/2 1 KM
0 1/2 1 MILE


"I'm going to move on Joe Johnston the day Grant telegraphs me he is going to hit Bobby Lee. And if you don't have my army supplied, we'll eat your mules up, sir; eat your mules up!"

*Remark made by Gen. Sherman to a quartermaster officer at the commencement of the Atlanta campaign.*

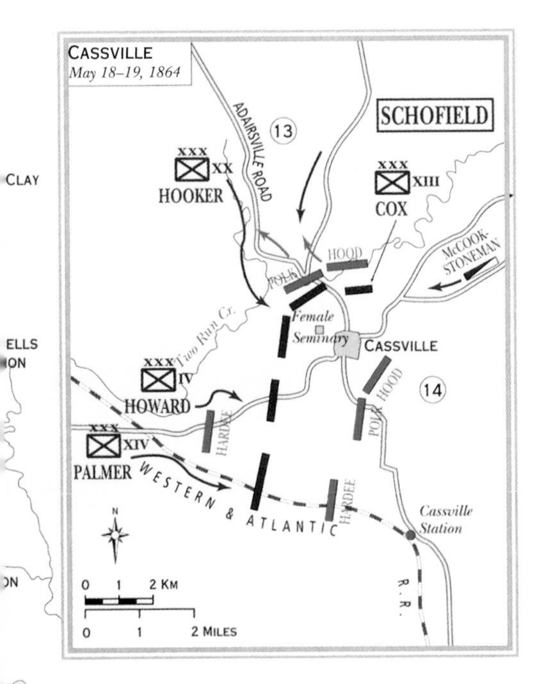
CASSVILLE
May 18–19, 1864
SCHOFIELD
ADAIRSVILLE ROAD
HOOKER
COX
HOOD
POLK
McCOOK-STONEMAN
Female Seminary
CASSVILLE
HOWARD
HARDEE
POLK
HOOD
PALMER
WESTERN & ATLANTIC
HARDEE
Cassville Station
R.R.
0 1 2 KM
0 1 2 MILES

*Sherman's armies, following Johnston south, converged on the Confederates in the Cassville area. Johnston ordered an attack on the separated units of Sherman's army, but the confederates were forced to yield and fall back to the Etowah River (see map left).*

"The Federal army, even if beaten, would have had a secure place of refuge at Chattanooga, while our only place of safety was Atlanta, a hundred miles off with three rivers intervening... I therefore decided to remain on the defensive... due to it's lengthening lines the numerical superiority of the Federal army would be reduced daily so that we might hope to cope with it on equal terms beyond the Chattahoochee."

*Gen. Johnston defending his strategy of withdrawal.*

*Gen. Sherman (above) commanded the Union forces in the western theater. Called by W.C. Gray "the most American looking man I ever saw", Sherman was respected by his troops, although he found it difficult to accept the personal popularity of his subordinates.*

*Union infantry advance uphill toward Confederate positions on the first day of the battle of Resaca (below). Sherman expressed disappointment at McPherson's failure to break the railroad and seize Resaca but readily admitted the difficulty of the task.*

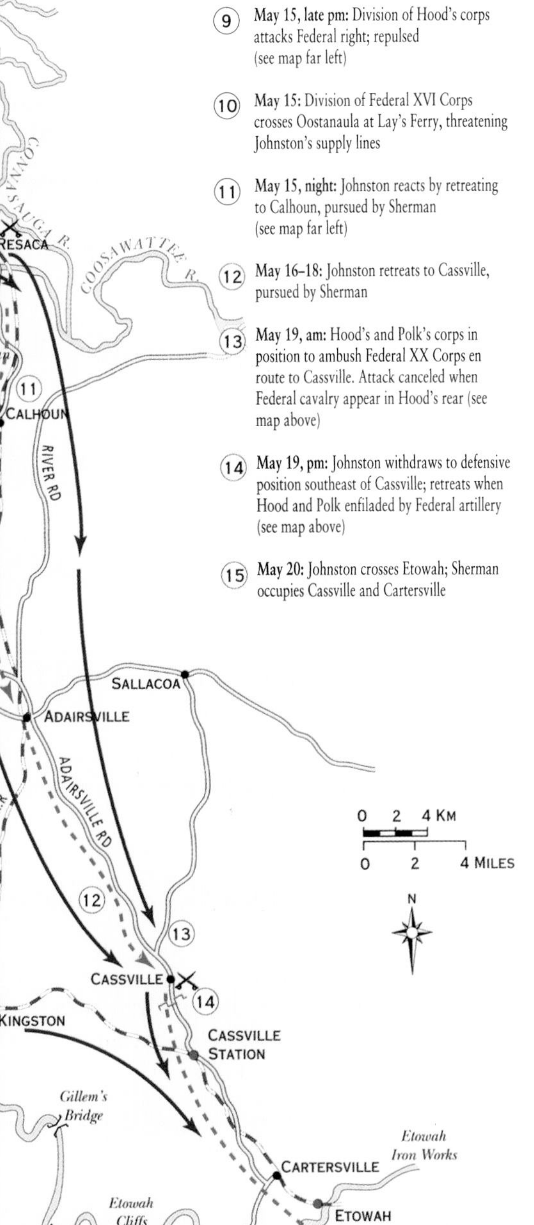
9 May 15, late pm: Division of Hood's corps attacks Federal right; repulsed (see map far left)
10 May 15: Division of Federal XVI Corps crosses Oostanaula at Lay's Ferry, threatening Johnston's supply lines
11 May 15, night: Johnston reacts by retreating to Calhoun, pursued by Sherman (see map far left)
12 May 16–18: Johnston retreats to Cassville, pursued by Sherman
13 May 19, am: Hood's and Polk's corps in position to ambush Federal XX Corps en route to Cassville. Attack canceled when Federal cavalry appear in Hood's rear (see map above)
14 May 19, pm: Johnston withdraws to defensive position southeast of Cassville; retreats when Hood and Polk enfiladed by Federal artillery (see map above)
15 May 20: Johnston crosses Etowah; Sherman occupies Cassville and Cartersville
RESACA
COOSAWATTEE R.
CALHOUN
RIVER RD
SALLACOA
ADAIRSVILLE
ADAIRSVILLE RD
0 2 4 KM
0 2 4 MILES
CASSVILLE
KINGSTON
CASSVILLE STATION
Gillem's Bridge
Etowah Iron Works
CARTERSVILLE
Etowah Cliffs
ETOWAH STATION
Milan's Bridge
ETOWAH R.

# Sherman's Atlanta Campaign MAY 21 – JULY 9 1864

After crossing the Etowah, Johnston took up another impregnable position at Allatoona. Sherman responded by swinging his army around to the west and south of the town, a move he expected would cause Johnston to fall back to the Chattahoochee River. Instead, Johnston blocked Serman's thrust at New Hope church on May 25, and at Pickett's Mill on May 27, but then himself suffered a reverse in a bungled assault at Dallas on May 28. Unable to advance, and suffering supply problems due to his separation from the railroad, Sherman withdrew to Acworth and the railroad early in June.

On June 10, having resupplied and reinforced his army, Sherman again advanced toward the Chattahoochee. Johnston slowly retreated until he reached Kennesaw Mountain near Marietta, an immensely strong position, where he again held Sherman in check. Frustrated, and fearing a stalemate in Georgia that would enable Johnston to reinforce Lee in Virginia, Sherman attempted to break through the Confederate center with a frontal assault on June 27, only to suffer a bloody repulse.

Meanwhile, Schofield's Army of the Ohio - which, in essence, consisted of XXII Corps

*Attacking in difficult, heavily wooded terrain, Howard's corps was repulsed at Pickett's Mill with heavy casualties (see map below).*

*At New Hope Church Sherman's formation advanced against Johnston's army. Several assaults by Hooker's Corps were repelled with heavy losses. Later, aiming to sabotage a Union movement to the left, Johnston ordered a reconnaissance in force against McPherson, near Dallas. Vicious fighting ensued, resulting in a Confederate withdrawal (see map below right).*

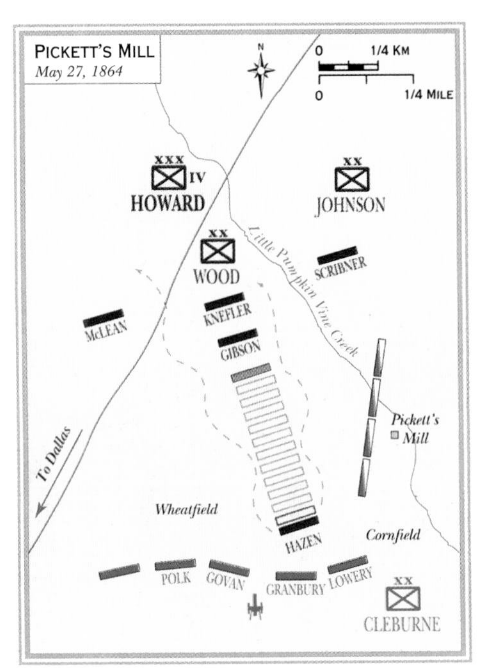
PICKETT'S MILL
May 27, 1864
0 1/4 KM
0 1/4 MILE
HOWARD
IV
JOHNSON
WOOD
SCRIBNER
KNEFLER
GIBSON
McLEAN
Little Pumpkin Vine Creek
Pickett's Mill
To Dallas
Wheatfield
HAZEN
Cornfield
POLK
GOVAN
GRANBURY
LOWERY
CLEBURNE

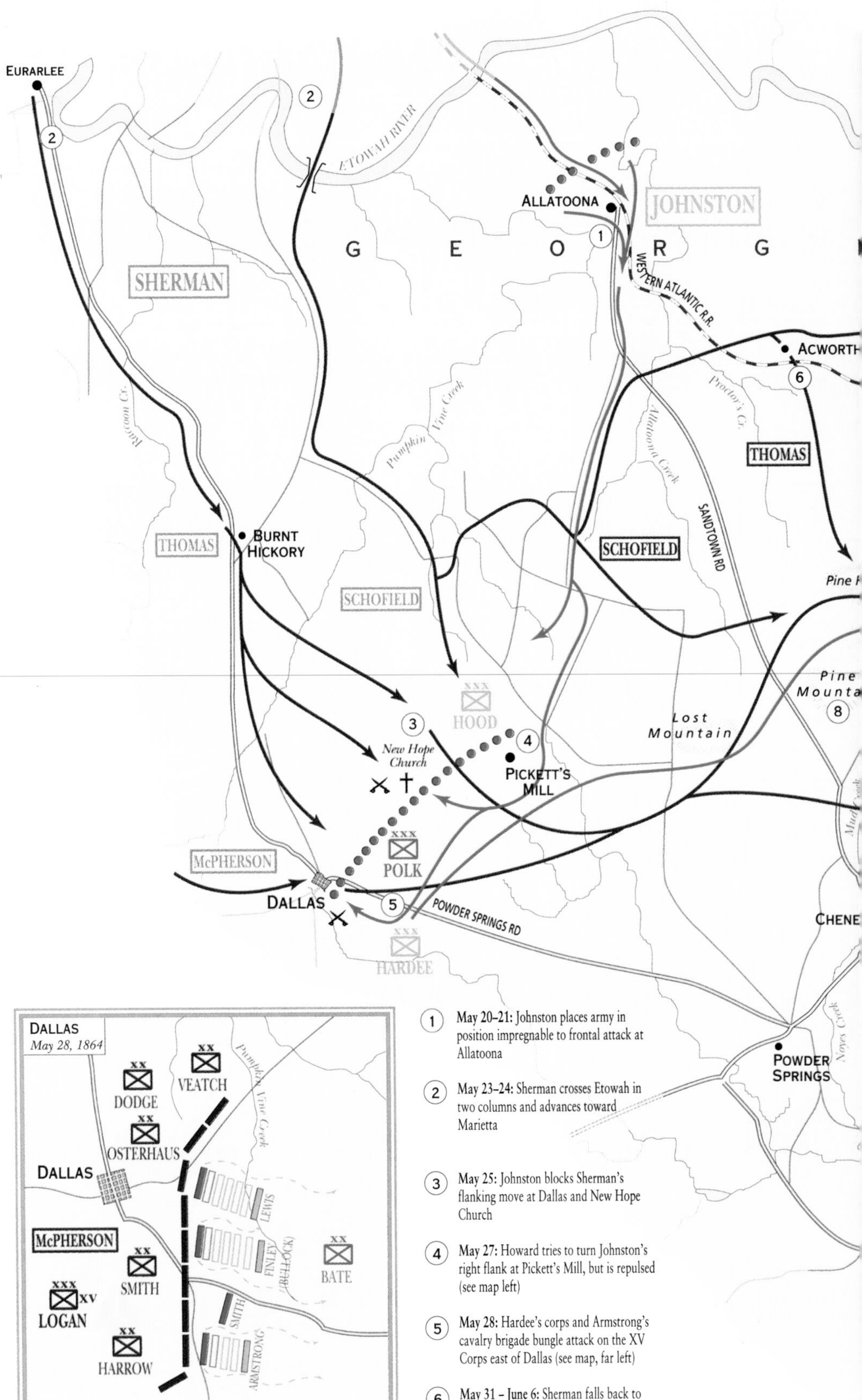
EURARLEE
ETOWAH RIVER
ALLATOONA
JOHNSTON
G E O R G
SHERMAN
WESTERN ATLANTIC R.R.
ACWORTH
THOMAS
Raccoon Cr.
Pumpkin Vine Creek
Allatoona Creek
Proctor's Cr.
SANDTOWN RD
BURNT HICKORY
THOMAS
SCHOFIELD
SCHOFIELD
Pine
HOOD
Pine Mountain
Lost Mountain
New Hope Church
PICKETT'S MILL
POLK
McPHERSON
DALLAS
POWDER SPRINGS RD
HARDEE
CHENE
POWDER SPRINGS
DALLAS
May 28, 1864
DODGE
VEATCH
OSTERHAUS
DALLAS
McPHERSON
SMITH
LOGAN
XV
HARROW
LEWIS
FINLEY
BATE
SMITH
ARMSTRONG
Pumpkin Vine Creek
0 1/4 1/2 KM
0 1/4 1/2 MILE
1 May 20–21: Johnston places army in position impregnable to frontal attack at Allatoona
2 May 23–24: Sherman crosses Etowah in two columns and advances toward Marietta
3 May 25: Johnston blocks Sherman's flanking move at Dallas and New Hope Church
4 May 27: Howard tries to turn Johnston's right flank at Pickett's Mill, but is repulsed (see map left)
5 May 28: Hardee's corps and Armstrong's cavalry brigade bungle attack on the XV Corps east of Dallas (see map, far left)
6 May 31 – June 6: Sherman falls back to Acworth and railroad
7 June 10–19: Sherman advances through rain and mud to area north and west of Kennesaw Mountain
8 June 12: Polk killed on Pine Mountain

9. **June 22:** Hood makes doomed attack on XX Corps and part of the XXIII Corps at Kolb's Farm
10. **June 27:** Sherman fails in an attempt to break through the Confederate center at Kennesaw Mountain (see map below right)
11. **June 27 – July 2:** XXIII Corps outflanks Confederate left south of Kennesaw Mountain, Johnston abandons his position
12. **July 5:** Johnston establishes a new line on the north bank of the Chattahoochee
13. **July 8–9:** Federal XXIII Corps and cavalry cross the Chattahoochee. Johnston withdraws to outskirts of Atlanta

*Union troops prepare to continue the assault upon Kennesaw Mountain, June 27 (above). The battle was brief, fierce and decisive. Union casualties were particularly high as the attack was directed uphill against well prepared Confederate fortifications, enabling the defenders to take careful and deadly aim.*

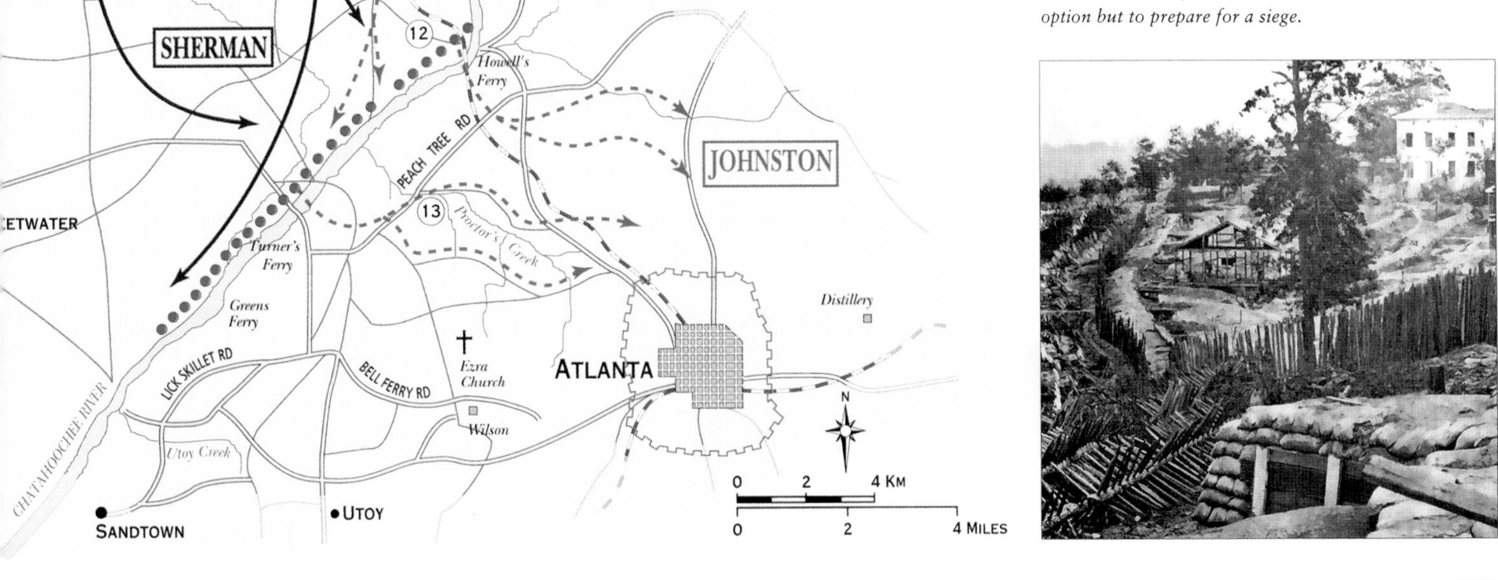
Brush Mountain
Black Jack Mountain
SHANTY
To Atlanta
HOOD
MARIETTA
Isham's Ford
Vinings Station
RUFF'S MILL
SHERMAN
Howell's Ferry
PEACH TREE RD
JOHNSTON
Proctor's Creek
Turner's Ferry
Greens Ferry
LICK SKILLET RD
BELL FERRY RD
Ezra Church
ATLANTA
Distillery
Wilson
Utoy Creek
CHATTAHOOCHEE RIVER
SANDTOWN
UTOY
0 2 4 KM
0 2 4 MILES

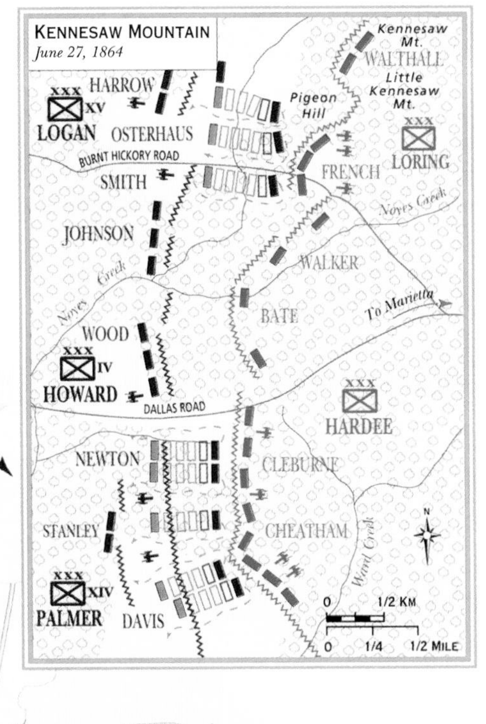
KENNESAW MOUNTAIN
June 27, 1864
Kennesaw Mt.
WALTHALL
Little Kennesaw Mt.
HARROW
XV
LOGAN
OSTERHAUS
Pigeon Hill
LORING
BURNT HICKORY ROAD
SMITH
FRENCH
Noyes Creek
JOHNSON
WALKER
BATE
To Marietta
WOOD
IV
HOWARD
DALLAS ROAD
HARDEE
NEWTON
CLEBURNE
STANLEY
CHEATHAM
XIV
PALMER
DAVIS
0 1/2 KM
0 1/4 1/2 MILE

only succeeded in turning Johnston's left flank. This accomplished, he retreated in early July to a previously prepared fortified line on the north bank of the Chattahoochee.

Once again, Sherman resorted to a flanking maneuver, ordering sections of his army to cross the Chattahoochee at various points above the Confederate position. Johnston responded by falling back across the river and withdrawing to the outskirts of Atlanta.

Sherman had breached the last major natural barrier between him and his objective; his army, despite hardships and hard fighting, was now stronger than the Confederate army. Thus, Johnston's professed strategy of wearing down the Federals until they were weak enough to be attacked successfully, was not succeeding.

*Attacking strong Confederate entrenchments at Kennesaw Mountain, the Federals were able to seize only the outposts, and continued assaults resulted in high casualties while relatively few were inflicted (see map left).*

"Within the space of about one acre... were over seven hundred dead men, all Yanks, and most of them within ten or fifteen yards of our line. Our men, as it seems, reserved their fire until they came within such deadly range that every shot took effect... I shudder as I think of the mutilated forms."

*A Confederate participant in the battle of New Hope Church.*

*Confederate fortifications surrounding Atlanta were formidable, consisting of a network of forts, breastworks, ditches and chevaux-de-frise (below). Unable to contemplate a direct assault, Sherman had no option but to prepare for a siege.*

# Battles for Atlanta JULY 20–28 1864

On July 17, Confederate President Jefferson Davis, displeased with Johnston's failure to stop Sherman's advance and doubting that he would make a determined effort to defend Atlanta, removed Johnston from command and replaced him with John B. Hood, who had a well-deserved reputation as a bold, aggressive fighter. Davis believed that the loss of Atlanta would be a devastating blow to the Confederacy, whereas if it could be held as Lee was holding Richmond the war-weary North would abandon its effort to force the South back into the Union.

On July 18–19, having crossed the Chatta-hoochee with his entire army, Sherman moved on Atlanta in two columns, with one (McPherson and Schofield) approaching it from the east and the other (Thomas) from the north. Hoping to exploit this division in the enemy forces, Hood on July 20, attacked Thomas at Peach Tree Creek with the corps of Hardee and Lieutenant General A.P. Stewart, now commanding in the place of Polk, killed on June 12. Although he caught the Union troops by surprise and for the most part unentrenched, Hood was repulsed with 2,500 casualties while inflicting 1,700.

Nothing daunted, Hood next sought to smash Sherman's other wing east of Atlanta by sending Hardee's corps, on the night of July 21, around the left flank of McPherson's Army of the Tennessee with the intent of striking it from the rear. This maneuver took Sherman unawares and only the fortuitous presence of the XVI Corps on the otherwise exposed Union left flank prevented a Federal disaster. As it was, McPherson was killed, the Army of Tennessee's left was bent back, and its center temporarily pene-

*The Confederate assault at Peach Tree Creek was nullified by two factors: the delay in launching the assault which enabled the majority of the troops to cross the creek, and the unexpectedly rapid advance of McPherson from the east which necessitated Hardee's withdrawal (see detail of map right).*

1. **July 20:** Hardee's and Stewart's corps attack Thomas and are repulsed with heavy losses in Battle of Peach Tree Creek
2. **July 20:** McPherson moves to within artillery range of Atlanta, his advance slowed by Wheeler's cavalry
3. **July 20:** Cheatham's (formerly Hood's) corps conducts delaying action against advance of Schofield and two divisions of IV Corps
4. **July 21:** Leggett's division, XVII Corps, captures Bald Hill
5. **July 21:** Hardee's corps and Wheeler's cavalry swing around Union left flank, Cheatham's corps deploys east of Atlanta, and Stewart's corps withdraws to fortifications on Atlanta's north side
6. **July 22, early afternoon:** Hardee's corps attacks Union left flank; Bate's and Walker's divisions are repulsed by the XVI Corps but Cleburne's and Maney's divisions force back the left of the XVII Corps
7. **July 22, early afternoon:** McPherson killed by Confederates advancing through gap in Union line; Major General John A. Logan is appointed acting commander of the Army of the Tennessee
8. **July 22, late afternoon:** Brown's division of Cheatham's corps penetrates center of XV Corps via a poorly defended railroad cut but is driven back by Union counterattack
9. **July 22, evening:** Hardee makes a final attempt to drive XVII Corps from its position on Union left, but his attack fails
10. **July 26 - 28:** Sherman sends Howard, now commanding the Army of the Tennessee, swinging west of Atlanta in an attempt to cut the railroad south of the city
11. **July 27 - 28:** Hood sends S.D. Lee's corps (formerly Cheatham's) and Stewart's corps to block Howard's movement, then to strike Howard's right flank and rear on July 29
12. **July 28:** Disregarding Hood's orders, Lee and Stewart attack Howard frontally at Ezra Church; Confederates repulsed with heavy losses

BATTLE OF EZRA CHURCH
*July 28, 1864*

*A contemporary woodcut of the attack on Ezra Church (below). Gen. Howard's defense of the Federal position did much to redeem his reputation which had been tarnished by the unpreparedness of his division at Chancellorsville in 1863.*

*Hood's plan at Ezra Church was to surprise the Army of the Tennessee as it moved west to cut the Macon & Western Railroad. Anticipating Hood's strategy Howard had ordered the construction of makeshift breastworks at Ezra Church from behind which Federal troops were able to repel the repeated and costly Confederate assaults (see detail of map above).*

trated by a frontal attack by Major General B.F. Cheatham's corps (formerly Hood's). Confederate losses at Bald Hill totaled about 5,500, Federal about 3,750.

Hood's final July attempt to defeat Sherman occurred on July 28 when he sent the corps of Lieutenant General S.D. Lee (formerly Cheatham's) and Stewart to strike the Army of the Tennessee, now headed by Major General O.O. Howard, in the flank and rear as it moved southward west of Atlanta. Instead, owing to Lee's impulsiveness, these corps attacked Howard frontally at Ezra Church, losing nearly 3,000 men while inflicting a mere 632 casualties.

Hood's attacks were well-conceived but lacked sufficient strength and skill in execution to succeed. Even so he held Sherman at bay at Atlanta, and at the end of July his cavalry routed two Union mounted divisions that raided the Macon & Western Railroad and attempted to free Federal prisoners at Macon and Andersonville.

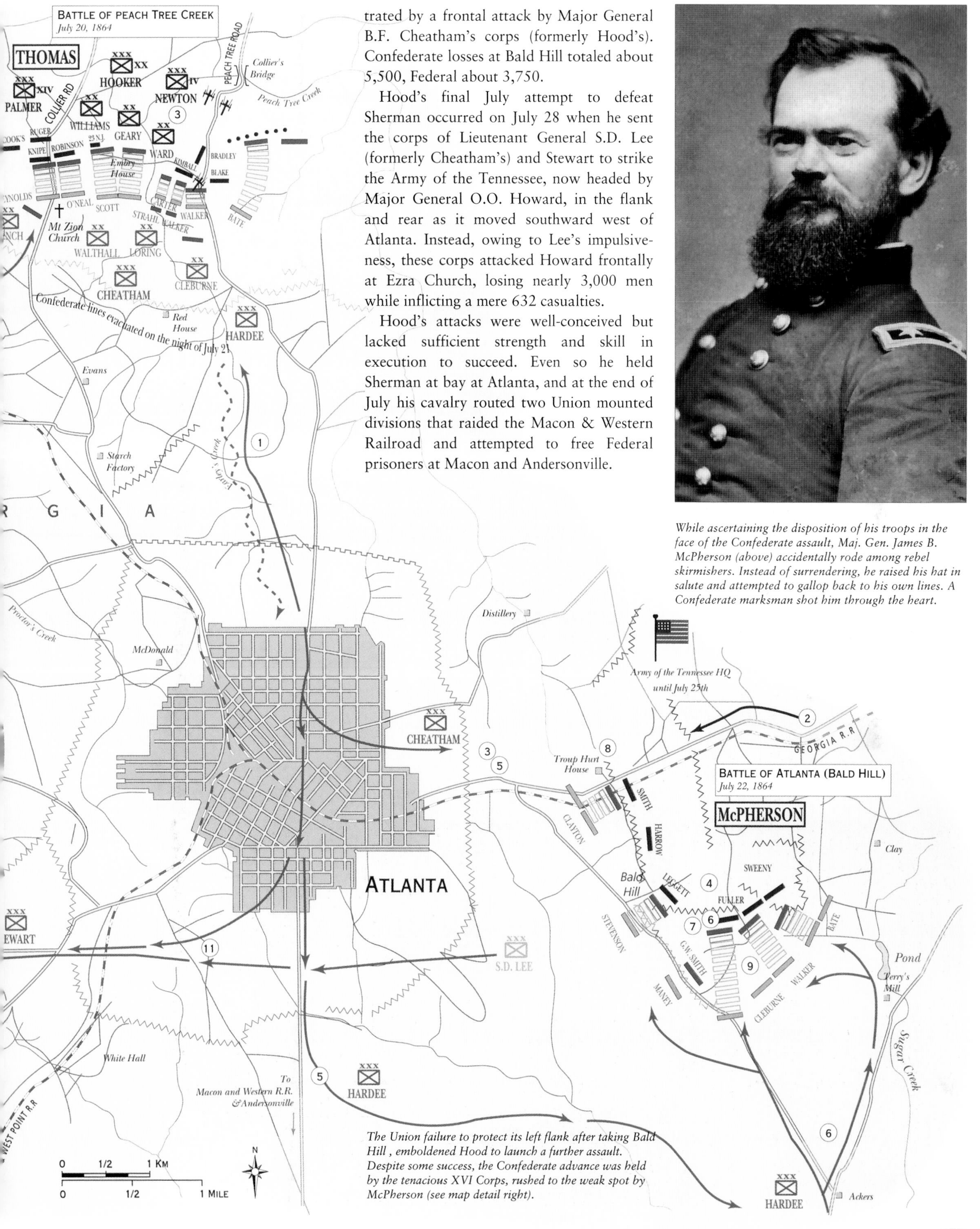
BATTLE OF PEACH TREE CREEK
July 20, 1864
THOMAS
PALMER
HOOKER
NEWTON
WILLIAMS
GEARY
WARD
RUGER
KNIPE
ROBINSON
KIMBALL
BRADLEY
BLAKE
Embry House
O'NEAL
SCOTT
CARTER
STRAHL
WALKER
BATE
Mt Zion Church
WALTHALL
LORING
CHEATHAM
CLEBURNE
HARDEE
Collier's Bridge
Peach Tree Creek
PEACH TREE ROAD
COLLIER RD
Confederate lines evacuated on the night of July 21
Red House
Evans
Starch Factory
Proctor's Creek
McDonald
ATLANTA
Distillery
Army of the Tennessee HQ until July 25th
GEORGIA R.R.
Troup Hurt House
BATTLE OF ATLANTA (BALD HILL)
July 22, 1864
McPHERSON
SMITH
HARROW
CLAYTON
Bald Hill
LEGGETT
SWEENY
FULLER
Clay
STEVENSON
G.W. SMITH
MANEY
CLEBURNE
WALKER
BATE
Pond
Terry's Mill
Sugar Creek
S.D. LEE
White Hall
To Macon and Western R.R. & Andersonville
WEST POINT R.R.
HARDEE
Ackers
EWART
1 KM
1 MILE
N

The Union failure to protect its left flank after taking Bald Hill, emboldened Hood to launch a further assault. Despite some success, the Confederate advance was held by the tenacious XVI Corps, rushed to the weak spot by McPherson (see map detail right).

*While ascertaining the disposition of his troops in the face of the Confederate assault, Maj. Gen. James B. McPherson (above) accidentally rode among rebel skirmishers. Instead of surrendering, he raised his hat in salute and attempted to gallop back to his own lines. A Confederate marksman shot him through the heart.*

# Atlanta Campaign: Final Phase AUGUST 1 – SEPTEMBER 2 1864

On August 10, hoping to take advantage of the virtual destruction of two Federal cavalry divisions at the end of July, Hood dispatched the bulk of Wheeler's cavalry across the Chattahoochee River into northern Georgia. Hood's aim was to force Sherman to retreat from Atlanta by wrecking his railroad supply line. Wheeler, however, failed to do any serious damage to the railroad, and soon had to flee into Tennessee, thereafter ceasing to be a factor in the campaign.

Meanwhile, Sherman again tried to turn Hood's left flank, west of Atlanta, only to be repulsed at Utoy Creek on August 6 and 7. Similarly, a bombardment of Atlanta achieved nothing but civilian casualties. A cavalry raid against the Macon & Western Railroad – Hood's sole remaining supply line – also failed. Sherman, therefore, once again resorted to a large-scale flanking maneuver.

Leaving XX Corps to guard the Chattahoochee railroad bridge, Sherman withdrew the rest of his army from the trenches around Atlanta and swung it south toward Jonesboro. Hood was unable to react until he ascertained Sherman's precise objective. Finally, on August 30, when he heard that a large Federal force was approaching Jonesboro, he sent the corps of Hardee and S.D. Lee to the town with orders to drive back the Federals. On August 31, Hardee assaulted Howard's Army of the Tennessee, but was easily repulsed. However, Hardee held against a Federal counterattack the following day, then withdrew during the night to Lovejoy's Station.

"As I became more and more convinced of our inability to successfully resist an advance of the Federal Army, I concluded to resume active operations, move upon Sherman's communications, and avert, if possible, impending disaster from the Confederacy."

*Hood outlining his plans after the evacuation of Atlanta.*

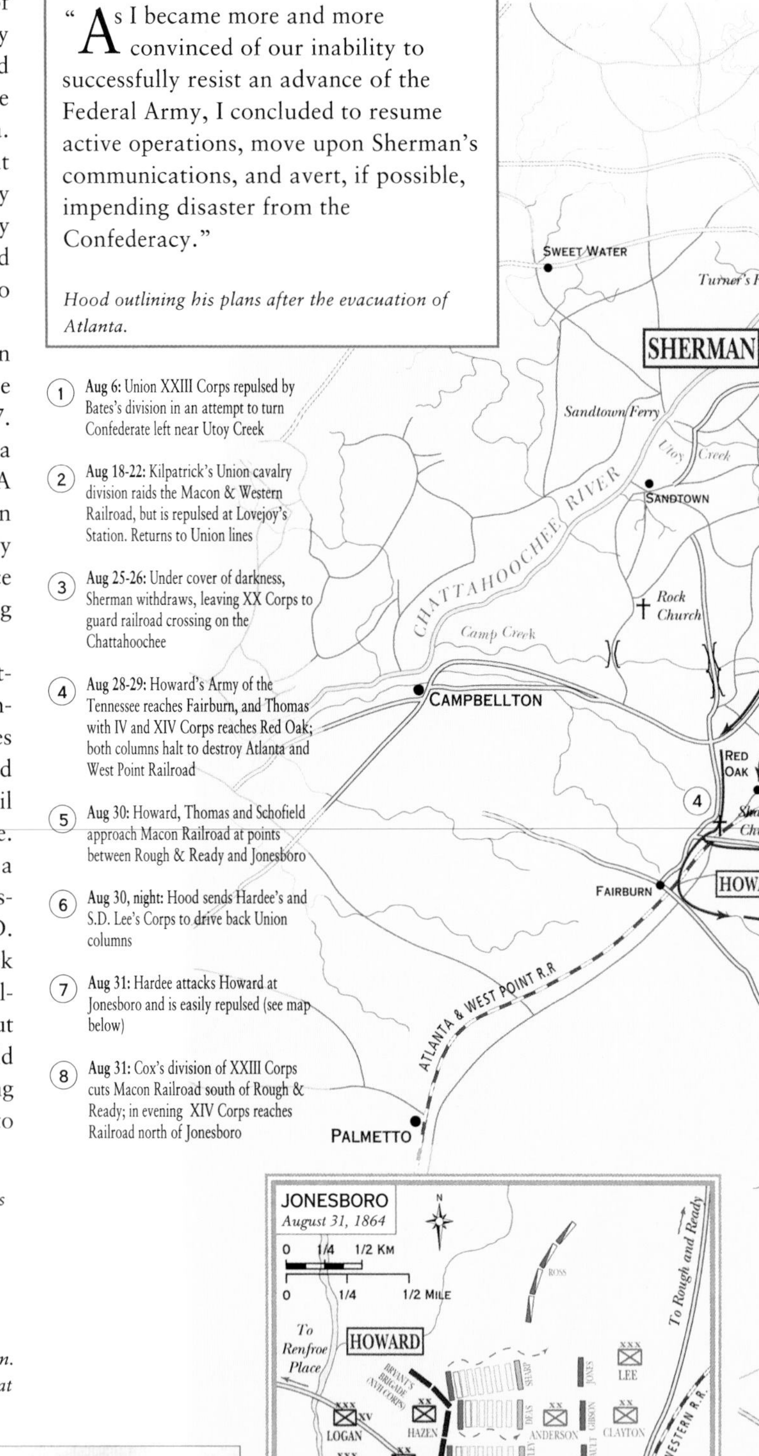
1 Aug 6: Union XXIII Corps repulsed by Bates's division in an attempt to turn Confederate left near Utoy Creek
2 Aug 18-22: Kilpatrick's Union cavalry division raids the Macon & Western Railroad, but is repulsed at Lovejoy's Station. Returns to Union lines
3 Aug 25-26: Under cover of darkness, Sherman withdraws, leaving XX Corps to guard railroad crossing on the Chattahoochee
4 Aug 28-29: Howard's Army of the Tennessee reaches Fairburn, and Thomas with IV and XIV Corps reaches Red Oak; both columns halt to destroy Atlanta and West Point Railroad
5 Aug 30: Howard, Thomas and Schofield approach Macon Railroad at points between Rough & Ready and Jonesboro
6 Aug 30, night: Hood sends Hardee's and S.D. Lee's Corps to drive back Union columns
7 Aug 31: Hardee attacks Howard at Jonesboro and is easily repulsed (see map below)
8 Aug 31: Cox's division of XXIII Corps cuts Macon Railroad south of Rough & Ready; in evening XIV Corps reaches Railroad north of Jonesboro
Vinings Station
Pace's Ferry
BUCK H
SLOCUM
WESTERN & ATLANTIC R.R.
PEACH TREE RD
SWEET WATER
Turner's Ferry
SHERMAN
LICK SKILLET
Ezra Church
BELL FERRY RD
Sandtown Ferry
Utoy Creek
SANDTOWN
UTOY
CHATTAHOOCHEE RIVER
Rock Church
Mt. Gilead Church
EAST PO
Camp Creek
CAMPBELLTON
SCHOFIELD
MARROW'S MILL
RED OAK
Shadna Church
Shoal Creek Church
THAMES
THO
FAIRBURN
HOWARD
BETHSAIDA
RENFROES
Pond Fork
ATLANTA & WEST POINT R.R.
Camp Creek
PALMETTO
FAYETTEVILLE
White Water Cr.
Flat Creek
WHITE WATER
GLENGROVE

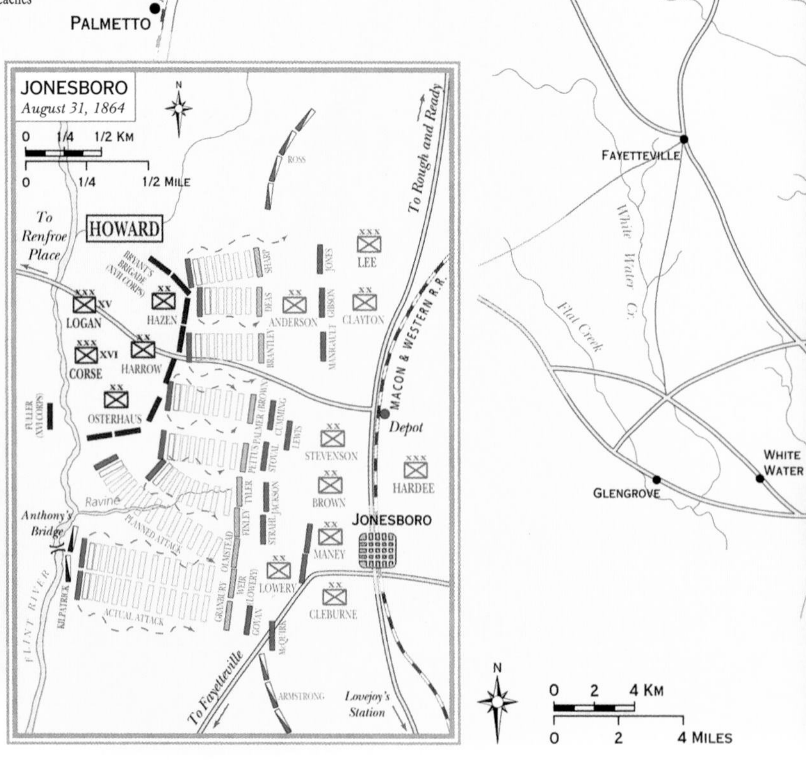
JONESBORO
August 31, 1864
0 1/4 1/2 KM
0 1/4 1/2 MILE
To Renfroe Place
HOWARD
LOGAN
HAZEN
CORSE
HARROW
OSTERHAUS
LEE
ANDERSON
CLAYTON
To Rough and Ready
MACON & WESTERN R.R.
Depot
STEVENSON
HARDEE
BROWN
JONESBORO
MANEY
LOWERY
CLEBURNE
Ravine
Anthony's Bridge
PLANNED ATTACK
ACTUAL ATTACK
FLINT RIVER
KILPATRICK
To Fayetteville
Lovejoy's Station
0 2 4 KM
0 2 4 MILES

*At Jonesboro Hardee's frontal assaults against Howard's Federal entrenchments resulted in ten times as many Confederate casualties as Union (see map, below right)*

*Immediately prior to their evacuation of Atlanta, Confederate troops ensured that their supplies and wagons did not fall into Union hands by destroying them. The explosion at first encouraged Sherman to believe that his advance parties had joined battle ( below).*

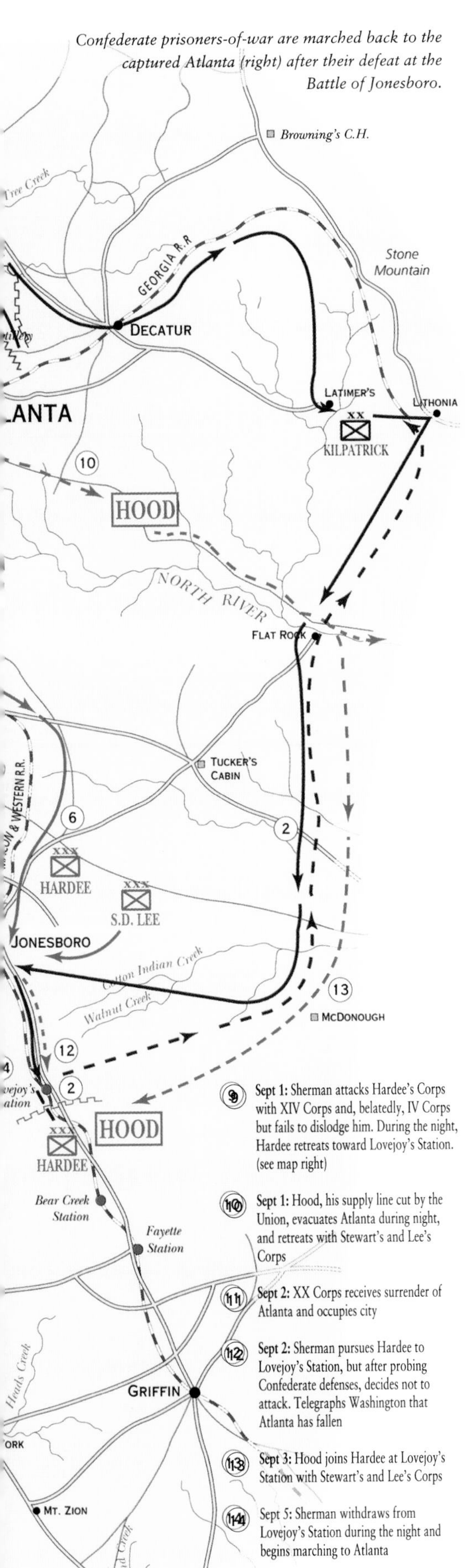
Browning's C.H.
Tree Creek
GEORGIA R.R.
Stone Mountain
DECATUR
LATIMER'S
LITHONIA
KILPATRICK
ANTA
10
HOOD
NORTH RIVER
FLAT ROCK
TUCKER'S CABIN
MACON & WESTERN R.R.
6
2
HARDEE
S.D. LEE
JONESBORO
Cotton Indian Creek
Walnut Creek
13
MCDONOUGH
12
2
HOOD
HARDEE
Bear Creek Station
Fayette Station
Heads Creek
GRIFFIN
MT. ZION
Sand Creek
9 Sept 1: Sherman attacks Hardee's Corps with XIV Corps and, belatedly, IV Corps but fails to dislodge him. During the night, Hardee retreats toward Lovejoy's Station. (see map right)
10 Sept 1: Hood, his supply line cut by the Union, evacuates Atlanta during night, and retreats with Stewart's and Lee's Corps
11 Sept 2: XX Corps receives surrender of Atlanta and occupies city
12 Sept 2: Sherman pursues Hardee to Lovejoy's Station, but after probing Confederate defenses, decides not to attack. Telegraphs Washington that Atlanta has fallen
13 Sept 3: Hood joins Hardee at Lovejoy's Station with Stewart's and Lee's Corps
14 Sept 5: Sherman withdraws from Lovejoy's Station during the night and begins marching to Atlanta

*Confederate prisoners-of-war are marched back to the captured Atlanta (right) after their defeat at the Battle of Jonesboro.*

That same night, the rest of the Confederate army evacuated the now untenable Atlanta, which was occupied by XX Corps on September 2. When he heard the news, Sherman telegraphed Washington: "Atlanta is ours, and fairly won," then fell back to the city, thus ending the campaign. During the campaign, both Federal and Confederate losses each totaled about 40,000.

By reviving sagging Union morale, the fall of Atlanta assured Lincoln's reelection in 1864. It also assured the Confederacy's ultimate defeat: at this stage of the war, it's only realistic hope of victory was that the Union would lose its will to continue the struggle.

*The railroad-wrecking expeditions of Sherman's cavalry (above), led by Maj. Gen. H. J. Fitzpatrick were limited in their effectiveness as the rebel engineers proved themselves as capable as their Union counterparts at repairing damaged tracks.*

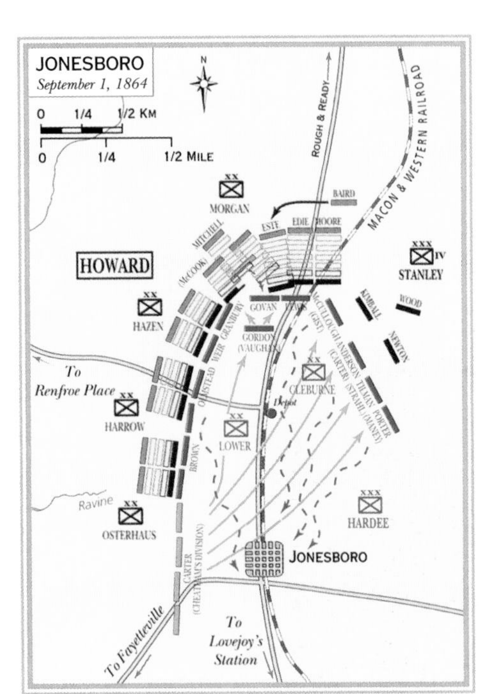
JONESBORO
September 1, 1864
0 1/4 1/2 KM
0 1/4 1/2 MILE
N
ROUGH & READY
MACON & WESTERN RAILROAD
MORGAN
BAIRD
ESTE
EDIE
MOORE
HOWARD
STANLEY
HAZEN
GOVAN
KIMBALL
WOOD
NEWTON
CLEBURNE
To Renfroe Place
HARROW
LOWER
Ravine
OSTERHAUS
HARDEE
JONESBORO
To Lovejoy's Station
To Fayetteville

*Furious fighting continued at Jonesboro on September 1. Confederates suffered severe casualties until Hardee pulled back to Lovejoy's Station at nightfall (map left).*

> "Not one instant did that line hesitate – it moved steadily forward to the enemy's works – over the works with a shout – over the cannon – over the rebels, and then commenced stern work with the bayonet, but the despairing cries of surrender soon stopped it, the firing ceased, and 1,000 rebels were... prisoners of war."
>
> *An Illinios major speaking of Jonesboro.*

# Forrest's Operations in Mississippi and Tennessee MARCH 16 – NOVEMBER 5 1864

In 1864 Nathan Bedford Forrest, exhibited the military genius and ferocity which made him one of the most outstanding cavalry leaders of the Civil War.

During March and April, Forrest's cavalry raided Federal installations in Kentucky and Tennessee. On April 12, he attacked the 580-man garrison at Fort Pillow. When the fort refused to surrender, 1,500 Confederates stormed the earthwork and quickly subdued the mixed garrison of white Tennessee "Unionists" and black soldiers. Forrest lost 100, but captured 226 Federals, killed 231, and wounded 100. Forrest was publicly accused in the North of allowing his men to murder black soldiers after they had surrendered.

Afraid that Forrest would raid his vital rail supply lines in middle Tennessee, General Sherman ordered General Sturgis into northern Mississippi to locate and eliminate Forrest's cavalry. On June 2, Sturgis left Memphis with 8,100 infantry, cavalry and 20 cannon.

In Alabama, en route to raid Tennessee railroads, Forrest was ordered back to meet Sturgis. On June 10, he intercepted Sturgis's 3,300 cavalry near Brice's Crossroads. By early afternoon, the Federal infantry, exhausted by a rapid march, arrived to relieve the cavalry. Fighting was intense in the dense underbrush. The Federal infantry, disastrously outflanked, panicked and fled across the Hatchie River, with Forrest in hot pursuit. Forrest's brilliant victory resulted in Federal losses of 617, with 1,618 captured, while the Confederates lost 492.

On July 5 Sherman sent Major General Andrew J. Smith with 14,000 Federal troops into Mississippi in a further attempt to intercept Forrest and protect Federal supply lines. Smith advanced south from La Grange, cutting a swath of destruction as he went. To halt Smith, General S.D. Lee joined Forrest, and with a total of 9,460 men, attacked Smith at Harrisburg near Tupelo on July 14. In several uncoordinated frontal assaults, the Confed-erates suffered a bloody repulse from the entrenched Federals. But, short of rations and ammunition, Smith withdrew the next day, pursued by Forrest. In a sharp rear-guard action along Old Town Creek, Forrest was shot in the foot, which put him out of action. Although he had not destroyed Forrest, Smith badly crippled Lee's Confederates and protected Sherman's supply line.

On August 21, fully recovered from his wound, and hoping to forestall another

(1) **March 16:** Forrest begins raids

(2) **March 24:** Confederate cavalry seizes Union City, capturing 500 Federal troopers and 300 horses

(3) **March 25:** Forrest demands surrender of Fort Anderson at Paducah, but Federals refuse. Cavalry force attacks fort, but is repulsed. Confederates withdraw

(4) **March 29:** At Bolivar, Confederate cavalry rout Federal cavalry, forcing it to flee to Memphis

(5) **April 12:** Forrest attacks and captures Fort Pillow (see map right)

(6) **June 2–9:** Gen. Sturgis leaves Memphis to locate Forrest. Arrives near Brice's Crossroads on the 9th

(7) **June 6–9:** In Alabama, Forrest learns of Sturgis's pursuit and returns to Tupelo; meets with his superior, Gen. S.D. Lee, at Booneville to make battle plans

(8) **June 10:** Battle of Brice's Crossroads (see map top right). Forrest pursues fleeing Federals until 12th

(9) **July 5–13:** Gen. Smith, sent to intercept Forrest, advances south from La Grange. Enters Tupelo on 13th

(10) **July 13:** Lee joins Forrest at Okolona and they overhaul Smith at Harrisburg. Federals spend night erecting breastworks

(11) **July 14:** Engagement at Tupelo (see map right)

(12) **Aug18-20:** Forrest masses his raiders at Oxford, then proceeds to Hernando

(13) **Aug 21:** Forrest raids Memphis

(14) **Oct 19 - November 3:** Forrest leaves Corinth for Fort Heiman and Paris Landing. Captures and burns Federal ships and installations. Federals retaliate

(15) **Nov 4:** Using artillery, Forrest bombards Johnsonville with devastating effectiveness. Withdraws on 5th (see map far right)

*By mid-morning on April 12, Forrest had hemmed in the Union garrison at Fort Pillow. In the final assault, fighting dismounted, Confederates stormed the fort. Union soldiers fled to the river and were caught in a murderous hail of small arms fire. Excessive killing occurred and Forrest was accused of murdering black prisoners (see map below). Nine days later, Forrest thundered into sleeping Memphis, narrowly missing capturing Gen. Hurlbut. Gen. Washburn also escaped by scampering out of the back of Federal headquarters in his nightshirt, and fleeing to Fort Pickering. Having stirred up a hornet's nest, Forrest withdrew.*

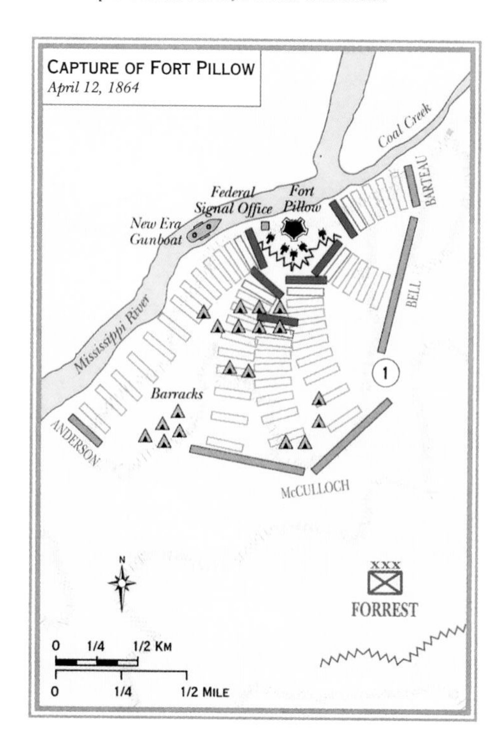
CAPTURE OF FORT PILLOW
April 12, 1864
Coal Creek
Federal Signal Office
Fort Pillow
New Era Gunboat
Mississippi River
BARTEAU
BELL
Barracks
ANDERSON
McCULLOCH
1
N
FORREST
0 1/4 1/2 KM
0 1/4 1/2 MILE

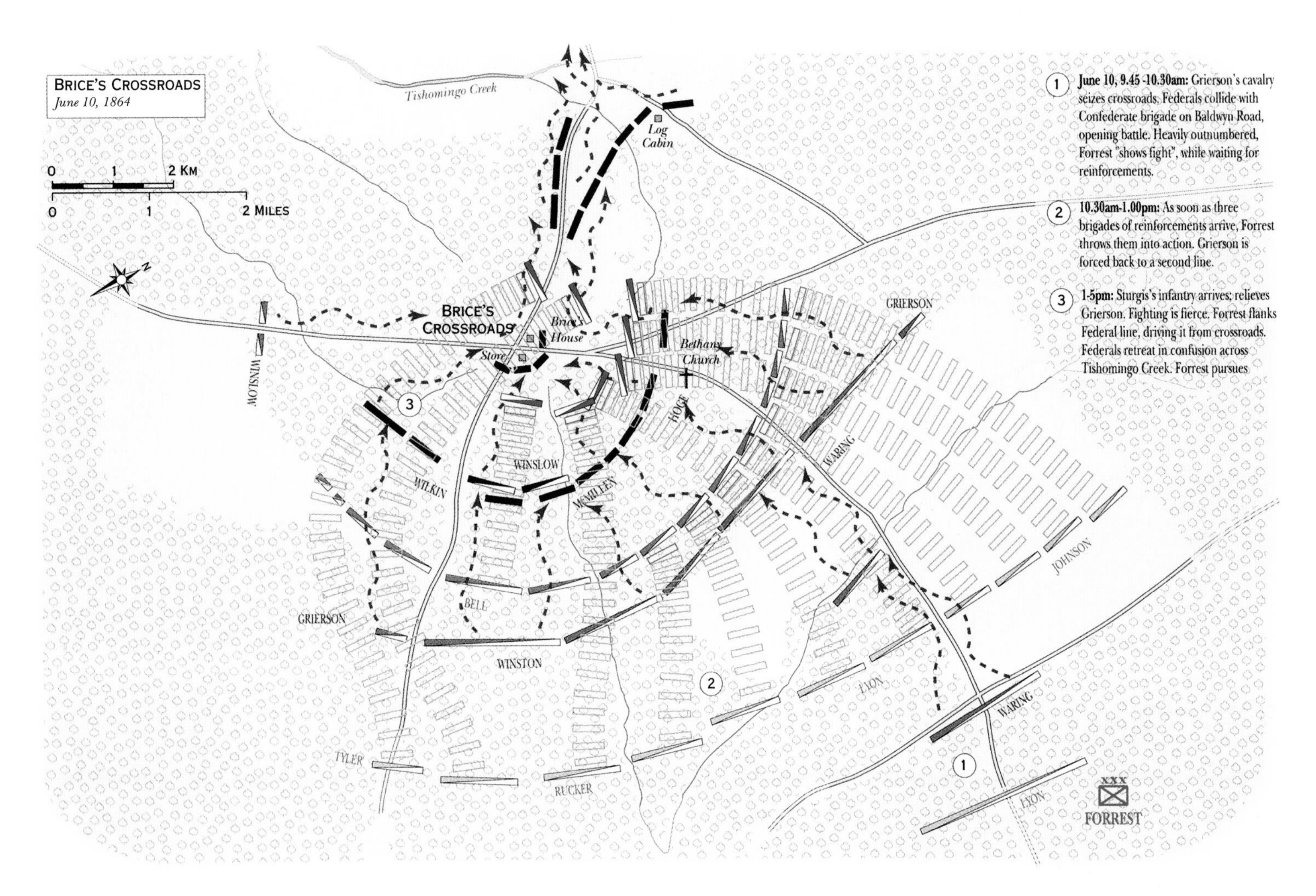
BRICE'S CROSSROADS
June 10, 1864
0 1 2 KM
0 1 2 MILES
Tishomingo Creek
Log Cabin
BRICE'S CROSSROADS
Brice's House
Store
Bethany Church
WINSLOW
GRIERSON
WILKIN
WINSLOW
MCMILLEN
HOGE
WARING
JOHNSON
BELL
WINSTON
TYLER
RUCKER
LYON
WARING
LYON
FORREST
1 June 10, 9.45-10.30am: Grierson's cavalry seizes crossroads. Federals collide with Confederate brigade on Baldwyn Road, opening battle. Heavily outnumbered, Forrest "shows fight", while waiting for reinforcements.
2 10.30am-1.00pm: As soon as three brigades of reinforcements arrive, Forrest throws them into action. Grierson is forced back to a second line.
3 1-5pm: Sturgis's infantry arrives; relieves Grierson. Fighting is fierce. Forrest flanks Federal line, driving it from crossroads. Federals retreat in confusion across Tishomingo Creek. Forrest pursues

advance by Smith into Mississippi, Forrest led 2,000 men in a daring raid into Memphis. The Confederates occupied part of the city and skirmished with the 5,000-man Federal garrison. In a savage rearguard skirmish, Forrest mortally wounded a Union cavalry officer – one of 30 Union soldiers that Forrest killed in hand-to-hand combat during the war. Forrest withdrew from Memphis, with only minor casualties.

Following a successful September raid on rail lines in Alabama and Tennessee, Forrest attacked Federal shipping on the Tennessee River. On October 29 and 30 Forrest used artillery to destroy and capture Federal steamboats near Fort Heiman, Kentucky. Adding to his reputation as a bold and innovative commander, Forrest put troopers aboard two of the captured boats, and used his makeshift "navy" to disrupt river traffic. The Federal navy, however, recaptured one of the boats, and forced the Confederates to burn the other.

Forrest next struck Johnsonville, an important depot used by the Federals to transfer supplies from steamboats for shipment by rail to Nashville. Secreting his artillery on the west bank, on November 4 Forrest shelled steamboats, warehouses and troops. The Federals panicked, set fire to their vessels, which in turn ignited their stores, causing damage estimated at $6,700,000. After shelling the depot next morning, Forrest withdrew into north Alabama to join General Hood's planned invasion of middle Tennessee.

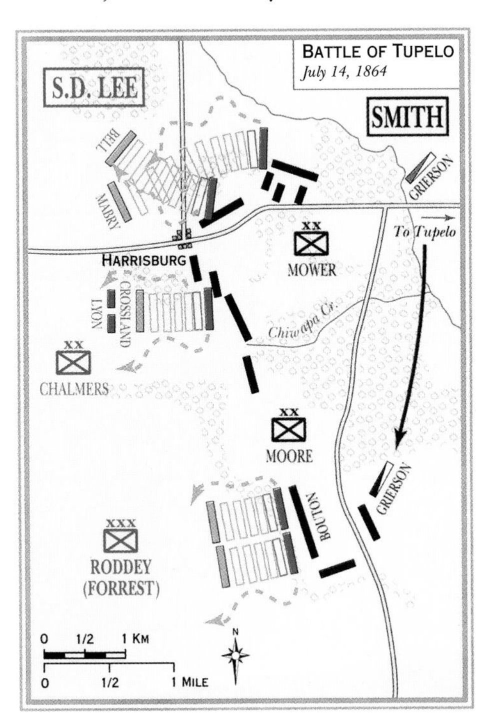
BATTLE OF TUPELO
July 14, 1864
S.D. LEE
SMITH
BELL
MABRY
GRIERSON
To Tupelo
HARRISBURG
MOWER
LYON
CROSSLAND
Chiwapa Cr.
CHALMERS
MOORE
BOUTON
GRIERSON
RODDEY (FORREST)
0 1/2 1 KM
0 1/2 1 MILE

*At Harrisburg, near Tupelo, the Confederates delivered a series of brave but uncoordinated frontal attacks on the Union positions. Heavy losses were incurred, and each assault was repulsed. That night, Smith set fire to Harrisburg. In a rare night attack, Forrest led a dismounted assault, but broke off when subjected to a hail of musketry. Smith withdrew north the following morning (see map left).*

*In a 28-day foray into West Tennessee, Forrest and his raiders blocked the Tennessee near Fort Heiman, then captured Federal shipping which they used to advance on Johnsonville. Despite Federal retaliation, the Confederates bombarded the town, causing extensive damage to Federal shipping and supplies (below).*

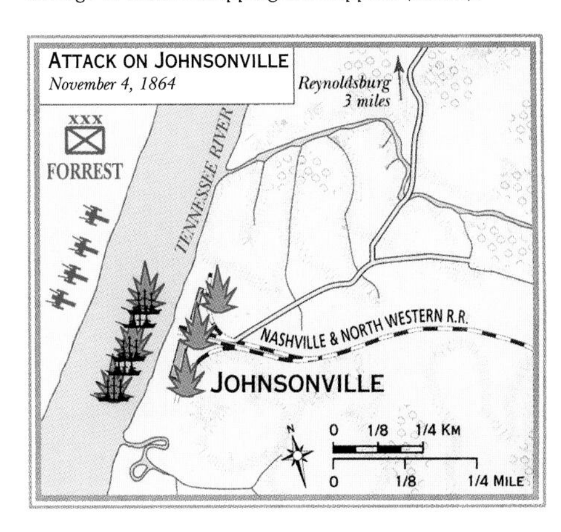
ATTACK ON JOHNSONVILLE
November 4, 1864
Reynoldsburg 3 miles
FORREST
TENNESSEE RIVER
NASHVILLE & NORTH WESTERN R.R.
JOHNSONVILLE
0 1/8 1/4 KM
0 1/8 1/4 MILE

# Mobile Bay Campaign AUGUST 5 – 23 1864

On his promotion to Federal General-in-Chief, Grant set himself a number of tasks, one of which was the capture of Mobile, Alabama. This would close one of the two major ports for blockade running left open to the Confederacy.

The attack on Mobile was aimed at the forts at the entrances to Mobile Bay, and on the Confederate naval squadron that included the ironclad ram *Tennessee*, considered to be invincible. The attack was a joint venture by the Federal army, led by Major General Gordon Granger, and navy, led by Rear Admiral David G. Farragut.

On August 3, Granger's soldiers landed on Dauphin Island, from which they opened fire on Fort Gaines on the 5th. That same morning, Farragut prepared to attack the two forts, the Confederate squadron, and its "infernal machines" – torpedoes. Eighteen ships, including four ironclad monitors, got underway. The monitors in the starboard column, led by *Tecumseh*, were to pass close to Fort Morgan. The port column consisted of 14 wooden warships lashed together in pairs, with *Brooklyn* in the lead, trailed by Farragut's flagship *Hartford*.

Confederate guns opened fire. As *Tecumseh* closed on Fort Morgan, she ran afoul of the Confederate torpedoes, exploded and sank with heavy loss of life. When *Brooklyn* seemed to hesitate, Farragut

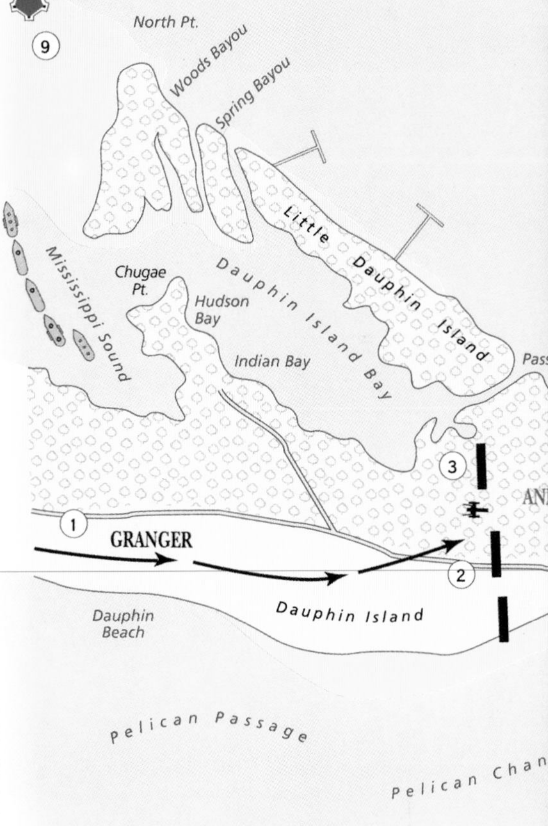
1 Aug 3, 1864 pm: Granger's Federal troops disembark on Dauphin Island
2 Aug 4, midnight: Granger's engineers emplace guns within 1,200 yards of Fort Gaines
3 Aug 5, dawn: Granger opens fire on Fort Gaines
4 Aug 5: Farragut's Federal squadron sails for Fort Morgan
Fort Powell
North Pt.
Woods Bayou
Spring Bayou
Little Dauphin Island
Mississippi Sound
Chugae Pt.
Hudson Bay
Dauphin Island Bay
Indian Bay
GRANGER
Dauphin Beach
Dauphin Island
Pelican Passage
Pelican Chan
5 Aug 5, 7.45am: Tecumseh runs afoul of torpedoes and sinks
6 Aug 5, 7.52am: Hartford takes lead and at 8.05am casts off Metacomet
7 Aug 5, 8.10-8.30am: First phase of battle ends with capture of one Confederate gunboat and one aground
Pelican Island
Tidal Flat

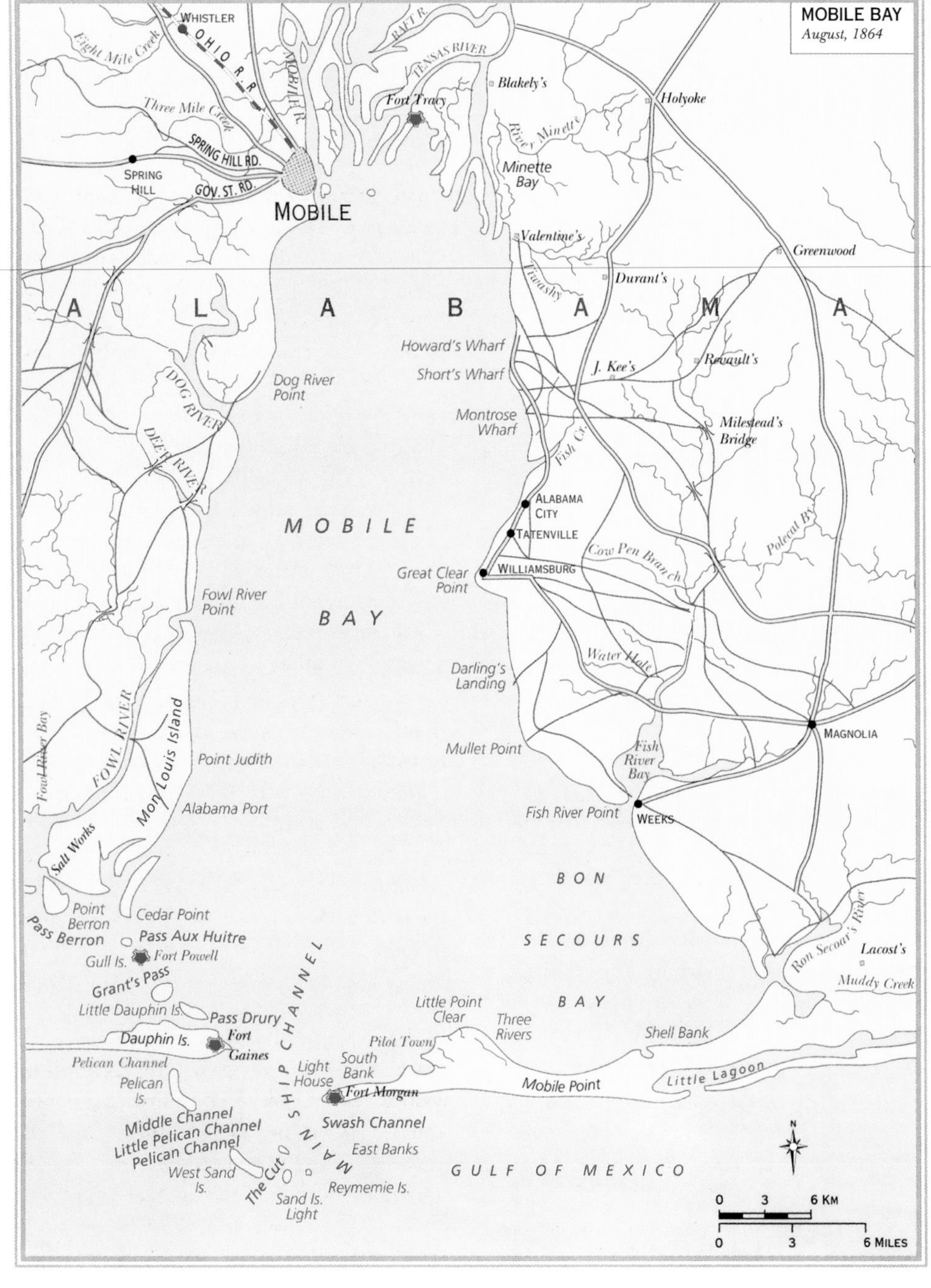
MOBILE BAY
August, 1864
WHISTLER
OHIO R.R.
MOBILE R.
TENSAS RIVER
Eight Mile Creek
Three Mile Creek
SPRING HILL RD.
GOV. ST. RD.
SPRING HILL
MOBILE
Fort Tracy
Blakely's
Holyoke
Minette Bay
Valentine's
Greenwood
Durant's
ALABAMA
Howard's Wharf
Short's Wharf
Dog River Point
J. Kee's
Montrose Wharf
Milestead's Bridge
DOG RIVER
DEER RIVER
MOBILE
BAY
ALABAMA CITY
TATENVILLE
WILLIAMSBURG
Great Clear Point
Cow Pen Branch
Fowl River Point
Darling's Landing
Water Hole
MAGNOLIA
Mullet Point
Fish River Bay
FOWL RIVER
Mon Louis Island
Point Judith
Alabama Port
Fish River Point
WEEKS
Salt Works
BON SECOURS BAY
Point Berron
Cedar Point
Pass Berron
Pass Aux Huitre
Gull Is.
Fort Powell
Grant's Pass
Little Dauphin Is.
Pass Drury
Dauphin Is.
Fort Gaines
MAIN SHIP CHANNEL
Little Point Clear
Three Rivers
Pilot Town
Shell Bank
Lacost's
Muddy Creek
Pelican Channel
Pelican Is.
Light House
South Bank
Fort Morgan
Mobile Point
Little Lagoon
Middle Channel
Little Pelican Channel
Pelican Channel
Swash Channel
East Banks
West Sand Is.
The Cut
Sand Is. Light
Reymemie Is.
GULF OF MEXICO
N
0 3 6 KM
0 3 6 MILES

"First you see the puff of white smoke upon the distant ramparts, and then you see the shot coming, looking exactly as if some gigantic hand has thrown in play a ball toward you."

*A Union Surgeon serving on board the* U.S.S. Lackawanna.

"Do the best you can, sir, and when all is done, surrender."

*Commodore Franklin Buchanan to his second-in-command during the Union assault on the* Tennessee.

*Having suffered intense bombardment from Federal land batteries, three monitors and, and other naval vessels, Fort Morgan surrendered on August 23 (see map right). The capture of the fort gave the Federals control of the port but the city continued to be held by the Confederates.*

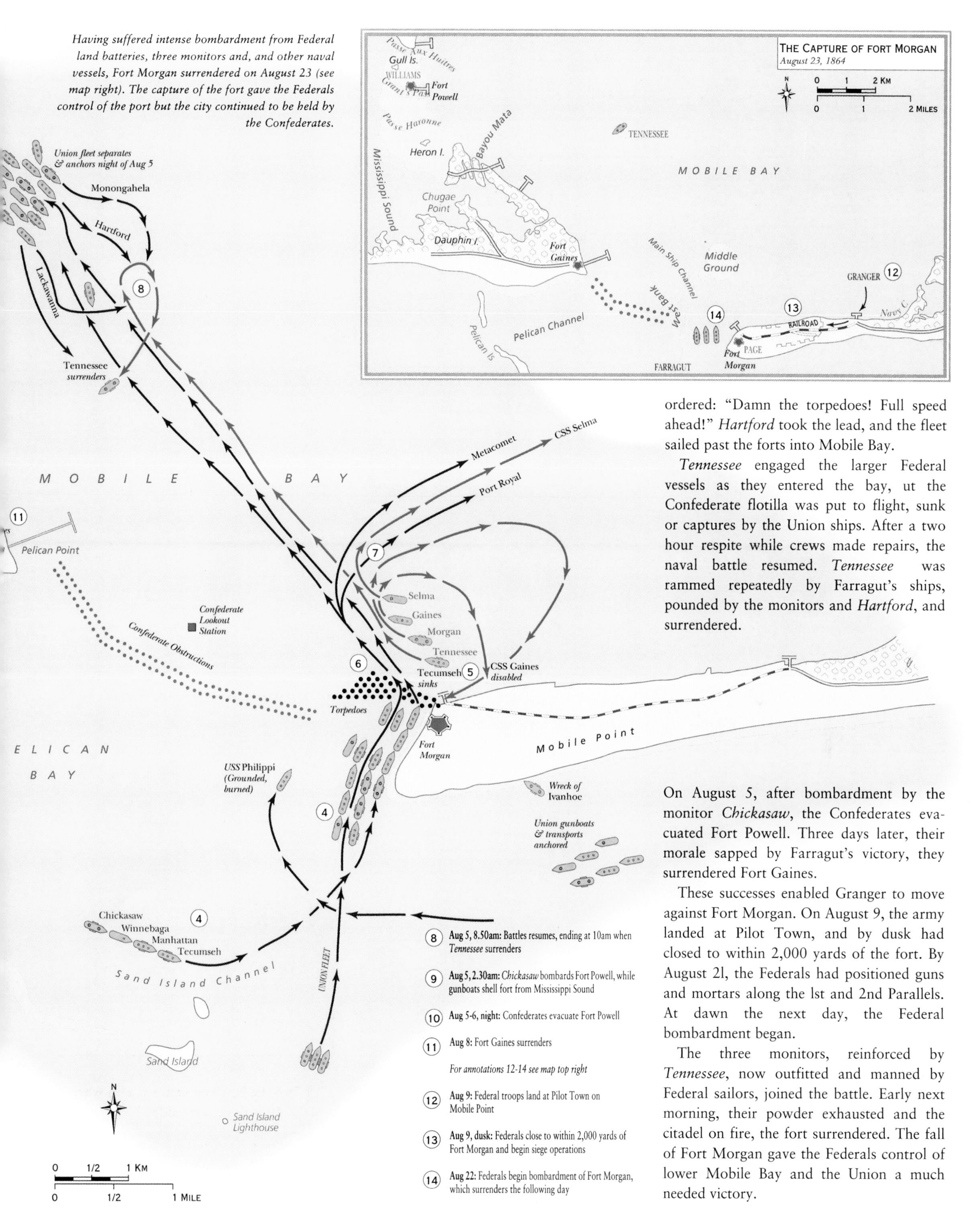
THE CAPTURE OF FORT MORGAN
August 23, 1864
0 1 2 KM
0 1 2 MILES
Gull Is.
Passe Aux Huitres
WILLIAMS
Grant's Pass
Fort Powell
Passe Haronne
Heron I.
Bayou Mata
Mississippi Sound
Chugae Point
Dauphin I.
Fort Gaines
TENNESSEE
MOBILE BAY
Main Ship Channel
West Bank
Middle Ground
Pelican Channel
Pelican Is
GRANGER
RAILROAD
Navy C.
FARRAGUT
Fort Morgan
PAGE
Union fleet separates & anchors night of Aug 5
Monongahela
Hartford
Lackawanna
Tennessee surrenders
MOBILE BAY
Pelican Point
Confederate Lookout Station
Confederate Obstructions
Metacomet
CSS Selma
Port Royal
Selma
Gaines
Morgan
Tennessee
Tecumseh sinks
CSS Gaines disabled
Torpedoes
Fort Morgan
Mobile Point
PELICAN BAY
USS Philippi (Grounded, burned)
Wreck of Ivanhoe
Union gunboats & transports anchored
Chickasaw
Winnebaga
Manhattan
Tecumseh
Sand Island Channel
UNION FLEET
Sand Island
Sand Island Lighthouse
0 1/2 1 KM
0 1/2 1 MILE
8 Aug 5, 8.50am: Battles resumes, ending at 10am when Tennessee surrenders
9 Aug 5, 2.30am: Chickasaw bombards Fort Powell, while gunboats shell fort from Mississippi Sound
10 Aug 5-6, night: Confederates evacuate Fort Powell
11 Aug 8: Fort Gaines surrenders
For annotations 12-14 see map top right
12 Aug 9: Federal troops land at Pilot Town on Mobile Point
13 Aug 9, dusk: Federals close to within 2,000 yards of Fort Morgan and begin siege operations
14 Aug 22: Federals begin bombardment of Fort Morgan, which surrenders the following day

ordered: "Damn the torpedoes! Full speed ahead!" *Hartford* took the lead, and the fleet sailed past the forts into Mobile Bay.

*Tennessee* engaged the larger Federal vessels as they entered the bay, ut the Confederate flotilla was put to flight, sunk or captures by the Union ships. After a two hour respite while crews made repairs, the naval battle resumed. *Tennessee* was rammed repeatedly by Farragut's ships, pounded by the monitors and *Hartford*, and surrendered.

On August 5, after bombardment by the monitor *Chickasaw*, the Confederates evacuated Fort Powell. Three days later, their morale sapped by Farragut's victory, they surrendered Fort Gaines.

These successes enabled Granger to move against Fort Morgan. On August 9, the army landed at Pilot Town, and by dusk had closed to within 2,000 yards of the fort. By August 2l, the Federals had positioned guns and mortars along the lst and 2nd Parallels. At dawn the next day, the Federal bombardment began.

The three monitors, reinforced by *Tennessee*, now outfitted and manned by Federal sailors, joined the battle. Early next morning, their powder exhausted and the citadel on fire, the fort surrendered. The fall of Fort Morgan gave the Federals control of lower Mobile Bay and the Union a much needed victory.

# Siege of Petersburg JUNE – OCTOBER 1864

As most of the railroads which served Lee's army and Richmond – the Confederate capital – passed through Petersburg, in June 1864 the city became the target for General grant's Army of the Potomac. When Grant's initial attempt to seize Petersburg by direct assault failed, he issued orders for the Army of the Potomac to entrench east of the city. However, Grant was unwilling to abandon the use of maneuver entirely. On June 22, he sent II and VI Corps southwest to extend the Federal lines beyond the Confederate flank and sever the railroad to Weldon, North Carolina. A vigorous Confederate counter-attack caused the Federals to recoil on June 23 with losses of 2,962 men.

Momentarily suspending his efforts to maneuver the enemy out of Petersburg, Grant permitted IX Corps to dig a tunnel that would undermine the Confederate works. The mine, an impressive engineering achievement, performed on July 30 as expected by blowing a huge crater 30ft deep and 70ft wide in the Confederate defenses. Unfortunately, the direct assault that followed the explosion was an abject failure, costing 3,798 Federal casualties but only 1,500 Confederate.

In August, in an effort to stretch Lee's forces to breaking point, Grant adopted a strategy of extending both his flanks simul taneously. On the Petersburg front on August 18, he sent V Corps westward in another attempt to break the railroad to Weldon. Federal infantrymen struck the tracks in the vicinity of Globe Tavern and began to wreck them. Immediate Confederate counterattacks forced the Federals to relenquish some ground, but they maintained their hold on the railroad. Federal losses totalled 4,455 men, while Confederate casualties numbered 1,600. Subsequent Federal efforts to damage the railroad south of Globe Tavern resulted in a defeat for II Corps at Reams Station on August 25. This action cost Grant an additional 2,742 men, while Lee's casualties were only 720.

On September 30, Grant again extended both flanks, gaining on the Petersburg front a salient around Popular Springs Church. as part of the same operation, on the richmond front Union troops captured Fort Harrison. Considering the fort a vital part of Richmond's defenses Lee ordered an assault the next day to recapture it, but the Federals repulsed the attack. These actions dented but did not break the left of Lee's line. In this action 2,889 Federals fell, along with approximately 900 Confederates. A month later, a similar Federal thrust near Hatcher's Run fell short of it's objectives, but again extended the federal line westward in exchange for 1,758 men. Winter weather now caused a cessation in major operations until the following year.

> "A slight tremor of the earth for a second, then... a vast column of earth and smoke shoots upward to a great height... having paused for a moment in mid- air, and then, hurtling down with a roaring sound, showers of stones, broken timbers and blackened human limbs, subsides."
>
> *A Union captain on the effects of the mine.*

*A Union mortar position in front of Petersburg. Each mortar would fire a 13-inch spheroid shell once every 30 minutes. To ease the tedium of their work, the gunners often chalked an imaginary Petersburg address on each of the shells.*

*While the Confederate defenders of Petersburg suffered considerable hardships during the siege, the Union troops enjoyed comparative luxury, their kitchens being well supplied and of a permanent construction (below). Both sides, however, became the victims of inactivity and boredom.*

1. **June 22–23:** in an attempt to cut Lee's supply lines, Grant sends Federal II and VI Corps southwest to sever the Weldon Railroad. Federals suffer major repulse by A.P. Hill's three divisions near Jerusalem Plank Road
2. **July 30:** Federals explode mine, which blows a huge crater in Confederate defenses. Subsequent Federal attack is a failure (see map top right)
3. **July 30:** Sheridan's cavalry circles Confederate defenses, but failure of the infantry attack at the crater causes him to halt
4. **Aug 18–21:** Federals begin to wreck railroad near Globe Tavern, but are forced back by Confederate counterattacks, although they maintain hold on the railroad
5. **Aug 25:** Further attacks on the railroad by Federal II Corps at Reams Station are repulsed (see map below right)
6. **Sep 30–Oct 1:** Grant's forces gain salient around Poplar Springs Church, forcing Lee to stretch his lines westward to Hatcher's Run (see map below right)

CHESTERFIELD COUNTY
PETERSBURG
APPOMATTOX RIVER
SOUTHSIDE R.R.
COX RD
BEAUREGAR
BOYDTON RD
HETH

SQUIRREL LEVEL RD
6
Poplar Springs Church
VAUGHAN RD
HETH
HALIFAX RD
WELDON AND PETERSBURG
V I R
D I N W I D D I E
C O U N T Y
5
Reams Station
Malone's Crossing
Rowanty Creek
Malone's Bridge
N
0 1/2 1 KM
0 1/2 1 MILE

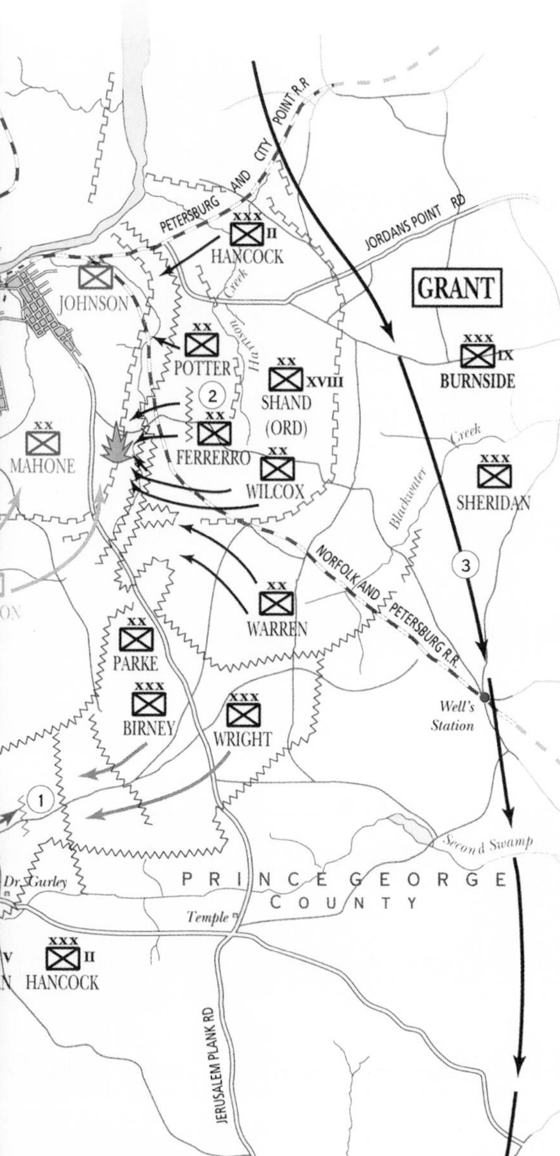

GRANT
PETERSBURG AND CITY POINT R.R.
JORDANS POINT RD
HANCOCK
JOHNSON
POTTER
SHAND (ORD)
FERRERRO
WILCOX
MAHONE
BURNSIDE
SHERIDAN
NORFOLK AND PETERSBURG R.R.
WARREN
PARKE
BIRNEY
WRIGHT
Well's Station
Second Swamp
PRINCE GEORGE COUNTY
Temple
Dr. Gurley
HANCOCK
JERUSALEM PLANK RD
Harwick Swamp
Lee's Mill
SHERIDAN


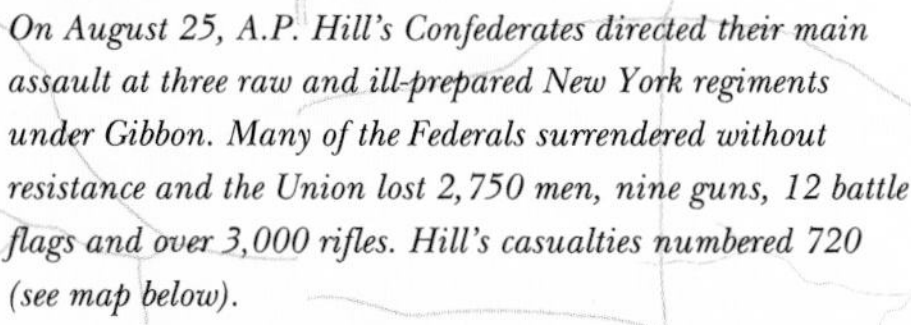

BATTLE OF THE CRATER
*July 30, 1864*

*In accordance with the plan of Lt. Col. Henry Pleasants, four tons of explosives were detonated under the Confederate Salient on July 30. The crater was 30-foot-deep, and 70-foot-wide. The confusion resulting from the explosion, coupled with poor leadership, resulted in many of the Federal attackers rushing into the crater rather than around it and many were slaughtered by the Confederates who had been allowed time to reorganize and man the defenses. Grant called the operation "the most stupendous failure," and in large part blamed Gen. Burnside and the division commander, Gen. James Ledlie (see map above).*

*On July 30, Burnside's IX Corps stormed the breach created by the four-ton mine (above). Burnside had intended that the assault be carried out by one of his African American divisions, but at the last moment Grant countermanded the order and relatively unprepared troops were used.*

*On August 25, A.P. Hill's Confederates directed their main assault at three raw and ill-prepared New York regiments under Gibbon. Many of the Federals surrendered without resistance and the Union lost 2,750 men, nine guns, 12 battle flags and over 3,000 rifles. Hill's casualties numbered 720 (see map below).*

*The wisdom of extending the Union flanks was proved at Poplar Springs Church on September 30. Although G.K. Warren's force suffered from fierce Confederate counterattacks, the pressure was beginning to tell, as it became necessary to rush Confederate troops from one threatened sector to another (see map below).*

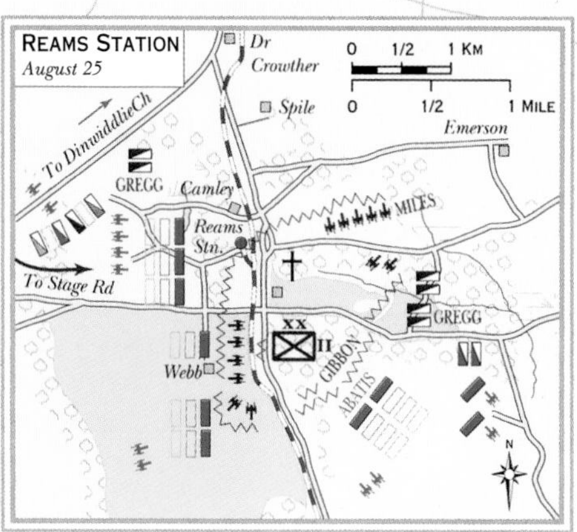

REAMS STATION
August 25
Dr Crowther
Spile
Emerson
To Dinwiddie Ch
GREGG
Camley
Reams Stn.
MILES
To Stage Rd
GREGG
Webb
GIBBON
ABATIS
0 1/2 1 KM
0 1/2 1 MILE


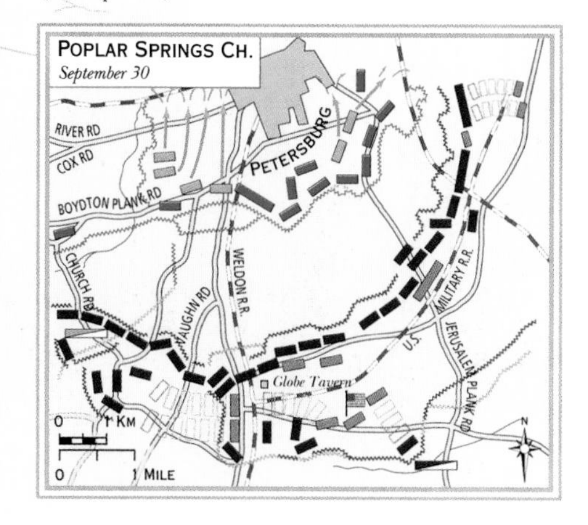

POPLAR SPRINGS CH.
September 30
RIVER RD
COX RD
BOYDTON PLANK RD
PETERSBURG
CHURCH RD
VAUGHN RD
WELDON R.R.
MILITARY R.R.
JERUSALEM PLANK RD
Globe Tavern
0 1 KM
0 1 MILE


"Our living is now very poor, nothing but corn-bread and poor beef – blue and tough – no vegetables, no coffee, sugar, tea or even molasses ... You would laugh, or cry, when you see me eating my supper... It is hard to maintain one's patriotism on ashcake and water."

*A Confederate captain in Petersburg.*

# The Siege of Richmond JULY – OCTOBER 1864

Although he remained focused upon Petersburg, General Grant perceived that Federal activity north of the James River might draw Confederate forces from the Petersburg trenches to permit a Federal breach. On July 27 he sent II Corps to attack through the X Corps bridgehead at Deep Bottom in a major demonstration. His goals were to weaken the Confederate garrison opposite the IX Corps mine at Petersburg and free a cavalry force to raid railroads near Richmond. The violent Confederate response did thin the Petersburg defenses, but it also prevented the cavalry raid. Two weeks later, on August 13, Grant again sent II and X Corps across the James to Deep Bottom. This time the goal was to turn the Confederates out of their defensive complex at Chaffin's Bluff. Again, Confederate reaction was greater than Grant anticipated, and after several days of indecisive fighting, he withdrew II Corps on August 20 to concentrate on the battle around Globe Tavern south of Petersburg.

To prevent Lee from reinforcing his units in the Shenandoah Valley, Grant once more sent a major force north of the James on September 28. This time X and XVIII Corps succeeded in cracking Confederate defenses at Fort Harrison. Confederate counterattacks over the next two days failed to restore their original position, and forced Lee to construct a new line. Furthermore, the fighting north of the James permitted Grant to gain ground southwest of Petersburg around Poplar Spring Church two days later. Federal losses numbered 3,327, with Confederate casualties being perhaps two thirds as great. After a brief initial success, a final Confederate effort on October 7 to regain part of the lost ground along the Darbytown road failed.

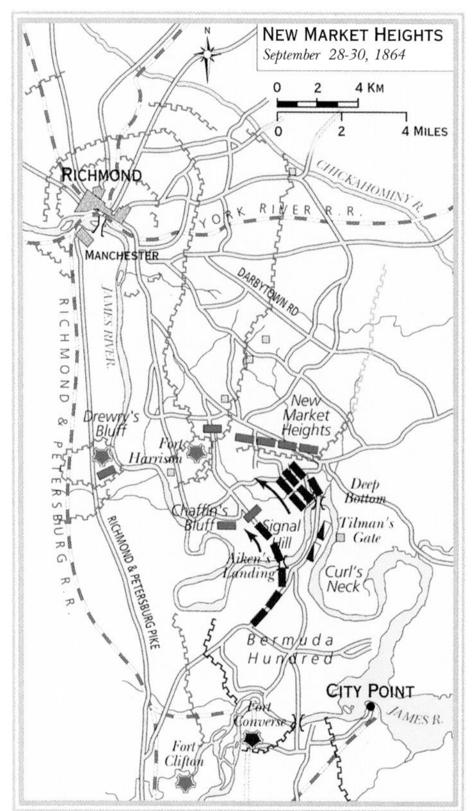

NEW MARKET HEIGHTS
September 28-30, 1864
0 2 4 KM
0 2 4 MILES
RICHMOND
MANCHESTER
CHICKAHOMINY R.
YORK RIVER R. R.
DARBYTOWN RD
JAMES RIVER
RICHMOND & PETERSBURG R. R.
RICHMOND & PETERSBURG PIKE
Drewry's Bluff
Fort Harrison
New Market Heights
Deep Bottom
Chaffin's Bluff
Signal Hill
Aiken's Landing
Tilman's Gate
Curl's Neck
Bermuda Hundred
CITY POINT
JAMES R.
Fort Converse
Fort Clifton


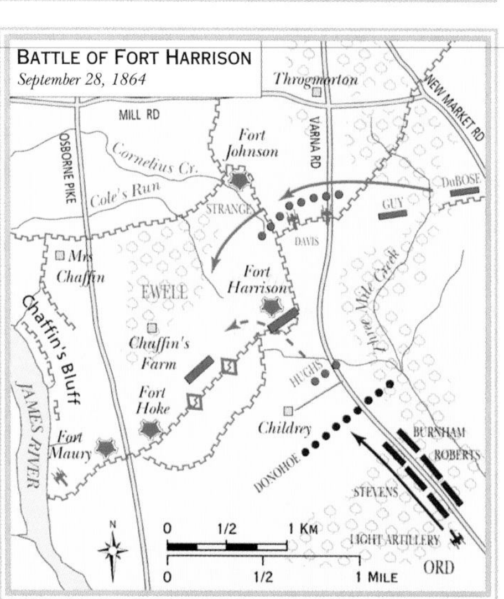

BATTLE OF FORT HARRISON
September 28, 1864
Throgmorton
MILL RD
OSBORNE PIKE
VARNA RD
NEW MARKET RD
Fort Johnson
Cornelius Cr.
Cole's Run
STRANGE
DuBOSE
GUY
DAVIS
Mrs Chaffin
EWELL
Fort Harrison
Chaffin's Farm
Chaffin's Bluff
Fort Hoke
HUGHS
Childrey
Fort Maury
JAMES RIVER
DONOHOE
BURNHAM
ROBERTS
STEVENS
LIGHT ARTILLERY
ORD
0 1/2 1 KM
0 1/2 1 MILE


*On September 28, Grant dispatched X and XVIII Corps north of the James River to carry out a two-pronged attack on Forts Harrison and Gilmer, works that formed part of Richmond's outer fortifications and covered Lee's Chaffin's Bluff defenses. Fighting was heaviest around New Market Heights (see map left) – which the Confederates were forced to abandon – and Fort Harrison. As two corps of infantry under Ord and Birney, together with Kautz's cavalry, approached the fort, the Confederates moved Guy and Dubose to reinforce the lines (see map below left).*

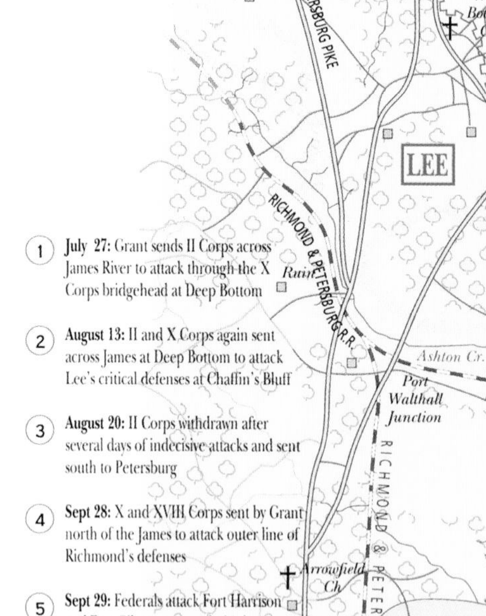

WILTON RD
Laurel Hill Ch
Fort Gilmor
MILL RD
OSBOURNE'S PIKE
Saw Mill
Cole's Run
Fort Harrison
Mrs Chaffin
Chaffin's Bluff
Kingsland Cr.
Signal Station
Graveyar
Mill
Howlett's Farm
RICHMOND & PETERSBURG PIKE
Ware Bottom Ch
LEE
RICHMOND & PETERSBURG R.R.
Ashton Cr.
Port Walthall Junction
Arrowfield Ch
Swift Run
1 July 27: Grant sends II Corps across James River to attack through the X Corps bridgehead at Deep Bottom
2 August 13: II and X Corps again sent across James at Deep Bottom to attack Lee's critical defenses at Chaffin's Bluff
3 August 20: II Corps withdrawn after several days of indecisive attacks and sent south to Petersburg
4 Sept 28: X and XVIII Corps sent by Grant north of the James to attack outer line of Richmond's defenses
5 Sept 29: Federals attack Fort Harrison and Fort Gilmer, two works in the Chaffin's Bluff defenses. Federals take Fort Harrison
6 Sept 30: Confederates attempt to retake Fort Harrison, but fail
7 Oct 7: Confederates attack Federal forces along Darbytown and New Market roads. After initial success, attack fails

*General Grant and his staff, photographed by Mathew Brady in June, 1864 (left).*

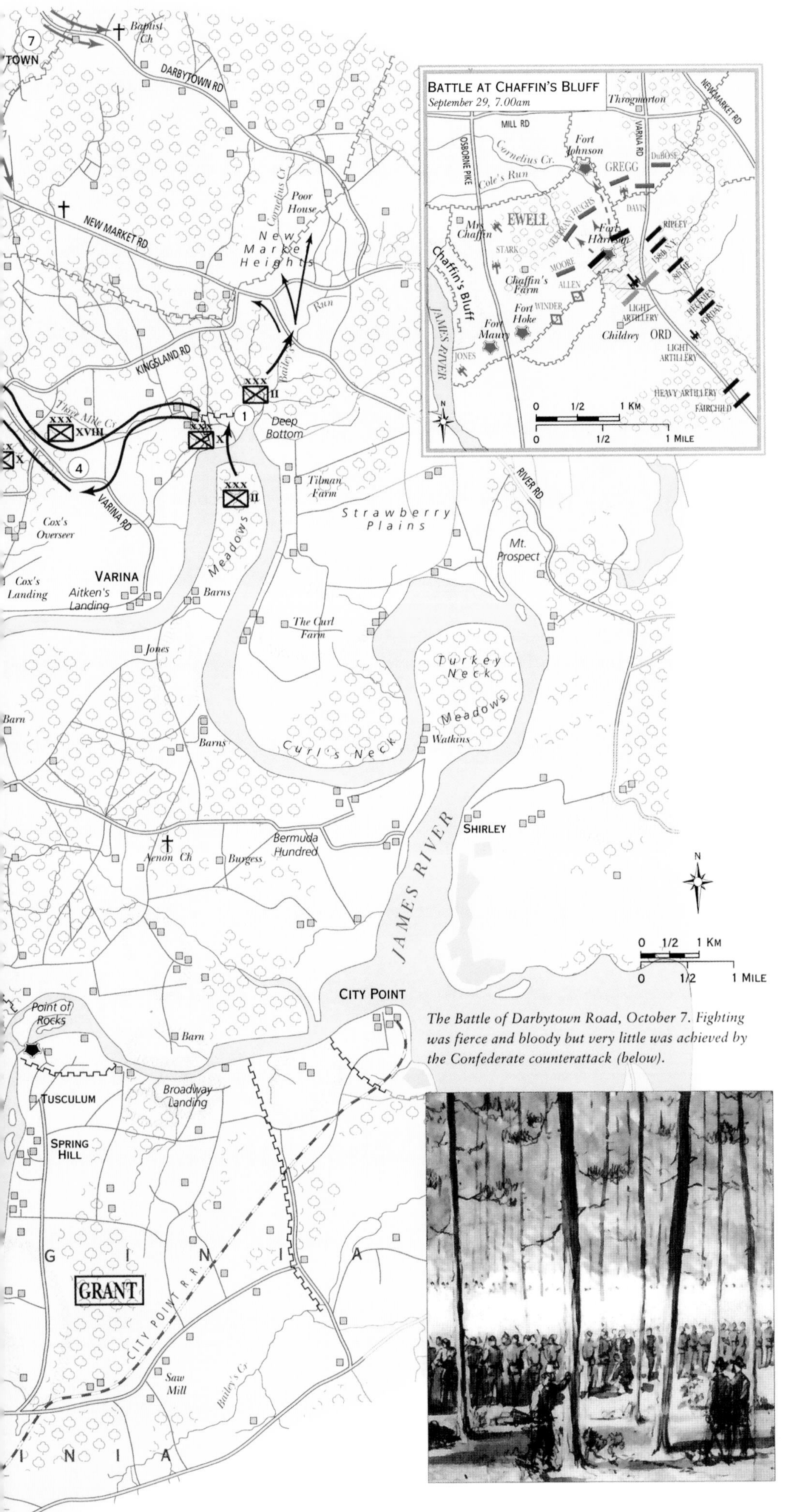
Baptist Ch
TOWN
DARBYTOWN RD
NEW MARKET RD
Poor House
New Market Heights
Cornelius Cr.
KINGSLAND RD
Bailey's Run
XXX II
XXX XVIII
XXX X
Deep Bottom
VARINA RD
Cox's Overseer
Cox's Landing
VARINA
Aitken's Landing
Tilman Farm
Strawberry Plains
RIVER RD
Mt. Prospect
Meadows
Barns
The Curl Farm
Jones
Turkey Neck
Meadows
Watkins
Barn
Barns
Curl's Neck
SHIRLEY
Bermuda Hundred
Nenon Ch
Burgess
JAMES RIVER
0 1/2 1 KM
0 1/2 1 MILE
CITY POINT
Point of Rocks
Barn
TUSCULUM
Broadway Landing
SPRING HILL
GRANT
CITY POINT R.R.
Saw Mill
Bailey's Cr.
G I N I A
I N I A
BATTLE AT CHAFFIN'S BLUFF
September 29, 7.00am
Throgmorton
NEW MARKET RD
MILL RD
OSBORNE PIKE
VARNA RD
Fort Johnson
Cornelius Cr.
Cole's Run
GREGG
DuBOSE
EWELL
Mrs Chaffin
STARK
DAVIS
RIPLEY
Fort Harrison
Chaffin's Farm
MOORE
ALLEN
Chaffin's Bluff
Fort Hoke
WINDER
Fort Maury
JAMES RIVER
JONES
LIGHT ARTILLERY
Childrey
ORD
LIGHT ARTILLERY
HECKMAN
JOURDAN
HEAVY ARTILLERY
FAIRCHILD
0 1/2 1 KM
0 1/2 1 MILE

The Battle of Darbytown Road, October 7. Fighting was fierce and bloody but very little was achieved by the Confederate counterattack (below).

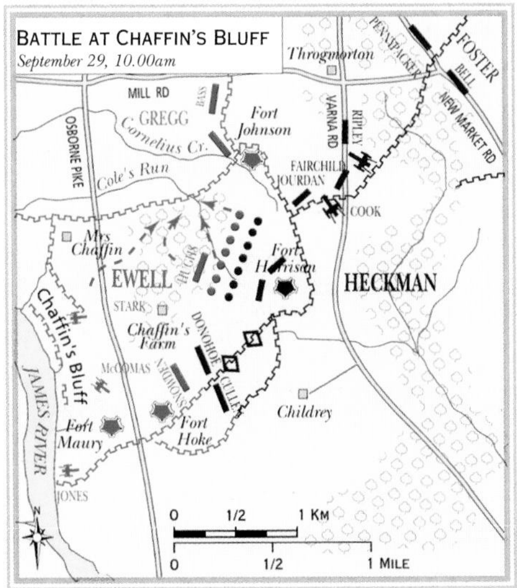
BATTLE AT CHAFFIN'S BLUFF
September 29, 10.00am
Throgmorton
PENNYPACKER
FOSTER
BELL
NEW MARKET RD
MILL RD
OSBORNE PIKE
VARNA RD
GREGG
Fort Johnson
Cornelius Cr.
Cole's Run
RIPLEY
FAIRCHILD
JOURDAN
COOK
Mrs Chaffin
EWELL
Fort Harrison
HECKMAN
STARK
Chaffin's Farm
Chaffin's Bluff
McCOMAS
Childrey
Fort Maury
Fort Hoke
JAMES RIVER
JONES
0 1/2 1 KM
0 1/2 1 MILE

*On September 29, appearing out of the early morning mist Ord stormed Fort Harrison and quickly overwhelmed its garrison (see map above left). Ord was severely wounded in the assault and had to be carried from the field. The ousted Confederates fell back in confusion toward Fort Johnson, where they regrouped for a counterattack which came the following day. The Federals, however, had rebuilt the fort's defenses overnight, and were able to counter the Confederate attack. Lee retrenched to the rear of Fort Harrison (see map above right).*

"The army is pursuing a course of masterly inactivity. Even the work of fortifying, which has been carried out with so much vigor during the past five months, is partially suspended. The hostile armies, separated by only a few rods of forbidden ground, are silently watching each other."

*A Michigan volunteer writing home from the Richmond siegeworks.*

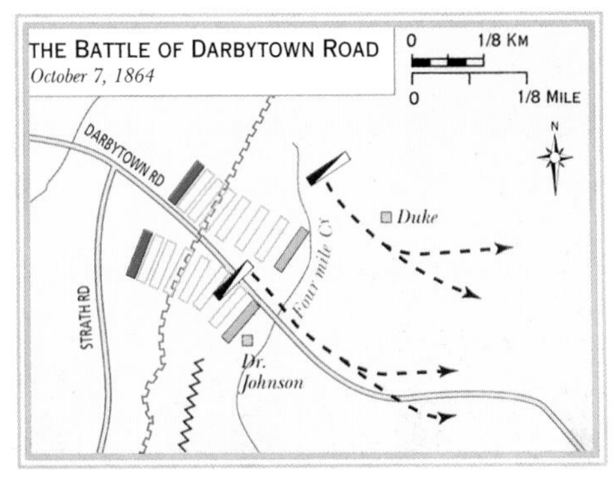
THE BATTLE OF DARBYTOWN ROAD
October 7, 1864
0 1/8 KM
0 1/8 MILE
DARBYTOWN RD
Duke
Four mile Cr.
STRATH RD
Dr. Johnson

*Attempting to push Union troops back from their threatening position near Richmond, the confederates attacked along the Darbytown and New Market roads. Although the attack began well, the Confederates were eventually repulsed (see map above).*

# Sheridan and Early in the Shenandoah Valley: Phase 1

## SEPTEMBER 19–22, 1864

General Philip Sheridan came to the Shenandoah Valley to eliminate Early's army, and wreck the breadbasket of the Confederacy. During early September, Sheridan's 20,000 Federals sparred cautiously with Early's 12,000 Confederates. With the presidential election approaching, Sheridan could not afford to lose. Jubal Early mistook Sheridan's reluctance to fight for timidity, and grew bolder. The Federals bided their time, waiting for an opportunity to strike.

Early confidently divided his army, spreading it from Martinsburg to Winchester. When Sheridan learned this, he advanced immediately against Winchester. Early barely had time to reconcentrate his army before the Federals attacked on September 19. Sheridan appeared to be stalemated – despite a series of heavy frontal assaults on the enemy – when the Federal cavalry crushed the Confederate left flank, forcing Early off the field. Sheridan lost 5,018 men to Early's 3,921.

Early tried to rally his defeated army and make a stand south of Strasburg, choosing the formidable heights of Fisher's Hill to position his army. A month previously Sheridan had elected not to attack this strong line, but now the Federals were flushed with success. Sheridan decided to flank Early's entrenched stronghold by marching to the west, along the face of Little North Mountain. The Confederates were shocked when, on September 22, the Federal troops descended on them, savaging their left flank. They fled their defenses, and Sheridan's soldiers overran the hill. The Federals sustained 528 casualties, inflicting 1,235 on the Confederates.

*The Second Division of the Federal XIX Corps in action at the battle of Winchester, September 19 (above). Faulty deployment at the beginning of the battle had threatened the Union army with defeat, despite its numerical superiority, and only by outflanking the Confederate left was disaster averted.*

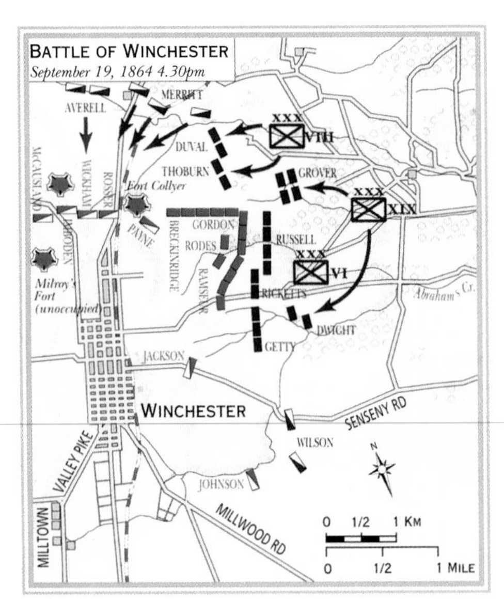
BATTLE OF WINCHESTER
September 19, 1864 4.30pm
MERRITT
AVERELL
DUVAL
THOBURN
Fort Collyer
GROVER
VIII
XIX
GORDON
RODES
RUSSELL
VI
RICKETTS
GETTY
DWIGHT
BRECKINRIDGE
RAMSEUR
PAYNE
WICKHAM
ROSSER
McCAUSLAND
JACKSON
JOHNSON
WILSON
Milroy's Fort (unoccupied)
Abraham's Cr.
WINCHESTER
SENSENY RD
MILLWOOD RD
VALLEY PIKE
MILLTOWN
N
0 1/2 1 KM
0 1/2 1 MILE

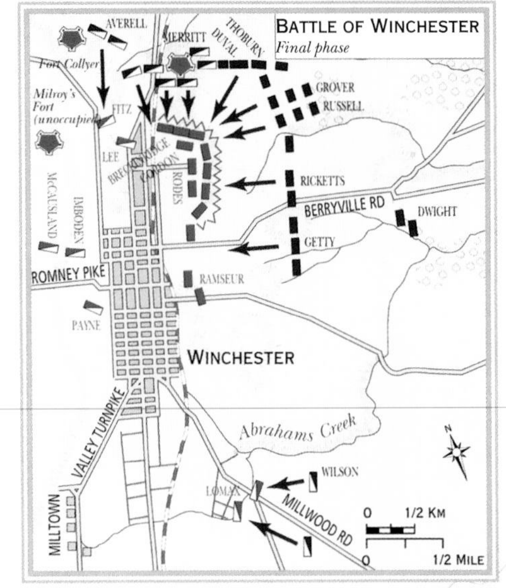
BATTLE OF WINCHESTER
Final phase
AVERELL
MERRITT
THOBURN
DUVAL
Fort Collyer
Milroy's Fort (unoccupied)
FITZ LEE
GROVER
RUSSELL
RICKETTS
BERRYVILLE RD
DWIGHT
GETTY
RODES
RAMSEUR
PAYNE
ROMNEY PIKE
WINCHESTER
Abrahams Creek
WILSON
MILLWOOD RD
VALLEY TURNPIKE
MILLTOWN
N
0 1/2 KM
0 1/2 MILE

*The Federals attacked just before dawn on September 19. After eight hours of attacks and counterattacks, four of Early's corps had been forced back into the town. After a prolonged lull, during which Early scented victory, at 4.30pm Sheridan moved Crook and Emory forward. Merritt's and Averell's cavalry divisions approached from the north to participate in the attack (see map left).*

*At about 5pm, Merritt's cavalry charged Breckinridge, and began a Confederate rout. The Federal infantry pressed the attack and by 6pm Early's army was, in the words of Sheridan's chief-of-staff, sent "whirling through Winchester" toward Fisher's hill (see map above).*

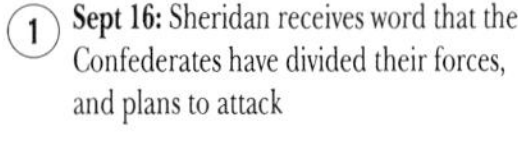
1 **Sept 16:** Sheridan receives word that the Confederates have divided their forces, and plans to attack

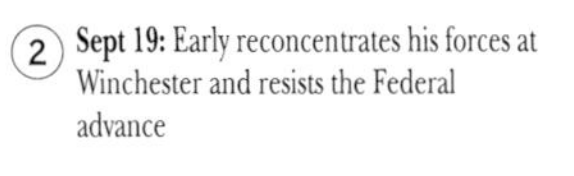
2 **Sept 19:** Early reconcentrates his forces at Winchester and resists the Federal advance

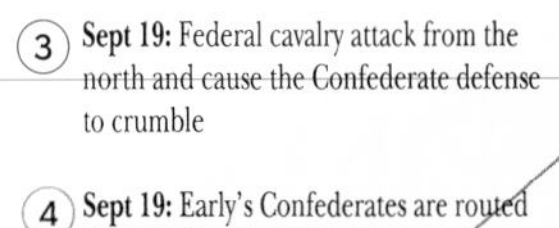
3 **Sept 19:** Federal cavalry attack from the north and cause the Confederate defense to crumble

4 **Sept 19:** Early's Confederates are routed through the city of Winchester

EARLY
N. W. TURNPIKE
WINCHESTER
EMORY
KERNSTOWN
NEWTOWN
Little North Mountain
BACK ROAD
Cedar Creek
WRIGHT
SHERIDAN
EMORY
CROOK
VALLEY TURNPIKE
MIDDLETOWN
NENEVEH
STRASBURG
BOWMAN'S FORD
BUCKTON
CEDARVILLE
Fisher's Hill
EARLY
Water Lick
TOMS BROOK
FRONT ROYAL
To Staunton
WOODSTOCK
Shenandoah River

5 **Sept 20:** Early establishes a strong Confederate defensive line at Fisher's Hill

6 **Sept 21:** Sheridan's Federals skirmish to test the Fisher's Hill defenses

7 **Sept 22:** Crook's Federals flank the Confederate fortifications, enabling Sheridan to overrun the Fisher's Hill defenses

8 **Sept 22:** Confederates retreat up the Valley, powerless to halt Sheridan's advance

Sheridan pursued the Confederates south as far as Staunton. Convinced that Early was finished, the Federal commander ordered his army to return down the Valley, systematically burning the crops as they went.

No longer could the Confederacy rely on the Shenandoah Valley to feed its armies. In the course of three days, Sheridan had achieved a double victory at Winchester and Fisher's Hill. These victories, together with the fall of Atlanta, ensured the reelection of Abraham Lincoln.

"The broad blue wave surged forward with a yell which lasted for minutes. In response there arose from the northern front of the wood a continuous, deafening wail of musketry... But the yell came steadily on and triumphed gloriously over the fusillade... Presently, looking to the left, we saw that the Vermonters were charging, and we jumped forward with a scream, the officers leading and the men hard after."

*John De Forest of the Twelfth Connecticut at the battle of Winchester.*

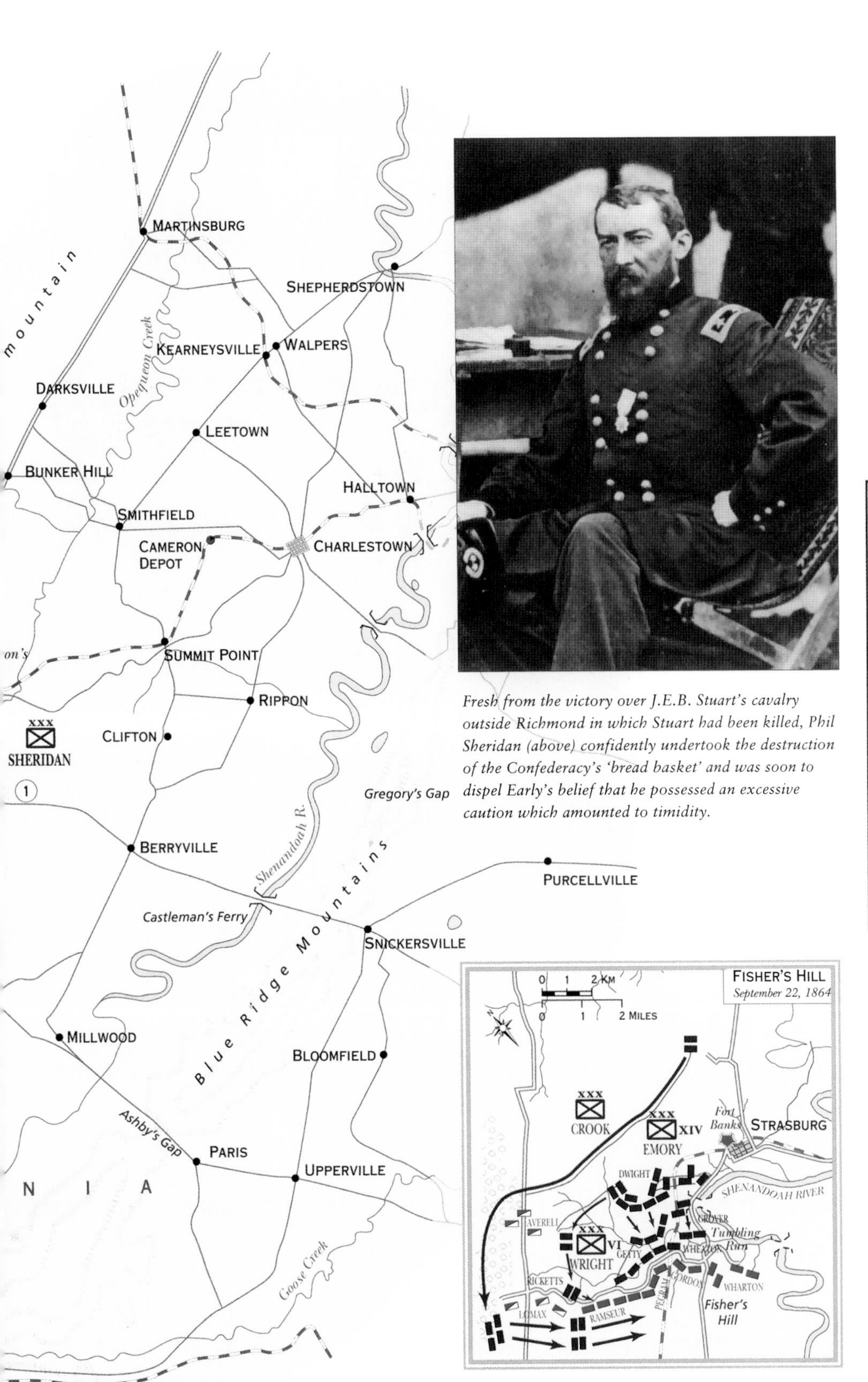
MARTINSBURG
SHEPHERDSTOWN
KEARNEYSVILLE
WALPERS
DARKSVILLE
Opequon Creek
LEETOWN
BUNKER HILL
HALLTOWN
SMITHFIELD
CAMERON DEPOT
CHARLESTOWN
SUMMIT POINT
RIPPON
CLIFTON
SHERIDAN
Gregory's Gap
BERRYVILLE
Shenandoah R.
PURCELLVILLE
Castleman's Ferry
SNICKERSVILLE
Blue Ridge Mountains
MILLWOOD
BLOOMFIELD
Ashby's Gap
PARIS
UPPERVILLE
N I A
Goose Creek
N
0 1 2 KM
0 1 2 MILES
FISHER'S HILL
September 22, 1864
CROOK
XIV
EMORY
Fort Banks
STRASBURG
DWIGHT
SHENANDOAH RIVER
AVERELL
VI
WRIGHT
GETTY
GROVER
Tumbling Run
WHARTON
RICKETTS
LOMAX
RAMSEUR
PEGRAM
GORDON
Fisher's Hill

*Fresh from the victory over J.E.B. Stuart's cavalry outside Richmond in which Stuart had been killed, Phil Sheridan (above) confidently undertook the destruction of the Confederacy's 'bread basket' and was soon to dispel Early's belief that he possessed an excessive caution which amounted to timidity.*

*Early retreated to a strong, defensive position on Fisher's Hill, hotly pursued by Sheridan. Two Federal corps made a feint attack against Early's entrenched line, descending into Tumbling Run Ravine and up Fisher's Hill while Crook's corps, aiming for the Confederate left, climbed the steep mountain paths. Appearing suddenly out of dense woodland, with the setting sun behind them, Crook's corps secured the Confederate entrenchments in their rear and flank. Once more, Early's Confederates were pushed southward (see map left).*

*Sheridan's wagon train advances up the Shenandoah Valley (below). Grant had ordered that the valley be stripped of it's wealth; everything that Sheridan could not carry away he must destroy, leaving nothing but "barren waste".*

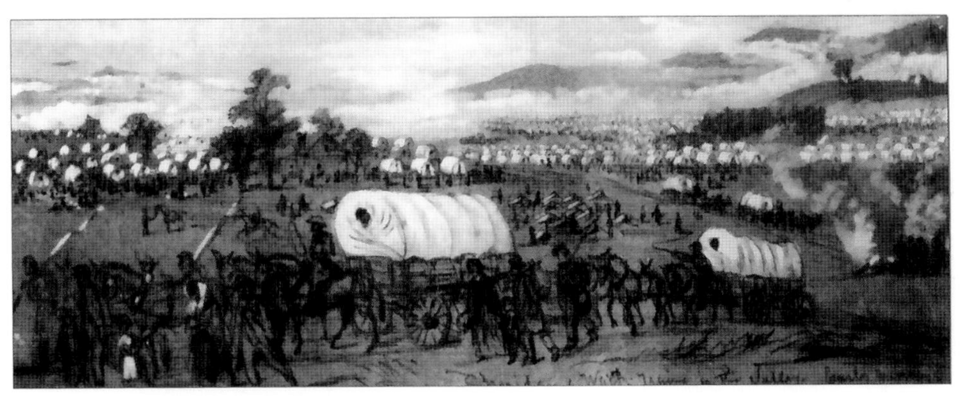

# Sheridan & Early in the Shenandoah Valley: Phase 2

## OCTOBER 9 1864 – MARCH 2 1865

Jubal Early's Confederates hovered near Sheridan's withdrawing army, looking for a chance to strike. The Federals marched north from Staunton, burning the Valley's farms, with the Confederate cavalry harrassing them continuously. Anxious to teach the Confederates a lesson, Sheridan ordered his cavalry to destroy Early's mounted force. The Federal horse soldiers hurled themselves at Early's cavalry at Tom's Brook on October 9, and routed it completely.

Sheridan's army of 32,000 pitched camped along Cedar Creek, below Middletown. Early followed with a reinforced army of 21,000, and surprised the Federals when they attacked the Union camp on October 19. Driven back beyond Middletown, the Federals retreated in confusion. Sheridan arrived and not only averted disaster, but turned the tide of battle with a massive counterattack. The Federals swept the Confederates back across Cedar Creek and inflicted a third major defeat on Jubal Early. The Federals had lost 5,672 men; the Confederates 2,910.

Early retreated to New Market and continued to skirmish with the Federals throughout November. Eventually, Early was forced to withdraw up the Valley to search for supplies. Sheridan ordered his army into winter quarters around Winchester. Throughout the winter, both

Confederate troops under the command of Maj. Gen. Thomas L. Rosser attack the rear of Sheridan's retreating army, October 8 (above). Although enthusiastically welcomed by Early, Rosser's reinforcements were under-nourished and ill-equipped.

"Our horses are worn down, and their is no source whence we can recruit. We have only pistols, sabers and old fashioned rifles, worn-out saddles, and none of the equipment in the way of portable furnaces, horse-shoes and other transportation requisite for efficient cavalry work; and above all, we have not enough food to keep the horses up."

*A Confederate cavalryman of Rosser's command.*

*Brig. Gen. John B. Gordon (above) who commanded the attack on the Union left flank at Cedar Creek in the Early hours of October 19. In fact, Gordon was, largely responsible for the formation of the Confederate plan of attack, and was so confident of success that he agreed to accept all responsibility in the event of failure.*

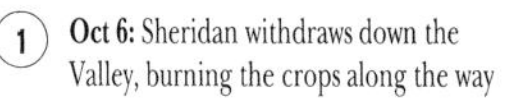

1. **Oct 6:** Sheridan withdraws down the Valley, burning the crops along the way
2. **Oct 9:** Sheridan's cavalry routs the Confederate horse soldiers at Tom's Brook
3. **Oct 19:** Early surprises the Federals and overruns their encampment along Cedar Creek
4. **Oct 19:** Sheridan rallys his army and counterattacks, driving the Confederates across Cedar Creek
5. **Feb 25:** Sheridan breaks camp at Winchester and heads south up the Valley
6. **March 2:** Sheridan brushes aside Early's cavalry at Mount Crawford
7. **March 3:** Sheridan crushes Early's defenses at Waynesboro and forces the Confederates out of the Valley

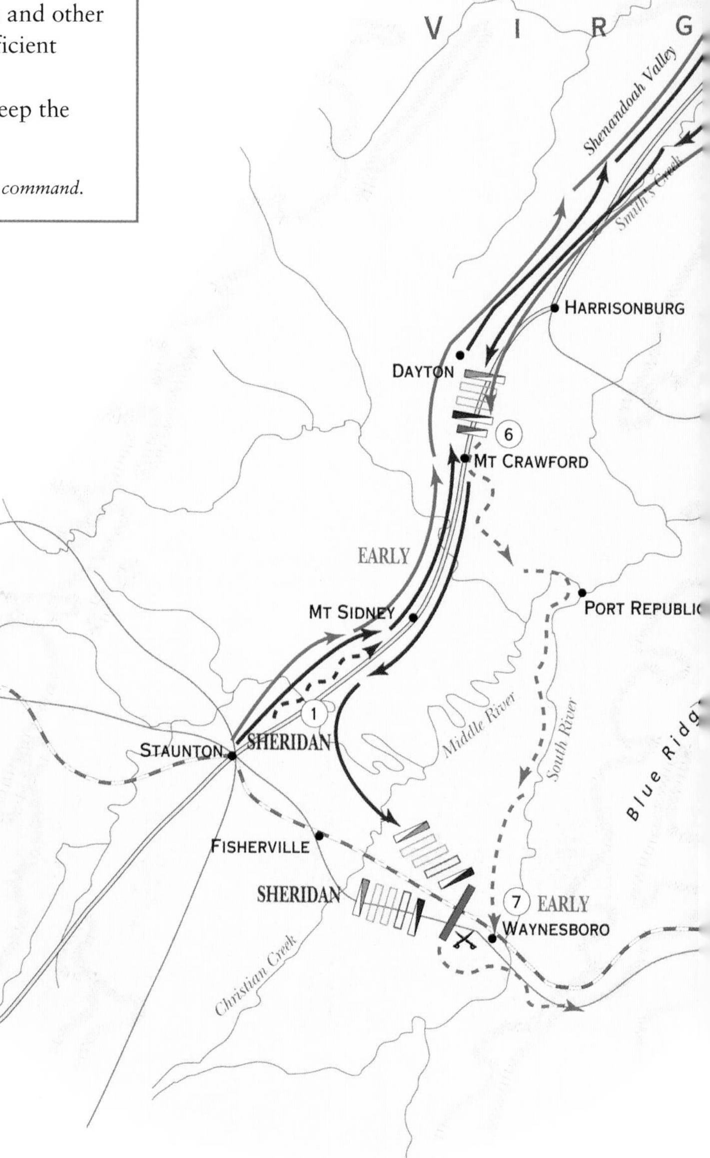
TIMBERVILLE
V I R G
Shenandoah Valley
Smith's Creek
HARRISONBURG
DAYTON
6
MT CRAWFORD
EARLY
PORT REPUBLIC
MT SIDNEY
1
STAUNTON
SHERIDAN
Middle River
South River
Blue Ridge
FISHERVILLE
SHERIDAN
7 EARLY
WAYNESBORO
Christian Creek

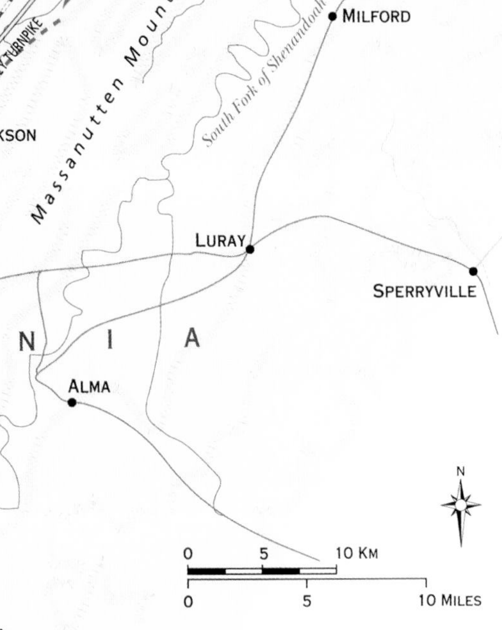

SHERIDAN
To Winchester
NEWTOWN
WEST VIRGINIA
Cedar Creek
MIDDLETOWN
NINEVEH
STRASBURG
Fishers Hill
CEDARVILLE
TOM'S BROOK
EARLY
FRONT ROYAL
ROSSER
WOODSTOCK
EDINBURG
VALLEY TURNPIKE
Massanutten Mountains
South Fork of Shenandoah R.
MILFORD
JACKSON
LURAY
SPERRYVILLE
N I A
ALMA
N
0 5 10 KM
0 5 10 MILES


> "With great pleasure, I tender to you... the thanks of the nation and my own personal admiration and gratitude for the month's operations in the Shenandoah Valley, and especially for the splendid work of October 19."
>
> *Lincoln to Sheridan, October 22.*

*Having crossed Cedar Creek during the night, Early's Confederates attacked Sheridan's camp at dawn on October 19 The astonished Federals dropped their weapons and fled north. Kershaw's division, meanwhile, routed Thoburn's two brigades a mile south of the main Federal encampment (see map below). The Confederates paused and the attack faltered. Federal VIII and XIX Corps withdrew in disorder to the north, with VI Corps covering the withdrawal (see map right). Having advanced beyond Middletown, Early ordered a halt to rest his exhausted army. Sheridan, absent when the battle began, now arrived and rallied his dispersed forces north of Middletown. At 4pm, Sheridan counterattacked. Although initially blocked, when the Federals flanked Gordon's division, the Confederate line began to crumble. The retreat, pushed by the cavalry of Custer and Merritt, became a rout and by nightfall, Early's defeated forces had reached New Market (see map below).*

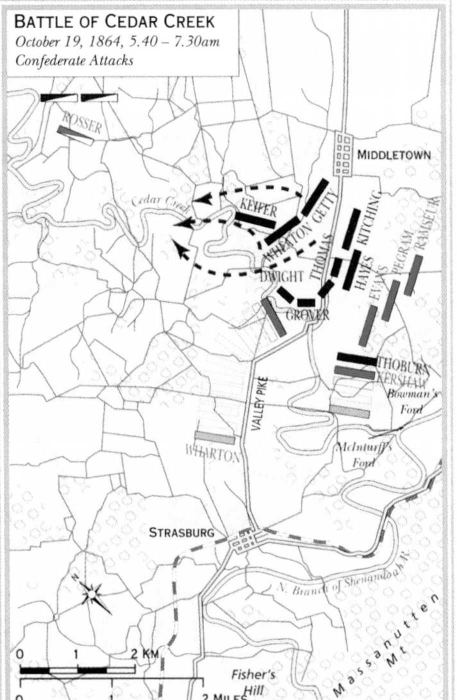

BATTLE OF CEDAR CREEK
October 19, 1864, 5.40 – 7.30am
Confederate Attacks
ROSSER
Cedar Creek
KEIFER
WHEATON
GETTY
THOMAS
KITCHING
HAYES
EVANS
PEGRAM
RAMSEUR
DWIGHT
GROVER
THOBURN
KERSHAW
Bowman's Ford
McInturff's Ford
WHARTON
VALLEY PIKE
MIDDLETOWN
STRASBURG
N. Branch of Shenandoah R.
Massanutten Mt
Fisher's Hill
0 1 2 KM
0 1 2 MILES


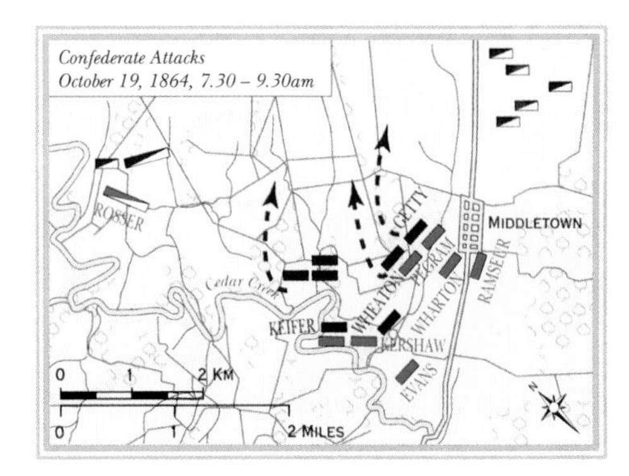

Confederate Attacks
October 19, 1864, 7.30 – 9.30am
ROSSER
Cedar Creek
GETTY
WHEATON
PEGRAM
WHARTON
RAMSEUR
KEIFER
KERSHAW
EVANS
MIDDLETOWN
0 1 2 KM
0 1 2 MILES


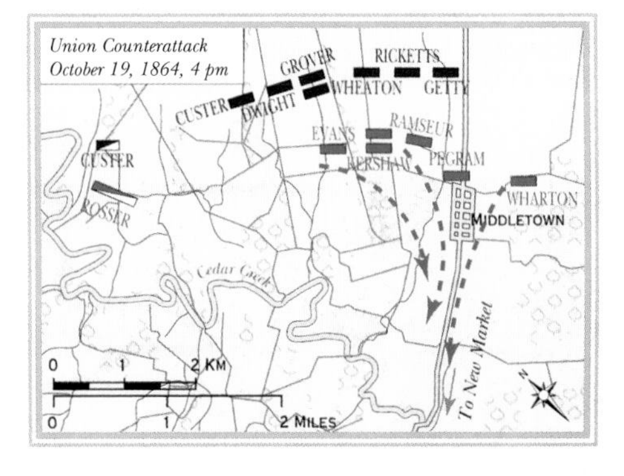

Union Counterattack
October 19, 1864, 4 pm
GROVER
RICKETTS
WHEATON
GETTY
CUSTER
DWIGHT
RAMSEUR
EVANS
KERSHAW
PEGRAM
CUSTER
ROSSER
WHARTON
MIDDLETOWN
Cedar Creek
To New Market
0 1 2 KM
0 1 2 MILES


*Sheridan's army pursuing Early up the Valley of the Shenandoah after the battle of Cedar Creek (below). Returning from Winchester, Sheridan was able to turn his demoralized troops and launch a devastating attack on the Confederates who had paused to loot the abandoned Union encampment.*

armies diminished in size; part of Early's command left to join Robert E. Lee in Petersburg; the rest dispersed to hunt for supplies. Sheridan's men were sent on various duties, until eventually he retained only two cavalry divisions.

Sheridan opened the final act of the Valley campaign when his cavalry broke camp on February 25, 1865, and started southward. Sheridan defeated Early's meager cavalry force on March 2, at Mount Crawford. The following day, the Federals destroyed the remnant of Early's army at Waynesboro. With this victory, Philip Sheridan had eliminated the last sizeable Confederate force in the Shenandoah Valley. His campaign was complete.

# Price's Raid in Missouri SEPTEMBER – OCTOBER 1864

To spark a general uprising against the Federal occupation of Missouri, General Sterling Price initiated the war's longest raid in September, 1864. Marching northeast from Arkansas with 12,000 men – mostly mounted infantry – Price learned that St. Louis was heavily defended. He instead captured Fort Davidson near Pilot Knob on September 27, losing ten per cent of his forces in costly assaults.

Price then proceeded to Jefferson City, intending to capture the state capital for political purposes, but skirmishes south of the city on October 7 revealed its defenses to be too strong. Consequently, he moved west to threaten Kansas City. Meeting little Federal resistance, Price captured Boonville, Lexington, and Independence, then split his forces and defeated the Federals simultaneously at Glasgow and Sedalia on October 15. Approaching Kansas City on October 22, he defeated Federal defenders at Byram's Ford on the Big Blue River. These men were part of the troops from Kansas and northwestern Missouri which General Samuel Curtis was concentrating against Price.

Supported by forces under Major General Alfred Pleasonton – who had been pursuing Price from the east – Curtis blocked Price's advance on Kansas City at Westport. A fierce battle began near dawn on October 23, continuing throughout the morning. Curtis had some 20,000 men, while Price's raiders had been reduced by cumulative losses and desertion to some 8,500 men. Curtis triumphed and Price was forced to retreat. The battle of Westport, the largest engagement west of the Mississippi River, had been a decisive Federal victory.

Crossing into Kansas, Price had camped along the Marais des Cygnes River when, on October 24, the pursuing Federals attacked his rear guard, then overtook his main force as it withdrew. Although outnumbered more than two-to-one, Pleasonton's Federals routed the Confederates. Price escaped by making a forced march to Carthage, Missouri, and reached Cane Hill, Arkansas, on November 1.

Price had inflicted millions of dollars in damage to government and private property, and had captured over 3,000 Federals – a significant accomplishment – but he lacked the logistical support system to hold any of the territory he liberated. Although it greatly alarmed the Federals, Price's raid, the last great Confederate offensive in the Trans-Mississippi, was no more than a diversion.

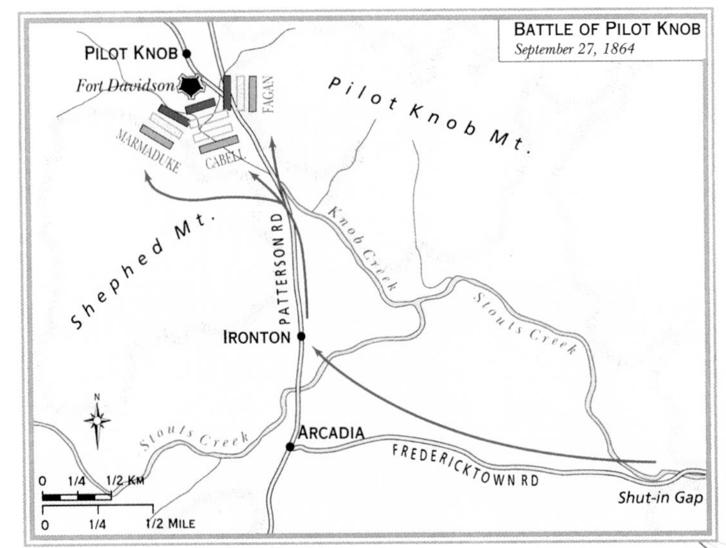
BATTLE OF PILOT KNOB
September 27, 1864
PILOT KNOB
Fort Davidson
FAGAN
MARMADUKE
CABELL
Pilot Knob Mt.
Shepherd Mt.
PATTERSON RD
Knob Creek
Stouts Creek
IRONTON
ARCADIA
FREDERICKTOWN RD
Shut-in Gap
N
0 1/4 1/2 KM
0 1/4 1/2 MILE

*Gen. Sterling Price's invasion of Missouri developed rapidly with skirmishing at Arcadia, Ironton and Mineral Point. His original plan to attack St. Louis – whose warehouses were stocked with useful goods – was dropped when Price learned of the strength of its defenses, manned by 8,000 Federals. Instead, Price decided to capture Fort Davidson, just south of Pilot Knob (above). On September 27, Price's 7,000 raiders charged the fort's thick walls. The Federals, commanded by Thomas Ewing, held off the attack with only 1,000 men and seven guns, inflicting 1,500 casualties on the Confederates at the cost of only 200. After refusing to surrender, Ewing secretly evacuated the fort during the night, leaving a slow fuse burning in the powder magazine which exploded in a sheet of flame when Ewing and his men were well on their way to Rolla.*

1 **Sept 19:** Price moves north from Pocahontas, in three columns converging near Pilot Knob
2 **Sept 27:** Price fails to capture Ewing's small garrison at Fort Davidson in a costly, inexpert assault
3 **Oct 6–8:** Price skirmishes with Federal forces at the Osage River. On Oct 7 Price drives a body of Federal cavalry back into Jefferson city, but impressed with its defenses, he moves west on October 8, intending to threaten Kansas City
4 **Oct 9:** Price defeats a small militia force at Boonville
MISSOURI RIVER
KANSAS CITY
INDEPENDENCE
WESTPORT
6
LEXINGTON
WARRENSBURG
7
Marais des Cygnes River
Big Blue River
Grand River
FORT SCOTT
OSECOLA
KANSAS
Shoal Creek
CARTHAGE
MT VERN
JOPLIN
INDIAN TERRITORY
NEESHO
CASSVILLE
MAYSVILLE
BENTONVILLE
FAYETTEVILLE
CANE HILL
8
A R K

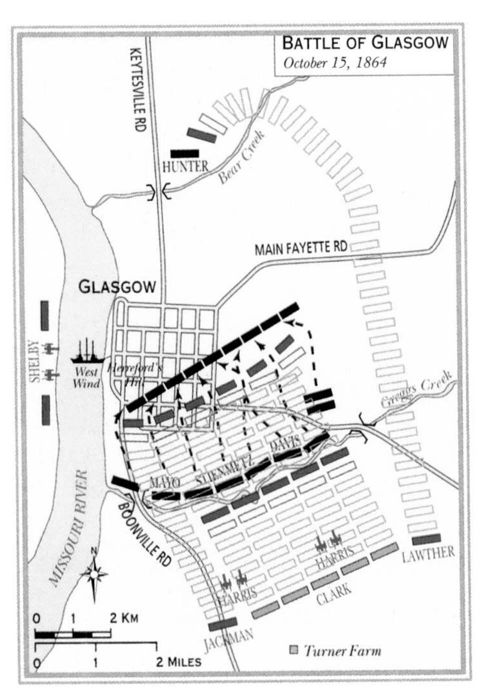
BATTLE OF GLASGOW
October 15, 1864
KEYTESVILLE RD
HUNTER
Bear Creek
MAIN FAYETTE RD
GLASGOW
SHELBY
West Wind
Hereford's Hill
Greggs Creek
DAVIS
MAYO
SHENMETZ
BOONVILLE RD
MISSOURI RIVER
HARRIS
LAWTHER
CLARK
JACKMAN
Turner Farm
0 1 2 KM
0 1 2 MILES

*Finding that Jefferson City, the state capital, was heavily defended, Gen. Price continued upriver toward Boonville and captured it. Splitting his forces, Price sent Jo Shelby's men to attack Sedatia, while Marmaduke's division attacked Glasgow on the same day, October 15 (left). Both attacks were successful and the towns captured.*

5. **Oct 15:** Splitting his forces, Price wins victories at Glasgow and Sedalia, capturing a substantial amount of supplies in the process. Casualties were slight

6. **Oct 22–23:** Price brushes aside the Federals blocking his approach to Westport at Byram's Ford on October 22. The following day he is defeated by Curtis and Pleasonton. With only 7,000 effectives, Price begins his retreat that same day

7. **Oct 25:** At a crossing of the Marais des Cygnes River, Pleasonton's forces rout Price's army in the last major combat of the campaign

8. **Nov 1:** Price reaches Cane Hill, ending the raid

> "Price entered Missouri fasting and furnishing his troops on the rich products and abundant spoils of the Missouri Valley, but crossed the Arkansas destitute, disarmed, disorganized, and avoiding starvation by eating raw corn and slippery-elm bark."
>
> *Gen. Samuel Ryan Curtis.*

*Gen. Samuel Ryan Curtis (right) was a graduate of West Point and a former congressman for the State of Iowa who had encountered and defeated Price in 1862 at Pea Ridge when the Confederate general had been under the overall command of Earl Van Dorn.*

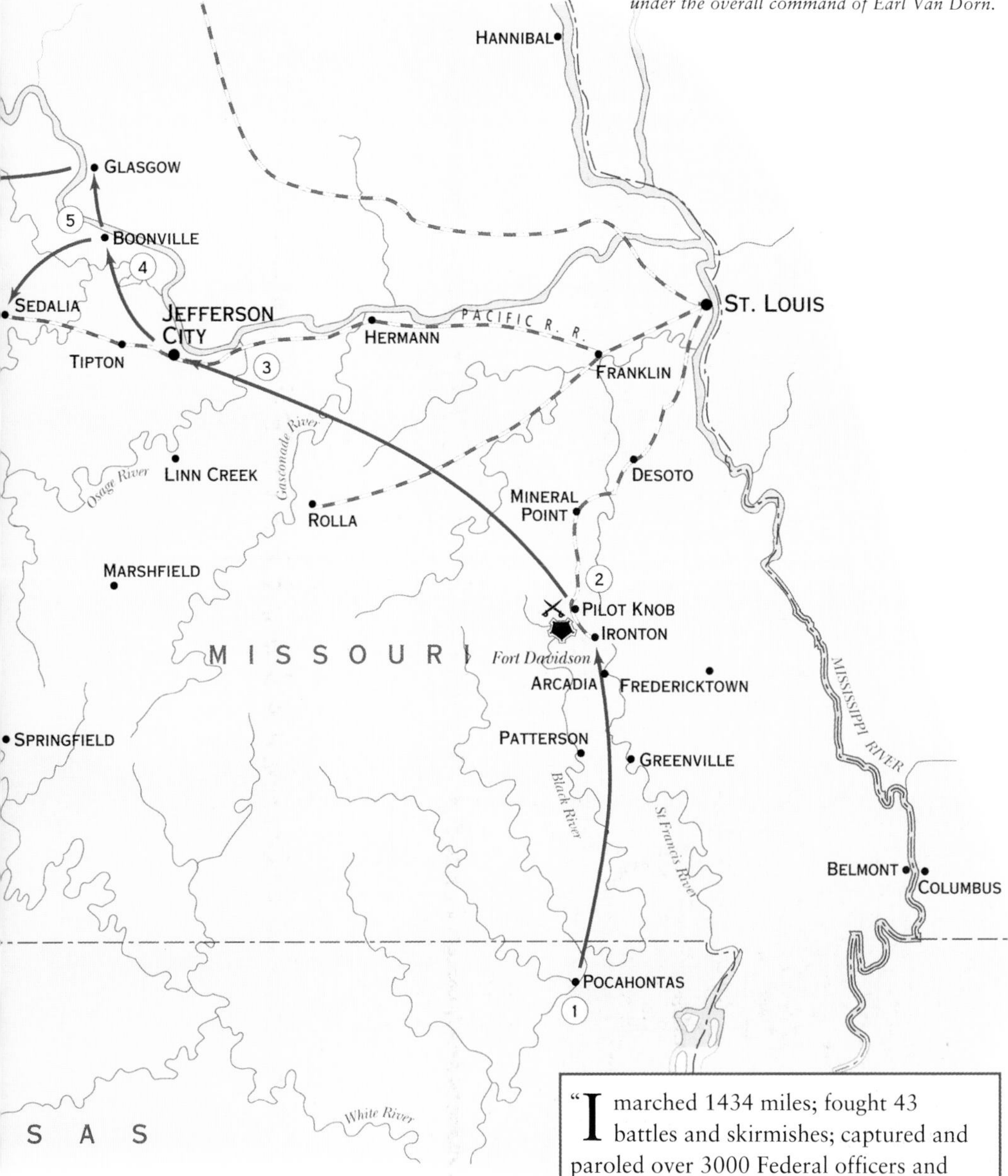
HANNIBAL
GLASGOW
BOONVILLE
SEDALIA
JEFFERSON CITY
TIPTON
HERMANN
PACIFIC R. R.
ST. LOUIS
FRANKLIN
Gasconade River
Osage River
LINN CREEK
ROLLA
MINERAL POINT
DESOTO
MARSHFIELD
PILOT KNOB
IRONTON
Fort Davidson
ARCADIA
FREDERICKTOWN
MISSOURI
SPRINGFIELD
PATTERSON
GREENVILLE
Black River
St Francis River
MISSISSIPPI RIVER
BELMONT
COLUMBUS
POCAHONTAS
White River
SAS
1
2
3
4
5

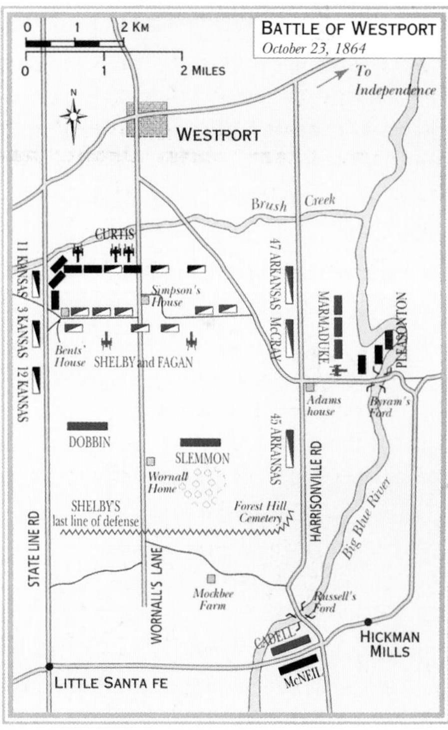
BATTLE OF WESTPORT
October 23, 1864
0 1 2 KM
0 1 2 MILES
N
To Independence
WESTPORT
Brush Creek
CURTIS
11 KANSAS
3 KANSAS
12 KANSAS
Simpson's House
Bents' House
SHELBY and FAGAN
47 ARKANSAS
McCRAY
MARMADUKE
PLEASONTON
Adams house
Byram's Ford
DOBBIN
SLEMMON
Wornall Home
45 ARKANSAS
SHELBY'S last line of defense
Forest Hill Cemetery
Big Blue River
STATE LINE RD
WORNALL'S LANE
HARRISONVILLE RD
Mockbee Farm
Russell's Ford
CABELL
McNEIL
HICKMAN MILLS
LITTLE SANTA FE

*Intending to defeat Gen. Curtis's Federals to his front at Westport, and then turn to face Pleasonton's force approaching from his rear, Price ordered Jo Shelby to attack shortly after daybreak on October 23. Shelby's assault began well, but the Federals countercharged, and the fighting swayed back and forth across Brush Creek. The Federals meanwhile found a route up a small ravine which enabled them to turn the Confederate left. Price now learned that part of his forces, under Marmaduke, had failed to hold Pleasonton at Byram's Ford, and the Federals were now coming up on Price's flank and rear. By early afternoon, Price was forced to withdraw all his Confederates south. Casualties were about 1,500 on each side (see map above).*

> "I marched 1434 miles; fought 43 battles and skirmishes; captured and paroled over 3000 Federal officers and men... and do not think I go beyond the truth when I state that I destroyed... property to the amount of $10,000,000 in value."
>
> *Report of Gen. Sterling Price.*

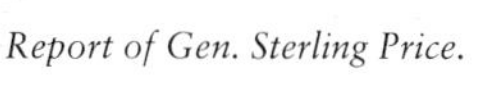

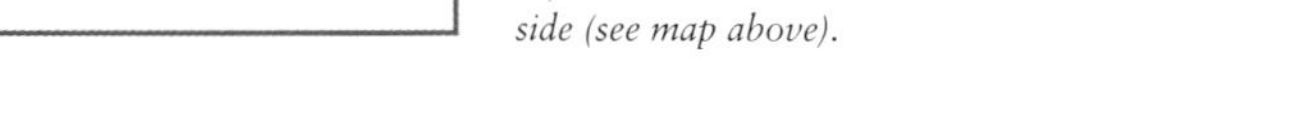

# Hood's Tennessee Campaign OCTOBER 1 – NOVEMBER 30 1864

Following his evacuation of Atlanta on September 2, 1864, Confederate General John Bell Hood withdrew to Palmetto, Georgia, to rebuild his army. This completed, he advanced into northern Georgia in early October with the object of destroying the Union lines of communication – the railroad between Chattanooga and Atlanta – thereby compelling Sherman to fall back to Tennessee. However, not only did he fail to wreck the railroad, but was forced to flee into northern Alabama to escape Sherman, who then returned to Atlanta.

Having learnt that Sherman was preparing to march from Atlanta to the sea, Hood planned to invade Tennessee, defeat the Union forces there, capture Nashville, and then join Lee in Virginia for an attack on Grant. On November 22 Hood entered Tennessee at the head of an army of 40,000. To confront him, Sherman sent General George H. Thomas with 55,000 troops, the majority of which were already at Nashville. The remainder of Thomas's forces, consisting of IV and XXIII Corps' were at Columbia, Tennessee, under Major General John M. Schofield.

On November 29 Hood attempted to prevent Schofield from joining Thomas at Nashville by swinging his army north of Columbia at Spring Hill. Schofield, however, arrived at Spring Hill first, repulsed a poorly-coordinated, half-hearted Confederate attack, then slipped away during the night without Hood and his generals trying to intercept him.

Early on November 30, Hood gave pursuit, and that afternoon came up on the Union forces south of Franklin. In an attempt to smash Schofield's army before it reached Nashville, Hood ignored advice to turn the enemy flank and instead ordered a frontal assault. For five hours Cheatham's and Stewart's corps assailed the well-fortified Union lines, but failed to break through. When night finally brought an end to the fighting, Schofield again retreated, this time to Nashville. Union losses numbered 2,236; the Confederate nearly 7,000, including twelve generals, six of whom were killed. In attempting to achieve the destruction of Schofield's army, Hood had only succeeded in destroying his own.

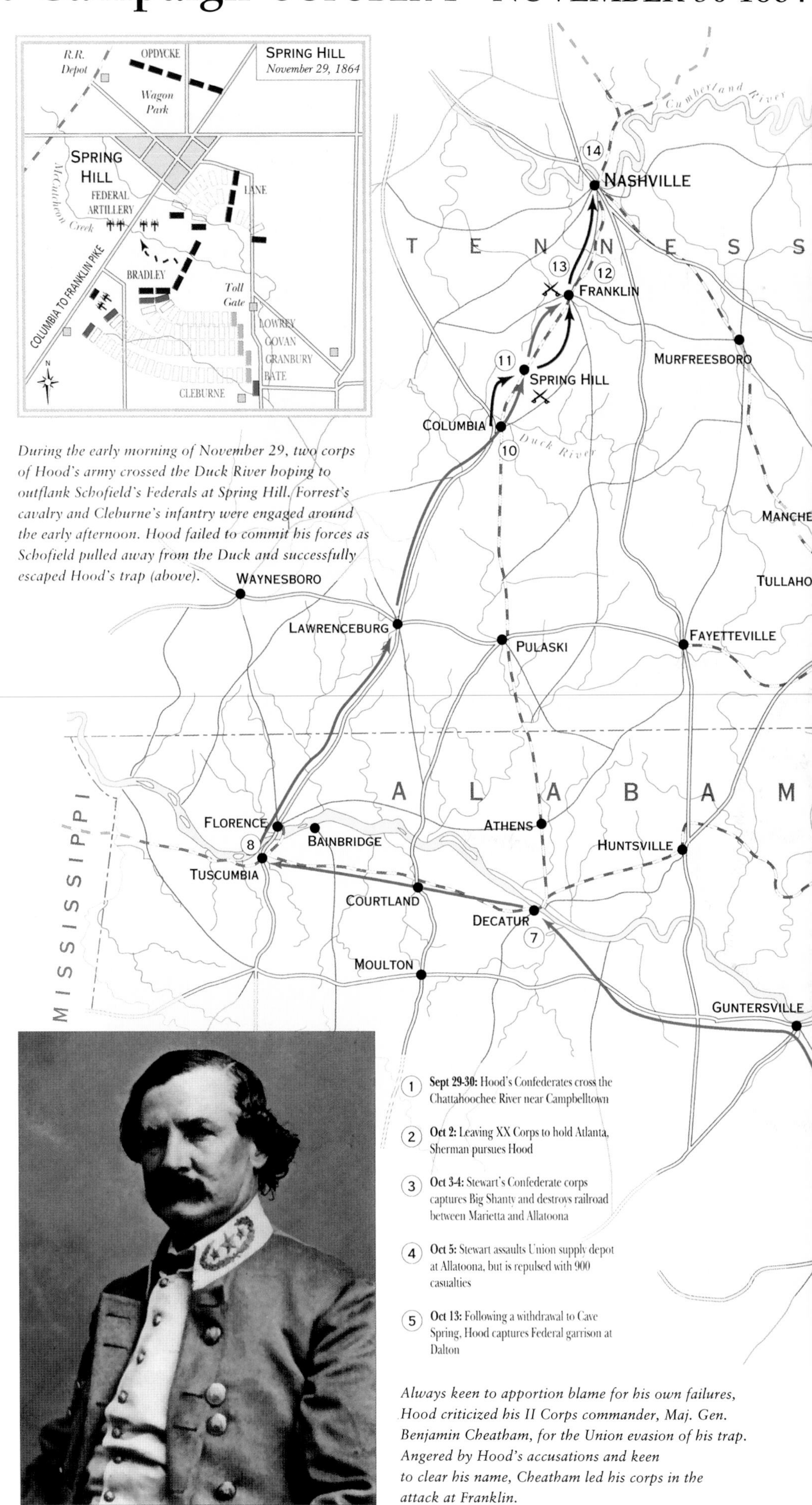

SPRING HILL
November 29, 1864
R.R. Depot
OPDYCKE
Wagon Park
SPRING HILL
FEDERAL ARTILLERY
LANE
McCutcheon Creek
COLUMBIA TO FRANKLIN PIKE
BRADLEY
Toll Gate
LOWREY
GOVAN
GRANBURY
BATE
CLEBURNE
N
Cumberland River
NASHVILLE
TENNESSEE
FRANKLIN
MURFREESBORO
SPRING HILL
COLUMBIA
Duck River
MANCHE
TULLAHO
WAYNESBORO
LAWRENCEBURG
PULASKI
FAYETTEVILLE
ALABAM
MISSISSIPPI
FLORENCE
BAINBRIDGE
ATHENS
HUNTSVILLE
TUSCUMBIA
COURTLAND
DECATUR
MOULTON
GUNTERSVILLE
14
13
12
11
10
8
7
1 Sept 29-30: Hood's Confederates cross the Chattahoochee River near Campbelltown
2 Oct 2: Leaving XX Corps to hold Atlanta, Sherman pursues Hood
3 Oct 3-4: Stewart's Confederate corps captures Big Shanty and destroys railroad between Marietta and Allatoona
4 Oct 5: Stewart assaults Union supply depot at Allatoona, but is repulsed with 900 casualties
5 Oct 13: Following a withdrawal to Cave Spring, Hood captures Federal garrison at Dalton


*During the early morning of November 29, two corps of Hood's army crossed the Duck River hoping to outflank Schofield's Federals at Spring Hill. Forrest's cavalry and Cleburne's infantry were engaged around the early afternoon. Hood failed to commit his forces as Schofield pulled away from the Duck and successfully escaped Hood's trap (above).*

*Always keen to apportion blame for his own failures, Hood criticized his II Corps commander, Maj. Gen. Benjamin Cheatham, for the Union evasion of his trap. Angered by Hood's accusations and keen to clear his name, Cheatham led his corps in the attack at Franklin.*

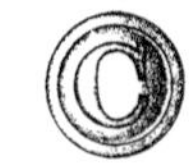

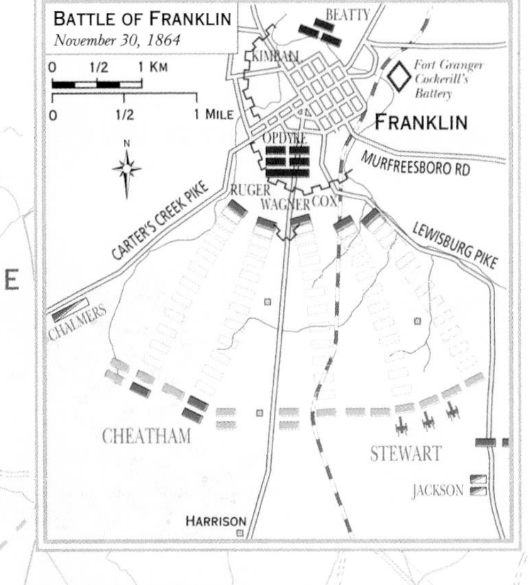

BATTLE OF FRANKLIN
November 30, 1864
0 1/2 1 KM
0 1/2 1 MILE
N
BEATTY
KIMBALL
Fort Granger
Cockerill's
Battery
FRANKLIN
OPDYKE
MURFREESBORO RD
RUGER
WAGNER
COX
CARTER'S CREEK PIKE
LEWISBURG PIKE
CHALMERS
CHEATHAM
STEWART
JACKSON
HARRISON


Arriving in Franklin at dawn, the vanguard of Schofield's retreating Federals formed a defensive line between the town and the Harpeth River. Hood ordered a frontal assault on the entrenchments and suffered a bloody repulse, despite having carried the outer works.

Once again proving the ability of a steady infantry line to resist cavalry charges, George D. Wagner's division opens fire on the afternoon of November 29 at Spring Hill (above).

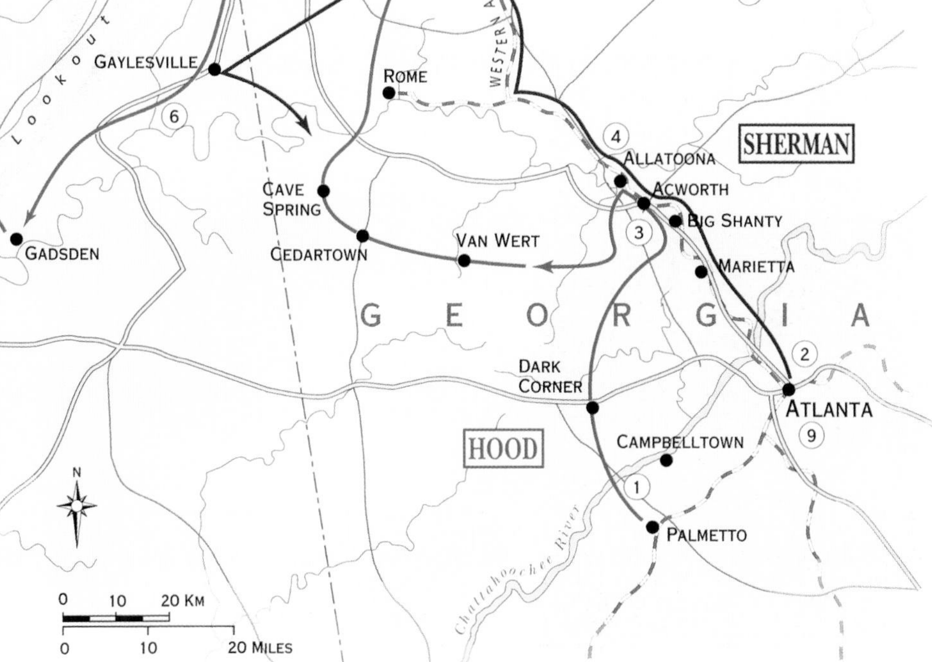

CHATTANOOGA
BRIDGEPORT
STEVENSON
RINGGOLD
DALTON
LAFAYETTE
VILLANOW
RESACA
SUMMERVILLE
Sand Mountain
Lookout Mountain
WESTERN AND ATLANTIC R.R.
GAYLESVILLE
ROME
SHERMAN
ALLATOONA
ACWORTH
BIG SHANTY
MARIETTA
CAVE SPRING
CEDARTOWN
VAN WERT
GADSDEN
GEORGIA
DARK CORNER
ATLANTA
CAMPBELLTOWN
HOOD
PALMETTO
Chattahoochee River
N
0 10 20 KM
0 10 20 MILES


6 **Oct 14-19:** Hood withdraws to Gadsden. Sherman pursues Hood to Gaylesville, then halts

7 **Oct 27 - Nov 2:** Hood vainly tries to cross Tennessee River at Decatur. Then resumes his western march

8 S.D. Lee's corps of Hood's army crosses Tennessee at Tuscumbia & occupies Florence

9 **Nov 15-16:** Sherman begins march from Atlanta to the sea

10 **Nov 21-27:** Hood marches to Columbia, which is held by Federal troops under Schofield

11 **Nov 28:** Fearful of being cut off from Nashville, Schofield retreats toward Spring Hill

12 **Nov 29:** Hood attempts, but fails, to intercept Schofield at Spring Hill; during the night Schofield retreats to Franklin

13 **Nov 30:** Hood pursues Schofield to Franklin, attacks and is repulsed

14 **Night:** Schofield withdraws to Nashville

Maj. Gen. John M. Schofield (above), Hood's mathematics coach at West Point, was a scholarly man of uncertain temper. Though a conscientious commander who had the respect of his troops, Schofield was never favoured by Sherman who found him slow to obey orders.

"The situation presented an occasion for one of those interesting and beautiful moves upon the chessboard of war, to perform which I had often desired an opportunity."

*Brig. John B. Hood upon his proposed entrapment of Schofield.*

"You could see a rebel's head falling off his horse on one side and his body on the other, and the horse running and nickering and looking for it's rider. Others you could see fall off with their feet caught in the stirrup, and the horse dragging and trampling them, dead or alive."

*A Wisconsin infantryman reporting on the effects of Union artillery at Spring Hill.*

# Battle of Nashville DECEMBER 15–16 1864

On December 2–3, Hood entrenched his army south of Nashville. Due to his appalling losses at Franklin, and the natural attrition of campaigning, he now had fewer than 30,000 men under his command. Opposing him were 55,000 well-fortified Union troops. Hood's plan, born of des-peration, was to await a Union attack and repulse it. Then, in his own words, "enter the city on the heels of the enemy."

The Union commander, General Thomas, intended to attack, but not until he could do so with maximum effect. In particular, he needed time to prepare his 12,000 cavalry, commanded by Brigadier General James H. Wilson. By December 9, Thomas was ready, but the onset of freezing rain rendered movement across the icy ground virtually impossible. Despite a barrage of orders from Grant in Virginia – who disliked and dis-trusted Thomas – to attack at once, Thomas postponed his offensive.

On December 15, the ground having thawed, Thomas advanced. While half his forces demonstrated against Hood's front, the other half, spearheaded by Wilson's cavalry fighting dismounted, turned Hood's left flank, forcing him to fall back. The following day Thomas renewed his assault, employing the same strategy with even more devastating results. Despite stubborn resistance, the entire Confederate left crumbled beneath the overwhelming Union pressure. This led in turn to the center fleeing in utter disorder to escape being cut off. Hood's losses totaled some 6,000, about 4,500 of them captured. Thomas' losses were slightly over 3,000.

Thomas pursued the Confederates until they crossed the Tennessee River in late December. Only a misdirected pontoon train, exhausted Union troops and horses, and a courageous rearguard action by Forrest, prevented the utter destruction of Hood's army. By the time it reached Mississippi, the Army of the Tennessee was little more than a demoralized remnant, no longer capable of playing a significant role in the war. Now only Lee's army in Virginia stood between the Confederacy and inevitable defeat.

*Stephen D. Lee (above) whose corps held the center of the Confederate line. The weight of fire directed toward his command on December 15 meant that his troops were unable to make any considerable contribution to the engagement.*

> "One long cheer relieved the pent up excitement of the reserves, and was wafted to the front as a testimonial of gratitude."
>
> *Union soldier reporting on the victory at Nashville.*

1. **Dec 15, c. 10am:** Wilson's cavalry corps and A. J. Smith's infantry corps begin to swing round to attack Confederate left
2. **c. 11am:** Steedman's Provisional Detachment attacks Confederate right and is repulsed
3. **c. 1pm:** Schofield's XXIII Corps advances to support Wilson's and Smith's attack
4. **c. 1pm:** Hood sends Johnson's division of Lee's corps to reinforce his left
5. **c. 2pm:** Kimball's and Beatty's divisions of Wood's IV Corps assail Confederate center
6. **Late pm:** Wilson's and Smith's attack drives Confederates from Hillsboro Pike. Hood withdraws to line covering the Granny White and Franklin Pikes
7. **Dec 16, 10am:** Steedman attacks Confederate right on Overton Hill, and is repulsed
8. **Noon:** Wilson reaches rear of Confederate left
9. **c. 3.30pm:** Beatty and Steedman make another unsuccessful Federal attack on Overton Hill
10. **4pm:** Elements of Smith's corps break Confederate line at Shy's Hill
11. **c. 4.30pm:** Wilson attacks Confederate left from south, Schofield from west, and Smith from north, routing Cheatham's and Stewart's corps
12. **Evening and night:** Remnants of Cheatham's and Smith's corps flee south along Granny Smith Pike

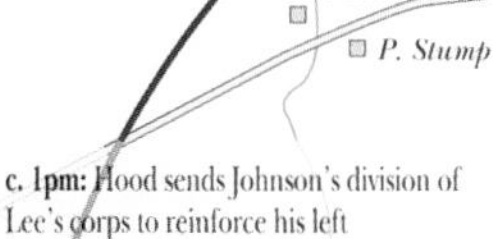
L. Lyon
B.F. Cook
CHARLOTTE PIKE
P. Stump

*Behind the fortifications of Nashville (below), Thomas's army suffered considerable hardships as the bitterly cold winter weather took its toll. Winds of up to 60 miles per hour swept through the camps and ice half an inch thick coated the cannon.*

> "For God's sake, drive the Yankee cavalry from our left and rear or all is lost."
>
> *Dispatch from Hood, December 15.*

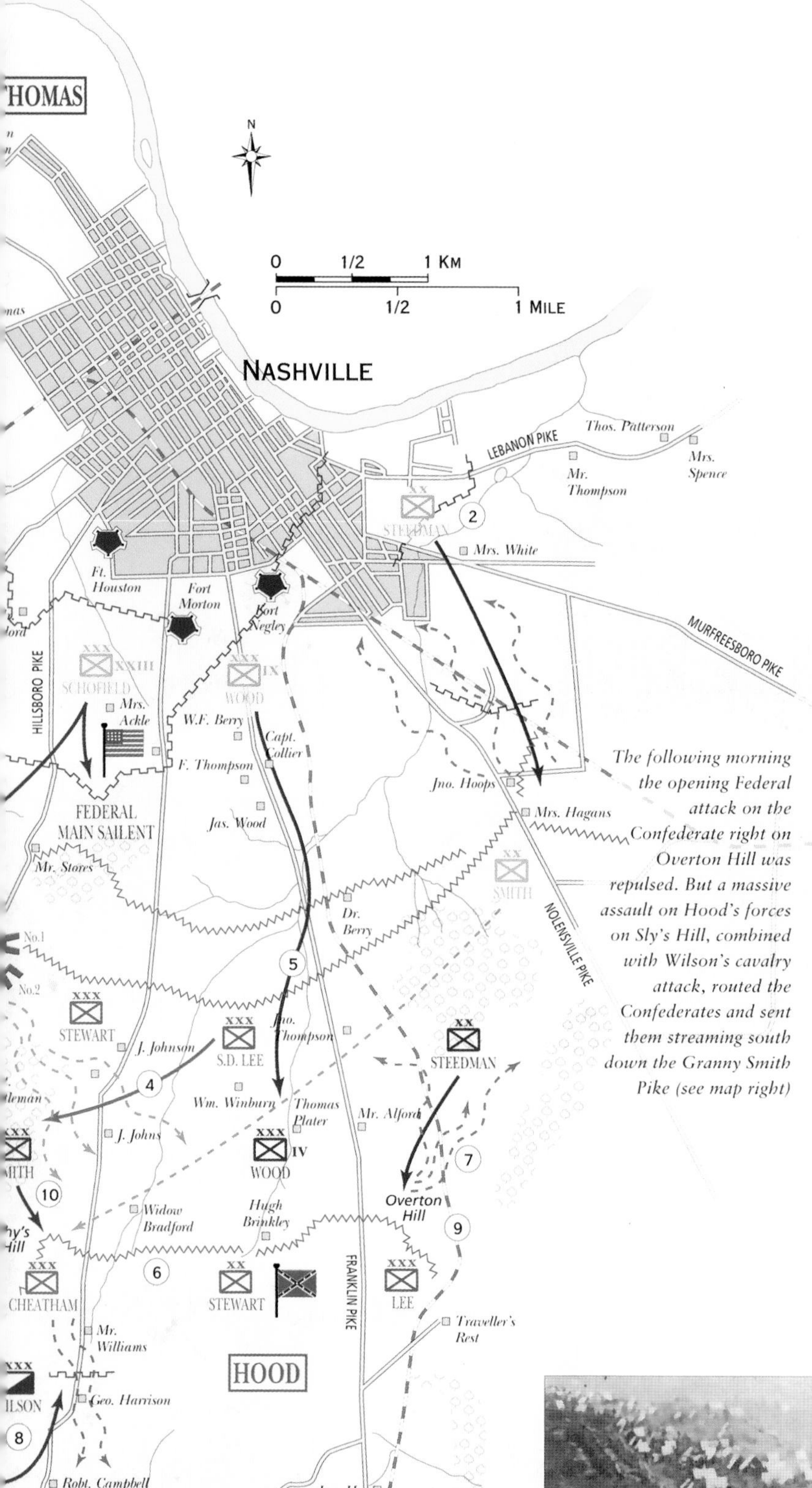

The following morning the opening Federal attack on the Confederate right on Overton Hill was repulsed. But a massive assault on Hood's forces on Sly's Hill, combined with Wilson's cavalry attack, routed the Confederates and sent them streaming south down the Granny Smith Pike (see map right)

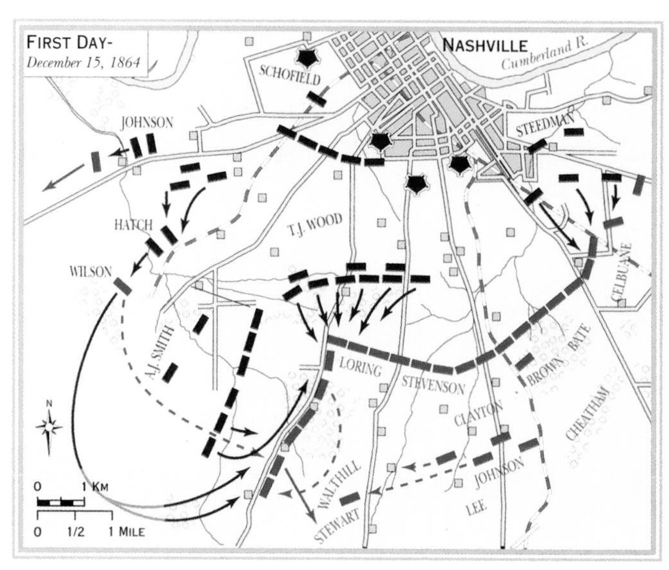

Gen. Thomas began his assault on Hood's Confederates with a diversionary attack by Steedman's division on the Confederate right. At midday, the bulk of the Federal army, led by Wood's corps enveloped Hood's left flank, sending the Confederates reeling back (see map above).

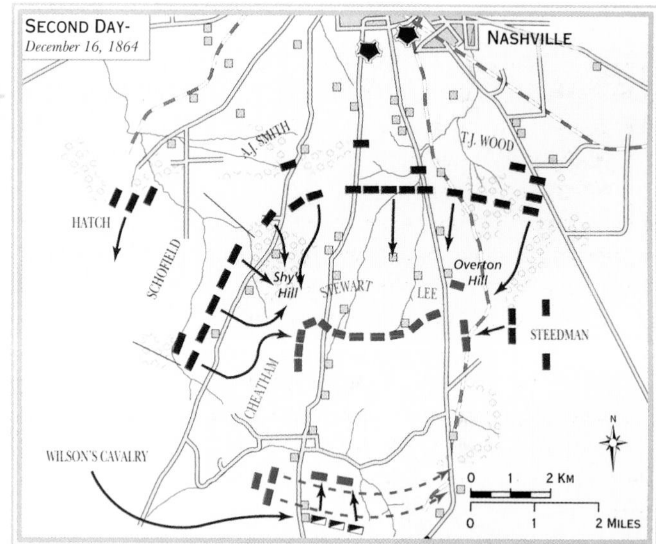

The Union encampment at Nashville, December 16, (below) after the Confederate defeat.

# Sherman's March from Atlanta to the Sea

## NOVEMBER 15 – DECEMBER 21 1864

After General Hood's Confederates evacuated Atlanta on September 2, Sherman's army occupied the city. Early in October, hoping to compel Sherman to relinquish Atlanta by destroying the Federal lines of communication – the railroad between the city and Chattanooga – Hood advanced into northern Georgia. Sherman rapidly pursued; although he prevented Hood from seriously damaging the railroad, he was unable to engage him in battle. Fearing that the strategic gains from the capture of Atlanta would be lost, Sherman received Grant's permission to march the bulk of his army from Atlanta to the Atlantic coast. He left a force in Tennessee sufficient to defend the state against invasion by Hood.

On November 15, Sherman left Atlanta with 62,000 men belonging to Howard's Army of the Tennessee, Slocum's Army of Georgia, and Kilpatrick's cavalry division. Sherman or-dered his men to "forage liberally on the country." This they did, plundering and burning, meeting with little opposition from the Confederates. In their wake the Federals left a sixty-mile wide swath of devastation. Hordes of blacks, lately freed, accompanied them.

On December 10 Sherman reached the outskirts of Savannah. Three days later an assault by a brigade of the Federal XV Corps captured Fort McAllister on the Ogeechee River, thereby establishing communications with the Federal navy under Dahlgren. On December 21 Sherman occupied Savannah after its garrison, hopelessly outnumbered, evacuated it. In a telegram to Lincoln, Sherman presented the city to the President "as a Christmas gift."

The importance of Sherman's march to the sea has been overrated. Sherman himself stated that it was merely a change of base, admitting that most of the destruction carried out by his soldiers was "sheer waste," without military value, and that probably Thomas's victory at Nashville made a stronger impact on Southern morale. Yet the march was one of the Civil War's most spectacular events, and demonstrated the truth of Sherman's de-claration that the Confederacy had become "a hollow shell."

> "Why don't you go over to South Carolina and serve them this way? They started it."
>
> *A Georgian farmer to Sherman.*

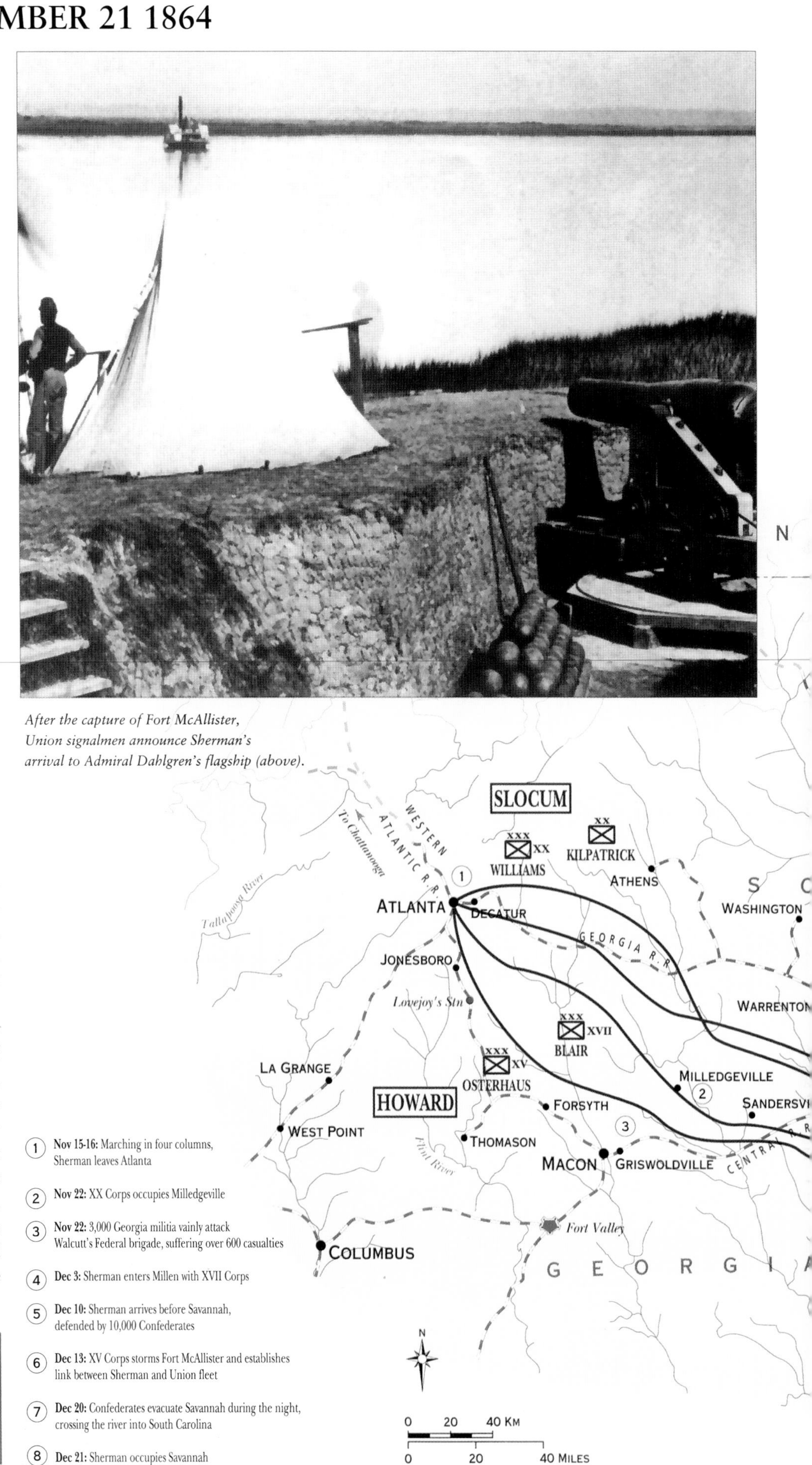
N
SLOCUM
XX
WILLIAMS
KILPATRICK
XVII
BLAIR
XV
OSTERHAUS
HOWARD
WESTERN ATLANTIC R.R.
To Chattanooga
Tallapoosa River
Flint River
ATLANTA
DECATUR
ATHENS
WASHINGTON
GEORGIA R.R.
JONESBORO
Lovejoy's Stn
WARRENTON
LA GRANGE
MILLEDGEVILLE
SANDERSVI
FORSYTH
WEST POINT
THOMASON
MACON
GRISWOLDVILLE
CENTRAL
Fort Valley
COLUMBUS
GEORGIA
1 Nov 15-16: Marching in four columns, Sherman leaves Atlanta
2 Nov 22: XX Corps occupies Milledgeville
3 Nov 22: 3,000 Georgia militia vainly attack Walcutt's Federal brigade, suffering over 600 casualties
4 Dec 3: Sherman enters Millen with XVII Corps
5 Dec 10: Sherman arrives before Savannah, defended by 10,000 Confederates
6 Dec 13: XV Corps storms Fort McAllister and establishes link between Sherman and Union fleet
7 Dec 20: Confederates evacuate Savannah during the night, crossing the river into South Carolina
8 Dec 21: Sherman occupies Savannah
0 20 40 KM
0 20 40 MILES

*After the capture of Fort McAllister, Union signalmen announce Sherman's arrival to Admiral Dahlgren's flagship (above).*

> "Whether called on to fight, to march, to wade streams... build bridges, make 'corduroy,' or tear up railroads, they have done it with alacrity and a degree of cheerfulness unsurpassed... on the whole they have supplied the wants of the army with as little violence as could be expected, and as little loss as I calculated."
>
> *Sherman expressing his pride in the achievements of his army.*

*Sherman's army parades before its general on Bay Street, Savannah (right).*

CHARLOTTE
YORKVILLE
SPARTANBURG
GREENVILLE
LAURENSVILLE
ALSTON
CAMDEN
FLORENCE
MARION
COLUMBIA
AUGUSTA
BRANCHVILLE
MCBEAN
XIV
WAYNESBORO
MILLEN
POCATALIGO
CHARLESTON
Fort Sumter
Charleston Harbor
Cape Romney
Bulls Bay
Honey Hill
BEAUFORT
STATESBORO
SAVANNAH
Fort Beauregard
Fort Walker
Fort Pulaski
Fort McAllister
Port Royal
Hilton Head I.
Savannah River
Tybee I.
Warsaw Sound
Ossabaw Sound
North Edisto R.
Edisto I.
South Edisto R.
St. Helena Sd.
Hunting I.
St. Helena I.
Saluda River
Broad River
Wateree River
Edisto River
Santee River
Black River
Combahee River
Savannah River
Ogeechee River
Canoochee

*Maj. Gen. H. J. Kilpatrick (right) who commanded Sherman's cavalry during the march to the sea. Kilpatrick allowed his men to loot but insisted that ammunition be reserved; the slaughter of livestock was to be performed with sabers not firearms.*

# 1865: The Triumph and the Tragedy

THE YEAR 1865 opened with the resolution of an issue that had generated bitter controversy – a bitterness that persisted for decades after the war – over the exchange of prisoners of war. In 1862, the Union and Confederate armed forces had signed a cartel for the exchange of prisoners captured in battle. The arrangement had worked reasonably well for a year, making large prison camps unnecessary. But when the Union army began to organize regiments of former slaves, the Confederate government announced that it would refuse to treat them as legitimate soldiers. If captured, they and their white officers would be put to death for the crime of inciting slave insurrections. In practice, however, the Confederate government did not enforce this policy, for Lincoln threatened retaliation against Confederate prisoners if it did so. But Confederate troops sometimes murdered black soldiers and their officers as they tried to surrender – most notably at Fort Pillow, a Union garrison on the Mississippi River north of Memphis, where cavalry commanded by Confederate General Nathan Bedford Forrest slaughtered scores of black (and some white) prisoners on April 12, 1864.

In most cases, though, Confederate officials returned captured black soldiers to slavery, or put them to hard labor on Southern fortifications. Expressing outrage at this treatment of soldiers wearing the uniform of the United States army, the Lincoln administration in 1863 suspended the exchange of prisoners until the Confederacy agreed to treat white and black prisoners alike. The Confederacy refused. The South would "die in the last ditch," said the Confederate exchange agent, before "giving up the right to send slaves back to slavery as property recaptured."

There matters stood as the heavy fighting of 1864 poured scores of thousands of captured soldiers into hastily built prison compounds. Due to overcrowding, poor sanitation, contaminated water, scanty rations, indadequate medical facilities, and the exposure of Northern prisoners to the heat of a deep South summer and of Southern prisoners to the cold of a Northern winter, these prisons rapidly became death camps. The suffering of Northern prisoners was especially acute because the deterioration of the Southern economy made it hard to feed and clothe even Confederate soldiers and civilians, let alone Yankee prisoners. Confederate prison officials did not provide any kind of shelter for captured enlisted men, who were forced to live in open stockades while Northern prisons housed captured Confederates in barracks or tents. Nearly 16% of all Union soldiers held in Southern prison camps died, compared with 12% of Confederate soldiers in Northern camps.

The Southern prison at Andersonville, Georgia, became the most notorious hellhole: a stockade camp of 26 acres designed to accommodate 15,000 prisoners, it held 33,000 in August 1864. During the worst weeks of that terrible summer of 1864 they died at the rate of a hundred or more a day. Altogether, 13,000 Northern soldiers died at Andersonville. For a generation after the war, many Northerners saw the graves of these soldiers as a symbol of Southern barbarism. As sectional passions cooled, they came to be seen as a symbol of the inhumanity of war.

The suffering of Union prisoners brought heavy pressure on the Lincoln administration to renew exchanges. Democrats made a political issue of this matter in the 1864 election. But the Confederates would not yield on the question of exchanging black soldiers. After the battles at New Market Heights, Fort Harrison, and Poplar Springs Church on the Richmond–Petersburg front in September 1864, General Lee proposed an informal exchange of prisoners. General Grant agreed, on condition that black soldiers captured in the fighting be included "the same as white soldiers." Lee replied that "negroes belonging to our citizens are not considered subjects of exchange and were not included in my proposition." No exchange, then, responded Grant; the Union government was "bound to secure to all persons received into her armies the rights due to soldiers." Lincoln backed this policy. Just as he would not recede from emancipation as a condition of peace, he would not sacrifice the principle of equal treatment of black prisoners, though Republican leaders warned that many Northerners "will work and vote against the President, because they think sympathy with a few negroes, also captured, is the cause of a refusal" to exchange prisoners.

Lincoln won reelection despite his political liabilities on this issue. Two months later, in January 1865, the Confederate government quietly abandoned its refusal to exchange black prisoners, and exchanges resumed. One reason for this reversal was a Confederate decision to recruit slaves to fight for the South. Two years earlier Jefferson Davis had denounced the North's arming of freed slaves as "the most execrable measure recorded in the history of guilty man." But by the beginning of 1865, Southern armies were desperate for manpower, and the slaves constituted the only remaining reserve. Supported by the powerful influence of Robert E. Lee, the Davis administration pressed the Confederate Congress to enact a bill for recruitment of black soldiers. The assumption that any slaves who fought for the South would have to be granted freedom as a reward generated angry opposition to this measure. "What did we go to war for, if not to protect our property?" asked a Virginia senator. By three votes in

*General Robert E. Lee, Commander-in-Chief of the Confederate armies. From a photograph by Mathew Brady taken on the back porch of the Lee home.*

the House and one in the Senate, Congress finally passed the bill on March 13, 1865. Whether any slaves would actually have fought for the Confederacy must remain forever a moot point. Before any black regiments could be formed, the war was over.

It ended not by a negotiated peace, but by the unconditional surrender of Confederate armies because that was how Lincoln intended it should end. In his annual message to Congress in December 1864, Lincoln referred to the abortive peace overtures during the preceding summer. In his stated conditions for peace, said Lincoln, Jefferson Davis did "not attempt to deceive us. He affords us no excuse to deceive ourselves. He cannot voluntarily reaccept the Union; we cannot voluntarily yield it. Between him and us the issue is distinct, simple, and inflexible. It is an issue which can only be tried by war, and decided by victory." Lincoln was prepared to fight on until unconditional victory was achieved, for "the national resources ... are unexhausted, and, as we believe, inexhaustible. We are *gaining* strength, and may, if need be, maintain the contest indefinitely." Lincoln's words expressed the mood of the Northern people, a mood described by the war correspondent of the London *Daily News*. The North, he wrote, was "silently, calmly, but desperately in earnest ... in a way the like of which the world never saw before. ... I am astonished the more I see and hear of the extent and depth of [this] determination ... to fight to the last."

By January 1865, it was clear to all but bitter-end Confederates, who vowed to die in the last ditch, that the downfall of the Confederacy was near. In mid-January, the largest fleet yet assembled in the war – 58 Union warships carrying 627 guns – rendezvoused near Fort Fisher at the mouth of the Cape Fear River below Wilmington. This fort protected the Confederate blockade runners that had made Wilmington the Confederacy's major port and virtually its only lifeline to the outside world by the end of 1864. The Union navy had long wanted to attack Fort Fisher, but the diversion of ships and troops to the long, futile campaign against Charleston had delayed the effort. In January 1865, two days of naval bombardment, followed by an attack of infantry and marines, captured Fort Fisher, thus coiling the tentacles of the Union anaconda ever more tightly around the South. Confederate Vice-President Alexander H. Stephens pronounced the fall of Fort Fisher "one of the greatest disasters which had befallen our Cause."

This was perhaps an exaggeration. But the fall of Fort Fisher did generate another Confederate gesture toward peace negotiations. Jefferson Davis appointed Stephens and two other officials to meet Union representatives "with a view to secure peace to the two countries." Lincoln responded with an expression of willingness to receive commissioners "with the view of securing peace to the people of our one common country."

This crucial difference in wording should have alerted Southerners to the futility of expecting peace terms short of unconditional surrender. Lincoln and Secretary of State Seward met personally with the Confederate envoys on board a Union ship at Hampton Roads, Virginia, on February 3. To every proposal for an armistice or preliminary terms, Lincoln replied that the Confederates must lay down their arms, give up slavery, and rejoin the Union. But these terms mean "unconditional submission," gasped the Confederate commissioners. Precisely, said Lincoln. The Southerners returned empty-handed to Richmond, where they reported glumly to Davis, who had expected this outcome. He used Lincoln's demands for "humiliating surrender" to stir up flagging Southern morale. Lincoln and Seward would soon find that "they had been speaking to their masters," Davis told a cheering crowd in Richmond. Confederate armies would yet "compel the Yankees, in less than twelve months, to petition us for peace on our own terms."

If William T. Sherman read these words, he must have uttered a sigh of exasperation. It was to break the back of such last-ditch resistance that he had made Georgia howl on his march from Atlanta to the sea. Now his 60,000 soldiers were moving north from Savannah on a sequel to that march, which would prove even more destructive. His men were determined to take revenge on South Carolina, which in their view had started the war. As one soldier declared as he entered South Carolina: "Here is where treason began and, by God, here is where it will end!" Sherman's avengers made

*The fall of Petersburg on April 2, 1865, after a five month siege. From a lithograph by Kurz and Allison*

even less distinction between civilian and military property in South Carolina than they had in Georgia. They left even less standing in Columbia (the state capital) than they had of Atlanta. Seemingly invincible, Sherman's army pushed into North Carolina and brushed aside a force that Joseph E. Johnston had assembled at Bentonville to try to stop them. The trail of devastation left in their wake appalled Confederates. "All is gloom, despondency, and inactivity," wrote a South Carolina physician. "Our army is demoralized and the people panic-stricken. To fight longer seems to be madness."

In his second inaugural address on March 4, 1865, Lincoln again made it clear that the bloodshed would not cease until Confederates soldiers laid down their arms. The best known words from that address urged a binding up of the nation's wounds "with malice toward none" and "charity for all." But more significant in the midst of the still raging conflict, were these words:

*American slavery is one of those offences which, in the providence of God ... He now wills to remove [through] this terrible war, as the woe due to those by whom the offence came. ... Fondly do we hope – fervently do we pray – that this mighty scourge of war may speedily pass away. Yet if God wills that it continue, until all the wealth piled by the bondman's two hundred and fifty years of unrequited toil shall be sunk, and until every drop of blood drawn with the lash, shall be paid by another drawn with the sword, as was said three thousand years ago, so still it must be said "the judgements of the Lord, are true and righteous altogether."*

Ulysses S. Grant did not intend the war to last that long. He knew that the Army of Northern Virginia – the only entity that kept the Confederacy alive – was on the verge of disintegration. Scores of soldiers were deserting every day. On April 1, Sheridan's cavalry (which had returned from the Shenandoah Valley) and an infantry corps smashed the right flank of Lee's line at Five Forks and cut off the last railroad into Petersburg. Next day, Grant attacked all along the line and forced Lee to abandon both Petersburg and Richmond. As the Confederate government fled its capital, its army set fire to all the military stores it could not carry. The fires spread and burned more of Richmond than Northern troops had burned of Atlanta or Columbia.

Lee's starving men limped westward, hoping to turn south and join the remnants of Johnston's army in North Carolina. But Sheridan's cavalry raced ahead and cut them off at Appomattox, 90 miles from Petersburg, on April 8. When the weary Confederates tried a breakout attack the next morning, their first probe revealed solid ranks of Union infantry arrayed behind the cavalry. It was the end. "There is

*The fall of Richmond, the Confederate capital, on the night of April 2, 1865. From a print by Currier and Ives*

nothing left for me to do," said Lee, "but to go and see General Grant, and I would rather die a thousand deaths." Go he did, to the house of Wilmer McLean, who in 1861 had lived near Manassas where a shell had crashed through his kitchen roof during the first battle of Bull Run. McLean had moved to the remote village of Appomattox to escape the war, only to have its final drama played out in his parlor. There the son of an Ohio tanner dictated surrender terms to a scion of one of Virginia's First Families.

Grant's terms were generous. They paroled the 30,000 captured Confederates and allowed them to go home on condition that they promise never again to take up arms against the United States. These terms served as the model for the surrender of the other Confederate armies over the next few weeks. In effect, Appomattox ended the war. After completing the surrender formalities on April 9, Grant introduced Lee to his staff, which included Colonel Ely Parker, a Seneca Indian. As Lee shook hands with Parker, he stared for a moment at Parker's dark features and said: "I am glad to see one real American here." Parker replied solemnly, "We are all Americans." And indeed, they now were.

Wild celebrations broke out in the North at news of the fall of Richmond, followed soon by news of Appomattox. In Washington, a 900-gun salute proclaimed the capture of the Confederate capital. "From one end of Pennsylvania Avenue to the other the air seemed to burn with the bright hues of

the flag," wrote a reporter. "Men embraced one another, 'treated' one another, made up old quarrels, renewed old friendships, marched arm-in-arm singing." The North barely recovered from this jubilation when news of Lee's surrender started it all over again. On Wall Street in New York "men embraced each other and hugged each other, *kissed* each other, retreated into doorways to dry their eyes and came out again to flourish their hats and hurrah," wrote a participant. Such intensity of feeling was "founded on memories of years of failure, all but hopeless, and the consciousness that national victory was at last secured."

In Washington, large groups of people went from house to house of prominent officals, serenading them and calling for speeches. Lincoln was ready for the crowd that came to the White House on April 11. The President spoke from a prepared text on the problem of reconstructing the Union in a manner to bind up the nation's wounds, but at the same time to insure justice for the freed slaves, including the right to vote for those who were qualified. At least one listener did not like this talk. "That means nigger citizenship," snarled John Wilkes Booth to a companion. "Now, by God, I'll put him through. That is the last speech he will ever make." An aspiring young actor overshadowed by the greater thespian fame of his father and older brother Edwin, Booth was a native of Maryland, a Confederate supporter, and a frustrated, unstable egotist who hated Lincoln for what he had done to Booth's beloved South. Three days after his speech on April 11, the careworn Lincoln sought to relax by attending a British comedy at Ford's Theater in Washington. In the middle of the play, Booth broke into the President's box and shot him fatally in the head. As he jumped from Lincoln's box to the stage and escaped out of a back door, Booth shouted Virginia's state motto at the stunned audience: "Sic semper tyrannis!" – "Thus be it ever to tyrants."

Thus the tragedy of the American Civil War, begun by the fanatical secessionist Edmund Ruffin – who pulled the lanyard on the first cannon fired at Fort Sumter on April 12, 1861 – ended with an assassin's bullet fired by another Southern fanatic on April 14, 1865. At the cost of more than 620,000 lives, the United States was preserved as one nation, indivisible, and slavery was abolished. But the only certainty as the nation looked ahead, was that the future would be radically different from the past.

*Lee surrenders to Grant at Appomattox Courthouse on April 9, 1865*

# Fort Fisher DECEMBER 8 – JANUARY 15 1865

By the end of 1864 Wilmington, North Carolina, was the last important port still in Confederate hands. Its approaches were guarded by Fort Fisher's powerful batteries, which were well-protected by mounds of sand. Fields of fire were well covered, and troops protected by bomb proofs.

On December 20, the largest number of Federal warships ever assembled in the war arrived off Fort Fisher, with Admiral Porter in command. General Butler was in charge of land forces.

As a preliminary to the assault, the navy packed the gunboat *Louisiana* with tons of explosives and on December 24, detonated her 450 yards from the fort. It accomplished nothing.

At midday on the 24th, Porter's fleet began the bombardment. For over five hours they laid down a volume of fire sometimes reaching 115 rounds per minute. Butler's troops landed on the beach north of the fort. As they neared the fort, the naval fire ceased. To his astonishment, Butler discovered the fort still intact and the defenders relatively unharmed. The weather now began to worsen, and concerned that a storm would isolate his forces from their naval support, Butler withdrew. The first assault on Fort Fisher was a failure.

On January 13, a Federal fleet with Porter again in command, and accompanied by

*Essential to Union victory was the strangulation of Lee's supply routes. Most of the supplies reaching his army were carried by blockade runners whose access to Wilmington was guarded by the enormous Fort Fisher (right). Built of sand and dirt over a log framework, the fort absorbed shells far more effectively than stone fortifications such as Fort Sumter.*

> "The effect of the explosion will be ... very severe, stunning men at a distance of three or four hundred yards ... and making them unable to stand for any length of time a fire from the ships ... I think that houses in Wilmington will tumble to the ground and I think if the rebels fight after the explosion they have more in them than I gave them credit for."
>
> *Admiral Porter predicting the effects of the* U.S.S. Louisiana's *detonation.*

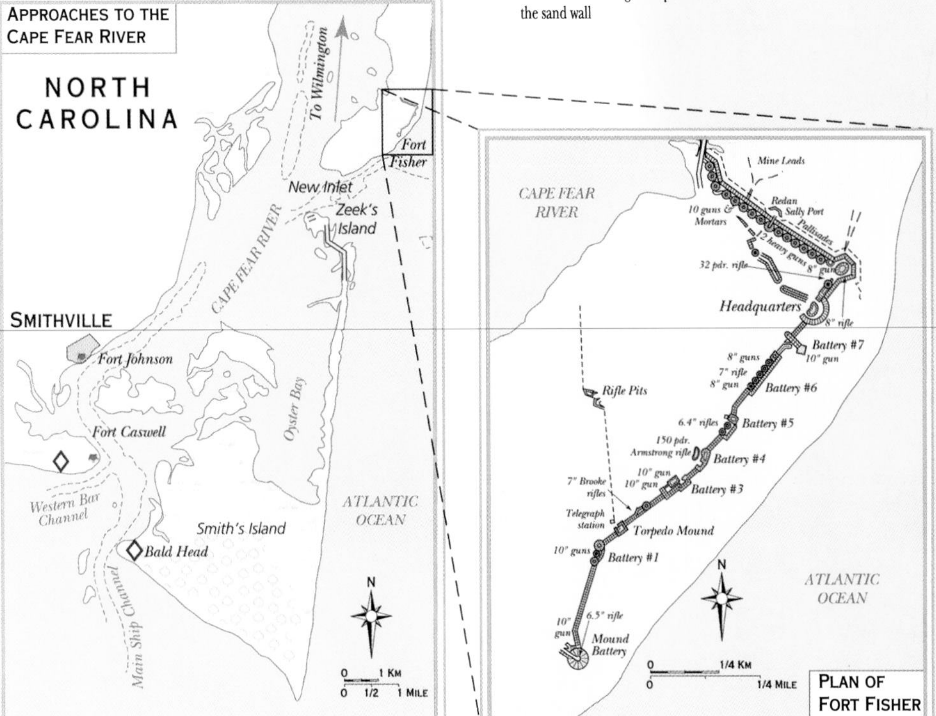

APPROACHES TO THE CAPE FEAR RIVER
NORTH CAROLINA
To Wilmington
Fort Fisher
New Inlet
Zeek's Island
CAPE FEAR RIVER
SMITHVILLE
Fort Johnson
Oyster Bay
Fort Caswell
Western Bar Channel
Smith's Island
Bald Head
Main Ship Channel
ATLANTIC OCEAN
N
0 1 KM
0 1/2 1 MILE
CAPE FEAR RIVER
Mine Leads
Redan
Sally Port
10 guns & Mortars
Palisades
12 heavy guns
32 pdr. rifle
8" gun
Headquarters
8" rifle
Battery #7
10" gun
8" guns
7" rifle
8" gun
Battery #6
Rifle Pits
6.4" rifles
Battery #5
150 pdr. Armstrong rifle
Battery #4
10" gun
10" gun
7" Brooke rifles
Battery #3
Telegraph station
Torpedo Mound
10" guns
Battery #1
N
ATLANTIC OCEAN
10" gun
6.5" rifle
Mound Battery
0 1/4 KM
0 1/4 MILE
PLAN OF FORT FISHER


1. **Dec 24 1.40 am:** *USS Louisiana* explodes causing negligible damage to the fort
2. **12 noon:** Union fleet bombards fort firing almost 115 shells per minute from 627 guns. The fleet fires almost 10,000 shells during the five hour bombardment
3. **Dec 25 2.30 pm:** The first Union attack on the fort is thrown back. The Union commander Brig. Gen. G.G. Weitzel also discovers approaches covered by torpedoes. (mines)
   **6 pm:** Weitzel withdraws and his troops are troops are taken off by Porter's boats
4. **Dec 27:** Col. Lamb repairs damage to fort correctly suspecting that the Union fleet will return
5. **Jan 12:** The Union fleet returns. Before dawn on Jan 13the ironclads move in close and open fire. Confederate guns return fire revealing their positions on the sand wall

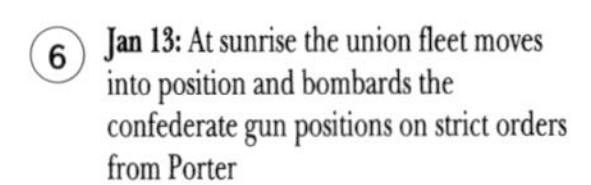

6. **Jan 13:** At sunrise the union fleet moves into position and bombards the confederate gun positions on strict orders from Porter
7. **Jan 13 8 am:** Gen. A. Terry lands his troops and prepares a defensive line facing north to block possible Confederate reinforcements, meanwhile Porter bombards all day and into the night
8. **Jan 14:** Lamb receives reinforcements: 700 soldiers and 50 sailors. Jan 15 a further 350 make it before their ship is driven off by Union bombardment
9. **Jan 15 3 pm:** The bombardment ceases and the land force attacks the North and N. East walls

CAPE FEAR RIVER

Battery Buchanan

*After the first assault on the fort, Benjamin Butler had complained that the naval gunnery had been ragged and ineffective. This failure was largely due to the attempts of Porter's gunners to knock down the Confederate flag by aiming high (left). Porter ensured that this fault was rectified in the subsequent assault.*

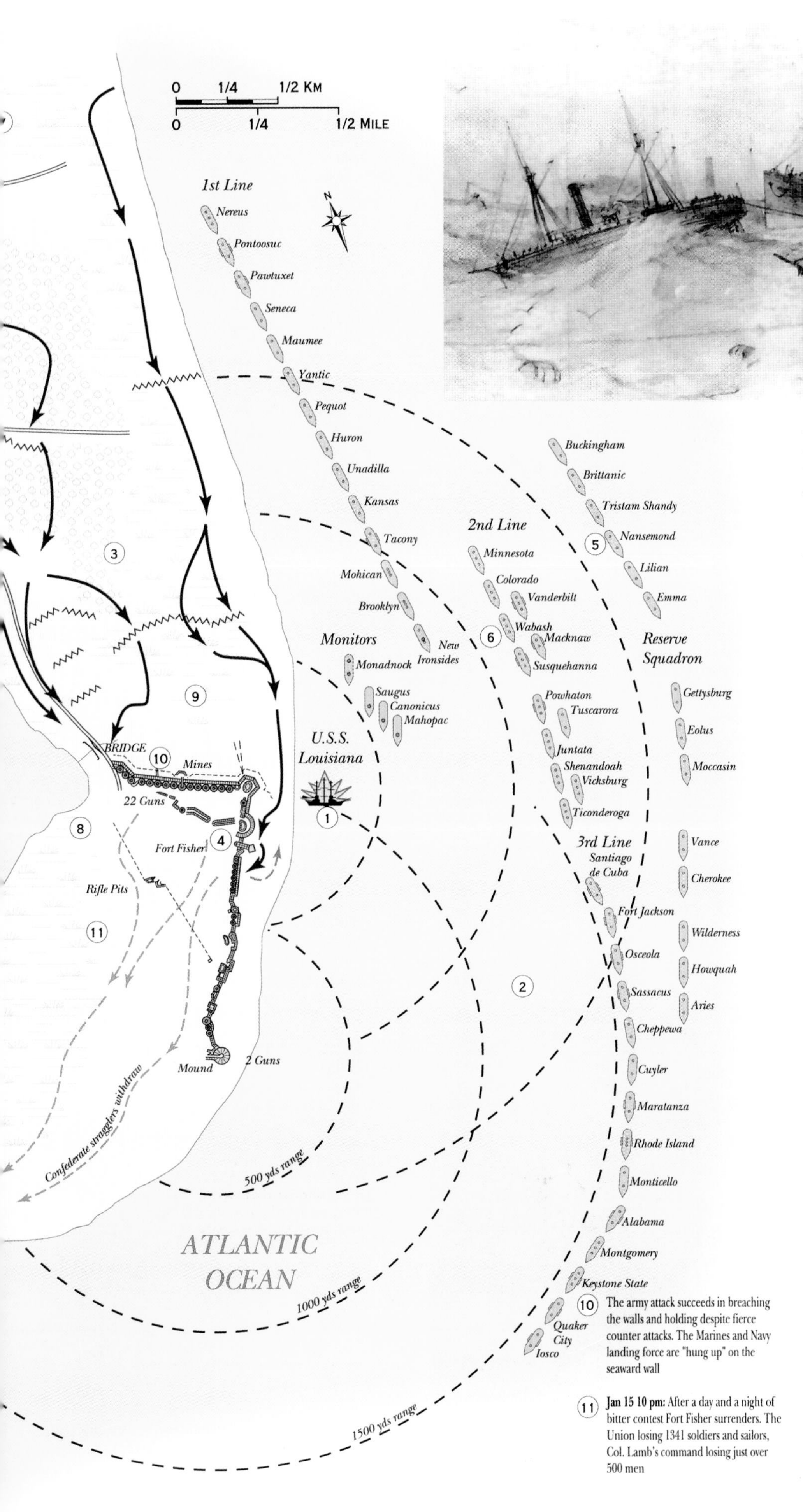
0 1/4 1/2 KM
0 1/4 1/2 MILE
1st Line
Nereus
Pontoosuc
Pawtuxet
Seneca
Maumee
Yantic
Pequot
Huron
Unadilla
Kansas
Tacony
Mohican
Brooklyn
New Ironsides
Monitors
Monadnock
Saugus
Canonicus
Mahopac
U.S.S. Louisiana
2nd Line
Minnesota
Colorado
Vanderbilt
Wabash
Macknaw
Susquehanna
Powhaton
Tuscarora
Juntata
Shenandoah
Vicksburg
Ticonderoga
3rd Line
Santiago de Cuba
Fort Jackson
Osceola
Sassacus
Cheppewa
Cuyler
Maratanza
Rhode Island
Monticello
Alabama
Montgomery
Keystone State
Quaker City
Iosco
Buckingham
Brittanic
Tristam Shandy
Nansemond
Lilian
Emma
Reserve Squadron
Gettysburg
Eolus
Moccasin
Vance
Cherokee
Wilderness
Howquah
Aries
BRIDGE
Mines
22 Guns
Fort Fisher
Rifle Pits
Mound
2 Guns
Confederate stragglers withdraw
500 yds range
1000 yds range
1500 yds range
ATLANTIC OCEAN
1
2
3
4
5
6
8
9
10
11
10 The army attack succeeds in breaching the walls and holding despite fierce counter attacks. The Marines and Navy landing force are "hung up" on the seaward wall
11 Jan 15 10 pm: After a day and a night of bitter contest Fort Fisher surrenders. The Union losing 1341 soldiers and sailors, Col. Lamb's command losing just over 500 men

*Porter's fleet (above) leaving Chesapeake Bay for Fort Fisher, December 13, 1864. Consisting of nearly sixty warships plus troop transports, the fleet was the largest assembled during the war. The attack was delayed when the fleet was scattered by strong winds on December 22. The squadron's seven monitors anchored and weathered the storm but the remainder of Porter's ships were dispersed over a wide area.*

12,000 soldiers led by General Alfred Terry, attacked Fort Fisher for the second time.

Porter had learned from his mistakes: this time he moved his vessels closer to the target and concentrated his fire on the fort's land face, where the main attack would take place. To assist in the assault, Porter sent marines and sailors ashore.

The bombardment continued throughout the 13th and into the night, thus preventing the defenders repairing the damage. Terry landed his force, and while the sailors and marines moved on the sea face of the fort, his men charged from the north. Although the attack on the sea face was repulsed, Terry broke through and entered the fort. Fierce hand to hand combat followed. The battle continued throughout the 14th until finally, at 10 p.m., the Confederates surrendered.

The fall of Fort Fisher ended Wilmington's usefulness as a Confederate port. On February 22 1865 the city itself fell to Federal forces.

> "Great cannon were broken in two, and over their ruins were lying the dead; others were partly buried in graves dug by the shells which had slain them if there has ever been a larger or more stubborn hand-to-hand encounter, I have failed to meet with it in history."
>
> *Col. W. Lamb, garrison commander of Fort Fisher.*

# Sherman's Carolinas Campaign FEBRUARY 1 – APRIL 26 1865

General Grant's original plan had been to bring Sherman's army – once it reached the Atlantic coast – by ship to Virginia, where it would join with Federal forces already there to crush Robert E. Lee. But after Sherman's occupation of Savannah, Grant changed his mind, authorizing Sherman to march through South and North Carolina to Virginia. Sherman was to destroy Lee's supply base and wage psychological war against the South.

On February 1 Sherman began the march through South Carolina, his army of 60,000 divided into two wings: the left, under General Slocum; the right under Howard. Regarded by many Unionists as "the cradle of secession," South Carolina was "punished" by the soldiers with vindictive delight; they looted, ransacked, and burned, meeting little Confederate resistance. The climax came at Columbia, the state capital, which was largely destroyed by fire on the night of February 17.

Early in March, Sherman entered North Carolina, where a makeshift force of about 21,000 Confederates, under General Joseph E. Johnston, opposed him. In an attempt to prevent Schofield's 30,000 Federals at Wilmington from joining Sherman, Johnston first sent 8,500 men under General Bragg to attack the vanguard of the Federal army at Kinston. The attack delayed its advance.

Johnston then tried to attack Sherman's left wing, under Slocum, as it moved toward Goldsboro. The result was the battle of Averasboro on March 16, in which 11,000 Confederates checked Slocum until forced to retreat to avoid being outflanked. This engagement was followed by the battle of

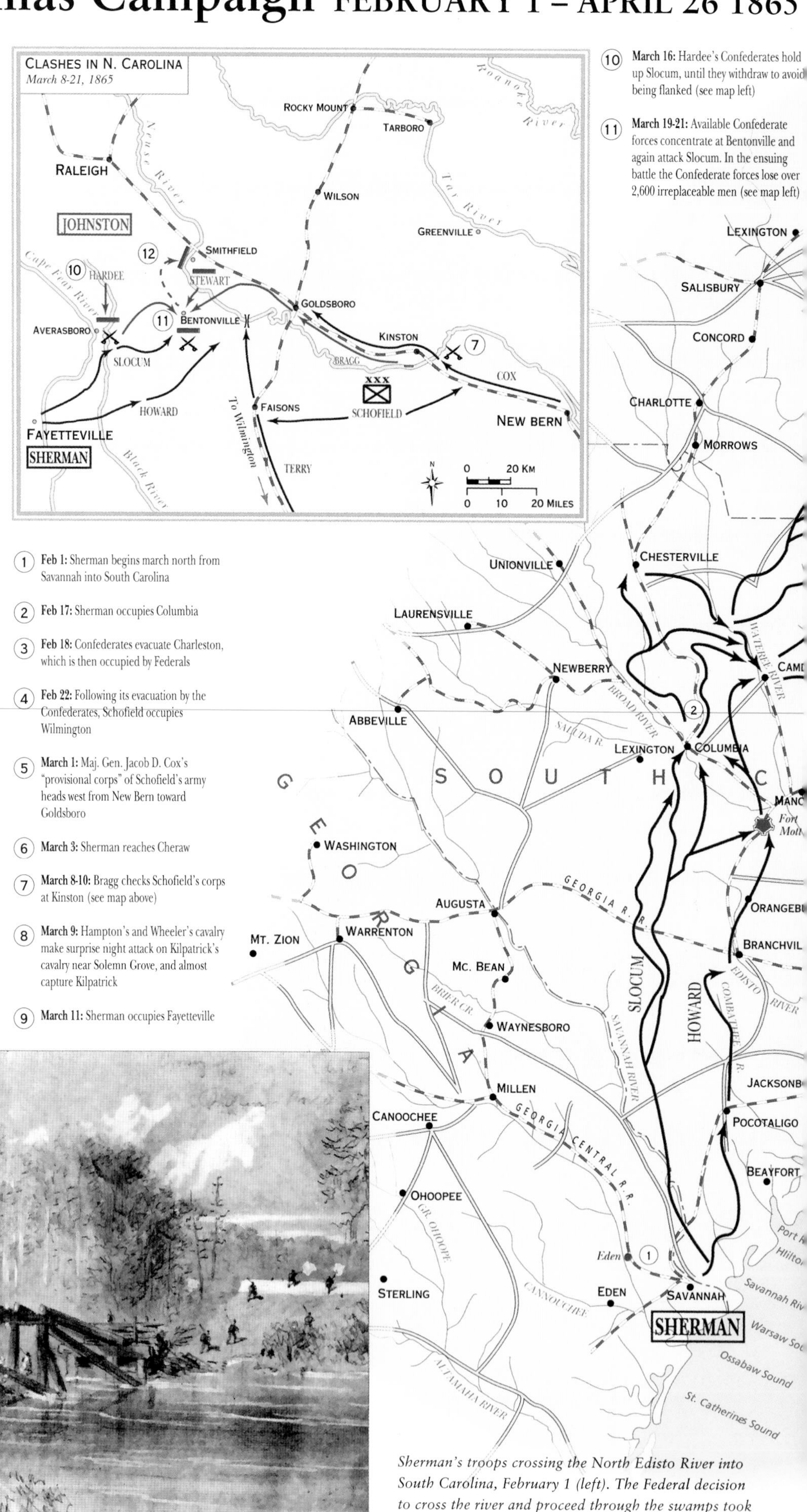
CLASHES IN N. CAROLINA
March 8-21, 1865
ROCKY MOUNT
TARBORO
Roanoke River
RALEIGH
Neuse River
Tar River
WILSON
JOHNSTON
GREENVILLE
SMITHFIELD
HARDEE
STEWART
Cape Fear River
GOLDSBORO
AVERASBORO
BENTONVILLE
KINSTON
SLOCUM
BRAGG
COX
HOWARD
SCHOFIELD
FAISONS
FAYETTEVILLE
SHERMAN
NEW BERN
To Wilmington
Black River
TERRY
0 20 KM
0 10 20 MILES
(10) March 16: Hardee's Confederates hold up Slocum, until they withdraw to avoid being flanked (see map left)
(11) March 19-21: Available Confederate forces concentrate at Bentonville and again attack Slocum. In the ensuing battle the Confederate forces lose over 2,600 irreplaceable men (see map left)
(1) Feb 1: Sherman begins march north from Savannah into South Carolina
(2) Feb 17: Sherman occupies Columbia
(3) Feb 18: Confederates evacuate Charleston, which is then occupied by Federals
(4) Feb 22: Following its evacuation by the Confederates, Schofield occupies Wilmington
(5) March 1: Maj. Gen. Jacob D. Cox's "provisional corps" of Schofield's army heads west from New Bern toward Goldsboro
(6) March 3: Sherman reaches Cheraw
(7) March 8-10: Bragg checks Schofield's corps at Kinston (see map above)
(8) March 9: Hampton's and Wheeler's cavalry make surprise night attack on Kilpatrick's cavalry near Solemn Grove, and almost capture Kilpatrick
(9) March 11: Sherman occupies Fayetteville
LEXINGTON
SALISBURY
CONCORD
CHARLOTTE
MORROWS
CHESTERVILLE
UNIONVILLE
LAURENSVILLE
NEWBERRY
ABBEVILLE
LEXINGTON
COLUMBIA
SOUTH C
GEORGIA
WASHINGTON
AUGUSTA
GEORGIA R.R.
MT. ZION
WARRENTON
MC. BEAN
WAYNESBORO
MILLEN
GEORGIA CENTRAL R.R.
CANOOCHEE
OHOOPEE
SLOCUM
HOWARD
ORANGEBURG
BRANCHVILLE
JACKSONBORO
POCOTALIGO
BEAUFORT
Eden
EDEN
SAVANNAH
SHERMAN
STERLING
Savannah River
Ossabaw Sound
St. Catherines Sound
Warsaw Sound
Altamaha River

*Sherman's troops crossing the North Edisto River into South Carolina, February 1 (left). The Federal decision to cross the river and proceed through the swamps took the Confederates completely by surprise.*

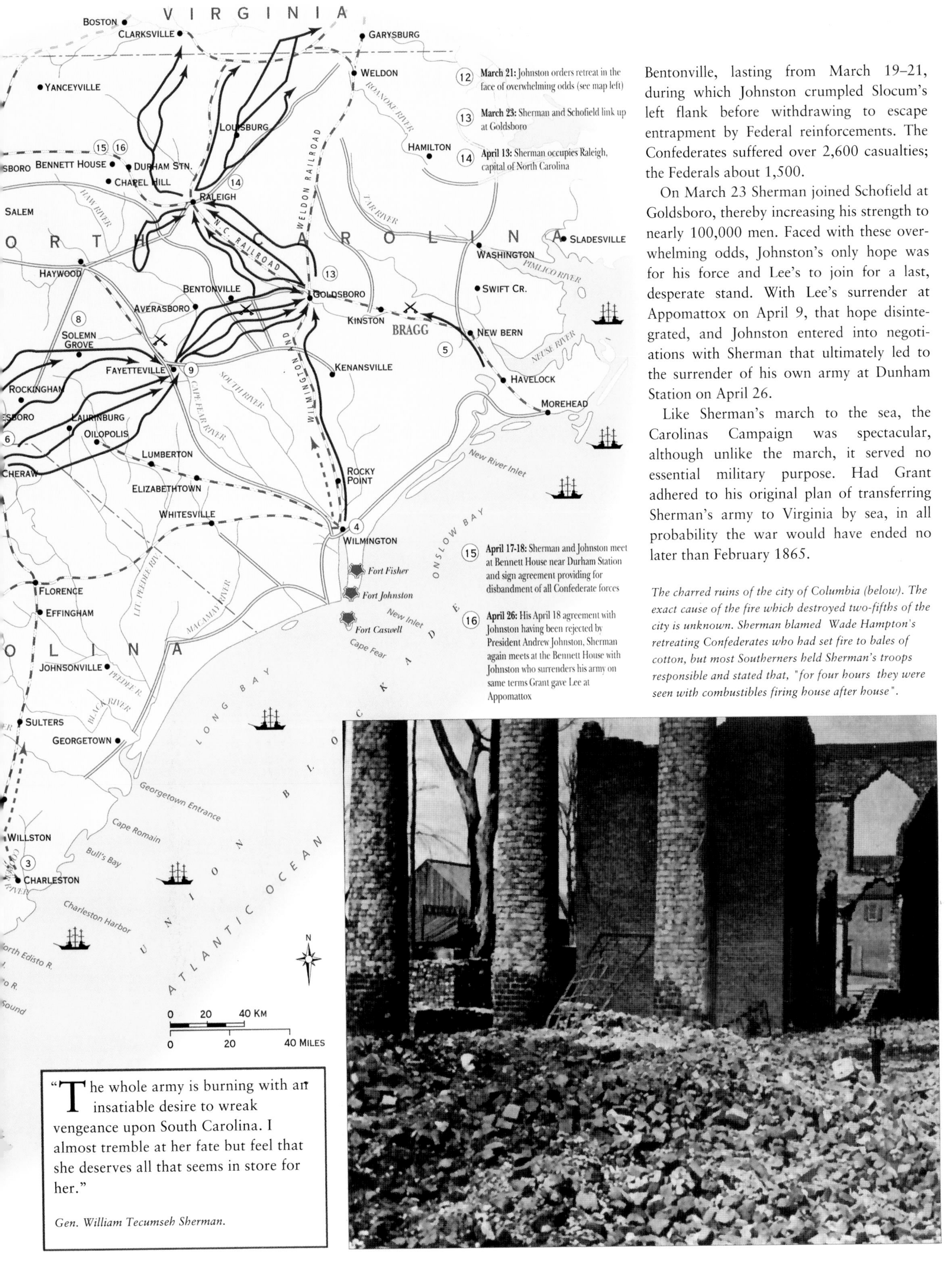
VIRGINIA
BOSTON
CLARKSVILLE
GARYSBURG
WELDON
YANCEYVILLE
LOUISBURG
HAMILTON
BENNETT HOUSE
DURHAM STN.
CHAPEL HILL
RALEIGH
SALEM
WELDON RAILROAD
N.C. RAILROAD
NORTH CAROLINA
SLADESVILLE
WASHINGTON
HAYWOOD
BENTONVILLE
GOLDSBORO
SWIFT CR.
AVERASBORO
KINSTON
BRAGG
NEW BERN
SOLEMN GROVE
FAYETTEVILLE
KENANSVILLE
HAVELOCK
ROCKINGHAM
MOREHEAD
LAURINBURG
OILOPOLIS
LUMBERTON
CHERAW
ROCKY POINT
ELIZABETHTOWN
WHITESVILLE
WILMINGTON
Fort Fisher
Fort Johnston
Fort Caswell
New Inlet
Cape Fear
New River Inlet
ONSLOW BAY
FLORENCE
EFFINGHAM
JOHNSONVILLE
SULTERS
GEORGETOWN
LONG BAY
Georgetown Entrance
Cape Romain
Bull's Bay
WILLSTON
CHARLESTON
Charleston Harbor
UNION BLOCKADE
ATLANTIC OCEAN
0 20 40 KM
0 20 40 MILES
12 March 21: Johnston orders retreat in the face of overwhelming odds (see map left)
13 March 23: Sherman and Schofield link up at Goldsboro
14 April 13: Sherman occupies Raleigh, capital of North Carolina
15 April 17-18: Sherman and Johnston meet at Bennett House near Durham Station and sign agreement providing for disbandment of all Confederate forces
16 April 26: His April 18 agreement with Johnston having been rejected by President Andrew Johnston, Sherman again meets at the Bennett House with Johnston who surrenders his army on same terms Grant gave Lee at Appomattox

Bentonville, lasting from March 19–21, during which Johnston crumpled Slocum's left flank before withdrawing to escape entrapment by Federal reinforcements. The Confederates suffered over 2,600 casualties; the Federals about 1,500.

On March 23 Sherman joined Schofield at Goldsboro, thereby increasing his strength to nearly 100,000 men. Faced with these overwhelming odds, Johnston's only hope was for his force and Lee's to join for a last, desperate stand. With Lee's surrender at Appomattox on April 9, that hope disintegrated, and Johnston entered into negotiations with Sherman that ultimately led to the surrender of his own army at Dunham Station on April 26.

Like Sherman's march to the sea, the Carolinas Campaign was spectacular, although unlike the march, it served no essential military purpose. Had Grant adhered to his original plan of transferring Sherman's army to Virginia by sea, in all probability the war would have ended no later than February 1865.

*The charred ruins of the city of Columbia (below). The exact cause of the fire which destroyed two-fifths of the city is unknown. Sherman blamed Wade Hampton's retreating Confederates who had set fire to bales of cotton, but most Southerners held Sherman's troops responsible and stated that, "for four hours they were seen with combustibles firing house after house".*

"The whole army is burning with an insatiable desire to wreak vengeance upon South Carolina. I almost tremble at her fate but feel that she deserves all that seems in store for her."

*Gen. William Tecumseh Sherman.*

# Fall of Petersburg and Richmond FEBRUARY 5 – APRIL 2 1865

During the winter both combatants remained quiescent in the trenches around Petersburg. On February 5, in the sole important movement, General Grant sent a cavalry and infantry force to disrupt Confederate traffic on the Boydton Plank Road near Hatcher's Run. A few wagons were intercepted, but the movement provoked Confederate counterattacks on February 6. Following the action, the Federal trenches were extended another two miles. Realizing that this gradual expansion of the Federal lines would ultimately doom his defense, Robert E. Lee orchestrated a final desperate effort to cripple Grant's army. He reasoned that a break in the Federal lines near their City Point base would cause them to contract, and permit Confederate movement to North Carolina. On March 25 Confederate forces attacked in the vicinity of Fort Stedman and momentarily achieved success. Massive Federal counterattacks soon restored the Federal position, however, ending the last offensive movement of the Army of Northern Virginia.

Seizing the initiative, Grant mounted a new thrust southwest of Petersburg with Sheridan's cavalry and two infantry corps. On March 31, Confederate counterattacks forced Sheridan back to Dinwiddie Court House, but were unable to prevent the Federal infantry from extending westward along the White Oak Road. The next day Sheridan's cavalry and V Corps advanced to Five Forks where they crushed George Pickett's command. This decisive action sealed Petersburg's fate. On April 2, Grant directed a massive assault on the Petersburg lines, breaking them in several places. By sacrificing several detachments in rear guard actions, Lee was able to evacuate both Richmond and Petersburg that evening with most of his forces.

*A Union battery in one of Petersburg's captured forts (above). The Union artillery had undoubtedly inflicted far more considerable damage than that of the Confederates, and the ground within the fortifications was found to be honeycombed with the effects of the shot.*

"On every hand we could see indications that Lee's army would soon melt away. Prisoners were pouring into our lines by thousands; baggage wagons, artillery, mortars, and baggage of all kinds lined the roads along which the rebels were fleeing."

*A Union veteran on the retreat of Lee's army from Petersburg.*

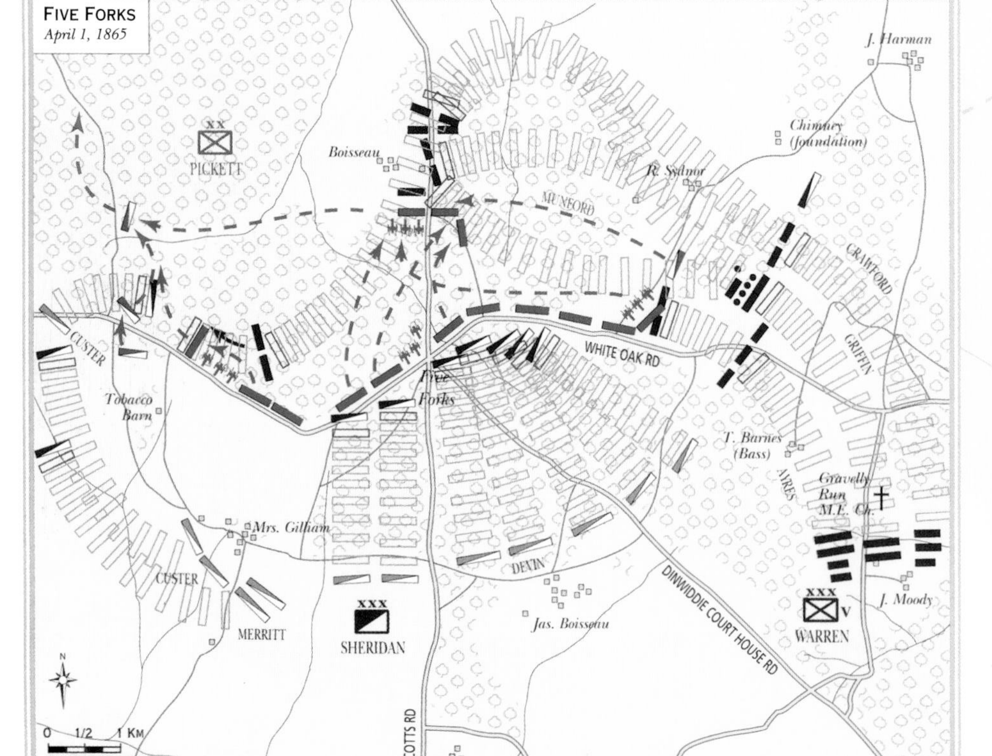
FIVE FORKS
April 1, 1865
PICKETT
Boisseau
J. Harman
Chimney (foundation)
R. Sydnor
MUNFORD
CRAWFORD
GRIFFIN
CUSTER
WHITE OAK RD
Five Forks
Tobacco Barn
T. Barnes (Bass)
AYRES
Gravelly Run M.E. Ch.
Mrs. Gilliam
CUSTER
MERRITT
SHERIDAN
DEVIN
Jas. Boisseau
DINWIDDIE COURT HOUSE RD
J. Moody
WARREN
SCOTTS RD
P. Scott
0 1/2 1 KM
0 1/2 1 MILE

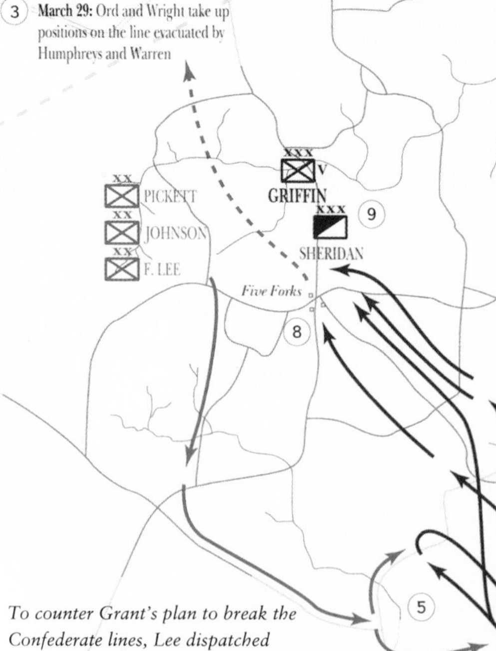
1 Feb 5–7: This part of the line gained by the Union in the battle of Hatcher's Run
2 March 25: Confederates attack near Fort Stedman. Federals counterattack and restore position
3 March 29: Ord and Wright take up positions on the line evacuated by Humphreys and Warren
PICKETT
JOHNSON
F. LEE
GRIFFIN
SHERIDAN
Five Forks
Dinwiddie C.H.

*To counter Grant's plan to break the Confederate lines, Lee dispatched George Pickett and W.H. Lee's cavalry. They met with Sheridan's cavalry and an infantry corps at Five Forks on April 1. Sheridan deployed Merritt to drive Pickett from his entrenched line, while Warren's V Corps assailed the Confederate left. The Federal attack collapsed Pickett's front and encircled his left (see map left).*

> "The condition of our army was daily becoming more desperate. Starvation, literal starvation was doing it's deadly work. So depleted and poisoned was the blood of many of Lee's men from insufficient and unsound food that a slight wound which would probably not have been reported at the beginning of the war would often cause blood-poisoning, gangrene and death."
>
> *Gen. Gordon at the siege of Petersburg.*

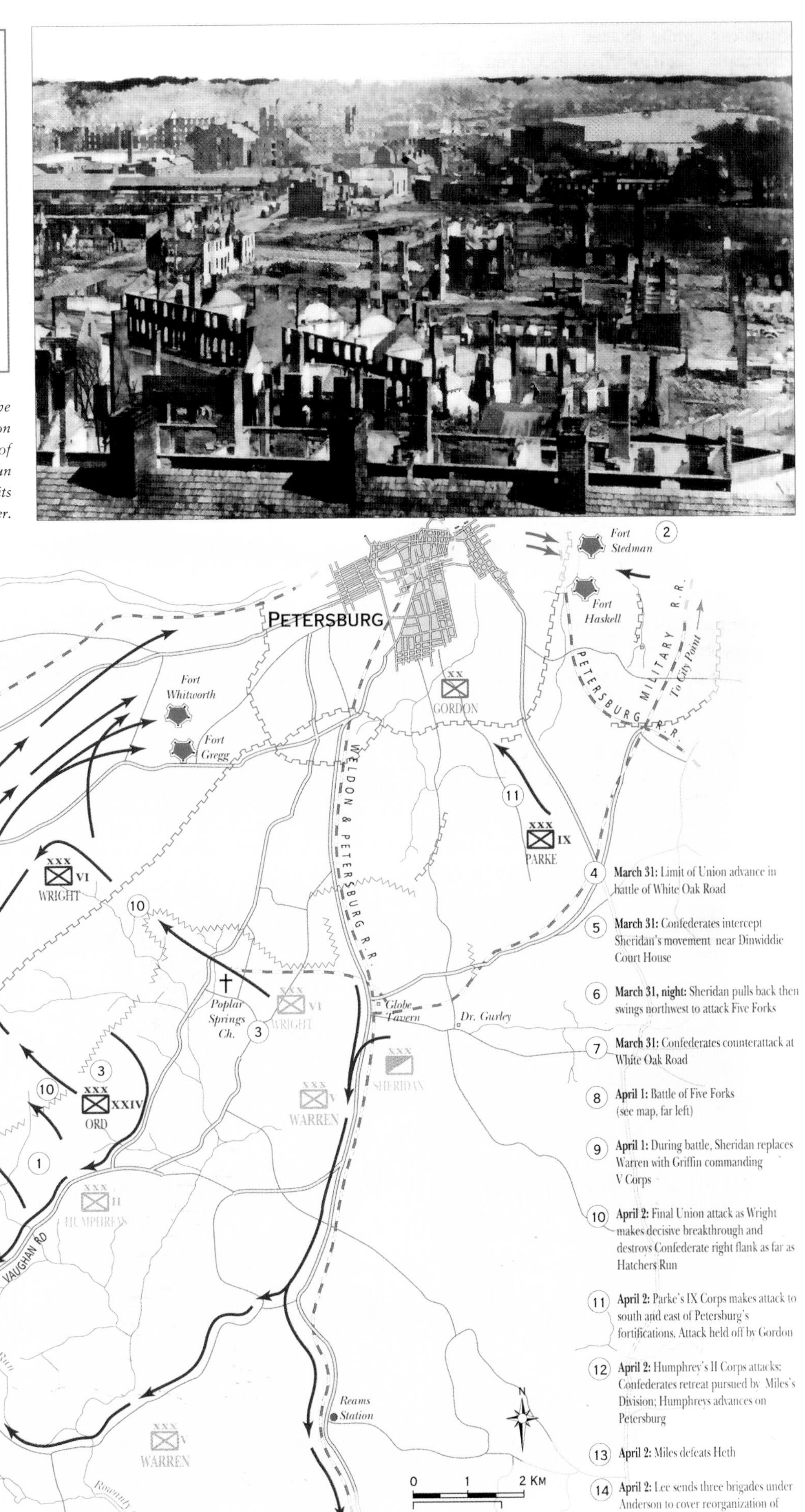
PETERSBURG
Fort Stedman
Fort Haskell
PETERSBURG MILITARY R.R.
To City Point
Fort Whitworth
Fort Gregg
GORDON
SOUTHSIDE R.R.
APPOMATTOX RIVER
PICKETT
JOHNSON
CLAIBORNE RD
MILES
A.P. HILL
BOYDTON PLANK ROAD
WRIGHT
WELDON & PETERSBURG R.R.
PARKE
HETH
WHITE OAK RD
Burgess
Poplar Springs Ch.
Globe Tavern
Dr. Gurley
ORD
HUMPHREYS
SHERIDAN
WARREN
Crow
Dabney's Mill
VAUGHAN RD
Gravelly Run
Reams Station
Rowanty Cr.
0 1 2 KM
0 1 MILE
4 March 31: Limit of Union advance in battle of White Oak Road
5 March 31: Confederates intercept Sheridan's movement near Dinwiddie Court House
6 March 31, night: Sheridan pulls back then swings northwest to attack Five Forks
7 March 31: Confederates counterattack at White Oak Road
8 April 1: Battle of Five Forks (see map, far left)
9 April 1: During battle, Sheridan replaces Warren with Griffin commanding V Corps
10 April 2: Final Union attack as Wright makes decisive breakthrough and destroys Confederate right flank as far as Hatchers Run
11 April 2: Parke's IX Corps makes attack to south and east of Petersburg's fortifications. Attack held off by Gordon
12 April 2: Humphrey's II Corps attacks; Confederates retreat pursued by Miles's Division; Humphreys advances on Petersburg
13 April 2: Miles defeats Heth
14 April 2: Lee sends three brigades under Anderson to cover reorganization of Pickett's command

*The ruins of Richmond (right), photographed after the city's fall. Upon the retreat from the capital, Jefferson Davis and his government had ordered the destruction of all ammunition dumps, but the fires thus begun eventually destroyed much of the city, including its commercial quarter.*

# The Road to Appomattox APRIL 2–9 1865

After evacuating Richmond and Petersburg during the night of April 2–3, Confederate forces moved on multiple routes toward a junction at Amelia Court House on the Richmond and Danville Railroad. There Lee hoped to be resupplied before marching south into North Carolina to join General Jospeph Johnston's forces. Reaching Amelia on April 5, Lee united his columns but found no rations. He also learned that Grant's vigorous pursuit had placed large forces across his path near Jetersville, eight miles west of Amelia. Looping around the Federal concentration in an exhausting night march, Lee continued westward toward Farmville on the South Side Railroad. There he hoped to be supplied by rail from Lynchburg. Grant followed, sending II and VI Corps behind the Confederates while the cavalry and the V Corps, plus elements of the Army of the James, attempted to turn Lee's southern flank. On April 6 at Sayler's Creek, east of Farmville, Lee's rearguard was smashed by II and VI Corps. This action cost the Confederates more than 7,000 casualties and many of their wagons; Federal losses amounted to 1,180 men.

Lee found rations at Farmville, but the rapid Federal pursuit prevented him from turning south toward North Carolina. Instead, he was forced to cross to the north bank of the Appomattox River and continue his retreat westward. Unable to prevent his pursuers from also crossing the stream, he was forced to halt briefly at Cumberland Church to repulse attacks by II Corps. Although necessary, this delay permitted Sheridan's cavalry and the Army of the James to gain ground freely south of the river. By the evening of April 8, Federal cavalry had reached Lee's rear at Appomattox Station and captured several trains filled with rations. On the following morning, Lee attempted to break through Sheridan's blockade, but failed when elements of the Army of the James arrived to support the cavalry. Simultaneously, units II Corps pressed the rear of Lee's dwindling army. Surrounded at last by overwhelming force, Lee opened negotiations with Grant which resulted in the surrender of the Army of Northern Virginia at Appomattox Court House on the afternoon of April 9, 1865. This surrender brought the war in Virginia to an end.

*Robert E. Lee photographed at his Richmond home in the uniform which he had worn when surrendering the Army of Northern Virginia a few days earlier (right). Lee disliked being photographed and the scarcity of likenesses had resulted in his being represented in newspapers throughout the war as dark haired and, apart from a black moustache, clean shaven. Such portrayals were based upon the general's appearance when he was a professor at West Point.*

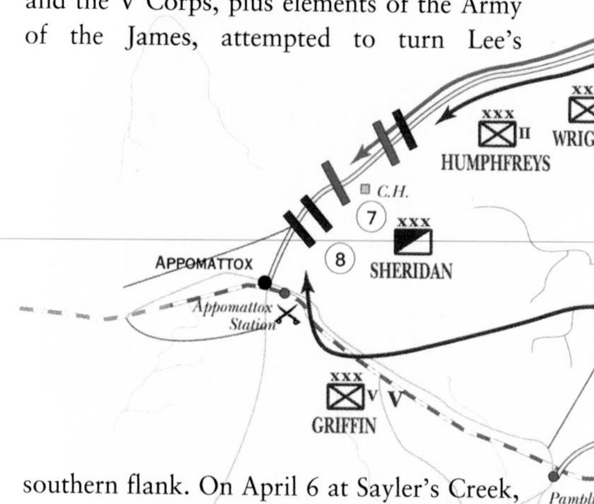
Cumberland C.H.
GRIFFIN
New Store
PARKE
WRIGHT
HUMPHREYS
PAINESVILLE
HUMPHREYS
HUMPHREYS
HUMPHREYS
C.H.
7
8
SHERIDAN
WRIGHT
APPOMATTOX
Cumberland Church
Sayler's Creek
5
DEATONSVILLE
JETERSVILLE
Appomattox Station
SOUTH SIDE R.R.
GRIFFIN
6
FARMVILLE
SHERIDAN
GRIFFIN
Buffalo River
PRINCE EDWARD
Rice's Station
WRIGHT
Pamplin's Station
SHERIDAN
Sayler's Creek
BURKEVILLE
ORD
4
BURKE
N
0 3 6 KM
0 3 6 MILES

*The McLean house at Appomatox . The house was owned by Wilmer McLean who soon regretted having lent his home since souvenir-hunters quickly stripped it after the signing of the peace (below).*

"Each successive division ... halts, the men face inward toward us across the road, twelve feet away; then carefully 'dress' their line... lastly... they tenderly fold their flags, battle-worn and torn, bloodstained, heart-holding colors, and lay them down; some frenziedly rushing from the ranks... pressing them to their lips with burning tears. And only the flag of the Union greets the sky!"

*Gen. Joshua L. Chamberlain describing the surrender of the Confederate army, April 12 , 1865.*

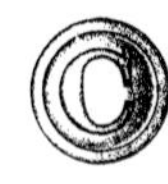

"After four years of arduous service marked by unsurpassed courage and fortitude, the Army of North Virginia has been compelled to yield to overwhelming numbers and resources... With an unceasing admiration of your constancy and devotion to your Country ... I bid you all an affectionate farewell."

*Gen. Lee, General Order Number Nine, April 10 , 1865.*

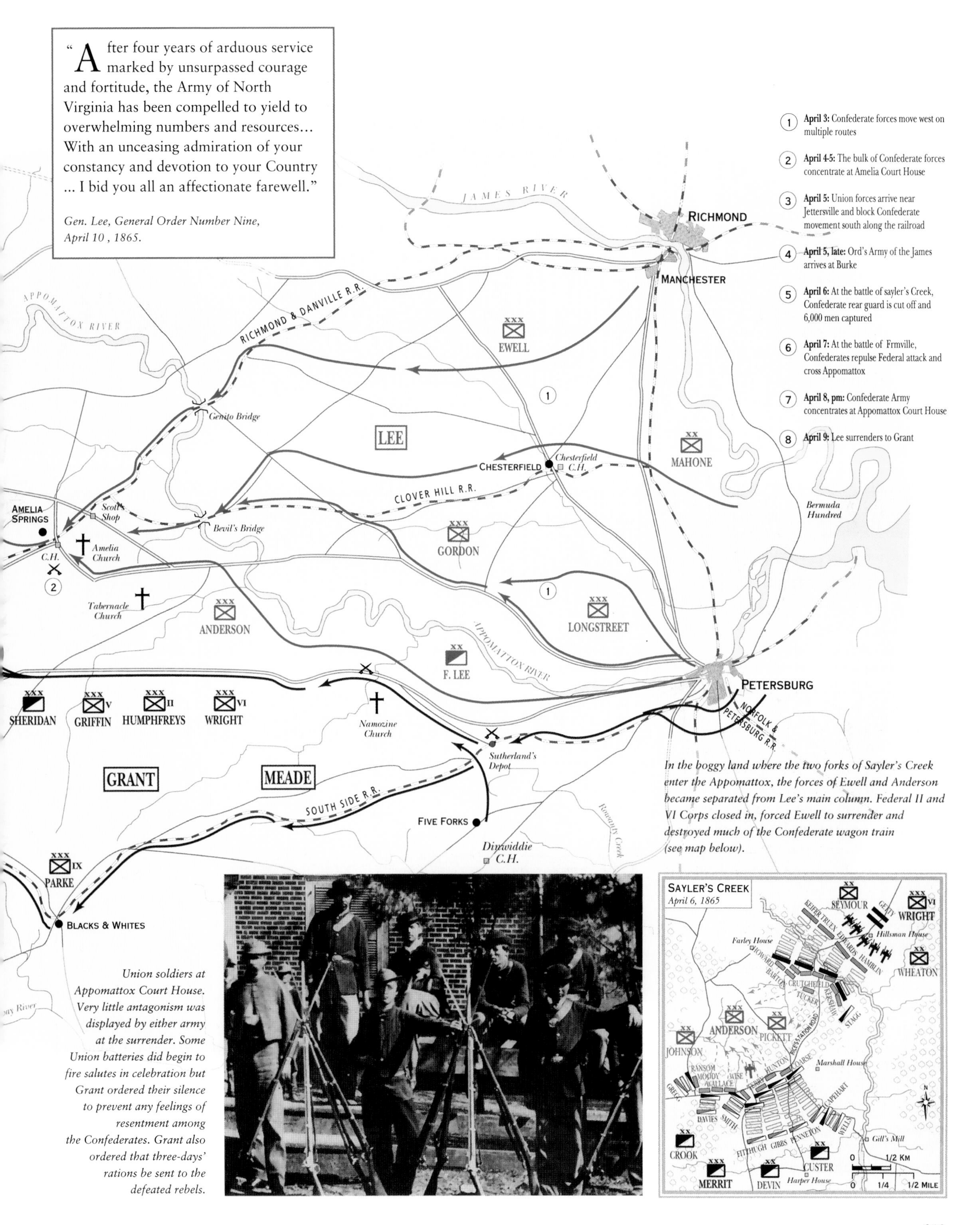
1 April 3: Confederate forces move west on multiple routes
2 April 4-5: The bulk of Confederate forces concentrate at Amelia Court House
3 April 5: Union forces arrive near Jettersville and block Confederate movement south along the railroad
4 April 5, late: Ord's Army of the James arrives at Burke
5 April 6: At the battle of sayler's Creek, Confederate rear guard is cut off and 6,000 men captured
6 April 7: At the battle of Frmville, Confederates repulse Federal attack and cross Appomattox
7 April 8, pm: Confederate Army concentrates at Appomattox Court House
8 April 9: Lee surrenders to Grant
JAMES RIVER
RICHMOND
MANCHESTER
APPOMATTOX RIVER
RICHMOND & DANVILLE R.R.
EWELL
Genito Bridge
LEE
CHESTERFIELD
Chesterfield C.H.
MAHONE
CLOVER HILL R.R.
Bermuda Hundred
AMELIA SPRINGS
Scott's Shop
Bevil's Bridge
Amelia Church
C.H.
GORDON
Tabernacle Church
ANDERSON
LONGSTREET
F. LEE
PETERSBURG
SHERIDAN
GRIFFIN
HUMPHFREYS
WRIGHT
Namozine Church
NORFOLK & PETERSBURG R.R.
Sutherland's Depot
GRANT
MEADE
SOUTH SIDE R.R.
FIVE FORKS
Dinwiddie C.H.
PARKE
BLACKS & WHITES
SAYLER'S CREEK
April 6, 1865
SEYMOUR
WRIGHT
WHEATON
Farley House
Hillsman House
ANDERSON
PICKETT
JOHNSON
Marshall House
CROOK
MERRIT
DEVIN
CUSTER
Harper House
Gill's Mill
1/2 KM
1/4
1/2 MILE

*In the boggy land where the two forks of Sayler's Creek enter the Appomattox, the forces of Ewell and Anderson became separated from Lee's main column. Federal II and VI Corps closed in, forced Ewell to surrender and destroyed much of the Confederate wagon train (see map below).*

*Union soldiers at Appomattox Court House. Very little antagonism was displayed by either army at the surrender. Some Union batteries did begin to fire salutes in celebration but Grant ordered their silence to prevent any feelings of resentment among the Confederates. Grant also ordered that three-days' rations be sent to the defeated rebels.*

# Epilogue

THE NATION WENT into profound mourning at Lincoln's death. Thousands paid their last respects at the White House on April 19, and General Grant wept openly by his catafalque. Seven million people lined railroad tracks from Washington to Springfield to view Lincoln's funeral train as it retraced the exact route Lincoln had taken to Washington four years earlier.

Union cavalry captured the fleeing Jefferson Davis in Georgia on May 10. The United States government discovered evidence of the role of Confederate agents in a plan late in the war to kidnap Lincoln, take him to Richmond and hold him as a bargaining counter in peace negotiations. It was the failure of this plan that had caused Booth and several co-conspirators to assassinate Lincoln, and attempt to assassinate Secretary of State Seward and Vice-President Andrew Johnson. The government thought it also had evidence of the involvement of high Confederate officials, including Jefferson Davis, in the kidnapping and – by implication – the assassination plots. But as it became clear that this evidence would not stand up in court, the government gradually dropped plans to try Davis. The only people tried and convicted were those directly involved in the assassination plots. Four of the conspirators were hanged on July 7, 1865. The other four were imprisoned. One died in prison; the other three received pardons from President Andrew Johnson in 1869. Booth had been killed by one of the soldiers who had trapped him in a burning barn in Virginia on April 26, two weeks after the assassination.

Jefferson Davis remained in prison for two years, while the government pondered whether to try him, not for involvement in the assassination, but for treason. The government finally decided to let him go, recognizing that a jury in Virginia (where, according to the Constitution, Davis would have to be tried) was unlikely to convict him. Davis returned to Mississippi, where he lived quietly until 1889, writing his memoirs and a history of the Confederacy, in which he insisted that the South was right, seccession was legal, and the Confederacy should have its independence.

But only a few die-hard, unreconstructed rebels paid much attention to Davis. The Civil War not only preserved the Union; it also transformed it. The old decentralized republic, in which the federal government had few direct contacts with the average citizen except through the post office, became a nation that taxed people directly, and created an internal revenue bureau to collect the taxes, drafted men into the army, increased the powers of federal courts, created a national currency and a national banking system, and confiscated at least three billion dollars of personal property by emancipating the four million slaves. Eleven of the first twelve amendments to the Constitution had limited the power of the national government; six of the next seven, beginning with the 13th Amendment in 1865, vastly increased national powers at the expense of the states.

The first three of these postwar amendments accomplished the most radical and rapid social and political change in American history: the abolition of slavery (13th), the granting of equal citizenship (14th), and voting rights (15th) to former slaves, all within a period of five years. This transformation of four million slaves into citizens with equal rights became the central issue of the troubled twelve-year Reconstruction period after the Civil war, during which the promise of equal rights was fulfilled for a brief time and then abandoned, to be revived again in the second half of the 20th century when it still remains one of the most important and divisive issues facing the American people.

The war tipped the sectional balance of power in favor of the North for half a century or more. Prior to the war, from the adoption of the Constitution in 1789 until 1861, slaveholders from states that joined the Confederacy had served as Presidents of the United States during 49 of the 72 years – more than two-thirds of the time. Twenty-three of the 36 speakers of the House, and 24 of the presidents pro tem of the Senate, had been Southerners. The Supreme Court always had a Southern majority before the Civil War; 20 of the 35 justices to 1861 had been appointed from the slave states. After the war, a century passed before a resident of an ex-Confederate state was elected president. For half a century, only one of the speakers of the House, and no president pro tem of the Senate came from the South, and only five of the 26 Supreme Court justices named during that half century were Southerners.

The United States went to war in 1861 to preserve the *Union*; it emerged from war in 1865 having created a *nation*. Before 1861, the two words "United States" were generally used as a plural noun: "the United States" *are* a republic. After 1865, the United States became a singular noun. The loose union of states became a single nation. Lincoln's wartime speeches marked this transition. In his first inaugural address he mentioned the "Union" 20 times but the "nation" not once. In his first message to Congress on July 4, 1861, Lincoln used the word Union 32 times and nation only three times. But in his Gettysburg Address in November 1863, he did not mention the Union at all, but spoke of the nation five times to invoke a new birth of freedom and nationhood.

Were the accomplishments of the war worth the cost in lives lost and maimed? Few of the four million emancipated slaves, who joyously celebrated the Day of Jubilee in 1865, doubted the answer. Nor did the millions of Northerners who celebrated the news of Lee's surrender. Even most Southerners eventually came to share the sentiments expressed by young Woodrow Wilson, who had experienced the war as a child in Georgia. When he was a student at the University of Virginia law school in 1880, Wilson said that "*because* I love the South, I rejoice in the failure of the Confederacy. ...Conceive of this Union divided into two separate and independent sovereignties! ...Slavery was ennervating our Southern society. ...[Nevertheless] I recognize and pay loving tribute to the virtues of the leaders of secession ... to the righteousness of the cause which they thought they were promoting – to the immortal courage of the soldiers of the Confederacy." These words expressed themes that would help reconcile generations of Southerners to defeat: their forebears had fought bravely for what they believed right; perhaps they deserved to win; but in the long run it was well that they had lost.

Why? Because the Civil War resolved two fundamental problems left unresolved by the American Revolution. The first was the question whether this new nation, this republic born in a world of kings, emperors, tyrants, and oligarchs, could survive. The republican experiment launched in 1776 was a fragile entity. The founding fathers – indeed all Americans for the next three generations – were fearful about the prospects for its survival. They knew that most republics through history had been overthrown by revolutions, or had collapsed into dictatorships or civil war. Americans alive in 1860 had twice seen French republics succumb to collapse into dictatorship or anarchy. The same fate, they feared, could await them.

The Civil War was the great "testing," as Lincoln expressed it at Gettysburg, whether a "government of the people, by the people, for the people" would survive or "perish from the earth." It did not perish. Northern victory preserved the nation created in 1776. Since 1865 no disaffected state or region has tried to secede from the United States. That question appears to have been settled.

At Gettysburg, Lincoln spoke also of a "new birth of freedom." This referred to the other problem left unresolved by the Revolution of 1776 – slavery. The Civil War settled that issue as well. The 13th Amendment declared that "neither slavery nor involuntary servitude ... shall exist within the United States." It was that achievement, above all others, that caused the Civil War to be labeled as the Second American Revolution. Mark Twain put it best, half a dozen years after the war. "That cataclysm," he wrote, "uprooted institutions that were centuries old, changed the politics of a people, and wrought so profoundly upon the entire national character that the influence cannot be measured short of two or three generations." Five generations have passed, and we are still measuring the influence of the Civil War.

# Select Bibliography

**Boatner, Mark,** *The Civil War Dictionary* (2nd ed., New York, 1991)

**Catton, Bruce,** *The Centennial History of the Civil War*, 3 vols. (Garden City, N.Y., 1961–1965)

Vol. I: *The Coming Fury*

Vol. II: *Terrible Swift Sword*

Vol. III: *Never Call Retreat*

**Catton, Bruce,** *Mr. Lincoln's Army* (Garden City, N.Y., 1956)

**Catton, Bruce,** *Glory Road: The Bloody Route from Fredericksburg to Gettysburg* (Garden City, N.Y., 1956)

**Catton, Bruce,** *A Stillness at Appomattox* (Garden City, N.Y., 1957)

**Catton, Bruce,** *Grant Moves South* (Boston, 1960)

**Catton, Bruce,** *Grant Takes Command* (Boston, 1969)

**Connelly, Thomas L.,** *Army of the Heartland: The Army of Tennessee, 1861–1862* (Baton Rouge, 1967)

**Connelly, Thomas L.,** *Autumn of Glory: The Army of Tennessee, 1862–1865* (Baton Rouge, 1971)

**Davis, William C.,** *Jefferson Davis: The Man and His Hour* (New York, 1991)

**Faust, Patricia L., ed.** *Historical Times Illustrated Encyclopedia of the Civil War* (New York, 1986)

**Foote, Shelby,** *The Civil War: A Narrative*, 3 vols. (New York, 1958–1974)

Vol. I: *Fort Sumter to Perryville*

Vol. II: *Fredericksburg to Meridian*

Vol. III: *Red River to Appomattox*

**Freeman, Douglas Southall,** *R.E. Lee: A Biography*, 4 vols. (New York, 1934–1935)

**Freeman, Douglas Southall,** *Lee's Lieutenants: A Study in Command*, 3 vols. (New York, 1942–1944)

**Jones, Virgil Carrington,** *The Civil War at Sea*, 3 vols. (New York, 1960–1962)

**Long, E.B.,** *The Civil War Day By Day: An Almanac, 1861–1865* (Garden City, N.Y., 1971)

**McPherson, James M.,** *Battle Cry of Freedom: The Civil War Era* (New York, 1988)

**Nevins, Allan,** *Ordeal of the Union*, 4 vols. (New York, 1947–1950)

**Nevins, Allan,** *The War for the Union*, 4 vols. (New York, 1959–1971)

**Randall, James G.,** *Lincoln the President*, 4 vols. 4th vol. completed by Richard N. Current (New York, 1946–1955)

**Williams, Kenneth P.,** *Lincoln Finds a General*, 5 vols. (New York, 1949–1959)

**Williams, T. Harry,** *Lincoln and His Generals* (New York, 1952)

**Woodworth, Steven E.,** *Jefferson Davis and His Generals* (Lawrence, Kansas, 1990)

# Acknowledgments

PICTURE CREDITS

Chicago Historical Society
Dr William J. Schultz
Frank and Marie-T. Wood Print Collections, Alexandria, Virginia
Historical Picture Service Society, Chicago
Houghton Library, Harvard University
Kean Archives, Philadelphia
Library of Congress, Washington D.C.
Museum of the Confederacy, Richmond, Virginia
Peter Batty
The National Archives
U.S.A.M.H.I.
Valentine Museum, Cook Collection
Weidenfeld and Nicolson Limited

PICTURE RESEARCH

Stephen Haddelsey

ADDITIONAL THANKS TO

Jean Cox
Andrea Fairbrass

# Index

Items set in *italic* are pictures, items set in **bold** are maps. A "Q" after the entry is a quote.

## GENERAL INDEX

**PLACE NAMES INDEX**